T0192491

FEYNMAN DIAGRAM TECHNIQUES IN CONDENSED MATTER PHYSICS

A concise introduction to Feynman diagram techniques, this book shows how they can be applied to the analysis of complex many-particle systems, and offers a review of the essential elements of quantum mechanics, solid-state physics, and statistical mechanics.

Alongside a detailed account of the method of second quantization, the book covers topics such as Green's and correlation functions, diagrammatic techniques, superconductivity, and contains several case studies. Some background knowledge in quantum mechanics, solid-state physics, and mathematical methods of physics is assumed.

Detailed derivations of formulas and in-depth examples and chapter exercises from various areas of condensed matter physics make this a valuable resource for both researchers and advanced undergraduate students in condensed-matter theory, many-body physics, and electrical engineering. Solutions to the exercises are made available online.

RADI A. JISHI is a Professor of Physics at California State University. His research interests center on condensed matter theory, carbon networks, superconductivity, and the electronic structure of crystals.

FEYNMAN DIAGRAM TECHNIQUES IN CONDENSED MATTER PHYSICS

RADI A. JISHI
California State University

CAMBRIDGE
UNIVERSITY PRESS

University Printing House, Cambridge CB2 8BS, United Kingdom

Cambridge University Press is part of the University of Cambridge.

It furthers the University's mission by disseminating knowledge in the pursuit of education, learning and research at the highest international levels of excellence.

www.cambridge.org
Information on this title: www.cambridge.org/9781107655331

© R. A. Jishi 2013

This publication is in copyright. Subject to statutory exception and to the provisions of relevant collective licensing agreements, no reproduction of any part may take place without the written permission of Cambridge University Press.

First published 2013
First paperback edition 2014

A catalogue record for this publication is available from the British Library

Library of Congress Cataloguing in Publication data
Jishi, Radi A., 1955–
Feynman diagram techniques in condensed matter physics / Radi A. Jishi, California State University.
pages cm
Includes bibliographical references and index.
ISBN 978-1-107-02517-2 (hardback)
1. Feynman diagrams. 2. Many-body problem. 3. Condensed matter. I. Title.
QC794.6.F4J57 2013
530.4'1 – dc23 2013005735

ISBN 978-1-107-02517-2 Hardback
ISBN 978-1-107-65533-1 Paperback

Cambridge University Press has no responsibility for the persistence or accuracy of URLs for external or third-party internet websites referred to in this publication, and does not guarantee that any content on such websites is, or will remain, accurate or appropriate.

To the memory of
my parents

Contents

Preface

In both theory and practice, condensed matter physics is concerned with the physical properties of materials that are comprised of complex many-particle systems. Modeling the systems' behavior is essential to achieving a better understanding of the properties of these systems and their practical use in technology and industry.

Maximal knowledge about a many-particle system is gained by solving the Schrödinger equation. However, an exact solution of the Schrödinger equation is not possible, so resort is made to approximation schemes based on perturbation theory. It is generally true that, in order to properly describe the properties of an interacting many-particle system, perturbation theory must be carried out to infinite order. The best approach we have for doing so involves the use of Green's function and Feynman diagrams. Furthermore, much of our knowledge about a given complex system is obtained by measuring its response to an external probe, such as an electromagnetic field, a beam of electrons, or some other form of perturbation; its response to this perturbation is best described in terms of Green's function.

Two years ago, I set out to put together a guide that would allow advanced undergraduate and beginning graduate students in physics and electrical engineering to understand how Green's functions and Feynman diagrams are used to more accurately model complicated interactions in condensed matter physics. As time went by and the book was taking form, it became clear that it had turned into a reference manual that would be useful to professionals and educators as well as students. It is a self-contained place to learn or review how Feynman diagrams are used to solve problems in condensed matter physics. Great care has been taken to show how to create them, use them, and solve problems with them, one step at a time. It has been a labor of love. My reward is the thought that it will help others to understand the subject.

The book begins with a brief review of quantum mechanics, followed by a short chapter on single-particle states. Taken together with the accompanying exercises, these two chapters provide a decent review of quantum mechanics and solid state physics. The method of second quantization, being of crucial importance, is discussed at length in Chapter 3, and applied to the jellium model in Chapter 4. Since Green's functions at finite temperature are defined in terms of thermal averages, a review of the basic elements of statistical mechanics is presented in Chapter 5, which, I hope, will be accessible to readers without extensive knowledge of the subject.

Real-time Green's functions are discussed in Chapter 6, and some applications of these functions are presented in Chapter 7. Imaginary-time functions and Feynman diagram techniques are dealt with in Chapters 8 and 9. Every effort has been made to provide a step-by-step derivation of all the formulas, in as much detail as is necessary. Rules for the creation of the diagrams and their translation into algebraic expressions are clearly delineated. Feynman diagram techniques are then applied to the interacting electron gas in Chapter 10, to electron–phonon and electron–photon interactions in Chapter 11, and to superconductivity in Chapter 12. These techniques are then extended to systems that are not in equilibrium in Chapter 13.

Many exercises are given at the end of each chapter. For the more difficult problems, some guidance is given to allow the reader to arrive at the solution. Solutions to many of the exercises, as well as additional material, will be provided on my website (www.calstatela.edu/faculty/rjishi).

Over the course of the two years that it took me to finish this book, I received help in various ways from many people. In particular, I would like to thank David Guzman for extensive help in preparing this manuscript, and Hamad Alyahyaei for reading the first five chapters. I am indebted to Linda Alviti, who read the whole book and made valuable comments. I am grateful to Professor I. E. Dzyaloshinski for reading Chapter 9 and for his encouraging words. I also want to thank Dr. John Fowler, Dr. Simon Capelin, Antoaneta Ouzounova, Fiona Saunders, Kirsten Bot, and Claire Poole from Cambridge University Press for their help, guidance, and patience. I would also like to express my gratitude to my wife and children for their encouragement and support. Permission to use the quote from Russell's *The Scientific Outlook* (2001) was provided by Taylor and Francis (Routledge). Copyright is owned by Taylor and Francis and The Bertrand Russell Foundation Ltd. Permission to use Gould's quote from *Ever Since Darwin* (1977) was provided by W.W. Norton & Company.

This book is dedicated to the memory of my parents, who, despite adverse conditions, did all they could to provide me with a decent education.

Los Angeles, California *R. A. J.*
July, 2012

1
A brief review of quantum mechanics

Come forth into the light of things,
Let nature be your teacher.
–*William Wordsworth,*
The Tables Turned

The main focus of this book is many-particle systems such as electrons in a crystal. Such systems are studied within the framework of quantum mechanics, with which the reader is assumed to be familiar. Nevertheless, a brief review of this subject will provide an opportunity to establish notation and collect results that will be used later on.

1.1 The postulates

Quantum mechanics is based on five postulates, listed below with some explanatory comments.

(I) The quantum state

The quantum state of a particle, at time t, is described by a continuous, single-valued, square-integrable wave function $\Psi(\mathbf{r}, t)$, where $\mathbf{r}$ is the position of the particle. In Dirac notation, the state is represented by a state vector, or ket, $|\Psi(t)\rangle$, which is an element of a vector space V. We define a dual vector space V* whose elements, called bras, are in one-to-one correspondence with the elements of V: ket $|\alpha\rangle \in \mathrm{V} \leftrightarrow$ bra $\langle\alpha| \in \mathrm{V}^*$, as illustrated in Figure 1.1. The bra corresponding to ket $c\,|\alpha\rangle$ is $c^*\,\langle\alpha|$, where c^* is the complex conjugate of c. The inner product of kets $|\alpha\rangle$ and $|\beta\rangle$ is denoted by $\langle\beta|\alpha\rangle$, and it is a complex number (c-number). Note that the inner product is obtained by combining a bra and a ket. By definition, $\langle\beta|\alpha\rangle = \langle\alpha|\beta\rangle^*$. The state vectors $|\Psi(t)\rangle$ and $c\,|\Psi(t)\rangle$, where c is any nonzero complex number ($c \in \mathbb{C} - \{0\}$), describe the same physical state; because of that, the state

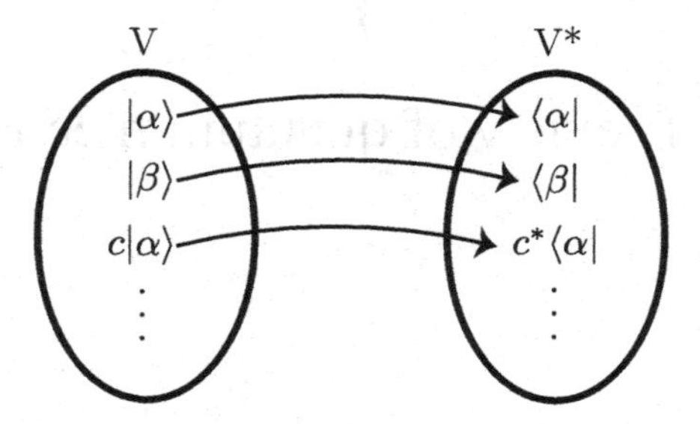

Figure 1.1 Vector space V of kets and the corresponding dual space V* of bras. A one-to-one correspondence exists between kets and bras.

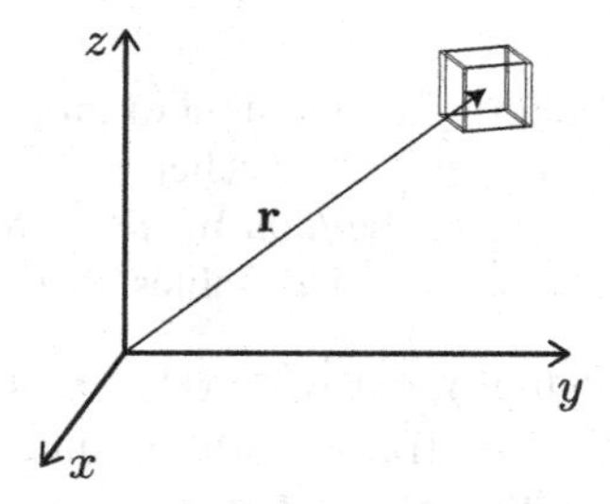

Figure 1.2 The probability of finding the particle, at time t, in the cube of volume d^3r, centered on $\mathbf{r}$, is $|\Psi(\mathbf{r}, t)|^2 d^3r$.

is usually taken to be normalized to unity: $\langle\Psi(t)|\Psi(t)\rangle = 1$. The normalized wave function has a probabilistic interpretation: $\Psi(\mathbf{r}, t)$ is the probability amplitude of finding the particle at position $\mathbf{r}$ at time t; this means that $|\Psi(\mathbf{r}, t)|^2 d^3r$ is the probability of finding the particle, at time t, in the infinitesimal volume d^3r centered on point $\mathbf{r}$ (see Figure 1.2).

Note that the description of a quantum state is completely different from the one used in classical mechanics, where the state of a particle is specified by its position $\mathbf{r}$ and momentum $\mathbf{p}$ at time t.

(II) Observables

An observable is represented by a linear, Hermitian operator acting on the state space. If A is an operator, being linear means that

$$A\,(c_1|\alpha\rangle + c_2|\beta\rangle) = c_1 A|\alpha\rangle + c_2 A|\beta\rangle, \quad |\alpha\rangle, |\beta\rangle \in \mathrm{V}, \quad c_1, c_2 \in \mathbb{C},$$

and being hermitian means that $A^\dagger = A$, where $A^\dagger$ is the adjoint of A, defined by the relation

$$\langle\beta|A^\dagger|\alpha\rangle = \langle\alpha|A|\beta\rangle^*.$$

In particular, the position of a particle is represented by the operator $\mathbf{r}$, its momentum $\mathbf{p}$ by $-i\hbar\boldsymbol{\nabla}$, and its energy by the Hamiltonian operator H,

$$H = -\frac{\hbar^2}{2m}\nabla^2 + V(\mathbf{r}, t). \tag{1.1}$$

$V(\mathbf{r}, t)$ is the operator that represents the potential energy of the particle, m is the particle's mass, and $\hbar$ is Planck's constant h divided by 2π.

As with states, the representation of observables in quantum mechanics is completely different from that of their classical counterparts, which are simply represented by their numerical values.

(III) Time evolution

The state $|\Psi(t)\rangle$ of a system evolves in time according to the Schrödinger equation

$$i\hbar\frac{\partial}{\partial t}\,|\Psi(t)\rangle = H\,|\Psi(t)\rangle. \tag{1.2}$$

If the Hamiltonian H does not depend explicitly on time, then

$$|\Psi(t)\rangle = e^{-iHt/\hbar}|\Psi(0)\rangle. \tag{1.3}$$

The operator $e^{-iHt/\hbar}$ is called the time evolution operator. Defining the stationary states $|\phi_n\rangle$ as the solutions of the eigenvalue equation, known as the time-independent Schrödinger equation,

$$H|\phi_n\rangle = E_n|\phi_n\rangle, \tag{1.4}$$

it is readily verified that $|\phi_n\rangle e^{-iE_nt/\hbar}$ is a solution of Eq. (1.2); the general solution of Eq. (1.2), when H is independent of t, is then given by

$$|\Psi(t)\rangle = \sum_n c_n|\phi_n\rangle e^{-iE_nt/\hbar}.$$

In contrast, the evolution of the classical state of a particle is determined by Hamilton's function H via Hamilton's equations of motion which, in one dimension, are

$$\dot{x} = \partial H/\partial p, \quad \dot{p} = -\partial H/\partial x. \tag{1.5}$$

(IV) Measurements

Let an observable be represented by the linear, Hermitian operator A, and consider the eigenvalue equation

$$A|\phi_n\rangle = a_n|\phi_n\rangle, \tag{1.6}$$

where $a_1, a_2, \ldots$ are the eigenvalues, and $|\phi_1\rangle, |\phi_2\rangle, \ldots$ the corresponding eigenvectors, or eigenkets. In general, there may be infinitely many eigenvalues and eigenkets. If k eigenkets correspond to the same eigenvalue a_l, then a_l is said to be k-fold degenerate. The following is postulated:

1. The outcome of any measurement of A is always one of its eigenvalues.
2. The eigenkets $|\phi_1\rangle, |\phi_2\rangle, \ldots$ form a complete set of states, i.e., they form a basis set that spans the state vector space.
3. If the state of a system is described by the normalized state vector $|\Psi(t)\rangle$, and if the states $|\phi_1\rangle, |\phi_2\rangle, \ldots$ are orthonormal, then the probability of finding the system in state $|\phi_n\rangle$ (in which case a measurement of observable A yields the eigenvalue a_n) at time t is given by $|\langle\phi_n|\Psi(t)\rangle|^2$. That is, $\langle\phi_n|\Psi(t)\rangle$ is the probability amplitude for a system, in state $|\Psi(t)\rangle$, to be found in state $|\phi_n\rangle$ at time t.
4. The state of a system, immediately following a measurement of A that gave the value a_n, collapses to the state $|\phi_n\rangle$ (if a_n is degenerate, the state collapses to the subspace spanned by the degenerate states corresponding to the eigenvalue a_n).

We note that the eigenvalues of a hermitian operator are real; hence, the outcome of any measurement of an observable is a real number, as it should be. Further, for a hermitian operator, the eigenkets corresponding to different eigenvalues are necessarily orthogonal. In the case of a k-fold degeneracy, where k eigenkets correspond to the same eigenvalue, every ket in the k-dimensional subspace that the eigenkets span is an eigenket of A with the same eigenvalue. It is always possible to choose within this subspace a set of k eigenkets that are orthogonal to each other. By normalizing the eigenkets, it is always possible to choose the eigenkets $|\phi_1\rangle, |\phi_2\rangle, \ldots$ so as to form a complete orthonormal basis that spans the vector space of state vectors. Orthonormality means that $\langle\phi_i|\phi_j\rangle = \delta_{ij}$ where

$$\delta_{ij} = \begin{cases} 0 & i \neq j \\ 1 & i = j \end{cases} \tag{1.7}$$

is the Kronecker delta, occasionally written as $\delta_{i,j}$ with a comma inserted between the indices if its absence could cause confusion. Completeness means that states $|\phi_1\rangle, |\phi_2\rangle, \ldots$ form a basis set: any state vector $|\Psi(t)\rangle \in \mathrm{V}$ can be expanded as

$$|\Psi(t)\rangle = \sum_n c_n(t)|\phi_n\rangle. \tag{1.8}$$

If the basis is chosen to be an orthonormal one, i.e., if $|\phi_1\rangle, |\phi_2\rangle, \ldots$ form an orthonormal set, then for an arbitrary state $|\Psi(t)\rangle$,

$$|\Psi(t)\rangle = \sum_n c_n(t)|\phi_n\rangle \Rightarrow \langle\phi_m|\Psi(t)\rangle = \sum_n c_n(t)\langle\phi_m|\phi_n\rangle = \sum_n c_n(t)\delta_{nm}$$
$$= c_m(t) \Rightarrow |\Psi(t)\rangle = \sum_n \langle\phi_n|\Psi(t)\rangle\,|\phi_n\rangle = \sum_n |\phi_n\rangle\langle\phi_n|\Psi(t)\rangle$$
$$\Rightarrow \sum_n |\phi_n\rangle\langle\phi_n| = 1. \tag{1.9}$$

Equation (1.9) expresses mathematically the property of completeness of the orthonormal states $|\phi_1\rangle, |\phi_2\rangle, \ldots$. Note that $|\phi_n\rangle\langle\phi_n|$ is an operator: it acts on a ket to yield another ket, and the 1 on the right hand side (RHS) of Eq. (1.9) is the identity operator.

An important complete orthonormal set of states is formed by the eigenkets of the position operator $\mathbf{r}$,

$$\mathbf{r}|\mathbf{r}\rangle = \mathbf{r}|\mathbf{r}\rangle. \tag{1.10}$$

On the left hand side (LHS), $\mathbf{r}$ is the position operator, sometimes written as $\hat{\mathbf{r}}$ or $\mathbf{r}_{op}$ to emphasize that it is an operator, while $\mathbf{r}$ on the RHS is the eigenvalue of the position operator. The ket $|\mathbf{r}\rangle$ is the state of a particle with a well defined position $\mathbf{r}$. Since the operator $\mathbf{r}$ is hermitian, the states $|\mathbf{r}\rangle$ form a complete orthonormal set. Because $\mathbf{r}$ is continuous, the orthonormality and completeness of the states now read

$$\langle\mathbf{r}|\mathbf{r}'\rangle = \delta(\mathbf{r}-\mathbf{r}') \text{ (orthonormality)}, \qquad \int |\mathbf{r}\rangle\langle\mathbf{r}|\, d^3r = 1 \text{ (completeness)}. \tag{1.11}$$

$\delta(\mathbf{r}-\mathbf{r}')$ is the Dirac-delta function, defined as follows:

$$\delta(\mathbf{r}) = \begin{cases} 0 & \mathbf{r} \neq 0 \\ \infty & \mathbf{r} = 0 \end{cases} \tag{1.12}$$

and

$$\int \delta(\mathbf{r})d^3r = 1, \tag{1.13}$$

the integration being over all space. In one dimension, $\delta(x-x')$ is represented graphically as in Figure 1.3.

One important property of $\delta(\mathbf{r}-\mathbf{r}')$ is the sifting property,

$$\int f(\mathbf{r})\delta(\mathbf{r}-\mathbf{r}')d^3r = f(\mathbf{r}'). \tag{1.14}$$

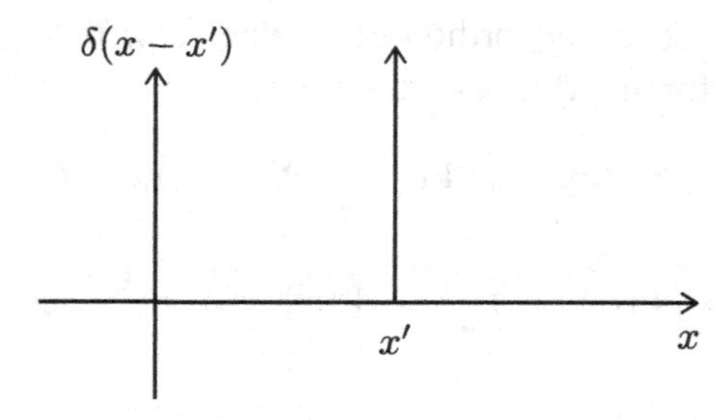

Figure 1.3 Dirac-delta function $\delta(x - x')$. It is zero for all values of x except for $x = x'$ where it is infinite. However, its integral over any interval containing x' is unity.

We also note that $\delta(\mathbf{r} - \mathbf{r}') = \delta(\mathbf{r}' - \mathbf{r})$ and $\delta(a\mathbf{r}) = \delta(\mathbf{r})/|a|^d$, where d is the dimension of space: $d = 3$ if $\mathbf{r}$ is a three-dimensional vector. A particularly useful representation of the Dirac-delta function is the following:

$$\delta(\mathbf{r}) = \frac{1}{(2\pi)^3}\int e^{\pm i\mathbf{k}.\mathbf{r}} d^3k. \tag{1.15}$$

Another useful representation of the Dirac-delta function is

$$\delta(x) = d\theta(x)/dx \tag{1.16}$$

where $\theta(x)$ is the step function:

$$\theta(x) = \begin{cases} 0 & x < 0 \\ 1 & x > 0. \end{cases} \tag{1.17}$$

Note that $d\theta(x)/dx = 0$ for $x \neq 0$, $d\theta(x)/dx = \infty$ for $x = 0$, and the integral of $d\theta(x)/dx$ over any interval that includes $x = 0$ is equal to 1.

Introducing a resolution of identity ($1 = \int |\mathbf{r}\rangle\langle\mathbf{r}| d^3r$), the state vector $|\Psi(t)\rangle$ may be written as

$$|\Psi(t)\rangle = \int |\mathbf{r}\rangle\langle\mathbf{r}\,|\Psi(t)\rangle\, d^3r.$$

This is the continuous analog of the discrete case for which $|\Psi(t)\rangle = \sum_n |\phi_n\rangle\langle\phi_n\,|\Psi(t)\rangle$.

$|\langle\phi_n\,|\Psi(t)\rangle|^2$ has been interpreted as the probability for a particle in state $|\Psi(t)\rangle$ to be found at time t in state $|\phi_n\rangle$. By analogy, should $|\langle\mathbf{r}\,|\Psi(t)\rangle|^2$ be interpreted as the probability for a particle in state $|\Psi(t)\rangle$ to be found at time t in state $|\mathbf{r}\rangle$, i.e., to be at position $\mathbf{r}$ at time t? Two problems beset this interpretation:

(a) $|\Psi(t)\rangle$ and $|\phi_n\rangle$ are dimensionless ($\langle\Psi(t)\,|\Psi(t)\rangle = 1$, $\langle\phi_n|\phi_n\rangle = 1$); hence, $|\langle\phi_n\,|\Psi(t)\rangle|^2$ is dimensionless and can be interpreted as a probability. However, the orthonormality and completeness relations for states $|\mathbf{r}\rangle$, as expressed in Eq. (1.11), reveal that states $|\mathbf{r}\rangle$ have dimension $1/\text{Length}^{3/2}$. Thus, $|\langle\mathbf{r}\,|\Psi(t)\rangle|^2$ has dimension 1/Volume, and it cannot be interpreted as a probability; rather, it is more properly interpreted as a probability density.

(b) Suppose that a particle is in state $|\Psi(t)\rangle$ and a measurement is carried out to determine its position. No detector could ever pinpoint the location of a particle to exactly one point; the best a detector could do is to "click" whenever the particle is in some small volume d^3r surrounding the position $\mathbf{r}$. Assuming that $\langle\mathbf{r}\,|\Psi(t)\rangle$ does not change appreciably within the volume d^3r, the probability that the detector clicks should be proportional to $|\langle\mathbf{r}\,|\Psi(t)\rangle|^2 d^3r$. The constant of proportionality is determined by requiring that the probability of finding the particle somewhere in space be unity. Noting that

$$\int |\langle\mathbf{r}\,|\Psi(t)\rangle|^2 d^3r = \int d^3r \langle\mathbf{r}\,|\Psi(t)\rangle^* \langle\mathbf{r}\,|\Psi(t)\rangle = \int d^3r \langle\Psi(t)|\mathbf{r}\rangle\langle\mathbf{r}\,|\Psi(t)\rangle$$
$$= \langle\Psi(t)\,|\Psi(t)\rangle = 1,$$

the proportionality constant is seen to be 1. $\langle\mathbf{r}\,|\Psi(t)\rangle|^2 d^3r$ is thus interpreted as the probability for a particle in state $|\Psi(t)\rangle$ to be found at time t in the infinitesimal volume d^3r centered on $\mathbf{r}$. Comparing this with the probabilistic interpretation of $\Psi(\mathbf{r}, t)$ given in postulate I, the following identification is made:

$$\langle\mathbf{r}\,|\Psi(t)\rangle = \Psi(\mathbf{r}, t). \tag{1.18}$$

The state vector $|\Psi(t)\rangle$ may now be written as

$$|\Psi(t)\rangle = \int |\mathbf{r}\rangle\langle\mathbf{r}\,|\Psi(t)\rangle\, d^3r = \int \Psi(\mathbf{r}, t)|\mathbf{r}\rangle d^3r.$$

In other words, the wave function $\Psi(\mathbf{r}, t)$ is the component of state vector $|\Psi(t)\rangle$ along $|\mathbf{r}\rangle$.

In the $\mathbf{r}$-representation, the orthonormality of states $|\phi_1\rangle$, $|\phi_2\rangle, \ldots$ reads

$$\delta_{ij} = \langle\phi_i|\phi_j\rangle = \int\langle\phi_i|\mathbf{r}\rangle\langle\mathbf{r}|\phi_j\rangle d^3r = \int\phi_i^*(\mathbf{r})\,\phi_j(\mathbf{r})\,d^3r, \tag{1.19}$$

and their completeness is expressed as

$$1 = \sum_n |\phi_n\rangle\langle\phi_n| = \sum_n \iint |\mathbf{r}\rangle\langle\mathbf{r}|\phi_n\rangle\langle\phi_n|\mathbf{r}'\rangle\langle\mathbf{r}'|\,d^3r\,d^3r'$$
$$= \iint \sum_n \phi_n(\mathbf{r})\,\phi_n^*(\mathbf{r}')|\mathbf{r}\rangle\langle\mathbf{r}'|\,d^3r\,d^3r'.$$

For the above to be true, it must be that

$$\sum_n \phi_n(\mathbf{r})\, \phi_n^*(\mathbf{r}') = \delta(\mathbf{r} - \mathbf{r}'). \tag{1.20}$$

This expresses the completeness property in the $\mathbf{r}$-representation.

We note that if operators A and B, representing two observables, commute ($AB = BA$), a complete set of states can be chosen so as to be simultaneous eigenstates of A and B; A and B may then be measured simultaneously.

So far, the fact that particles have spin has been ignored. To specify the state of a particle, its spin state must also be specified. For example, an electron has spin $s = 1/2$, and the z-component S_z of the spin operator has eigenvalues $+\hbar/2$ and $-\hbar/2$,

$$S_z\,|\uparrow\rangle = \frac{\hbar}{2}\,|\uparrow\rangle, \quad S_z\,|\downarrow\rangle = -\frac{\hbar}{2}\,|\downarrow\rangle. \tag{1.21}$$

The spin-up state $|\uparrow\rangle$ is also denoted by $|1/2\rangle$, or $|+\rangle$, or α, while the spin-down state may also be written as $|-1/2\rangle$, or $|-\rangle$, or β. A general spin state, denoted by $|\chi\rangle$, is a linear combination of the basis states $|\uparrow\rangle$ and $|\downarrow\rangle$,

$$|\chi\rangle = a\,|\uparrow\rangle + b\,|\downarrow\rangle$$

where $a = \langle\uparrow|\chi\rangle$ and $b = \langle\downarrow|\chi\rangle$. If $|\chi\rangle$ is normalized ($\langle\chi|\chi\rangle = 1$), the probability of finding the spin up is $|a|^2$ and that of finding it down is $|b|^2$.

The spin states $|\uparrow\rangle$ and $|\downarrow\rangle$ span a two-dimensional complex vector space, the spin space V_{spin}: they form an orthonormal basis for V_{spin},

$$\langle\uparrow|\uparrow\rangle = \langle\downarrow|\downarrow\rangle = 1, \quad \langle\uparrow|\downarrow\rangle = 0, \quad |\uparrow\rangle\langle\uparrow| + |\downarrow\rangle\langle\downarrow| = 1. \tag{1.22}$$

The above equations express, respectively, normalization, orthogonality, and completeness of the spin states. In general, for a particular spin s, the spin projection $\sigma = -s, -s+1, \ldots, s$; the spin space is a $(2s+1)$-dimensional complex vector space. The orthonormality and completeness relations are

$$\langle\sigma|\sigma'\rangle = \delta_{\sigma\sigma'}, \quad \sum_\sigma |\sigma\rangle\langle\sigma| = 1 \tag{1.23}$$

where $\sigma, \sigma' = -s, -s+1, \ldots, s$.

On the other hand, the states $|\phi_n\rangle$, which are eigenstates of a linear hermitian operator that depends on spatial coordinates, span a spatial vector space $\mathrm{V}_{\text{spatial}}$. The states $|\phi_\nu\rangle = |\phi_n\rangle \otimes |\sigma\rangle$, $\sigma = -s, -s+1, \ldots, s$ and $n = 1, 2, \ldots$ form an orthonormal basis for the direct product space $\mathrm{V} = \mathrm{V}_{\text{spatial}} \otimes \mathrm{V}_{\text{spin}}$, known as the Hilbert space. The state of a particle is a vector $|\Psi(t)\rangle \in \mathrm{V}$; hence, it can be expanded in the basis states,

$$|\Psi(t)\rangle = \sum_{n\sigma} c_{n\sigma}(t)\, |\phi_n\rangle \otimes |\sigma\rangle = \sum_{\nu} c_\nu(t)|\phi_\nu\rangle. \tag{1.24}$$

Here, ν is a collective index that specifies the spatial and spin quantum numbers. For example, four quantum numbers specify the eigenstates of a hydrogen atom: the principal quantum number n that determines the energy of the state, l which determines the value of L^2 (the square of the orbital angular momentum), m which determines the value of L_z (the z-component of the orbital angular momentum), and σ which is either $\uparrow$ or $\downarrow$. In this case $\nu = [nlm\sigma]$, while the index n in $|\phi_n\rangle$ stands for the spatial quantum numbers $[nlm]$. The ket $|\phi_\nu\rangle = |\phi_n\rangle \otimes |\sigma\rangle$, being a direct product of an orbital (spatial) state and a spin state, is called a spin orbital.

The orthonormality and completeness of the states $|\phi_\nu\rangle$ mean that

$$\langle\phi_\nu|\phi_{\nu'}\rangle = \langle\phi_n|\phi_{n'}\rangle\langle\sigma|\sigma'\rangle = \delta_{nn'}\delta_{\sigma\sigma'} = \delta_{\nu\nu'} \tag{1.25}$$

$$\sum_\nu |\phi_\nu\rangle\langle\phi_\nu| = \sum_n |\phi_n\rangle\langle\phi_n| \otimes \sum_\sigma |\sigma\rangle\langle\sigma| = 1_{\text{spatial}} \otimes 1_{\text{spin}} = 1. \tag{1.26}$$

Here, 1_{spatial} (1_{spin}) is the identity operator in V_{spatial} (V_{spin}), and 1 on the RHS is the identity operator in the Hilbert space (the direct product space).

So far, we have restricted the discussion to a one-particle system. We now consider a system comprised of N identical particles. Identical particles, such as electrons, are truly indistinguishable in quantum mechanics. The stationary states (eigenfunctions of the Hamiltonian H) of a system of N identical particles will be written as $\Psi(1, 2, \ldots, N)$, depending on the spatial and spin coordinates of the particles. Because of the indistinguishability of the particles, the Hamiltonian remains unchanged if any two particles are interchanged. This means that H commutes with P_{ij}, the permutation operator which interchanges particles i and j. It follows that the eigenfunctions of H can be chosen to be simultaneously eigenfunctions of P_{ij}. Denoting the eigenvalues of P_{ij} by λ, we can write

$$P_{ij}\Psi(1, \ldots, i, \ldots, j, \ldots, N) = \lambda\Psi(1, \ldots, i, \ldots, j, \ldots, N).$$

Applying P_{ij} to both sides of the above equation, and noting that $P_{ij}^2 = 1$, we obtain

$$\Psi(1, \ldots, i, \ldots, j, \ldots, N) = \lambda^2\Psi(1, \ldots, i, \ldots, j, \ldots, N).$$

Thus, $\lambda^2 = 1 \Rightarrow \lambda = \pm 1$. For $\lambda = +1(-1)$, the wave function is symmetric (antisymmetric) under the exchange of coordinates (spatial and spin) of any two particles. In nature, particles with integral spin (0, 1, 2, ...), known as bosons, have symmetric wave functions under the exchange of the coordinates of two particles, and they obey Bose–Einstein statistics. On the other hand, particles with half integral spin (1/2, 3/2, ...), known as fermions, have antisymmetric wave functions under the exchange of the coordinates of two particles, and they obey Fermi–Dirac statistics. The Pauli exclusion principle is a direct consequence of this antisymmetry of the fermionic wave function. The last postulate of quantum mechanics follows.

(V) Wave function of a system of identical particles

Under the interchange of all coordinates (spatial and spin) of one particle with those of another, the wave function of a collection of identical particles must be symmetric if the particles are bosons, and antisymmetric if the particles are fermions:

$$\Psi(1, \ldots, j, \ldots, i, \ldots, N) = \begin{cases} \Psi(1, \ldots, i, \ldots, j, \ldots, N) & \text{Bosons} \\ -\Psi(1, \ldots, i, \ldots, j, \ldots, N) & \text{Fermions.} \end{cases} \tag{1.27}$$

We close this section by remarking that some exotic quasiparticles, known as anyons, which arise as excitations of a two-dimensional electron gas in a magnetic field, are believed to obey some fractional statistics, which are neither Bose–Einstein nor Fermi–Dirac statistics (Wilczek, 1982).

1.2 The harmonic oscillator

We briefly review the solution of the harmonic oscillator problem in quantum mechanics. For a particle of mass m confined to a harmonic potential, the Hamiltonian is given by

$$H = \frac{p^2}{2m} + \frac{1}{2}m\omega^2 x^2, \tag{1.28}$$

where ω is the oscillator frequency. We introduce two new operators

$$a = \left(\frac{m\omega}{2\hbar}\right)^{1/2}\left(x + \frac{i}{m\omega}p\right), \quad a^\dagger = \left(\frac{m\omega}{2\hbar}\right)^{1/2}\left(x - \frac{i}{m\omega}p\right). \tag{1.29}$$

Since x and p are hermitian, $a^\dagger$ is the adjoint of a, and vice versa. The operators x and $p = -i\hbar d/dx$ do not commute: $xp \neq px$. We define the commutator of any two operators A and B by

$$[A, B] = AB - BA. \tag{1.30}$$

By letting the commutator $[x, p]$ act on an arbitrary differentiable function $f(x)$, it is found that $[x, p] = i\hbar$. It follows that $[a, a^\dagger] = 1$. In terms of the new operators,

$$H = \hbar\omega(N + 1/2), \quad N = a^\dagger a. \tag{1.31}$$

The hermitian operator N is called the number operator. Let the eigenvalues of N be denoted by n and the corresponding eigenkets by $|n\rangle$,

$$N|n\rangle = n|n\rangle. \tag{1.32}$$

The relation $[AB, C] = A[B, C] + [A, C]B$, easily verified, implies that

$$[N, a] = -a, \quad [N, a^\dagger] = a^\dagger.$$

With the help of these commutation relations we can easily show that

$$Na|n\rangle = (n-1)a|n\rangle, \quad Na^\dagger|n\rangle = (n+1)a^\dagger|n\rangle.$$

Hence, we can write

$$a|n\rangle = c|n-1\rangle, \quad a^\dagger|n\rangle = c'|n+1\rangle$$

where c and c' are constants. The first relation implies that $\langle n|a^\dagger = c^*\langle n-1|$. Therefore $\langle n|a^\dagger a|n\rangle = c^*c\langle n-1|n-1\rangle$. But $\langle n|a^\dagger a|n\rangle = n\langle n|n\rangle$; if we thus require that the eigenkets be normalized ($\langle n|n\rangle = \langle n-1|n-1\rangle = 1$), then $c^*c = n$, and we may choose $c = \sqrt{n}$. Similar considerations yield $c' = \sqrt{n+1}$. Thus,

$$a|n\rangle = \sqrt{n}|n-1\rangle, \quad a^\dagger|n\rangle = \sqrt{n+1}|n+1\rangle, \quad H|n\rangle = (n+1/2)\hbar\omega|n\rangle. \tag{1.33}$$

From the above equation we find that $a|0\rangle = 0$.

Let $|\beta\rangle = a|n\rangle$, then $\langle\beta| = \langle n|a^\dagger$, and $\langle\beta|\beta\rangle = \langle n|a^\dagger a|n\rangle = n\langle n|n\rangle = n$. But $\langle\beta|\beta\rangle \geq 0$; hence $n \geq 0$. Starting with a ket $|n\rangle$, we can apply the operator a repeatedly, each time lowering n by 1. If n is not an integer, we will end up with kets $|n\rangle$ having negative values of n, which is not allowed since $n \geq 0$. If n is an integer, then upon repeatedly applying the operator a we end up with the ket $|0\rangle$; further application of a gives $a|0\rangle = 0$. Therefore n must be an integer, and the eigenvalues of the Hamiltonian are $(n+1/2)\hbar\omega$, where $n = 0, 1, 2, \ldots$; the energy is quantized in steps of $\hbar\omega$. Since $a^\dagger$ increases n by 1, it increases the energy by one quantum; in other words, it creates a quantum of energy and it acquires the name "creation operator." In contrast, a annihilates (destroys) one quantum of energy, and it is called the annihilation, or destruction, operator. The ground state wave function is obtained from the equation

$$a|0\rangle = 0 \Rightarrow \left(\frac{m\omega}{2\hbar}\right)^{1/2}\left(x + \frac{i}{m\omega}p\right)|0\rangle = 0,$$

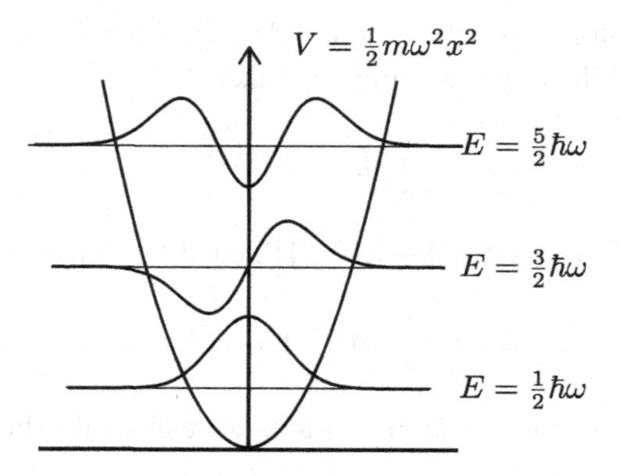

Figure 1.4 Eigenfunctions corresponding to the three lowest energy levels of a one-dimensional harmonic oscillator.

which, upon replacing p by $-i\hbar d/dx$, translates to

$$\left(x + \frac{\hbar}{m\omega}\frac{d}{dx}\right)\phi_0 = 0$$

$$\Longrightarrow \phi_0(x) = \left(\frac{m\omega}{\pi\hbar}\right)^{1/4} \exp\left(-\frac{m\omega}{2\hbar}x^2\right). \tag{1.34}$$

The first excited state $|1\rangle$ is given by $|1\rangle = a^\dagger|0\rangle$; we find

$$\phi_1(x) = \left(\frac{b}{2\sqrt{\pi}}\right)^{1/2} (2bx)\exp(-b^2x^2/2) \tag{1.35}$$

where $b = (m\omega/\hbar)^{1/2}$. We can continue in this fashion; in general,

$$\phi_n(x) = A_n H_n(bx)\exp(-b^2x^2/2). \tag{1.36}$$

A_n is a normalization factor and H_n is the Hermite polynomial of order n,

$$H_n(\xi) = (-1)^n \exp(\xi^2)\,\frac{d^n}{d\xi^n}\exp(-\xi^2). \tag{1.37}$$

The eigenfunctions corresponding to a few of the lowest energy levels are shown in Figure 1.4. Note that for n even (odd) the eigenfunctions are even (odd) functions of x.

In the traditional approach to the harmonic oscillator problem, one instead employs a power series solution to the time-independent Schrödinger equation, which is a second-differential equation. It is through the requirement that the wave function vanish at infinity that a truncation of the series is brought about. This, in turn, leads to the quantization of energy. One might ask where the boundary conditions (the wave function vanishes at $x = \pm\infty$) were used in the above discussion. They were used indirectly when it was required that the stationary states $|n\rangle$ be normalized to unity. This can only be true if the corresponding eigenfunctions vanish at infinity.

Further reading

Griffiths, D.J. (2005). *Introduction to Quantum Mechanics*, 2nd edn. Upper Saddle River, NJ: Pearson Education, Inc.

Sakurai, J.J. (1994). *Modern Quantum Mechanics*, rev. edn. Reading, MA: Addison-Wesley Publishing Company, Inc.

Problems

1.1 *Operators.*
  (a) Evaluate the commutators $[x, p]$, $[x^2, p]$, and $[p, V(x)]$.
  (b) Show that $(AB)^\dagger = B^\dagger A^\dagger$.
  (c) Show that $Tr(ABC) = Tr(CAB)$, where Tr is the trace.
  (d) Show that by setting $\mathbf{S} = (\hbar/2)\boldsymbol{\sigma}$, the Pauli spin matrices

$$\sigma_x = \begin{bmatrix} 0 & 1 \\ 1 & 0 \end{bmatrix}, \quad \sigma_y = \begin{bmatrix} 0 & -i \\ i & 0 \end{bmatrix}, \quad \sigma_z = \begin{bmatrix} 1 & 0 \\ 0 & -1 \end{bmatrix}$$

  provide a valid representation for the spin operator $\mathbf{S}$.
  (e) Find $S_x|\uparrow\rangle$, $S_x|\downarrow\rangle$, $S_y|\uparrow\rangle$, and $S_y|\downarrow\rangle$.

1.2 *Delta-function representation.* Show that $\int_{-\infty}^{\infty} e^{ikx}dk = 2\pi\delta(x)$.

1.3 *Another delta function representation.* Show that

$$\lim_{a\to\infty} \frac{\sin^2(ax)}{\pi a x^2} = \delta(x).$$

1.4 *Periodic boundary conditions.* An electron is confined to a cube of side L. Assume that the eigenfunctions obey periodic boundary conditions,

$$\phi(x, y, z) = \phi(x + L, y, z) = \phi(x, y + L, z) = \phi(x, y, z + L).$$

Under these boundary conditions, opposite faces of the cube are identified; for example, the faces $x = 0$ and $x = L$ are the same. The ranges of values of x, y, and z are $0 \le x < L$, $0 \le y < L$, $0 \le z < L$. Find the normalized eigenfunctions and show by explicit calculation that they form a complete orthonormal set of states.

1.5 *Singlet and triplet states.* Consider a two-electron system (for example, the two electrons in a helium atom). The total spin $\mathbf{S} = \mathbf{S}_1 + \mathbf{S}_2$, where $\mathbf{S}_1$ and $\mathbf{S}_2$ are the spin operators of electrons 1 and 2. Consider the singlet state $\frac{1}{\sqrt{2}}[\alpha(1)\beta(2) - \beta(1)\alpha(2)]$ and the three triplet states $\alpha(1)\alpha(2)$, $\frac{1}{\sqrt{2}}[\alpha(1)\beta(2) + \beta(1)\alpha(2)]$, and $\beta(1)\beta(2)$. Show that these are eigenstates of S^2 and S_z.

1.6 *A particle bound by a delta-function potential.* A particle of mass m, moving in one dimension, is bound by the delta-function potential $V(x) = -\lambda\delta(x)$ where λ is a positive constant.
(a) Determine the energy of the bound state (for a bound state, in this case, $E < 0$).
(b) The potential energy is suddenly changed from $-\lambda\delta(x)$ to $-b\lambda\delta(x)$, where b is a dimensionless positive constant. What is the probability that the particle remains bound?

1.7 *Harmonic oscillator.* For a one-dimensional harmonic oscillator of mass m and frequency ω, in the ground state, show that $\langle p^2/2m\rangle = \langle \frac{1}{2}m\omega^2x^2\rangle = \hbar\omega/4$.

1.8 *Coherent states.* For a one-dimensional harmonic oscillator, show that:
(a) The operator $a^\dagger$ does not have any eigenstates.
(b) The state $|z\rangle = e^{-z^*z/2}e^{za^\dagger}|0\rangle$, for any complex number z, is a normalized eigenstate of the annihilation operator a with eigenvalue z.

1.9 *Time-independent perturbation.* Suppose that for a given system the Hamiltonian is $H = H_0 + V$, where H_0 and V have no explicit time dependence. Assume that the solutions of the eigenvalue equation $H_0|n\rangle = E_n|n\rangle$ are known, and that the energy levels are nondegenerate. If V is small, in the sense that the shift in the energy of the states brought about by the presence of V is small compared to the energy difference between neighboring states, i.e., $|\Delta E_n| \ll |E_{n\pm1} - E_n|$, one can show that

$$\Delta E_n = \langle n|V|n\rangle + \sum_{m\neq n} \frac{|\langle m|V|n\rangle|^2}{E_n - E_m} + \cdots .$$

Consider a one-dimensional harmonic oscillator subjected to a perturbation $V = \lambda x$.
(a) Calculate the shift in the energy of the ground state to second order in V.
(b) Solve the problem exactly.

1.10 *Heisenberg picture of quantum mechanics.* In the Schrödinger picture, the usual picture of quantum mechanics, the state evolves in time, but the operators representing observables are time-independent. Assuming that the Hamiltonian does not depend explicitly on time, the state evolves in time according to

$$|\psi_S(t)\rangle = e^{-iH(t-t_0)/\hbar}|\psi_S(t_0)\rangle.$$

Setting $t_0 = 0$, for simplicity, $|\psi_S(t)\rangle = e^{-iHt/\hbar}|\psi_S(0)\rangle$. The expectation value of an operator A varies with time according to

$$\langle A\rangle(t) = \langle\psi_S(t)|A|\psi_S(t)\rangle = \langle\psi_S(0)|e^{iHt/\hbar}Ae^{-iHt/\hbar}|\psi_S(0)\rangle.$$

The above suggests a second approach to quantum mechanics, the Heisenberg picture, in which the state is frozen at what it was at $t = 0$, but the operator A evolves in time. In this picture,

$$|\psi_H\rangle = |\psi_S(0)\rangle, \qquad A_H(t) = e^{iHt/\hbar}Ae^{-iHt/\hbar}.$$

(a) Derive the Heisenberg equation of motion

$$\frac{d}{dt}A_H(t) = \frac{i}{\hbar}[H, A_H(t)].$$

(b) Show that for the harmonic oscillator of frequency ω,

$$a(t) = a(0)e^{-i\omega t}, \qquad a^\dagger(t) = a^\dagger(0)e^{i\omega t}$$

where $a(t)$ and $a^\dagger(t)$ are the annihilation and creation operators in the Heisenberg picture.

1.11 *The interaction picture.* Let the Hamiltonian for a system be given by

$$H = H_0 + V(t)$$

where H_0 is time-independent. The state in the Schrödinger picture is $|\psi_S(t)\rangle$. In the interaction picture, the state is defined by

$$|\psi_I(t)\rangle = e^{iH_0t/\hbar}|\psi_S(t)\rangle.$$

If, in the Schrödinger picture, an observable is represented by the operator A, the corresponding operator in the interaction picture is

$$A_I(t) = e^{iH_0t/\hbar}Ae^{-iH_0t/\hbar}.$$

(a) Show that $\frac{d}{dt}A_I(t) = \frac{i}{\hbar}[H_0, A_I(t)]$.
(b) Show that $i\hbar\frac{\partial}{\partial t}|\psi_I(t)\rangle = V_I(t)|\psi_I(t)\rangle$.

(c) The evolution operator in the interaction picture, $U_I(t, t_0)$, is defined by $|\psi_I(t)\rangle = U_I(t, t_0)|\psi_I(t_0)\rangle$. What is the differential equation satisfied by $U_I(t, t_0)$?

(d) Show that

$$U_I(t, t_0) = 1 - (i/\hbar)\int_{t_0}^{t} dt_1 V_I(t_1) + (-i/\hbar)^2 \int_{t_0}^{t} dt_1 \int_{t_0}^{t_1} dt_2 V_I(t_1) V_I(t_2) + \cdots .$$

(e) If at time t_0 the system is in an eigenstate $|i\rangle$ of H_0, then at time t the state will be $U(t, t_0)|i\rangle$, where $U(t, t_0) = e^{-iH(t-t_0)/\hbar}$ is the time evolution operator in the Schrödinger picture. The probability of finding the system in an eigenstate $|f\rangle$ of H_0 at time t is $P_{i\to f} = |\langle f|U(t, t_0)|i\rangle|^2$. Show that $P_{i\to f} = |\langle f|U_I(t, t_0)|i\rangle|^2$.

1.12 *Fermi golden rule.* The Hamiltonian for a system is given by $H = H_0 + V(t)$ where

$$V(t) = \begin{cases} 0 & t < 0 \\ V & t \geq 0. \end{cases}$$

V has no explicit dependence on time, but it may depend on position, momentum, and spin. At $t = 0$, the system is in an eigenstate $|i\rangle$ of H_0. The probability of finding the system at time t in an eigenstate $|f\rangle$ of H_0 is $P_{i\to f}(t) = |\langle f|U_I(t)|i\rangle|^2$, as shown in the previous problem. By expanding $U_I(t)$ in V_I, we can calculate $P_{i\to f}$ to various orders of the perturbation. Let $\omega_{fi} = (E_f - E_i)/\hbar$.

(a) By expanding $U_I(t)$ to first order in V, show that

$$P_{i\to f}(t) = \frac{4|\langle f|V|i\rangle|^2}{\hbar^2} \frac{sin^2(\omega_{fi}t/2)}{\omega_{fi}^2}.$$

(b) Now let $t \to \infty$ (steady state). The transition rate, $w_{i\to f}$, is defined as $w_{i\to f} = \frac{d}{dt} \lim_{t\to\infty} P_{i\to f}(t)$. Prove the Fermi golden rule,

$$w_{i\to f} = \frac{2\pi}{\hbar} |\langle f|V|i\rangle|^2 \delta(E_f - E_i).$$

1.13 *Harmonic perturbation.* The Hamiltonian for a system is $H = H_0 + V(t)$, where $V(t)$ is a harmonic perturbation turned on at $t = 0$,

$$V(t) = Ae^{i\omega t} + A^\dagger e^{-i\omega t} \qquad t \geq 0.$$

A is a time-independent operator. For $t < 0$, the system is in state $|i\rangle$, where $H_0|i\rangle = E_i|i\rangle$. Expanding $U_I(t)$ to first order in the perturbation:

(a) Show that the probability of finding the system in state $|f\rangle$, also an eigenstate of H_0 with energy $E_f = E_i + \hbar\omega_{fi}$, is

$$P_{i\to f} = \frac{1}{\hbar^2}\left|\frac{1-e^{i(\omega_{fi}+\omega)t}}{\omega_{fi}+\omega}\langle f|A|i\rangle + \frac{1-e^{i(\omega_{fi}-\omega)t}}{\omega_{fi}-\omega}\langle f|A^\dagger|i\rangle\right|^2 .$$

(b) Show that, as $t \to \infty$, the transition rate is given by

$$w_{i\to f} = \frac{2\pi}{\hbar}\left[|\langle f|A|i\rangle|^2\delta(E_f - E_i + \hbar\omega) + |\langle f|A^\dagger|i\rangle|^2\delta(E_f - E_i - \hbar\omega)\right].$$

2

Single-particle states

Good order is the foundation of all good things.
–*Edmund Burkes,* Reflections on the
Revolution in France

2.1 Introduction

Let us consider a system of N identical, interacting particles whose Hamiltonian is

$$H(1, 2, \ldots, N) = \sum_{i=1}^{N} h(i) + \frac{1}{2} \sum_{i \neq j}^{N} v(i, j). \tag{2.1}$$

$h(i)$ is the sum of the kinetic energy and potential energy, due to some external field, of particle i,

$$h(i) = -\frac{\hbar^2}{2m} \nabla^2 + v(i). \tag{2.2}$$

For example, if the interacting particles are the electrons in an atom, then $v(i)$ is the potential energy of electron i due to its interaction with the nucleus. For electrons in a crystal, $v(i)$ is the interaction of electron i with the ionic lattice. The last term in Eq. (2.1) represents the interaction between the particles, taken as a sum over pairs (i, j). The summation is carried over both indices i and j, but terms with $i = j$ are excluded. The factor $1/2$ ensures that each pair is counted only once. In general, H may depend on the spatial as well as the spin coordinates of the particles.

Recall, from the fourth postulate of quantum mechanics, that the wave function of a particle can be expanded in terms of a complete set of states $|\phi_\nu\rangle$. Similarly, the N-particle wave function can be expanded in terms of a complete set of N-particle states. These are constructed as properly symmetrized products of single-particle states (SPSs), as we will show in Chapter 3. Although any complete set of SPSs is

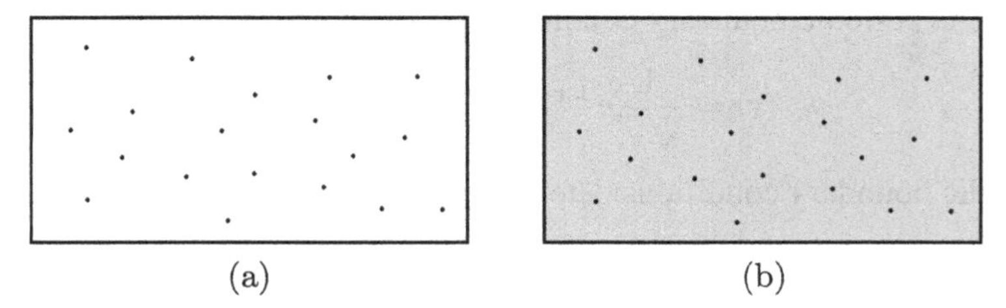

Figure 2.1 (a) The free electron model: the interactions of the electrons with each other and with the ions are ignored; each electron moves freely within the crystal and its wave function obeys periodic boundary conditions. (b) The jellium model: the electrons interact with each other and with a uniform positive background.

adequate for this purpose, a more convenient set is generated by the single-particle Hamiltonian h through the eigenvalue equation

$$h|\phi_\nu\rangle = \epsilon_\nu|\phi_\nu\rangle. \tag{2.3}$$

Each SPS $|\phi_\nu\rangle$ is characterized by a set of quantum numbers that are collectively denoted by ν. In the following sections we describe convenient SPSs for free electrons, electrons in a periodic potential, and electrons in a two-dimensional system that is in the presence of a magnetic field.

2.2 Electron gas

Consider a system of N electrons confined to a cube of side L and volume $V = L^3$. In the description of a metal within the free electron model, the interactions of the electrons with each other, and with the ions, are ignored. The Hamiltonian is then simply the sum of the kinetic energies of the electrons. In the so-called jellium model, the lattice ions are replaced by a uniform positive background, i.e., the positive charges on the ions are smeared so as to fill the crystal in such a way that the charge density is constant, and the electron–electron, electron–background, and background–background interactions are taken into account. The two models are illustrated in Figure 2.1. For either model, the convenient SPSs are those of a free electron confined to a cube of side L, with periodic boundary conditions,

$$\phi(x, y, z) = \phi(x + L, y, z) = \phi(x, y + L, z) = \phi(x, y, z + L). \tag{2.4}$$

The SPSs are found by solving the Schrödinger equation

$$-\frac{\hbar^2}{2m}\nabla^2\phi = \epsilon\phi, \tag{2.5}$$

subject to the periodic boundary conditions given above. We obtain

$$\phi_{\mathbf{k}\sigma}(\mathbf{r}) = \frac{1}{\sqrt{V}} e^{i\mathbf{k}.\mathbf{r}} |\sigma\rangle, \quad \epsilon_{\mathbf{k}\sigma} = \hbar^2 \mathbf{k}^2 / 2m. \tag{2.6}$$

The periodic boundary conditions determine the allowed values of **k**,

$$k_x, k_y, k_z = 0, \pm 2\pi/L, \pm 4\pi/L, \cdots = 2n\pi/L, \quad n \in \mathbb{Z}. \tag{2.7}$$

The spin ket $|\sigma\rangle$ is either $|\uparrow\rangle$ or $|\downarrow\rangle$. In Dirac notation, the SPSs are denoted by $|\mathbf{k}\sigma\rangle$. An SPS is thus described by four quantum numbers: k_x, k_y, k_z, and σ. The SPSs form an orthonormal set,

$$\langle \mathbf{k}'\sigma' | \mathbf{k}\sigma \rangle = \frac{1}{V} \int e^{i(\mathbf{k}-\mathbf{k}').\mathbf{r}} d^3 r \langle \sigma' | \sigma \rangle = \delta_{\mathbf{k}\mathbf{k}'} \delta_{\sigma\sigma'}, \tag{2.8}$$

and the set is complete,

$$\sum_{\mathbf{k}\sigma} \phi_{\mathbf{k}\sigma}(\mathbf{r}) \phi^*_{\mathbf{k}\sigma}(\mathbf{r}') = \frac{1}{V} \frac{V}{(2\pi)^3} \int d^3 k e^{i\mathbf{k}.(\mathbf{r}-\mathbf{r}')} \sum_{\sigma} |\sigma\rangle\langle\sigma| = \delta(\mathbf{r} - \mathbf{r}'). \tag{2.9}$$

In the above equation, we have made the replacement

$$\sum_{\mathbf{k}} F(\mathbf{k}) \rightarrow \frac{V}{(2\pi)^3} \int d^3 k \, F(\mathbf{k}).$$

This is justified as follows. Consider a volume $d^3 k$ in **k**-space, which is large compared to $(2\pi)^3/V$, but sufficiently small on the scale of variation of $F(\mathbf{k})$, i.e., $F(\mathbf{k})$ has almost the same value within the cube $d^3 k$ centered on **k**. Since the volume in **k**-space occupied by one **k**-point, as deduced from Eq. (2.7), is $(2\pi)^3/V$, the volume $d^3 k$ contains $V d^3 k/(2\pi)^3$ **k**-points. We have also used $\sum_\sigma |\sigma\rangle\langle\sigma| = 1$ and the representation of the Dirac-delta function given in Eq. (1.15).

In the ground state at zero temperature, electrons fill the states of lowest energy. Within the free electron model, where $\epsilon_{\mathbf{k}\sigma} = \hbar^2 \mathbf{k}^2/2m$, electrons fill states inside a sphere in **k**-space, known as the Fermi sphere, which is depicted in Figure 2.2. The volume of the sphere is $4\pi k_F^3/3$, where k_F, the Fermi wave vector (or wave number), is the radius of the Fermi sphere. Since each **k**-point occupies a volume $(2\pi)^3/V$ in **k**-space, the number of **k**-points within the Fermi sphere is $4\pi k_F^3/3/(2\pi)^3/V$. Each **k**-point can accommodate two electrons, one with spin up and one with spin down. It follows that

$$N = 2(\text{number of } \mathbf{k}\text{-points within the Fermi sphere}) = V k_F^3 / 3\pi^2.$$

In terms of the electron number density $n = N/V$, the Fermi wave vector is thus given by

$$k_F = (3\pi^2 n)^{1/3}. \tag{2.10}$$

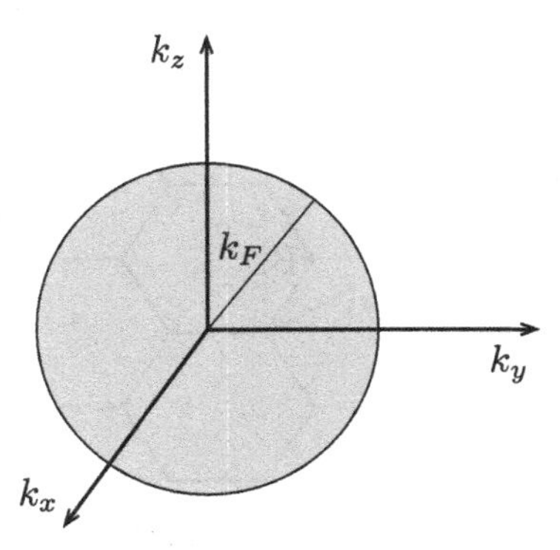

Figure 2.2 Fermi sphere of radius k_F. At zero temperature the states within the sphere are all occupied, while the states outside the sphere are all empty.

The mean energy of an electron is $\bar{\epsilon} = E/N$, where E is the total energy,

$$E = \sum_{\mathbf{k}\sigma} \hbar^2 k^2/2m = 2\sum_{\mathbf{k}} \hbar^2 k^2/2m.$$

The factor 2 arises from summing over σ ($\uparrow$ or $\downarrow$). The sum over $\mathbf{k}$ runs over all vectors within the Fermi sphere; replacing the sum by an integral,

$$E = 2\frac{V}{(2\pi)^3}\int d^3k\, \hbar^2 k^2/2m = \frac{V\hbar^2}{2\pi^2 m}\int_0^{k_F} k^4\, dk = \frac{V\hbar^2}{10\pi^2 m}k_F^5 = \frac{V}{5\pi^2}k_F^3 E_F$$

where $E_F = \hbar^2 k_F^2/2m$ is the Fermi energy. Since $k_F^3 = 3\pi^2 N/V$ (see Eq. [2.10]),

$$\bar{\epsilon} = E/N = 3E_F/5. \tag{2.11}$$

2.3 Bloch states

In the free electron model, lattice ions are ignored. In the jellium model, a uniform positive background replaces the ions. These models cannot explain why some crystals are metals while others are insulators. In reality, the ions vibrate about their equilibrium positions, which form a periodic structure. If the ions were fixed at their equilibrium positions, they would produce a fixed potential in which an electron would move, and the resulting stationary states of the electron would be the Bloch states. Ionic vibrations result in a time-dependent potential that causes scattering among the stationary states.

A three-dimensional lattice, generated by three noncoplanar vectors $\mathbf{a}_1$, $\mathbf{a}_2$, $\mathbf{a}_3$, and one point in space, is the set of all points that can be obtained from this point by all translations by vectors $\mathbf{R}_{n_1n_2n_3} = n_1\mathbf{a}_1 + n_2\mathbf{a}_2 + n_3\mathbf{a}_3$, where n_1, n_2, n_3 are integers ($n_1, n_2, n_3 \in \mathbb{Z}$). That is, if we start from any lattice point and undergo a displacement $\mathbf{R}_{n_1n_2n_3}$, for any integers n_1, n_2, and n_3, we will encounter another

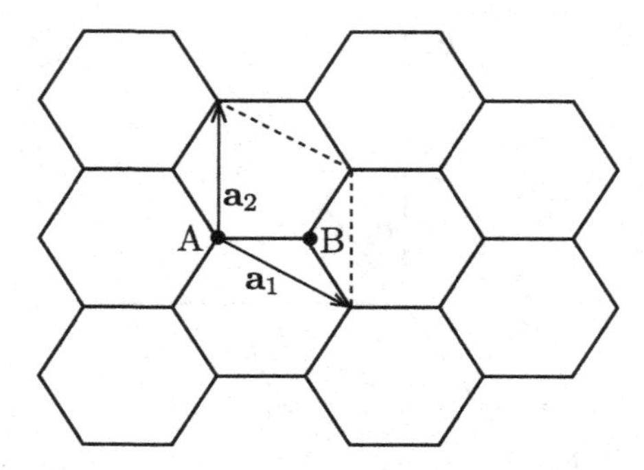

Figure 2.3 Graphene, a two-dimensional crystal. The primitive lattice vectors are $\mathbf{a}_1$ and $\mathbf{a}_2$, and the basis consists of two carbon atoms, A and B. The parallelogram formed by $\mathbf{a}_1$ and $\mathbf{a}_2$ is the unit cell.

lattice point. The vectors $\mathbf{a}_1$, $\mathbf{a}_2$, and $\mathbf{a}_3$ are called the primitive lattice vectors, and the parallelepiped they form is the primitive, or unit, cell. A crystal is obtained if one or more atoms, called the basis, are placed in each unit cell. We may thus write

$$\text{Crystal} = \text{Lattice} + \text{Basis}.$$

For example, the two-dimensional crystal graphene (Figure 2.3) has a honeycomb structure, with carbon atoms occupying the hexagonal corners. The primitive lattice vectors are $\mathbf{a}_1$ and $\mathbf{a}_2$, and the basis consists of two carbon atoms, A and B. The unit cell is the parallelogram formed by $\mathbf{a}_1$ and $\mathbf{a}_2$.

From the definition of a lattice, we infer that when the ions are at their equilibrium positions, the environment surrounding any point P in the crystal is identical to that surrounding any other point Q which is separated from P by a lattice vector $\mathbf{R}_{n_1n_2n_3}$ (see Figure 2.4). It follows that, if the ions sit at their equilibrium sites, the potential energy $V(\mathbf{r})$ of an electron (due to its interaction with the ions) has the same value at points P and Q. $V(\mathbf{r})$ is thus a periodic function of position, its periodicity being that of the lattice.

In a crystal, Bloch states form a convenient set of single-particle states (SPSs). These are solutions to the Schrödinger equation

$$h\phi_\nu = [-(\hbar^2/2m)\nabla^2 + V(\mathbf{r})]\phi_\nu = \epsilon_\nu\phi_\nu. \tag{2.12}$$

Here, $V(\mathbf{r})$ is a periodic potential with periodicity $\mathbf{R}$,

$$V(\mathbf{r} + \mathbf{R}) = V(\mathbf{r}) \tag{2.13}$$

where $\mathbf{R} = n_1\mathbf{a}_1 + n_2\mathbf{a}_2 + n_3\mathbf{a}_3,\ \ n_1, n_2, n_3 \in \mathbb{Z}$.

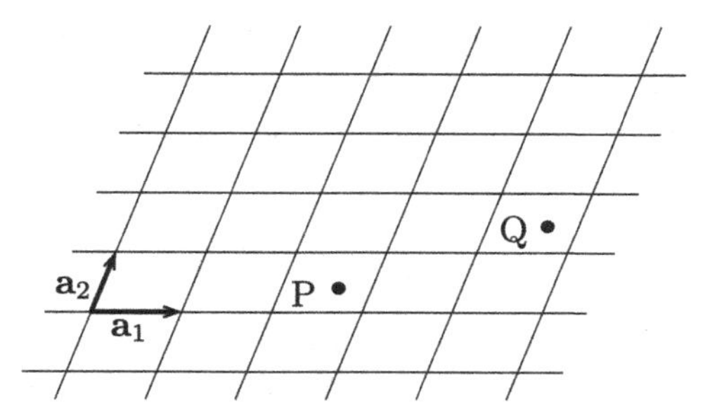

Figure 2.4 Two-dimensional crystal with primitive lattice vectors $\mathbf{a}_1$ and $\mathbf{a}_2$. The points P and Q are separated by a lattice vector equal to $2\mathbf{a}_1 + \mathbf{a}_2$. The potential felt by an electron at point P, due to its interaction with the ions, is identical to that felt by the electron at point Q.

An elegant way to solve the Schrödinger equation is by introducing the translation operator $T_{\mathbf{R}}$, defined by its action on an arbitrary function $f(\mathbf{r})$,

$$T_{\mathbf{R}}\, f(\mathbf{r}) = f(\mathbf{r} + \mathbf{R}). \tag{2.14}$$

Note that

$$T_{\mathbf{R}} T_{\mathbf{R}'} f(\mathbf{r}) = T_{\mathbf{R}} f(\mathbf{r} + \mathbf{R}') = f(\mathbf{r} + \mathbf{R} + \mathbf{R}') = T_{\mathbf{R}+\mathbf{R}'} f(\mathbf{r}) \Rightarrow T_{\mathbf{R}} T_{\mathbf{R}'} = T_{\mathbf{R}+\mathbf{R}'}.$$

Now consider

$$T_{\mathbf{R}} h(\mathbf{r}) f(\mathbf{r}) = T_{\mathbf{R}}[h(\mathbf{r}) f(\mathbf{r})] = h(\mathbf{r} + \mathbf{R}) f(\mathbf{r} + \mathbf{R}) = h(\mathbf{r}) T_{\mathbf{R}}\, f(\mathbf{r}). \tag{2.15}$$

The periodicity of the Hamiltonian, namely that $h(\mathbf{r} + \mathbf{R}) = h(\mathbf{r})$, results from (i) $\nabla^2_{\mathbf{r}+\mathbf{R}} = \nabla^2_{\mathbf{r}}$ and (ii) $V(\mathbf{r} + \mathbf{R}) = V(\mathbf{r})$. Since $f(\mathbf{r})$ is arbitrary, Eq. (2.15) implies that h and $T_{\mathbf{R}}$ commute. Furthermore,

$$T_{\mathbf{R}}\, T_{\mathbf{R}'}\, f(\mathbf{r}) = T_{\mathbf{R}}\, f(\mathbf{r} + \mathbf{R}') = f(\mathbf{r} + \mathbf{R}' + \mathbf{R}) = T_{\mathbf{R}'}\, f(\mathbf{r} + \mathbf{R}) = T_{\mathbf{R}'}\, T_{\mathbf{R}}\, f(\mathbf{r}).$$

Hence, $T_{\mathbf{R}}$ and $T_{\mathbf{R}'}$ commute. We conclude that $\{h, T_{\mathbf{R}}, T_{\mathbf{R}'}, \dots\}$ is a set of commuting operators. The eigenstates of h can thus be chosen so as to be simultaneously eigenstates of the translation operator $T_{\mathbf{R}}$ for every lattice vector $\mathbf{R}$. Let the eigenvalues of $T_{\mathbf{R}}$ be $\lambda(\mathbf{R})$: $T_{\mathbf{R}}\phi = \lambda(\mathbf{R})\phi$. Since

$$T_{\mathbf{R}'}\, T_{\mathbf{R}}\phi = T_{\mathbf{R}'}\lambda(\mathbf{R})\phi = \lambda(\mathbf{R})\, T_{\mathbf{R}'}\phi = \lambda(\mathbf{R})\lambda(\mathbf{R}')\phi,$$

and

$$T_{\mathbf{R}'}\, T_{\mathbf{R}}\phi = T_{\mathbf{R}+\mathbf{R}'}\phi = \lambda(\mathbf{R} + \mathbf{R}')\phi,$$

it follows that $\lambda(\mathbf{R})$ satisfies the equation

$$\lambda(\mathbf{R})\lambda(\mathbf{R}') = \lambda(\mathbf{R} + \mathbf{R}').$$

This holds if $\lambda(\mathbf{R}) = e^{i\mathbf{k}.\mathbf{R}}$ for some vector $\mathbf{k}$. For different values of $\mathbf{k}$, we get different values of the eigenvalue $\lambda(\mathbf{R})$, and correspondingly different eigenfunctions $\phi(\mathbf{r})$. Thus, $T_{\mathbf{R}}\phi_{\mathbf{k}}(\mathbf{r}) = e^{i\mathbf{k}.\mathbf{R}}\phi_{\mathbf{k}}(\mathbf{r})$. From the definition of the translation operator, it follows that, for any lattice vector $\mathbf{R}$,

$$\phi_{\mathbf{k}}(\mathbf{r} + \mathbf{R}) = e^{i\mathbf{k}.\mathbf{R}}\,\phi_{\mathbf{k}}(\mathbf{r}). \tag{2.16}$$

We have not yet indicated what values $\mathbf{k}$ may assume. These are determined by the periodic boundary conditions. If there are N_1 primitive cells along the direction of lattice vector $\mathbf{a}_1$, N_2 primitive cells along $\mathbf{a}_2$, and N_3 primitive cells along $\mathbf{a}_3$, the periodic boundary conditions take the form:

$$\phi_{\mathbf{k}}(\mathbf{r} + N_i\mathbf{a}_i) = \phi_{\mathbf{k}}(\mathbf{r}), \quad i = 1, 2, 3. \tag{2.17}$$

These boundary conditions are adopted under the assumption that the bulk properties of a crystal do not depend on the choice of boundary conditions on its surface. Periodic boundary conditions are more convenient for mathematical analysis than fixed boundary conditions, for which the wave function vanishes on the surface of the crystal. From Eq. (2.16), we can write

$$\phi_{\mathbf{k}}(\mathbf{r} + N_i\mathbf{a}_i) = e^{iN_i\mathbf{k}.\mathbf{a}_i}\,\phi_{\mathbf{k}}(\mathbf{r}), \quad i = 1, 2, 3.$$

Combined with the periodic boundary conditions, the above equation gives

$$e^{iN_i\mathbf{k}.\mathbf{a}_i} = 1, \quad i = 1, 2, 3. \tag{2.18}$$

In order to solve for $\mathbf{k}$, we introduce the reciprocal lattice vectors $\mathbf{b}_1$, $\mathbf{b}_2$, and $\mathbf{b}_3$ defined by

$$\mathbf{b}_1 = 2\pi\mathbf{a}_2 \times \mathbf{a}_3/\Omega, \quad \mathbf{b}_2 = 2\pi\mathbf{a}_3 \times \mathbf{a}_1/\Omega, \quad \mathbf{b}_3 = 2\pi\mathbf{a}_1 \times \mathbf{a}_2/\Omega. \tag{2.19}$$

Here, $\Omega = |\mathbf{a}_1.\mathbf{a}_2 \times \mathbf{a}_3|$ is the volume of the primitive cell. The vectors $\mathbf{b}_1$, $\mathbf{b}_2$, and $\mathbf{b}_3$ have the dimension of 1/Length. It is readily checked that

$$\mathbf{b}_i.\mathbf{a}_j = 2\pi\delta_{ij}. \tag{2.20}$$

If we write $\mathbf{k} = \beta_1\mathbf{b}_1 + \beta_2\mathbf{b}_2 + \beta_3\mathbf{b}_3$, then $\mathbf{k}.\mathbf{a}_i = 2\pi\beta_i$, and Eq. (2.18) becomes $exp(2\pi i\beta_i N_i) = 1 \Rightarrow \beta_i = m_i/N_i$, where m_i is an integer. Therefore

$$\mathbf{k} = \frac{m_1}{N_1}\mathbf{b}_1 + \frac{m_2}{N_2}\mathbf{b}_2 + \frac{m_3}{N_3}\mathbf{b}_3, \quad m_1, m_2, m_3 \in \mathbb{Z}. \tag{2.21}$$

Equation (2.16), with $\mathbf{k}$ given in Eq. (2.21), is the first form of Bloch's theorem.

An alternative and useful form of Bloch's theorem is obtained as follows. Let $u_{\mathbf{k}}(\mathbf{r}) = e^{-i\mathbf{k}.\mathbf{r}}\,\phi_{\mathbf{k}}(\mathbf{r})$. Using the first form of Bloch's theorem, we find

$$u_{\mathbf{k}}(\mathbf{r}) = e^{-i\mathbf{k}.\mathbf{r}}e^{-i\mathbf{k}.\mathbf{R}}\phi_{\mathbf{k}}(\mathbf{r} + \mathbf{R}) = e^{-i\mathbf{k}.(\mathbf{r}+\mathbf{R})}\phi_{\mathbf{k}}(\mathbf{r} + \mathbf{R}) = u_{\mathbf{k}}(\mathbf{r} + \mathbf{R}).$$

Hence, $u_{\mathbf{k}}(\mathbf{r})$ is a periodic function with the same periodicity as the lattice. The stationary states are given by

$$\phi_{\mathbf{k}}(\mathbf{r}) = e^{i\mathbf{k}.\mathbf{r}} u_{\mathbf{k}}(\mathbf{r}). \tag{2.22}$$

The stationary states are thus plane waves modulated by a function that has the periodicity of the lattice; this is the second form of Bloch's theorem. The Schrödinger equation is now written as

$$[-(\hbar^2/2m)\nabla^2 + V(\mathbf{r})]e^{i\mathbf{k}.\mathbf{r}} u_{\mathbf{k}}(\mathbf{r}) = \epsilon_{\mathbf{k}} e^{i\mathbf{k}.\mathbf{r}} u_{\mathbf{k}}(\mathbf{r}).$$

Noting that

$$\begin{aligned}\nabla[e^{i\mathbf{k}.\mathbf{r}} u_{\mathbf{k}}(\mathbf{r})] &= u_{\mathbf{k}}(\mathbf{r})\nabla e^{i\mathbf{k}.\mathbf{r}} + e^{i\mathbf{k}.\mathbf{r}}\nabla u_{\mathbf{k}}(\mathbf{r}) \\ &= u_{\mathbf{k}}(\mathbf{r})(i\mathbf{k})e^{i\mathbf{k}.\mathbf{r}} + e^{i\mathbf{k}.\mathbf{r}}\nabla u_{\mathbf{k}}(\mathbf{r}) = e^{i\mathbf{k}.\mathbf{r}}(\nabla + i\mathbf{k})u_{\mathbf{k}}(\mathbf{r}),\end{aligned}$$

we obtain

$$\left[-\frac{\hbar^2}{2m}(\nabla + i\mathbf{k})^2 + V(\mathbf{r})\right] u_{\mathbf{k}}(\mathbf{r}) = \epsilon_{\mathbf{k}} u_{\mathbf{k}}(\mathbf{r}). \tag{2.23}$$

This is viewed as an eigenvalue equation for $u_{\mathbf{k}}(\mathbf{r})$, to be solved within a primitive cell, subject to the periodic boundary conditions

$$u_{\mathbf{k}}(\mathbf{r}) = u_{\mathbf{k}}(\mathbf{r} + \mathbf{a}_i), \quad i = 1, 2, 3.$$

For each $\mathbf{k}$ there are infinitely many eigenvalues $\epsilon_{1\mathbf{k}}, \epsilon_{2\mathbf{k}}, \ldots$ with corresponding eigenfunctions $u_{1\mathbf{k}}(\mathbf{r})$, $u_{2\mathbf{k}}(\mathbf{r})$, ... The periodic functions should thus be written as $u_{n\mathbf{k}}(\mathbf{r})$, where $n = 1, 2, \ldots$ is called the band index. Because the separation between nearby $\mathbf{k}$-points is extremely small in comparison with the magnitude of the reciprocal lattice vectors, the eigenvalues $\epsilon_{n\mathbf{k}}$ may be considered as continuous functions of $\mathbf{k}$. For example, in a cubic crystal, where $\mathbf{a}_1, \mathbf{a}_2$, and $\mathbf{a}_3$ have the same magnitude a and are at right angles with each other, the reciprocal lattice vectors $\mathbf{b}_1, \mathbf{b}_2$, and $\mathbf{b}_3$ are also at right angles with each other, having the magnitude $b = 2\pi/a$. In contrast, the separation between nearby $\mathbf{k}$-points is $2\pi/L$; for a crystal with 10^{24} atoms, a/L is 10^{-8}. Thus, we see that the separation between adjacent $\mathbf{k}$-points is much smaller than the size of the reciprocal lattice vector.

According to Eq. (2.21), the set of allowed values for $\mathbf{k}$ is infinite. However, there is redundancy in the resulting set of energies and eigenfunctions, and the values of $\mathbf{k}$ can be restricted to a finite set. We prove this assertion as follows. The reciprocal lattice vectors $\mathbf{b}_1, \mathbf{b}_2$, and $\mathbf{b}_3$, treated as primitive vectors, generate a lattice in reciprocal space where a general lattice vector is

$$\mathbf{G} = m_1\mathbf{b}_1 + m_2\mathbf{b}_2 + m_3\mathbf{b}_3, \quad m_1, m_2, m_3 \in \mathbb{Z}.$$

We remark that for any reciprocal lattice vector $\mathbf{G}$ and any real lattice vector $\mathbf{R}$, the relation $e^{i\mathbf{G}.\mathbf{R}} = 1$ is satisfied. Equation (2.23) implies that

$$\left[-\frac{\hbar^2}{2m}(\nabla + i\mathbf{k} + i\mathbf{G})^2 + V(\mathbf{r})\right] u_{\mathbf{k}+\mathbf{G}}(\mathbf{r}) = \epsilon_{\mathbf{k}+\mathbf{G}}\, u_{\mathbf{k}+\mathbf{G}}(\mathbf{r}). \tag{2.24}$$

Let $f_{\mathbf{k}\mathbf{G}} = e^{i\mathbf{G}.\mathbf{r}} u_{\mathbf{k}+\mathbf{G}}(\mathbf{r})$. Following some algebraic manipulations, Eq. (2.24) reduces to

$$\left[-\frac{\hbar^2}{2m}(\nabla + i\mathbf{k})^2 + V(\mathbf{r})\right] f_{\mathbf{k}\mathbf{G}} = \epsilon_{\mathbf{k}+\mathbf{G}}\, f_{\mathbf{k}\mathbf{G}}.$$

Since $e^{i\mathbf{G}.\mathbf{R}} = 1$, the boundary condition $u_{\mathbf{k}+\mathbf{G}}(\mathbf{r}+\mathbf{R}) = u_{\mathbf{k}+\mathbf{G}}(\mathbf{r})$ implies that $f_{\mathbf{k}\mathbf{G}}(\mathbf{r}+\mathbf{R}) = f_{\mathbf{k}\mathbf{G}}(\mathbf{r})$. Noting that $f_{\mathbf{k}\mathbf{G}}$ satisfies the same eigenvalue equation as $u_{\mathbf{k}}(\mathbf{r})$, and that it obeys the same boundary conditions, the theorem regarding existence and uniqueness of the solutions of differential equations asserts that $f_{\mathbf{k}\mathbf{G}}(\mathbf{r}) = u_{\mathbf{k}}(\mathbf{r})$ and $\epsilon_{\mathbf{k}+\mathbf{G}} = \epsilon_{\mathbf{k}}$. Moreover,

$$\phi_{\mathbf{k}+\mathbf{G}}(\mathbf{r}) = e^{i(\mathbf{k}+\mathbf{G}).\mathbf{r}} u_{\mathbf{k}+\mathbf{G}}(\mathbf{r}) = e^{i\mathbf{k}.\mathbf{r}} f_{\mathbf{k}\mathbf{G}}(\mathbf{r}) = e^{i\mathbf{k}.\mathbf{r}} u_{\mathbf{k}}(\mathbf{r}) = \phi_{\mathbf{k}}(\mathbf{r}).$$

Inserting the band index, we can write

$$\phi_{n\mathbf{k}+\mathbf{G}}(\mathbf{r}) = \phi_{n\mathbf{k}}(\mathbf{r}), \quad \epsilon_{n\mathbf{k}+\mathbf{G}} = \epsilon_{n\mathbf{k}}. \tag{2.25}$$

This is the redundancy we mentioned earlier. The above relations allow us to restrict the values of $\mathbf{k}$ to one primitive cell in reciprocal space, since any $\mathbf{k}$-point outside the primitive cell can be reached from a $\mathbf{k}$-point inside the primitive cell by adding some reciprocal lattice vector $\mathbf{G}$. It is conventional to choose a Wigner–Seitz cell, known as the first Brillouin zone (FBZ), as the primitive cell in reciprocal space. It is constructed by drawing all reciprocal lattice vectors that emanate from a chosen point in the reciprocal lattice, and then drawing the perpendicular bisector planes of these vectors. The volume bounded by these planes, and centered on the chosen point, is the FBZ. This procedure is illustrated in Figure 2.5 for a two-dimensional square lattice.

Taking into account the spin state of the electron, the single-particle states, expressed as Bloch functions, are given by

$$\phi_{n\mathbf{k}\sigma}(\mathbf{r}) = e^{i\mathbf{k}.\mathbf{r}} u_{n\mathbf{k}}(\mathbf{r})|\sigma\rangle. \tag{2.26}$$

Five quantum numbers characterize the single-particle states: the band index $n = 1, 2, \ldots$, the three components k_x, k_y, k_z, of the wave vector $\mathbf{k} \in$ FBZ, and the spin index $\sigma = \uparrow$ or $\downarrow$.

From the above discussion, we see how bands arise once the static potential produced by the ions at their equilibrium positions is taken into account. It takes $2N_1N_2N_3$ electrons, i.e., twice the number of primitive cells in the crystal, to fill

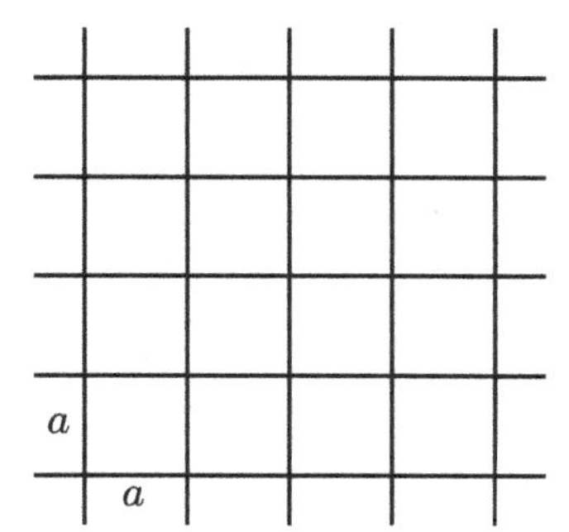

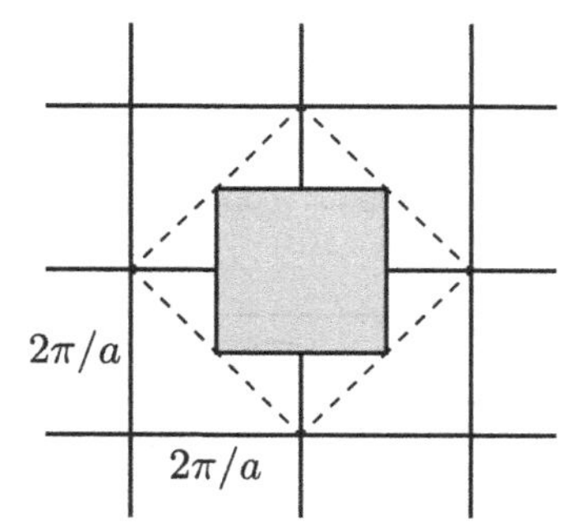

Figure 2.5 A square lattice, in real space, of side a (left figure), and the reciprocal lattice, also a square lattice, of side $2\pi/a$ (right figure). The shaded area is the first Brillouin zone.

one band, as deduced from the following argument. From Eq. (2.21), neighboring $\mathbf{k}$-points along the $\mathbf{b}_i$ direction ($i = 1, 2, 3$) are separated by $\mathbf{b}_i/N_i$. The volume in reciprocal space occupied by one $\mathbf{k}$-point is thus $\mathbf{b}_1.\mathbf{b}_2 \times \mathbf{b}_3/N_1N_2N_3$, i.e., the volume of a primitive cell in reciprocal space divided by $N_1N_2N_3$. Hence, the FBZ, whose volume is equal to that of a primitive cell in reciprocal space, contains $N_1N_2N_3$ $\mathbf{k}$-points. Since each state with quantum numbers n and $\mathbf{k}$ can accommodate two electrons, with opposite spin projections, each band can accommodate up to $2N_1N_2N_3$ electrons. In the ground state, at zero temperature, electrons fill the lowest energy states. If we end up with a situation where some bands are completely filled while the rest are empty, the crystal will be an insulator. If we do not, the crystal will be a metal.

Finally, we note that, in a metal with a partially filled band, it is generally the case that to a good approximation, we may set $u_{\mathbf{k}}(\mathbf{r}) = 1/\sqrt{V}$, and take $\epsilon_{\mathbf{k}} = \hbar^2k^2/2m^*$, where m^* is an effective electron mass. This is known as the effective mass approximation. In many metals, $m^* \approx m$, the free electron mass. A similar situation occurs in semiconductors with partially occupied bands, due to either thermal excitations or doping. In this case, however, m^* may be very different from m; in GaAs, for example, the effective electron mass is $m^* = 0.067m$.

2.4 Example: one-dimensional lattice

Consider a chain of N identical atoms. The equilibrium separation between the atoms is a. If a is large, the chain will be a collection of isolated atoms. Each atom has its own orbitals: 1s, 2s, 2p, ... with the lowest energy orbitals being occupied by electrons. As the atoms are brought closer together so that atomic wave functions begin to overlap, electrons tunnel from one atom to another, becoming delocalized, and the overlapping orbitals form bands. For example, whereas the 3s orbitals have

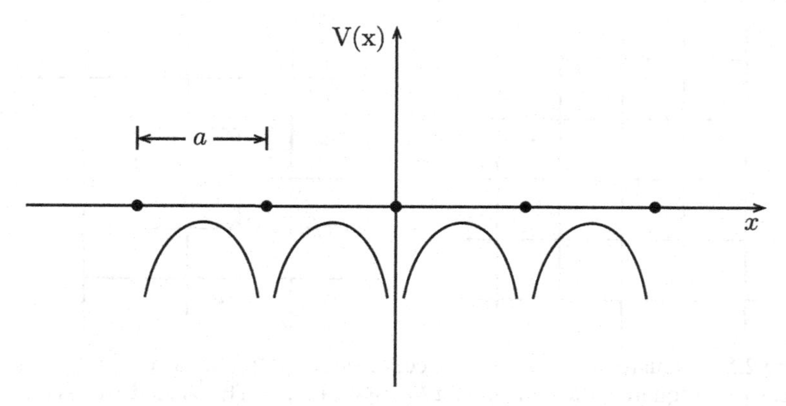

Figure 2.6 A line of identical atoms where the separation between adjacent atoms is a. The periodic potential seen by an electron is $V(x)$, and it is produced by the ions sitting at their equilibrium sites.

a well-defined energy in isolated atoms, they broaden into a band when atoms are brought closer together. The one-electron Hamiltonian is $H = p^2/2m + V(x)$, where $V(x) = V(x + a)$ is the periodic potential seen by the electron. This is sketched in Figure 2.6.

Now consider the band formed by the broadening of one type of atomic orbital, e.g., the 3s orbitals. Let $|\phi_m\rangle$ be the atomic orbital centered on atom number m, located at $x = ma,\ m = 1, \ldots, N$. We assume that $\langle\phi_m|H|\phi_m\rangle = \epsilon$, and that for $n \neq m$, $\langle\phi_n|H|\phi_m\rangle = -t\delta_{n,m\pm1}$, i.e., we assume that the overlap of atomic wave functions is appreciable only between nearest-neighbor atoms. We take t to be real. Our goal is to find the energy dispersion E_k for this band.

We want to solve the eigenvalue equation $H|\Psi_k\rangle = E_k|\Psi_k\rangle$. We take the N atomic orbitals centered on the N atoms as the basis states in which $|\Psi_k\rangle$ is expanded,

$$|\Psi_k\rangle = \sum_m c_{mk}|\phi_m\rangle \Rightarrow \Psi_k(x) = \sum_{m=1}^{N} c_{mk}\phi(x - ma).$$

The coefficients c_{mk} are not arbitrary; they are chosen so that $\Psi_k(x)$ is a Bloch function, being a stationary state of an electron in a periodic potential. We thus require that $\Psi_k(x + a) = e^{ika}\Psi_k(x)$; this, in turn, implies that

$$\sum_m c_{mk}\phi(x + a - ma) = e^{ika}\sum_m c_{mk}\phi(x - ma). \tag{2.27}$$

The LHS of the above equation is

$$LHS = \sum_m c_{mk}\phi[x - (m-1)a] = \sum_n c_{n+1,k}\phi(x - na) = \sum_m c_{m+1,k}\phi(x - ma). \tag{2.28}$$

Note that, strictly speaking, in the summation over n, n ranges from 0 to $N-1$, and the series is

$$\sum_n c_{n+1,k}\phi(x-na) = c_{1k}\phi(x) + c_{2k}\phi(x-a) + \cdots + c_{Nk}\phi[x-(N-1)a].$$

Periodic boundary conditions, however, mean that $\phi(x) = \phi(x-Na)$. Identifying $c_{N+1,k}$ with c_{1k}, the index n in the summation over n may be taken to range from 1 to N. From Eqs (2.27) and (2.28), we can write

$$\sum_{m=1}^{N} c_{m+1,k}\, \phi(x-ma) = e^{ika} \sum_{m=1}^{N} c_{mk}\, \phi(x-ma) \Rightarrow c_{m+1,k} = e^{ika}\, c_{mk}.$$

Therefore, $c_{mk} = e^{imka}$. Summarizing, the Bloch state is given by

$$|\Psi_k\rangle = \frac{1}{\sqrt{N}} \sum_m e^{imka}\, |\phi_m\rangle.$$

The factor $\frac{1}{\sqrt{N}}$ is a normalization factor. The eigenvalue equation now reads

$$\sum_m e^{imka} H\, |\phi_m\rangle = \sum_m e^{imka} E_k\, |\phi_m\rangle.$$

Multiplying by $\langle\phi_n|$ on both sides, we obtain

$$\sum_m e^{imka} \langle\phi_n|H|\phi_m\rangle = \sum_m e^{imka} E_k\, \langle\phi_n|\phi_m\rangle. \tag{2.29}$$

On the LHS, the matrix element vanishes unless $m = n, n-1$, or $n+1$,

$$LHS = \epsilon\, e^{inka} - t\, e^{i(n-1)ka} - t\, e^{i(n+1)ka}.$$

On the RHS of Eq. (2.29), $\langle\phi_n|\phi_m\rangle$ is the overlap of the atomic orbitals centered on sites n and m. Although it is nonzero, we will neglect it if $n \neq m$ and take $\langle\phi_n|\phi_m\rangle = \delta_{nm}$. The RHS thus reduces to $E_k e^{inka}$. Therefore,

$$E_k = \epsilon - t\, e^{-ika} - t\, e^{ika} = \epsilon - 2t\cos(ka). \tag{2.30}$$

2.5 Wannier states

For electrons subjected to the periodic potential produced by a lattice of ions, we have considered in Section 2.3 the basis set of Bloch states $|n\mathbf{k}\sigma\rangle$ characterized by a band index n, wave vector $\mathbf{k} \in$ FBZ, and spin projection σ. These are modulated plane waves that extend throughout the crystal. Another basis set of states, consisting of localized orbitals centered on lattice sites, may be constructed. For a given band index n, lattice site $\mathbf{R}_i$, and spin projection σ, consider the states

$$|ni\sigma\rangle = \frac{1}{\sqrt{N}} \sum_{\mathbf{k}\in\mathrm{FBZ}} e^{-i\mathbf{k}.\mathbf{R}_i} |n\mathbf{k}\sigma\rangle. \tag{2.31}$$

These are called Wannier states; they have the following properties:

- The Wannier function $\phi_{ni\sigma}(\mathbf{r}) = \langle\mathbf{r}|ni\sigma\rangle$ is centered on $\mathbf{R}_i$; hence it is written as $\phi_{n\sigma}(\mathbf{r}-\mathbf{R}_i)$.
- The Wannier states form a complete, orthonormal set.
- The Wannier function $\phi_{n\sigma}(\mathbf{r}-\mathbf{R}_i)$ is localized on the lattice site i.

The first property follows directly from the second form of Bloch's theorem. From the definition of the Wannier state, we can write

$$\phi_{ni\sigma}(\mathbf{r}) = \frac{1}{\sqrt{N}} \sum_{\mathbf{k}\in\mathrm{FBZ}} e^{-i\mathbf{k}.\mathbf{R}_i} e^{i\mathbf{k}.\mathbf{r}} u_{n\mathbf{k}}(\mathbf{r})|\sigma\rangle.$$

Since $u_{n\mathbf{k}}$ has the periodicity of the lattice, we can rewrite the above as

$$\phi_{ni\sigma}(\mathbf{r}) = \frac{1}{\sqrt{N}} \sum_{\mathbf{k}\in\mathrm{FBZ}} e^{i\mathbf{k}.(\mathbf{r}-\mathbf{R}_i)} u_{n\mathbf{k}}(\mathbf{r}-\mathbf{R}_i)|\sigma\rangle.$$

This shows that the Wannier function is a function of $\mathbf{r}-\mathbf{R}_i$, and we can write it as $\phi_{n\sigma}(\mathbf{r}-\mathbf{R}_i)$; it is centered on $\mathbf{R}_i$.

Since the Bloch states are orthonormal: $\langle n'\mathbf{k}'\sigma'|n\mathbf{k}\sigma\rangle = \delta_{nn'}\delta_{\mathbf{k}\mathbf{k}'}\delta_{\sigma\sigma'}$,

$$\begin{aligned}\langle n'j\sigma'|ni\sigma\rangle &= \frac{1}{N}\sum_{\mathbf{k}\mathbf{k}'} e^{i\mathbf{k}'.\mathbf{R}_j} e^{-i\mathbf{k}.\mathbf{R}_i} \langle n'\mathbf{k}'\sigma'|n\mathbf{k}\sigma\rangle \\ &= \delta_{nn'}\delta_{\sigma\sigma'}\frac{1}{N}\sum_{\mathbf{k}} e^{i\mathbf{k}.(\mathbf{R}_j-\mathbf{R}_i)} = \delta_{nn'}\delta_{\sigma\sigma'}\delta_{ij}.\end{aligned} \tag{2.32}$$

In the last step, we used the results of Problem 2.1. Wannier states are thus orthonormal. Furthermore, since the number of lattice sites is N, which is the same as the number of $\mathbf{k}$-points in the FBZ, the set of states $|ni\sigma\rangle$ is complete: it forms a basis for the expansion of any state.

To see that $\phi_{n\sigma}(\mathbf{r}-\mathbf{R}_i)$ is localized on the lattice site $\mathbf{R}_i$, we consider a one-dimensional lattice and assume that $u_{n\mathbf{k}}(x) = 1/\sqrt{L}$, so that the Bloch states are plane waves. For any lattice vector $\mathbf{R}$ of magnitude $R = ma$, where m is an integer,

$$\phi_{n\sigma}(x-ma) = \frac{1}{\sqrt{NL}}\sum_k e^{-imka} e^{ikx}|\sigma\rangle = \frac{1}{\sqrt{NL}}\sum_k e^{ik(x-ma)}|\sigma\rangle.$$

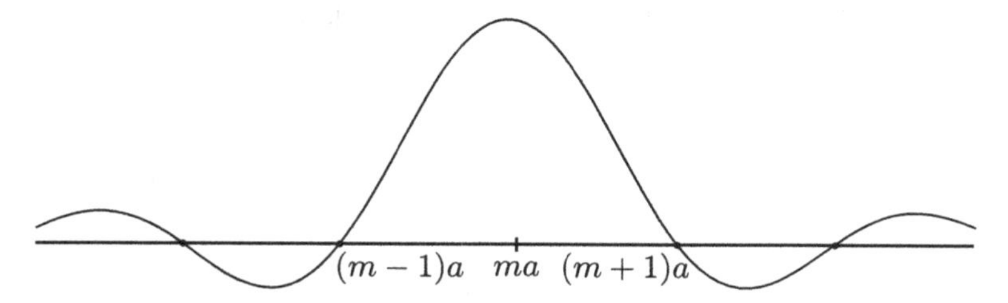

Figure 2.7 Wannier function centered on site m in a one-dimensional crystal.

Replacing the sum over k by an integral, the spatial part becomes

$$\phi_n(x - ma) = \frac{1}{\sqrt{NL}} \frac{L}{2\pi} \int_{-\pi/a}^{\pi/a} dk\; e^{ik(x-ma)} = \frac{\sqrt{L}}{\pi\sqrt{N}} \frac{sin[\pi(x - ma)/a]}{x - ma}.$$

A plot of $\phi_n(x - ma)$ shows that most of the weight of the function is at $x = ma$ (see Figure 2.7). The localization of the Wannier function is not as strong as that of an atomic orbital. The damped oscillations are necessary for Wannier functions centered on different sites to be orthogonal.

2.6 Two-dimensional electron gas in a magnetic field

A two-dimensional electron gas (2-DEG) is produced at semiconductor interfaces and in metal–oxide–semiconductor (MOS) structures. Electrons move freely in the x-y plane but are localized in the z-direction. Absent a magnetic field, the most convenient single-particle states are plane waves $|\mathbf{k}\sigma\rangle$, which are characterized by three quantum numbers: k_x, k_y, and σ. In the presence of a magnetic field, however, these states are not very convenient.

Consider an electron gas, confined to a rectangular sheet of length L_x and width L_y, i.e., in the presence of a static, uniform magnetic field $\mathbf{B}$ in the z-direction. What is the Hamiltonian for a charged particle in a magnetic field $\mathbf{B}$? To answer this question we go back to classical mechanics. The force on a particle of charge q and velocity $\mathbf{v}$ is, in cgs units, $\mathbf{F} = (q/c)\mathbf{v} \times \mathbf{B}$; in SI units, $q/c \to q$. Defining the vector potential $\mathbf{A}$ by $\mathbf{B} = \nabla \times \mathbf{A}$, it is not difficult to check that the Lagrangian L, given by

$$L = \frac{1}{2}mv^2 + \frac{q}{c}\mathbf{v}.\mathbf{A}, \tag{2.33}$$

does indeed produce the correct equation for the force. The proof consists in using the Euler–Lagrange equation of motion

$$\frac{d}{dt}\frac{\partial L}{\partial \dot{x}} = \frac{\partial L}{\partial x}$$

with similar equations for y and z, along with the following two equations:

$$\mathbf{v} \times \mathbf{B} = \mathbf{v} \times (\nabla \times \mathbf{A}) = \nabla(\mathbf{v}.\mathbf{A}) - (\mathbf{v}.\nabla)\mathbf{A}$$

$$\frac{d\mathbf{A}}{dt} = \frac{\partial \mathbf{A}}{\partial t} + \frac{\partial \mathbf{A}}{\partial x}\frac{dx}{dt} + \frac{\partial \mathbf{A}}{\partial y}\frac{dy}{dt} + \frac{\partial \mathbf{A}}{\partial z}\frac{dz}{dt} = (\mathbf{v}.\nabla)\mathbf{A}.$$

The first equation is checked easily. In the second, $\partial \mathbf{A}/\partial t = 0$ because $\mathbf{A}$ does not depend explicitly on time, since $\mathbf{B}$ is a static field.

From the Lagrangian, we construct the canonical momentum,

$$p_x = \partial L/\partial \dot{x} = m\dot{x} + (q/c)A_x,$$

along with similar equations for p_y and p_z. It follows that $m\mathbf{v} = \mathbf{p} - (q/c)\mathbf{A}$, and the kinetic energy is given by

$$T = \frac{1}{2}mv^2 = \frac{1}{2m}\left(\mathbf{p} - \frac{q}{c}\mathbf{A}\right)^2.$$

Thus, for a charged particle in a magnetic field, the kinetic energy portion of the Hamiltonian is obtained by $\mathbf{p} \rightarrow \mathbf{p} - q\mathbf{A}/c$.

For a given $\mathbf{B}$, there are infinitely many choices for $\mathbf{A}$, because for any $\mathbf{A}$, the vector potential $\mathbf{A}' = \mathbf{A} + \nabla f$, for an arbitrary function f, is an equally valid choice (since $\nabla \times \nabla f = 0$). Making a particular choice for $\mathbf{A}$ is called "fixing the gauge." For a uniform magnetic field $\mathbf{B}$ in the z-direction, we may choose $\mathbf{A} = (-By, 0, 0)$; this is the Landau gauge.

The Hamiltonian for the 2-DEG is $H = \sum_i h(i) + \frac{1}{2}\sum_{i \neq j} v(i, j)$. The second term describes the electron–electron interaction. The single-particle Hamiltonian is given by

$$h = \frac{1}{2m}\left(\mathbf{p} + \frac{e}{c}\mathbf{A}\right)^2 + (g\mu_B/\hbar)\mathbf{S}.\mathbf{B}, \tag{2.34}$$

where $-e$ is the charge of the electron, g is the gyromagnetic factor for the electron spin ($g \simeq 2$), $\mu_B = e\hbar/2mc$ is the Bohr magneton, and $\mathbf{S}$ is the electron spin operator. The second term in h may be written as $-\boldsymbol{\mu}.\mathbf{B}$, and it is the potential energy of a magnetic moment $\boldsymbol{\mu}$ in a magnetic field $\mathbf{B}$; for an electron, $\boldsymbol{\mu} = -(g\mu_B/\hbar)\mathbf{S}$. In actual systems realized at interfaces, m should be replaced by m^*, the effective mass of the electron.

A convenient set of single-particle states is formed by the eigenstates of h. Ignoring the spin part for the time being, we have, in the Landau gauge,

$$h = \frac{1}{2m}\left(p_x - \frac{eB}{c}y\right)^2 + \frac{1}{2m}p_y^2. \tag{2.35}$$

Inserting $\phi(x, y) = f(y)e^{ikx}/\sqrt{L_x}$ into the Schrödinger equation $h\phi = \epsilon\phi$, we obtain

$$\left[\frac{1}{2m}\left(\hbar k - \frac{eB}{c}y\right)^2 + \frac{1}{2m}p_y^2\right] f(y) = \epsilon f(y).$$

This may be rewritten, in terms of $\omega = eB/mc$, the cyclotron frequency, and with $y_0 = \hbar ck/eB$, as

$$\left[\frac{1}{2m}p_y^2 + \frac{1}{2}m\omega^2(y - y_0)^2\right] f(y) = \epsilon f(y). \tag{2.36}$$

This is the eigenvalue equation for a harmonic oscillator of frequency ω, centered at y_0. The eigenvalues are $\epsilon_n = (n + 1/2)\hbar\omega,\ n = 0, , 1, 2, \ldots,$ and the corresponding eigenfunctions are $A_n H_n(a(y - y_0))\exp(-a^2(y - y_0)^2/2)$, where H_n is the Hermite polynomial of degree n, A_n is a normalization constant, and $a = (m\omega/\hbar)^{1/2}$. The eigenfunctions of h are given by

$$\phi_{nk}(x, y) = \frac{1}{\sqrt{L_x}} A_n e^{ikx} H_n(a(y - y_0))e^{-a^2(y-y_0)^2/2}. \tag{2.37}$$

The periodic boundary conditions determine the allowed values of k: $k = 0, \pm 2\pi/L_x, \pm 4\pi/L_x, \ldots$. The energy levels ϵ_n, known as the Landau levels, do not depend on k; hence, they are degenerate. The orbital degeneracy of a Landau level, N_L, is the number of allowed values of k. This is determined by the requirement that y_0, the harmonic oscillator center, lies between 0 and L_y, which means that k lies between 0 and $eBL_y/\hbar c$. With a separation of $2\pi/L_x$ between consecutive values of k, the number of allowed values of k is $(eBL_y/\hbar c)/(2\pi/L_x)$; hence, $N_L = eBA/hc$, where $A = L_x L_y$ is the sample area, and h is Planck's constant. An equivalent expression is $N_L = \Phi/\Phi_0$, where $\Phi = BA$ is the magnetic flux through the sample, and $\Phi_0 = hc/e$ is the flux quantum.

Taking into account the spin part of the single-particle Hamiltonian, the single-particle states are $\phi_{nk\sigma} = \phi_{nk}|\sigma\rangle$, where $\phi_{nk}(x, y)$ is given in Eq. (2.37), and the corresponding energies are $\epsilon_{nk\sigma} = (n + 1/2)\hbar\omega + g\mu_B B\sigma$, where $\sigma = -1/2, +1/2$. Here, single-particle states are described by three quantum numbers: n, k, and σ.

Further reading

Ashcroft, N.W. and Mermin, N.D. (1976). *Solid State Physics*. Philadelphia: Saunders College.

Kittel, C. (2005). *Introduction to Solid State Physics*, 8th edn. New York: Wiley.

Omar, M.A. (1993) *Elementary Solid State Physics: Principles and Applications*, revised printing. Boston: Addison-Wesley.

Problems

2.1 *Important sums.*

(a) $\forall \mathbf{k} \in$ FBZ, show that $\sum_n e^{i\mathbf{k}.\mathbf{R}_n} = N\delta_{\mathbf{k},\mathbf{0}}$, where N is the number of primitive cells and the sum runs over all lattice sites.

(b) For every lattice vector $\mathbf{R}$, show that $\sum_{\mathbf{k}\in\text{FBZ}} e^{i\mathbf{k}.\mathbf{R}} = N\delta_{\mathbf{R},\mathbf{0}}$.

2.2 *Free electron model at zero temperature.* Consider a system of N free electrons confined to a cube of volume V, at $T = 0$ (ground state). Define the dimensionless parameter r_s through the relation $(4\pi/3)(r_s a_0)^3 = V/N$, where a_0 is the Bohr radius.

(a) Express the mean energy per electron in terms of r_s.

(b) Show that $d_\sigma(\epsilon_F)$, the density of states per unit volume, per spin orientation, at the Fermi energy, is given by $mk_F/2\pi^2\hbar^2$.

2.3 *Free electron model in lower dimensions.* Consider the free electron model in one and two dimensions at $T = 0$.

(a) Show that the Fermi wave vector is given by

$$k_F = \begin{cases} \sqrt{2\pi n} & 2\text{D} \\ \pi n/2 & 1\text{D} \end{cases}$$

where n is the electron number density (in 2D, n is the number of electrons per unit area, while in 1D, it is the number of electrons per unit length).

(b) Show that the mean energy per electron is given by $\bar{\epsilon} = \epsilon_F d/(d+2)$, where d is the dimension of space and ϵ_F is the Fermi energy.

2.4 *Graphene bands.* Graphene (see Figure 2.3) has two atoms per unit cell, denoted by A and B. The x- and y-axes are chosen such that $\mathbf{a}_1 = a(\sqrt{3}/2, -1/2)$, $\mathbf{a}_2 = a(0, 1)$, where $a = 0.246$ nm is the lattice constant. An isolated carbon atom has the electronic configuration $1s^2 2s^2 2p^2$. To form graphene, one electron is excited from $2s$ to $2p$, and the new configuration is $1s^2 2s^1 2p_x^1 2p_y^1 2p_z^1$. The $2s$, $2p_x$, and $2p_y$ orbitals get hybridized (mixed) and form three sp^2 orbitals that are oriented in the x–y plane at 120° with each other. The sp^2 orbitals on nearby atoms form strong bonds in the plane, giving rise to a honeycomb structure, and they broaden into the σ-bands that lie very low in energy. The p_z orbital on each atom is perpendicular to the graphene plane and is occupied by one electron. The p_z orbitals broaden into two π-bands (there are two atoms, hence two p_z orbitals, per unit cell).

Let us take atom A to sit at the origin of coordinates. The p_z orbital on this atom is $\phi(\mathbf{r})$, and that on any atom of type A, that can be reached from A by a lattice vector $\mathbf{R}_n$, is $\phi(\mathbf{r} - \mathbf{R}_n)$. The p_z orbital on B is $\phi(\mathbf{r} - \boldsymbol{\delta})$, where $\boldsymbol{\delta}$ is the vector from A to B, and that on any atom of type B, separated from B by $\mathbf{R}_n$, is $\phi(\mathbf{r}-\boldsymbol{\delta}-\mathbf{R}_n)$. We have

$$\int \phi^*(\mathbf{r}-\mathbf{R}_n)H\phi(\mathbf{r}-\mathbf{R}_n)d^3r = \int \phi^*(\mathbf{r}-\boldsymbol{\delta}-\mathbf{R}_n)H\phi(\mathbf{r}-\boldsymbol{\delta}-\mathbf{R}_n)d^3r = \epsilon.$$

We may shift the zero of energy and set $\epsilon = 0$. The equality of the matrix elements in the above equation is due to the symmetry of graphene under reflection in a plane that is a perpendicular bisector of the bond connecting atoms A and B. For simplicity, we make two assumptions:

- Only nearest-neighbor atoms interact; the matrix element of the Hamiltonian between orbitals on neighboring atoms is $-t$ ($t \simeq 3eV$): $\int \phi^*(\mathbf{r}-\boldsymbol{\delta})H\phi(\mathbf{r})d^3r = -t$.
- The ovelap between p_z orbitals on different sites is ignored, i.e., we assume that $\int \phi^*(\mathbf{r}-\boldsymbol{\delta})\phi(\mathbf{r})d^3r = 0$.

From the p_z orbitals on atoms of type A and B, the Bloch functions

$$\psi_{\mathbf{k}}^A(\mathbf{r}) = \frac{1}{\sqrt{N}}\sum_n e^{i\mathbf{k}.\mathbf{R}_n}\phi(\mathbf{r}-\mathbf{R}_n), \quad \psi_{\mathbf{k}}^B(\mathbf{r}) = \frac{1}{\sqrt{N}}\sum_n e^{i\mathbf{k}.\mathbf{R}_n}\phi(\mathbf{r}-\boldsymbol{\delta}-\mathbf{R}_n)$$

are constructed. To solve the Schrödinger equation $H\Psi_{\mathbf{k}}(\mathbf{r}) = E_{\mathbf{k}}\Psi_{\mathbf{k}}(\mathbf{r})$, we try a solution of the form $\Psi_{\mathbf{k}}(\mathbf{r}) = a\psi_{\mathbf{k}}^A(\mathbf{r}) + b\psi_{\mathbf{k}}^B(\mathbf{r})$.

(a) Find the primitive reciprocal lattice vectors $\mathbf{b}_1$ and $\mathbf{b}_2$ and draw the first Brillouin zone.

(b) Show that $E_{\mathbf{k}} = \pm t|g_{\mathbf{k}}|$, where

$$g_{\mathbf{k}} = 1+\exp\left[i\left(-\sqrt{3}k_x a/2+k_y a/2\right)\right]+\exp\left[-i\left(\sqrt{3}k_x a/2+k_y a/2\right)\right].$$

(c) Reduce $E_{\mathbf{k}}$ to the form:

$$E_{\mathbf{k}} = \pm t\left[3+4\cos\left(\sqrt{3}k_x a/2\right)\cos\left(k_y a/2\right)+2\cos\left(k_y a\right)\right]^{1/2}.$$

(d) In the vicinity of the points K$(2\pi/\sqrt{3}a, 2\pi/3a)$ and K$'(0, 4\pi/\sqrt{3}a)$ in the first Brillouin zone, show that $E_{\mathbf{k}} = \pm\hbar v_F k$ where $v_F = \sqrt{3}ta/2\hbar$ is the Fermi velocity, $k = |\mathbf{k} - (2\pi/\sqrt{3}a, 2\pi/3a)|$ (near point K), or $k = |\mathbf{k} - (0, 4\pi/\sqrt{3}a)|$ (near point K$'$). The energy dispersion is thus linear in the vicinity of K and K$'$.

(e) Assuming linear dispersion, show that the density of states per unit area is $d(E) = 8|E|/(3\pi a^2 t^2)$.

2.5 *More on graphene.* Assume that the p_z orbital on each site in graphene is described by the wave function $\phi(\mathbf{r}) = Ar\cos\theta\exp(-Zr/2a_0)$, where A is a normalization constant, a_0 is the Bohr radius, θ is the angle $\mathbf{r}$ makes with the c-axis (the axis perpendicular to the graphene plane), and Z is the effective charge on the nucleus (the nuclear charge is screened by the two core electrons in the $1s$ orbital, and to a lesser extent by the valence electrons; $Z \approx 3$). Show that

$$\int \phi^*(\mathbf{r})e^{-i\mathbf{q}.\mathbf{r}}\phi(\mathbf{r})d^3r = \left[1 + (qa_0/Z)^2\right]^{-3}.$$

2.6 *Matrix elements in graphene.*

(a) Using the results of Problems 4 and 5, evaluate $\langle\psi_{\mathbf{k}}^A|X|\psi_{\mathbf{k+q}}^A\rangle$, $\langle\psi_{\mathbf{k}}^B|X|\psi_{\mathbf{k+q}}^B\rangle$, $\langle\psi_{\mathbf{k}}^{\mathrm{v}}|X|\psi_{\mathbf{k+q}}^{\mathrm{v}}\rangle$, and $\langle\psi_{\mathbf{k}}^{\mathrm{c}}|X|\psi_{\mathbf{k+q}}^{\mathrm{v}}\rangle$, where $X = e^{-i\mathbf{q}.\mathbf{r}}$, and v(c) stands for the valence (conduction) band.

(b) Let $F_{ss'}(\mathbf{k},\mathbf{q}) = \left|\langle\psi_{\mathbf{k}}^s|e^{-i\mathbf{q}.\mathbf{r}}|\psi_{\mathbf{k+q}}^{s'}\rangle\right|^2$. Show that

$$F_{ss'}(\mathbf{k},\mathbf{q}) = \frac{1}{2}\left(1 + ss'\,\frac{k + q\cos\phi}{|\mathbf{k}+\mathbf{q}|}\right)$$

where $\cos\phi = \mathbf{k}.\mathbf{q}/kq$, and $s, s' = -1(+1)$ if $s, s' =$ v(c). For more details, see (Shung, 1986).

2.7 *Density of states $D(\epsilon)$.* The total number of states within a shell in $\mathbf{k}$-space bounded by the two constant energy surfaces $E_{\mathbf{k}} = \epsilon$ and $E_{\mathbf{k}} = \epsilon + d\epsilon$ is $D(\epsilon)d\epsilon$. The number of states within this shell is twice the number of $\mathbf{k}$-points within the shell because of spin degeneracy. Therefore,

$$D(\epsilon)d\epsilon = 2\frac{V}{(2\pi)^3}\int_{\text{shell}} d^3k = 2\frac{V}{(2\pi)^3}\int dS_\epsilon dk_\perp$$

where dS_ϵ is an area element on the inner surface and $dk_\perp$ is the perpendicular distance between the two surfaces of the shell. Show that

$$D(\epsilon) = 2\frac{V}{(2\pi)^3}\int \frac{dS_\epsilon}{|\nabla_{\mathbf{k}}E_{\mathbf{k}}|_{E_{\mathbf{k}}=\epsilon}}.$$

3

Second quantization

Nothing can be made out of nothing.
–*William Shakespeare,* King Lear

Historically, quantization of the motion of particles was developed first. The state was described by a wave function and observables by operators. When dealing with interactions between particles and fields, such as the electromagnetic field, the fields were treated classically. Classical field equations look like the quantum mechanical equations for the wave function of the field quanta. For example, the Klein–Gordon classical field equation is similar to the quantum mechanical wave equation for a relativistic spinless particle. Quantizing the fields, leading to quantum field theory, appears to be quantizing a theory that has already been quantized; hence the name "second quantization." In reality, there is only one quantization and one quantum theory.

The method of second quantization is important in the study of many-particle systems. It enables us to express many-body operators in terms of creation and annihilation operators, thus rendering calculations less cumbersome. Moreover, the method makes it possible to treat systems with a variable number of particles; that is why the method initially emerged in the context of quantum field theory.

In Chapter 1 we indicated that any one-particle wave function may be expanded in a complete set of states. In this chapter, we show that products of single-particle states, when properly symmetrized, form an orthonormal basis for the expansion of the wave function of an N-particle system. We then introduce creation and annihilation operators and show how to express one-body and two-body operators in those terms.

3.1 N-particle wave function

Suppose that we have a complete, orthonormal set of single-particle states $|\phi_\nu\rangle$, where ν is an index that represents all the quantum numbers that characterize the state. Orthonormality and completeness mean that

$$\langle \phi_\nu | \phi_{\nu'} \rangle = \delta_{\nu\nu'} \text{ (orthonormality)} \quad \sum_\nu |\phi_\nu\rangle\langle\phi_\nu| = 1 \text{ (completeness).} \tag{3.1}$$

We will show that the N-particle wave function $\Psi(1, 2, \ldots, N)$ can be expanded in terms of products of the single-particle states. We may proceed as follows. Suppose that we fix the spatial and spin coordinates of particles $2, 3, \ldots, N$. Then $\Psi(1, 2, \ldots, N)$ is a function of the coordinates of particle 1 alone; hence, we can expand it in a complete set of states $\phi_\nu(1)$,

$$\Psi(1, 2, \ldots, N) = \sum_{\nu_1} A_{\nu_1}(2, 3, \ldots, N)\phi_{\nu_1}(1).$$

If we now allow the coordinates of particle 2 to vary, $A_{\nu_1}(2, 3, \ldots, N)$ becomes a function of these coordinates, and we may expand it as

$$A_{\nu_1}(2, 3, \ldots, N) = \sum_{\nu_2} B_{\nu_1\nu_2}(3, 4, \ldots, N)\phi_{\nu_2}(2).$$

Continuing in this fashion, we end up with

$$\Psi(1, 2, \ldots, N) = \sum_{\nu_1\nu_2\ldots\nu_N} C_{\nu_1\nu_2\ldots\nu_N}\phi_{\nu_1}(1)\phi_{\nu_2}(2)\ldots\phi_{\nu_N}(N).$$

There is an alternative way to arrive at this result. States $|\phi_{\nu_1}\rangle_1$, for all values of ν_1, form an orthonormal basis for vector space V_1, the Hilbert space of the states of particle 1. States $|\phi_{\nu_2}\rangle_2$ form an orthonormal basis for V_2, the vector space of the states of particle 2, and so on. The state vector $|\Psi\rangle$ of the N-particle system belongs to the direct product space $V^{(N)} = V_1 \otimes V_2 \otimes \cdots \otimes V_N$, whose orthonormal basis consists of the direct product states $|\phi_{\nu_1}\rangle_1 \otimes |\phi_{\nu_2}\rangle_2 \otimes \cdots \otimes |\phi_{\nu_N}\rangle_N$. It follows that

$$|\Psi\rangle = \sum_{\nu_1\nu_2\ldots\nu_N} C_{\nu_1\nu_2\ldots\nu_N}|\phi_{\nu_1}\rangle_1 \otimes |\phi_{\nu_2}\rangle_2 \otimes \cdots \otimes |\phi_{\nu_N}\rangle_N. \tag{3.2}$$

3.2 Properly symmetrized products as a basis set

Although the products $\phi_{\nu_1}(1)\phi_{\nu_2}(2)\ldots\phi_{\nu_N}(N)$ of single-particle states may serve as a basis for the expansion of the N-particle wave function, they are not useful as such. This is because $\Psi(1, \ldots, i, \ldots, j, \ldots, N)$ must be symmetric (antisymmetric) under the exchange of i and j if the N identical particles are bosons (fermions). The product $\phi_{\nu_1}(1)\phi_{\nu_2}(2)\ldots\phi_{\nu_N}(N)$ lacks this property, and the symmetry (antisymmetry) property must be buried in the constants $C_{\nu_1\nu_2\ldots\nu_N}$. It is far more convenient to incorporate the appropriate symmetry into the product of the functions, so that $C_{\nu_1\nu_2\ldots\nu_N}$ will be completely symmetric upon the exchange of any two indices. For bosons, we can achieve this by summing the product over the $N!$ permutations of $1, 2, \ldots, N$; the basis states are thus given by

$$\Phi^B_{\nu_1\nu_2\ldots\nu_N}(1, 2, \ldots, N) = \frac{1}{\prod_\mu \sqrt{n_\mu!}} \frac{1}{\sqrt{N!}} \sum_P \phi_{\nu_1}[P(1)]\phi_{\nu_2}[P(2)] \ldots \phi_{\nu_N}[P(N)]. \tag{3.3}$$

Here $P(1), P(2), \ldots, P(N)$ is a permutation of $1, 2, \ldots, N$, and n_μ is the number of times the index μ appears in the product. The factor before the summation ensures that Φ^B is normalized.

For fermions, a similar expression for the basis states is used, except for the following two modifications. First, n_μ is either 0 or 1 (Pauli exclusion principle), so that $n_\mu! = 1$ ($0! = 1$ and $1! = 1$). Second, we must insert a minus sign whenever $P(1), P(2), \ldots, P(N)$ is an odd permutation of $1, 2, \ldots, N$. The fermionic basis functions are given by

$$\Phi^F_{\nu_1\nu_2\ldots\nu_N}(1, 2, \ldots, N) = \frac{1}{\sqrt{N!}} \sum_P (-1)^P \phi_{\nu_1}[P(1)]\phi_{\nu_2}[P(2)] \ldots \phi_{\nu_N}[P(N)]. \tag{3.4}$$

Equivalently, we may permute the indices instead of the coordinates

$$\Phi^F_{\nu_1\nu_2\ldots\nu_N}(1, 2, \ldots, N) = \frac{1}{\sqrt{N!}} \sum_P (-1)^P \phi_{P(\nu_1)}(1)\phi_{P(\nu_2)}(2) \ldots \phi_{P(\nu_N)}(N). \tag{3.5}$$

The above expression for Φ^F may be written in the form of a determinant,

$$\Phi^F_{\nu_1\nu_2\ldots\nu_N}(1, 2, \ldots, N) = \frac{1}{\sqrt{N!}} \begin{vmatrix} \phi_{\nu_1}(1) & \phi_{\nu_1}(2) & \ldots & \phi_{\nu_1}(N) \\ \phi_{\nu_2}(1) & \phi_{\nu_2}(2) & \ldots & \phi_{\nu_2}(N) \\ \vdots & & & \\ \phi_{\nu_N}(1) & \phi_{\nu_N}(2) & \ldots & \phi_{\nu_N}(N) \end{vmatrix}. \tag{3.6}$$

This is the Slater determinant. We make the following remarks:

1. The antisymmetry property is built into the determinant. Interchanging i and j amounts to the interchange of two columns, which changes the determinant's sign.
2. If particles i and j occupy the same state, i.e., $\nu_i = \nu_j$, then rows i and j become identical and the determinant vanishes, as it should (Pauli exclusion principle).
3. In the first row, only the single-particle state ϕ_{ν_1} appears; in the second row, only ϕ_{ν_2} appears, and so on. Therefore, there is no confusion in representing the Slater determinant by the ket $|\phi_{\nu_1}\phi_{\nu_2} \ldots \phi_{\nu_N}\rangle$.

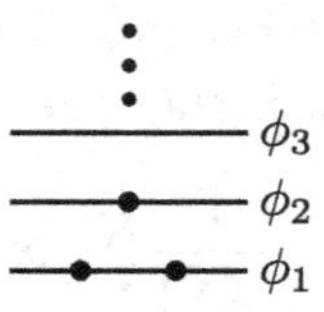

Figure 3.1 A system of three noninteracting bosons. Two bosons occupy the single-particle state ϕ_1 and one boson occupies ϕ_2.

In terms of the basis functions $\Phi^{B,F}$, the N-particle wave function is now expanded as

$$\Psi^{B,F}(1,\ldots,N) = \sum_{\nu_1\ldots\nu_N} A_{\nu_1\ldots\nu_N}\Phi^{B,F}_{\nu_1\ldots\nu_N}(1,\ldots,N). \tag{3.7}$$

3.3 Three examples

1. A system consists of three identical bosons. Denote the single-particle states by $\phi_1, \phi_2, \phi_3 \ldots$. Two bosons occupy the state ϕ_1, and one occupies the state ϕ_2 (see Figure 3.1). In this case, $N = 3$, $n_1 = 2$, $n_2 = 1$, and $n_3 = n_4 = \cdots = 0$. There are $3! = 6$ permutations of 1 2 3; they are 1 2 3, 1 3 2, 2 1 3, 2 3 1, 3 1 2, and 3 2 1. Therefore,

$$\begin{aligned}
\Phi^B_{112}(1,2,3) &= \frac{1}{\sqrt{2!1!0!0!\cdots}}\frac{1}{\sqrt{3!}}\sum_P \phi_1[P(1)]\phi_1[P(2)]\phi_2[P(3)] \\
&= \frac{1}{\sqrt{12}}\,[\phi_1(1)\phi_1(2)\phi_2(3) + \phi_1(1)\phi_1(3)\phi_2(2) + \phi_1(2)\phi_1(1)\phi_2(3) \\
&\quad + \phi_1(2)\phi_1(3)\phi_2(1) + \phi_1(3)\phi_1(1)\phi_2(2) + \phi_1(3)\phi_1(2)\phi_2(1)] \\
&= \frac{1}{\sqrt{3}}\,[\phi_1(1)\phi_1(2)\phi_2(3) + \phi_1(1)\phi_1(3)\phi_2(2) + \phi_1(2)\phi_1(3)\phi_2(1)]\,.
\end{aligned}$$

2. Three noninteracting electrons in a box of volume V occupy the states $\phi_{\mathbf{k}\uparrow}(\mathbf{r}) = \frac{1}{\sqrt{V}}e^{i\mathbf{k}.\mathbf{r}}|\uparrow\rangle$, $\phi_{\mathbf{k}\downarrow}(\mathbf{r}) = \frac{1}{\sqrt{V}}e^{i\mathbf{k}.\mathbf{r}}|\downarrow\rangle$, and $\phi_{\mathbf{k}'\uparrow}(\mathbf{r}) = \frac{1}{\sqrt{V}}e^{i\mathbf{k}'.\mathbf{r}}|\uparrow\rangle$. The wave function for the system is the Slater determinant

$$\Psi_{SD}(1,2,3) = \frac{1}{\sqrt{3!V^3}}\begin{vmatrix} e^{i\mathbf{k}.\mathbf{r}_1}|\uparrow\rangle_1 & e^{i\mathbf{k}.\mathbf{r}_2}|\uparrow\rangle_2 & e^{i\mathbf{k}.\mathbf{r}_3}|\uparrow\rangle_3 \\ e^{i\mathbf{k}.\mathbf{r}_1}|\downarrow\rangle_1 & e^{i\mathbf{k}.\mathbf{r}_2}|\downarrow\rangle_2 & e^{i\mathbf{k}.\mathbf{r}_3}|\downarrow\rangle_3 \\ e^{i\mathbf{k}'.\mathbf{r}_1}|\uparrow\rangle_1 & e^{i\mathbf{k}'.\mathbf{r}_2}|\uparrow\rangle_2 & e^{i\mathbf{k}'.\mathbf{r}_3}|\uparrow\rangle_3 \end{vmatrix}. \tag{3.8}$$

3. A system consists of two noninteracting electrons. The Hamiltonian is $H = h(1) + h(2)$. Let us assume that h is spin-independent. Being spin-independent, the Hamiltonian H commutes with S^2 and S_z,

$$[H, S^2] = [H, S_z] = [S^2, S_z] = 0.$$

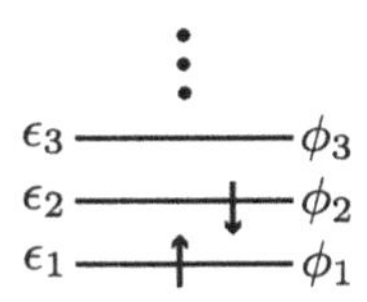

Figure 3.2 A system of two noninteracting electrons. One electron, with spin up, occupies the single-particle state ϕ_1. Another electron, with spin down, occupies ϕ_2. The energy of the system is $\epsilon_1 + \epsilon_2$.

Here, $\mathbf{S}$ is the total spin operator, and S_z is its projection on the z-axis,

$$\mathbf{S} = \mathbf{S}_1 + \mathbf{S}_2, \quad S_z = S_{1z} + S_{2z}.$$

The single-particle states are solutions of $h\phi_{n\sigma} = \epsilon_n \phi_{n\sigma}$, where n is the set of spatial quantum numbers and $\sigma = \uparrow$ or $\downarrow$. Suppose that one electron occupies the state $\phi_{1\uparrow} = \phi_1(\mathbf{r})\alpha$, where $\alpha = |\uparrow\rangle$, while the other electron occupies the state $\phi_{2\downarrow} = \phi_2(\mathbf{r})\beta$, where $\beta = |\downarrow\rangle$ (see Figure 3.2). The state of the system is given by the Slater determinant

$$\Psi_{SD} = \frac{1}{\sqrt{2!}} \begin{vmatrix} \phi_1(\mathbf{r}_1)\alpha(1) & \phi_1(\mathbf{r}_2)\alpha(2) \\ \phi_2(\mathbf{r}_1)\beta(1) & \phi_2(\mathbf{r}_2)\beta(2) \end{vmatrix} .$$

Expanding the matrix, we find

$$\Psi_{SD} = \frac{1}{\sqrt{2}} \left[\phi_1(\mathbf{r}_1)\phi_2(\mathbf{r}_2)\alpha(1)\beta(2) - \phi_1(\mathbf{r}_2)\phi_2(\mathbf{r}_1)\alpha(2)\beta(1) \right] .$$

The energy of the state is $\epsilon_1 + \epsilon_2$. This particular example allows us to discuss the following point. Since the Hamiltonian is spin-independent, we can write the stationary states as the product of a spatial function and a spin function. However, the Slater determinant given above is not amenable to such a factorization. Is something wrong? The answer is no. The problem is that Ψ_{SD}, even though it is an eigenfunction of both H, with eigenvalue $\epsilon_1 + \epsilon_2$, and S_z, with eigenvalue 0, nevertheless is not an eigenfunction of S^2. However, since H, S^2, and S_z commute among themselves, stationary states can be chosen that are eigenstates of all three operators simultaneously. We may construct two degenerate, antisymmetric eigenfunctions of H, with energy $\epsilon_1 + \epsilon_2$, which are also eigenfunctions of S^2 and S_z. Consider

$$\Psi(1,2) = \frac{1}{\sqrt{2}} \left[\phi_1(\mathbf{r}_1)\phi_2(\mathbf{r}_2) + \phi_1(\mathbf{r}_2)\phi_2(\mathbf{r}_1) \right] \frac{1}{\sqrt{2}} \left[\alpha(1)\beta(2) - \alpha(2)\beta(1) \right]$$

$$\Psi'(1,2) = \frac{1}{\sqrt{2}} \left[\phi_1(\mathbf{r}_1)\phi_2(\mathbf{r}_2) - \phi_1(\mathbf{r}_2)\phi_2(\mathbf{r}_1) \right] \frac{1}{\sqrt{2}} \left[\alpha(1)\beta(2) + \alpha(2)\beta(1) \right] .$$

$\Psi(1,2)$ is an eigenstate of S^2 and S_z with $s = 0$ and $m_s = 0$; it is a spin singlet. On the other hand, $\Psi'(1,2)$ is an eigenstate of S^2 and S_z with $s = 1$ and $m_s = 0$; it is the $m_s = 0$ component of the spin triplet (see Problem 1.5). It is readily verified that

$$\Psi_{SD}(1,2) = \frac{1}{\sqrt{2}}\left[\Psi(1,2) + \Psi'(1,2)\right].$$

Since $\Psi(1,2)$ and $\Psi'(1,2)$ are degenerate stationary states, Ψ_{SD} is also a stationary state with the same energy. If we take the difference of Ψ and Ψ', we obtain

$$\Psi'_{SD}(1,2) = \frac{1}{\sqrt{2}}\left[\Psi(1,2) - \Psi'(1,2)\right].$$

It is easy to verify that $\Psi'_{SD}(1,2)$ is the Slater determinant which describes the configuration where the electron in orbital ϕ_1 has spin down while the electron in orbital ϕ_2 has spin up. This is also a stationary state of $H = h(1) + h(2)$, with energy $\epsilon_1 + \epsilon_2$. In other words, we may choose $\Psi(1,2)$ and $\Psi'(1,2)$ as the two degenerate stationary states; each is expressed as the product of a spatial part and a spin part. Since $\Psi(1,2)$ and $\Psi'(1,2)$ are degenerate, the Slater determinants $\Psi_{SD}(1,2)$ and $\Psi'_{SD}(1,2)$, which are linear combinations of $\Psi(1,2)$ and $\Psi'(1,2)$, are also stationary states with the same energy, even though they cannot be factored into the product of a spatial part and a spin part.

3.4 Creation and annihilation operators

Dealing with determinants or with sums of the permutations of products of single-particle states is very cumbersome. It is worthwhile to try to encode the symmetry properties of the basis states into the algebraic properties of operators. We do this by introducing creation and annihilation operators. We treat the case of fermions in detail, and briefly give the corresponding results for bosons.

3.4.1 Fermions

Each single-particle state $|\phi_\nu\rangle$ is associated with a creation operator $c_\nu^\dagger$, defined by

$$c_\nu^\dagger |\phi_{\nu_1} \cdots \phi_{\nu_N}\rangle = |\phi_\nu \phi_{\nu_1} \cdots \phi_{\nu_N}\rangle. \tag{3.9}$$

The operator $c_\nu^\dagger$ thus creates a fermion in the single-particle state $|\phi_\nu\rangle$; it adds a row to the Slater determinant, which becomes the first row of the new $(N+1)\times(N+1)$ determinant. The action of the creation operator is illustrated in Figure 3.3. The action of $c_\nu^\dagger$ on a Slater determinant yields 0 if ν coincides with any of the indices

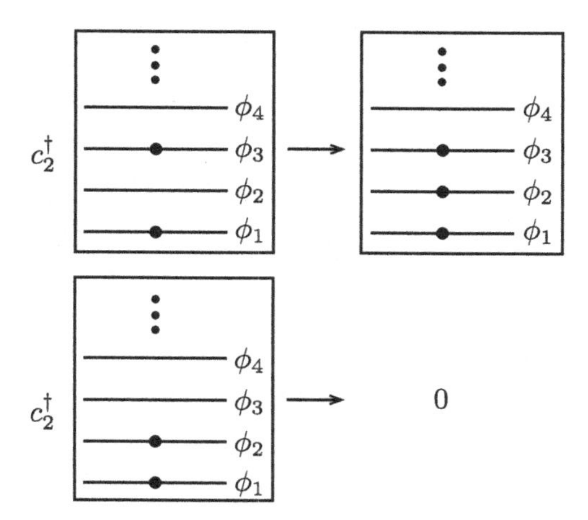

Figure 3.3 The action of the creation operator $c_2^\dagger$ on a system of fermions: if the state ϕ_2 is empty, the operator creates a particle in that state, but if the state is occupied by one particle, the result of the action of $c_2^\dagger$ is zero.

$\nu_1 \ldots \nu_N$, for then the resulting determinant would have two identical rows. Stated differently, we cannot create a fermion in a state that is already occupied (Pauli exclusion principle).

For an arbitrary Slater determinant $|\phi_{\nu_1}\phi_{\nu_2}\ldots\phi_{\nu_N}\rangle$ (arbitrary in the sense that the single-particle state indices $\nu_1 \ldots \nu_N$ are arbitrary), consider

$$c_\nu^\dagger c_{\nu'}^\dagger |\phi_{\nu_1}\cdots\phi_{\nu_N}\rangle = c_\nu^\dagger |\phi_{\nu'}\phi_{\nu_1}\cdots\phi_{\nu_N}\rangle = |\phi_\nu\phi_{\nu'}\phi_{\nu_1}\cdots\phi_{\nu_N}\rangle$$
$$c_{\nu'}^\dagger c_\nu^\dagger |\phi_{\nu_1}\ldots\phi_{\nu_N}\rangle = |\phi_{\nu'}\phi_\nu\phi_{\nu_1}\ldots\phi_{\nu_N}\rangle = -|\phi_\nu\phi_{\nu'}\phi_{\nu_1}\ldots\phi_{\nu_N}\rangle.$$

The minus sign results from the interchange of the first two rows. Since $|\phi_{\nu_1}\ldots\phi_{\nu_N}\rangle$ is arbitrary, adding the above two equations gives us

$$c_\nu^\dagger c_{\nu'}^\dagger + c_{\nu'}^\dagger c_\nu^\dagger = 0 \Longrightarrow \{c_\nu^\dagger, c_{\nu'}^\dagger\} = 0,$$

where, for any operators A and B, we define the anticommutator $\{A, B\}$ by

$$\{A, B\} = AB + BA. \tag{3.10}$$

Note, in particular, that if $\nu = \nu'$, we have $(c_\nu^\dagger)^2 = 0$: we cannot put two fermions in the same state, as Figure 3.3 illustrates.

Next, we define an annihilation operator c_ν that annihilates a particle in state $|\phi_\nu\rangle$,

$$c_\nu|\phi_\nu\phi_i\phi_j\cdots\rangle = |\phi_i\phi_j\cdots\rangle. \tag{3.11}$$

The annihilated state must be on the left, i.e., it must be the first row in the Slater determinant. If ϕ_ν is not on the left, then it must be moved to the leftmost position,

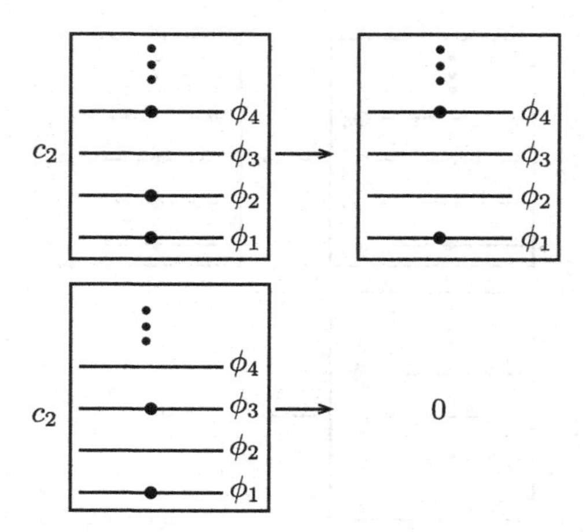

Figure 3.4 The action of the annihilation operator c_2 on a system of identical fermions. If ϕ_2 is occupied by one particle, c_2 renders the state empty. If the state is empty, the action of c_2 yields zero.

introducing a minus sign every time it is interchanged with another state. For example,

$$c_\nu|\phi_i\phi_\nu\phi_j\ldots\rangle = -c_\nu|\phi_\nu\phi_i\phi_j\ldots\rangle = -|\phi_i\phi_j\ldots\rangle.$$

Clearly, the state occupied by the particle to be annihilated must be among the collection of states in the Slater determinant; otherwise the action of c_ν is defined to yield zero,

$$c_\nu|\phi_{\nu_1}\ldots\phi_{\nu_N}\rangle = 0 \quad \text{if} \quad \nu \notin \{\nu_1,\ldots,\nu_N\}.$$

The action of an annihilation operator on a system of identical fermions is depicted in Figure 3.4.

The notation we have adopted suggests that c_ν is the adjoint (hermitian conjugate) of $c_\nu^\dagger$ and vice versa. This is indeed the case, as shown by the following argument. Consider the ket $|\Psi\rangle = |\phi_\nu\phi_{\nu_1}\ldots\rangle = c_\nu^\dagger|\phi_{\nu_1}\ldots\rangle$. Then the bra $\langle\Psi|$ is equal to $\langle\phi_{\nu_1}\ldots|(c_\nu^\dagger)^\dagger$. It follows that

$$1 = \langle\Psi|\Psi\rangle = \langle\phi_{\nu_1}\ldots|(c_\nu^\dagger)^\dagger|\phi_\nu\phi_{\nu_1}\ldots\rangle.$$

For this to be true, the following must hold

$$(c_\nu^\dagger)^\dagger|\phi_\nu\phi_{\nu_1}\ldots\rangle = |\phi_{\nu_1}\ldots\rangle,$$

which shows that $(c_\nu^\dagger)^\dagger = c_\nu$. Taking the adjoint of $\{c_\nu^\dagger, c_{\nu'}^\dagger\} = 0$, we obtain

$$\begin{aligned}0 &= (c_\nu^\dagger c_{\nu'}^\dagger + c_{\nu'}^\dagger c_\nu^\dagger)^\dagger = (c_\nu^\dagger c_{\nu'}^\dagger)^\dagger + (c_{\nu'}^\dagger c_\nu^\dagger)^\dagger = (c_{\nu'}^\dagger)^\dagger(c_\nu^\dagger)^\dagger + (c_\nu^\dagger)^\dagger(c_{\nu'}^\dagger)^\dagger \\ &= c_{\nu'}c_\nu + c_\nu c_{\nu'} = \{c_\nu, c_{\nu'}\}.\end{aligned}$$

In particular, if $\nu = \nu'$, we obtain $c_\nu^2 = 0$: a fermion cannot be annihilated twice; once it has been annihilated, it is no longer there, and a nonexistent particle cannot be annihilated.

What about $\{c_\nu, c_\nu^\dagger\}$? Consider an arbitrary Slater determinant $|\Psi\rangle = |\phi_{\nu_1} \dots \phi_{\nu_N}\rangle$. If the single-particle state $|\phi_\nu\rangle$ is not occupied,

$$\begin{aligned}(c_\nu c_\nu^\dagger + c_\nu^\dagger c_\nu)|\Psi\rangle &= (c_\nu c_\nu^\dagger + c_\nu^\dagger c_\nu)|\phi_{\nu_1} \dots \phi_{\nu_N}\rangle = c_\nu|\phi_\nu \phi_{\nu_1} \dots \phi_{\nu_N}\rangle + 0 \\ &= |\phi_{\nu_1} \dots \phi_{\nu_N}\rangle = |\Psi\rangle.\end{aligned}$$

Now suppose that the state $|\phi_\nu\rangle$ is occupied, say with $\nu = \nu_{j+1}$. Then

$$\begin{aligned}&(c_\nu c_\nu^\dagger + c_\nu^\dagger c_\nu)|\Psi\rangle = (c_\nu c_\nu^\dagger + c_\nu^\dagger c_\nu)|\phi_{\nu_1} \dots \phi_{\nu_j}\phi_\nu\phi_{\nu_{j+2}} \dots \phi_{\nu_N}\rangle \\ &\quad = 0 + (-1)^j c_\nu^\dagger c_\nu|\phi_\nu\phi_{\nu_1} \dots \phi_{\nu_j}\phi_{\nu_{j+2}} \dots \phi_{\nu_N}\rangle \\ &\quad = (-1)^j c_\nu^\dagger|\phi_{\nu_1} \dots \phi_{\nu_j}\phi_{\nu_{j+2}} \dots \phi_{\nu_N}\rangle = (-1)^j|\phi_\nu\phi_{\nu_1} \dots \phi_{\nu_j}\phi_{\nu_{j+2}} \dots \phi_{\nu_N}\rangle \\ &\quad = (-1)^j(-1)^j|\phi_{\nu_1} \dots \phi_{\nu_j}\phi_\nu\phi_{\nu_{j+2}} \dots \phi_{\nu_N}\rangle = |\phi_{\nu_1} \dots \phi_{\nu_N}\rangle = |\Psi\rangle.\end{aligned}$$

Note that in order to move ϕ_ν to the leftmost position, j interchanges are carried out, hence the first $(-1)^j$ factor. To move ϕ_ν back to its original position, j more interchanges are undertaken. We thus see that in both cases, whether $|\phi_\nu\rangle$ is vacant or occupied, the action of $\{c_\nu, c_\nu^\dagger\}$ leaves an arbitrary Slater determinant unaltered. We conclude that

$$\{c_\nu, c_\nu^\dagger\} = 1.$$

We now calculate $\{c_\nu, c_{\nu'}^\dagger\}$ for $\nu \neq \nu'$. Consider $(c_\nu c_{\nu'}^\dagger + c_{\nu'}^\dagger c_\nu)|\phi_{\nu_1} \dots \phi_{\nu_N}\rangle$. This is equal to zero unless $\nu \in \{\nu_1, \dots, \nu_N\}$ and $\nu' \notin \{\nu_1, \dots, \nu_N\}$. Let us assume that this is indeed the case. Then

$$\begin{aligned}&(c_\nu c_{\nu'}^\dagger + c_{\nu'}^\dagger c_\nu)|\phi_{\nu_1}\phi_{\nu_2} \dots \phi_\nu \dots \phi_{\nu_N}\rangle = -(c_\nu c_{\nu'}^\dagger + c_{\nu'}^\dagger c_\nu)|\phi_\nu\phi_{\nu_2} \dots \phi_{\nu_1} \dots \phi_{\nu_N}\rangle \\ &\quad = -c_\nu|\phi_{\nu'}\phi_\nu\phi_{\nu_2} \dots \phi_{\nu_1} \dots \phi_{\nu_N}\rangle - c_{\nu'}^\dagger|\phi_{\nu_2} \dots \phi_{\nu_1} \dots \phi_{\nu_N}\rangle \\ &\quad = c_\nu|\phi_\nu\phi_{\nu'}\phi_{\nu_2} \dots \phi_{\nu_1} \dots \phi_{\nu_N}\rangle - |\phi_{\nu'}\phi_{\nu_2} \dots \phi_{\nu_1} \dots \phi_{\nu_N}\rangle \\ &\quad = |\phi_{\nu'}\phi_{\nu_2} \dots \phi_{\nu_1} \dots \phi_{\nu_N}\rangle - |\phi_{\nu'}\phi_{\nu_2} \dots \phi_{\nu_1} \dots \phi_{\nu_N}\rangle = 0.\end{aligned}$$

In the first step, the interchange of ϕ_{ν_1} and ϕ_ν introduces the minus sign. We thus see that whichever way ν and ν' are related to the indices $\nu_1, \dots, \nu_N$, the action of $\{c_\nu, c_{\nu'}^\dagger\}$, for $\nu \neq \nu'$, on an arbitrary Slater determinant, yields zero. Hence, $\{c_\nu, c_{\nu'}^\dagger\} = 0$ for $\nu \neq \nu'$. Below we summarize our results,

$$\{c_\nu, c_{\nu'}\} = \{c_\nu^\dagger, c_{\nu'}^\dagger\} = 0, \quad \{c_\nu, c_{\nu'}^\dagger\} = \delta_{\nu\nu'}. \tag{3.12}$$

Let us conclude this subsection by considering the following question: what space do creation and annihilation operators act upon? Suppose that we have a complete set of single-particle states $|\phi_1\rangle, |\phi_2\rangle, \ldots$ that are ordered in some fashion, e.g., $\epsilon_1 \leq \epsilon_2 \leq \ldots$. The Slater determinant lists the occupied states; for example, $|\phi_1\phi_3\rangle$ represents a configuration where one particle occupies $|\phi_1\rangle$ and another particle occupies $|\phi_3\rangle$. We can represent this state as $|1\ 0\ 1\ 0\ 0\cdots\rangle$, which tells us that states $|\phi_1\rangle$ and $|\phi_3\rangle$ are occupied, each by one particle, while all the other states are empty. The states $|\phi_1\phi_3\rangle$ and $|1\ 0\ 1\ 0\ 0\cdots\rangle$ carry exactly the same information. In general, a state of noninteracting particles, where n_1 particles occupy $|\phi_1\rangle$, n_2 particles occupy $|\phi_2\rangle$, and so on, may be represented as $|n_1 n_2\cdots\rangle$. For fermions, $n_i = 0$ or 1, but for bosons, n_i can vary from 0 to N, the total number of particles. Representation of states in this fashion is known as number-representation. The vacuum state, with no particles at all, is written as $|0\rangle$, and is defined by $c_\nu|0\rangle = 0$ for all ν. For $N = 1$, the states $|\phi_1\rangle, |\phi_2\rangle, \ldots$ span the Hilbert space $\mathrm{V}^{(1)}$ of the quantum states of the one-particle system, as do the states $|1\ 0\ 0\ 0\ \cdots\rangle, |0\ 1\ 0\ 0\ \cdots\rangle, \ldots$. For $N = 2$, the basis states that span $\mathrm{V}^{(2)}$ (the vector space of the quantum states of the two-particle system) are $|1\ 1\ 0\ 0\ \cdots\rangle, |1\ 0\ 1\ 0\ \cdots\rangle, |0\ 1\ 1\ 0\ \cdots\rangle, |0\ 1\ 0\ 1\ \cdots\rangle, \ldots$. We can continue in this fashion for any value of N.

Let us consider an extended Hilbert space, called the Fock space, which is obtained as a direct sum,

$$\mathrm{F} = \mathrm{V}^{(0)} \oplus \mathrm{V}^{(1)} \oplus \mathrm{V}^{(2)} \oplus \cdots .$$

Here, $\mathrm{V}^{(0)}$ is the Hilbert space (vector space) for $N = 0$, i.e., it contains only the vacuum state $|0\rangle$; $\mathrm{V}^{(1)}$ is the vector space for a one-particle system, and so on. The operator $c_\nu^\dagger$, by creating a particle in state $|\phi_\nu\rangle$, increases the number of particles by 1; hence, if $|\Psi\rangle \in \mathrm{V}^{(k)}$, then $c_\nu^\dagger|\Psi\rangle \in \mathrm{V}^{(k+1)}$, while $c_\nu|\Psi\rangle \in \mathrm{V}^{(k-1)}$. The vector spaces $\mathrm{V}^{(k)}$, $\mathrm{V}^{(k+1)}$, and $\mathrm{V}^{(k-1)}$, are parts of F; hence, creation and annihilation operators act upon the Fock space.

Finally, we note that, in the number-representation, $c_\nu^\dagger$ and c_ν act in the following way

$$c_\nu^\dagger|n_1\cdots n_\nu\cdots\rangle = (-1)^{n_1+n_2+\cdots+n_{\nu-1}}(1-n_\nu)|n_1\cdots n_\nu+1\cdots\rangle \qquad (3.13)$$

$$c_\nu|n_1\cdots n_\nu\cdots\rangle = (-1)^{n_1+n_2+\cdots+n_{\nu-1}} n_\nu|n_1\cdots n_\nu-1\cdots\rangle. \qquad (3.14)$$

Since $n_\nu = 0$ or 1, these relations are easily verified.

3.4.2 Bosons

We only give a brief account of creation and annihilation operators for the case of bosons. We define a creation operator $a_\nu^\dagger$ by the following relation,

$$a_\nu^\dagger|n_1\cdots n_\nu\cdots\rangle = \sqrt{n_\nu+1}\,|n_1\cdots n_\nu+1\cdots\rangle. \qquad (3.15)$$

The operator $a_\nu^\dagger$ creates a particle in state $|\phi_\nu\rangle$. Similarly, an annihilation operator a_ν, which annihilates a particle in state $|\phi_\nu\rangle$, is defined by

$$a_\nu|n_1 \cdots n_\nu \cdots\rangle = \sqrt{n_\nu}\,|n_1 \cdots n_\nu - 1 \cdots\rangle. \tag{3.16}$$

If the state $|\phi_\nu\rangle$ is vacant ($n_\nu = 0$), the action of a_ν yields zero. By using an argument similar to the one used in the case of fermions, one shows that a_ν is the adjoint (hermitian conjugate) of $a_\nu^\dagger$. The symmetry of the state of identical bosons, under the exchange of coordinates of any two particles, leads to the following commutation relation between the creation and annihilation operators

$$[a_\nu, a_{\nu'}^\dagger] = \delta_{\nu\nu'}. \tag{3.17}$$

Equations (3.15–3.17) should be familiar from the study of the quantum harmonic oscillator.

3.5 One-body operators

The Hamiltonian for a system of N identical, interacting particles is generally the sum of a one-body operator $\sum_{i=1}^N h(i)$ and a two-body operator $(1/2)\sum_{i\neq j} v(i, j)$. For now, we will focus on the one-body operator and give its expression in terms of creation and annihilation operators.

Let $H_0 = \sum_{i=1}^N h(i)$, where $h(i)$ is an operator that depends on the coordinates of particle i. For example, $h(i)$ could be the kinetic energy $-(\hbar^2/2m)\nabla_i^2$ of particle i, or it could be the sum of the kinetic energy and the potential energy $v(i)$ produced by some external field. In general, h may depend on both spatial and spin coordinates.

Suppose that $|\phi_1\rangle, |\phi_2\rangle, \ldots$ constitute a complete, orthonormal set of single-particle states. For example, if for a system of electrons $|\phi\rangle = |\mathbf{k}\sigma\rangle$, the complete set of single-particle states will be $|\mathbf{k}_1 \uparrow\rangle, |\mathbf{k}_1 \downarrow\rangle, |\mathbf{k}_2 \uparrow\rangle, \ldots$.

We can express the operator H_0 in terms of the creation and annihilation operators $c_\nu^\dagger$ and c_ν. The derivation of such an expression is somewhat lengthy; it is given in Appendix A. Here, we merely state the result:

$$H_0 = \sum_{\nu\nu'} \langle\phi_{\nu'}|h|\phi_\nu\rangle c_{\nu'}^\dagger c_\nu. \tag{3.18}$$

This is the second quantized form of H_0, and it holds true for both fermions and bosons. The expression is plausible: a one-body operator is the sum of single-particle operators $h(1), h(2), \ldots, h(N)$. The effect of a single-particle operator is to scatter a particle from a state $|\phi_\nu\rangle$ into a state $|\phi_{\nu'}\rangle$. The scattering process can be viewed as the annihilation of a particle in state $|\phi_\nu\rangle$, followed by the creation of a particle in state $|\phi_{\nu'}\rangle$. The amplitude for this process is the matrix element $\langle\phi_{\nu'}|h|\phi_\nu\rangle$.

3.6 Examples

In the following we give a few examples that illustrate how to express one-body operators in second quantized form.

3.6.1 Kinetic energy of a system of N electrons

The kinetic energy of a system of N electrons is

$$T = \sum_{i=1}^{N} p_i^2/2m = \sum_{i=1}^{N} -(\hbar^2/2m)\nabla_i^2.$$

The second quantized form of the operator depends on the basis set of single-particle states. Let us choose the plane waves $|\mathbf{k}\sigma\rangle$ as basis states. The electrons are assumed to move within a box of volume $V = L^3$. Assuming periodic boundary conditions, we have

$$\phi_{\mathbf{k}\sigma}(\mathbf{r}) = \langle \mathbf{r}|\mathbf{k}\sigma\rangle = \frac{1}{\sqrt{V}} e^{i\mathbf{k}.\mathbf{r}}|\sigma\rangle,$$

where $\sigma = \uparrow$ or $\downarrow$ $(+1/2$ or $-1/2)$, and $k_x, k_y, k_z = 0, \pm 2\pi/L, \pm 4\pi/L, \ldots$. Being a one-body operator, T can be written as

$$T = \sum_{\mathbf{k}\sigma}\sum_{\mathbf{k}'\sigma'} \left\langle \mathbf{k}'\sigma' \left| -\frac{\hbar^2}{2m}\nabla^2 \right| \mathbf{k}\sigma \right\rangle c^{\dagger}_{\mathbf{k}'\sigma'} c_{\mathbf{k}\sigma}.$$

Since $-(\hbar^2/2m)\nabla^2|\mathbf{k}\sigma\rangle = (\hbar^2k^2/2m)|\mathbf{k}\sigma\rangle$ and $\langle \mathbf{k}'\sigma'|\mathbf{k}\sigma\rangle = \delta_{\mathbf{k}\mathbf{k}'}\delta_{\sigma\sigma'}$, the second quantized form of the kinetic energy is

$$T = \sum_{\mathbf{k}\sigma} \frac{\hbar^2 k^2}{2m} c^{\dagger}_{\mathbf{k}\sigma} c_{\mathbf{k}\sigma}. \tag{3.19}$$

3.6.2 External potential

The potential energy of a system of N particles due to interaction with an external field is

$$V_{\text{ext}} = \sum_{i=1}^{N} v(i)$$

In a crystal, $v(i)$ could be the interaction of electron i with a periodic lattice of ions. In general, $v(i)$ may depend on the spin of particle i; e.g., $v(i)$ may include spin-orbit coupling. Using a basis set of plane waves,

$$V_{\text{ext}} = \sum_{\mathbf{k}'\sigma'} \sum_{\mathbf{k}\sigma} \langle \mathbf{k}'\sigma'|v|\mathbf{k}\sigma\rangle c^{\dagger}_{\mathbf{k}'\sigma'} c_{\mathbf{k}\sigma}.$$

The matrix elements are given by

$$\langle \mathbf{k}'\sigma'|v|\mathbf{k}\sigma\rangle = \frac{1}{V}\int e^{i(\mathbf{k}-\mathbf{k}').\mathbf{r}} \langle \sigma'|v|\sigma\rangle d^3r.$$

If v is spin-dependent, the matrix element $\langle \sigma'|v|\sigma\rangle$ is evaluated first; the result will be a function of $\mathbf{r}$, and the $\mathbf{r}$ integration is then carried out. In the simpler case where v is spin-independent, $\langle \sigma'|v|\sigma\rangle = v(\mathbf{r})\delta_{\sigma\sigma'}$, and

$$\langle \mathbf{k}'\sigma'|v|\mathbf{k}\sigma\rangle = \delta_{\sigma\sigma'}\frac{1}{V}\int e^{-i(\mathbf{k}'-\mathbf{k}).\mathbf{r}} v(\mathbf{r}) d^3r = \frac{1}{V} v_{\mathbf{k}'-\mathbf{k}} \delta_{\sigma\sigma'}.$$

Here, $v_{\mathbf{q}}$ is the Fourier transform of $v(\mathbf{r})$,

$$v_{\mathbf{q}} = \int e^{-i\mathbf{q}.\mathbf{r}} v(\mathbf{r}) d^3r. \tag{3.20}$$

We note, in passing, that the inverse Fourier transform is

$$v(\mathbf{r}) = \frac{1}{V}\sum_{\mathbf{q}} e^{i\mathbf{q}.\mathbf{r}} v_{\mathbf{q}}. \tag{3.21}$$

In conclusion, the second quantized expression for V_{ext} is

$$V_{\text{ext}} = \frac{1}{V}\sum_{\mathbf{k}\mathbf{k}'\sigma} v_{\mathbf{k}'-\mathbf{k}} c^{\dagger}_{\mathbf{k}'\sigma} c_{\mathbf{k}\sigma} = \frac{1}{V}\sum_{\mathbf{k}\mathbf{q}\sigma} v_{\mathbf{q}} c^{\dagger}_{\mathbf{k}+\mathbf{q}\sigma} c_{\mathbf{k}\sigma}. \tag{3.22}$$

3.6.3 Particle-number density

If a system consists of one particle at position $\mathbf{r}'$, what is the particle-number density $n(\mathbf{r})$? Since $n(\mathbf{r}) = 0$ if $\mathbf{r} \neq \mathbf{r}'$, and the integral over all space of the density, $\int n(\mathbf{r})d^3r$, must give the total number of particles, which is 1, it follows that $n(\mathbf{r}) = \delta(\mathbf{r} - \mathbf{r}')$, the Dirac-delta function. In a system of N particles at positions $\mathbf{r}'_1, \mathbf{r}'_2, \ldots, \mathbf{r}'_N$, the particle-number density is

$$n(\mathbf{r}) = \sum_{i=1}^{N} \delta(\mathbf{r} - \mathbf{r}'_i). \tag{3.23}$$

This is a one-body operator of the form $\sum_{i=1}^{N} f(i)$. In the basis $|\mathbf{k}\sigma\rangle$,

$$\begin{aligned} n(\mathbf{r}) &= \sum_{\mathbf{k}'\sigma'}\sum_{\mathbf{k}\sigma}\frac{1}{V}\int e^{-i\mathbf{k}'.\mathbf{r}'}\delta(\mathbf{r}-\mathbf{r}')e^{i\mathbf{k}.\mathbf{r}'}d^3r'\langle\sigma'|\sigma\rangle c^\dagger_{\mathbf{k}'\sigma'}c_{\mathbf{k}\sigma} \\ &= \sum_{\mathbf{k}\sigma}\sum_{\mathbf{k}'}\frac{1}{V}e^{i(\mathbf{k}-\mathbf{k}').\mathbf{r}}c^\dagger_{\mathbf{k}'\sigma}c_{\mathbf{k}\sigma} = \frac{1}{V}\sum_{\mathbf{q}}e^{i\mathbf{q}.\mathbf{r}}\sum_{\mathbf{k}'\sigma}c^\dagger_{\mathbf{k}'\sigma}c_{\mathbf{k}'+\mathbf{q}\sigma} \\ &= \frac{1}{V}\sum_{\mathbf{q}}e^{i\mathbf{q}.\mathbf{r}}n_{\mathbf{q}}. \end{aligned} \tag{3.24}$$

We have introduced $n_{\mathbf{q}}$, the Fourier transform of $n(\mathbf{r})$, and it is given by

$$n_{\mathbf{q}} = \sum_{\mathbf{k}\sigma}c^\dagger_{\mathbf{k}\sigma}c_{\mathbf{k}+\mathbf{q}\sigma}. \tag{3.25}$$

3.7 Two-body operators

Consider the two-body operator $H' = (1/2)\sum_{i\neq j}v(i, j)$. The sum extends over both i and j, but terms with $i = j$ are excluded. H' represents the pairwise interaction between particles, such as the Coulomb interaction between electrons. Let us assume that we have a complete set $|\phi_1\rangle, |\phi_2\rangle, \ldots$ of orthonormal single-particle states. A detailed derivation of the second quantized form of the two-body operator is given in Appendix A. Here we only state the result, which holds equally true for both fermions and bosons,

$$H' = \frac{1}{2}\sum_{klmn}\langle\phi_k\phi_l|v|\phi_m\phi_n\rangle c^\dagger_k c^\dagger_l c_n c_m. \tag{3.26}$$

In the above equation, we have introduced $|\phi_k\phi_l\rangle$ and $\langle\phi_k\phi_l|$ defined by

$$|\phi_k\phi_l\rangle = |\phi_k\rangle\otimes|\phi_l\rangle \equiv |\phi_k\rangle|\phi_l\rangle, \qquad \langle\phi_k\phi_l| = \langle\phi_k|\otimes\langle\phi_l| \equiv \langle\phi_k|\langle\phi_l|.$$

Similar definitions apply to $|\phi_m\phi_n\rangle$ and $\langle\phi_m\phi_n|$.

Adopting a simplified notation, the Hamiltonian given by

$$H = \sum_i h(i) + \frac{1}{2}\sum_{i\neq j}v(i, j)$$

is written in second quantized form as

$$H = \sum_{kl}\langle k|h|l\rangle c^\dagger_k c_l + \frac{1}{2}\sum_{klmn}\langle kl|v|mn\rangle c^\dagger_k c^\dagger_l c_n c_m. \tag{3.27}$$

Notice that the order of n and m in the matrix element differs from the order in the operators.

3.8 Translationally invariant system

In a translationally invariant system, the interaction between two particles at $\mathbf{r}_1$ and $\mathbf{r}_2$ depends only on $\mathbf{r}_1 - \mathbf{r}_2$ and not on $\mathbf{r}_1$ and $\mathbf{r}_2$ separately: $v(\mathbf{r}_1, \mathbf{r}_2) = v(\mathbf{r}_1 - \mathbf{r}_2)$. The system acquires its name because if two particles within it at positions $\mathbf{r}_1$ and $\mathbf{r}_2$ are translated by the same vector $\mathbf{R}$ to new positions $\mathbf{r}'_1 = \mathbf{r}_1 + \mathbf{R}$ and $\mathbf{r}'_2 = \mathbf{r}_2 + \mathbf{R}$, their interaction energy does not change: $v(\mathbf{r}'_1 - \mathbf{r}'_2) = v(\mathbf{r}_1 - \mathbf{r}_2)$. For N particles, the total interaction energy is

$$V_{\text{int}} = \frac{1}{2} \sum_{i \neq j} v(\mathbf{r}_i - \mathbf{r}_j).$$

Using the basis states $|\mathbf{k}\sigma\rangle$, the second quantized form of V_{int} is

$$V_{\text{int}} = \frac{1}{2} \sum_{\mathbf{k}_1\sigma_1} \sum_{\mathbf{k}_2\sigma_2} \sum_{\mathbf{k}_3\sigma_3} \sum_{\mathbf{k}_4\sigma_4} \langle \mathbf{k}_1\sigma_1\mathbf{k}_2\sigma_2 | v | \mathbf{k}_3\sigma_3\mathbf{k}_4\sigma_4 \rangle c^\dagger_{\mathbf{k}_1\sigma_1} c^\dagger_{\mathbf{k}_2\sigma_2} c_{\mathbf{k}_4\sigma_4} c_{\mathbf{k}_3\sigma_3}.$$

Assuming, as is often the case, that v is spin-independent, the matrix element $M = \langle \mathbf{k}_1\sigma_1\mathbf{k}_2\sigma_2 | v | \mathbf{k}_3\sigma_3\mathbf{k}_4\sigma_4 \rangle$ is given by

$$\begin{aligned} M &= \frac{1}{V^2} \int d^3r_1 \int d^3r_2 \, e^{-i\mathbf{k}_1.\mathbf{r}_1} e^{-i\mathbf{k}_2.\mathbf{r}_2} v(\mathbf{r}_1 - \mathbf{r}_2) e^{i\mathbf{k}_3.\mathbf{r}_1} e^{i\mathbf{k}_4.\mathbf{r}_2} \langle \sigma_1 | \sigma_3 \rangle \langle \sigma_2 | \sigma_4 \rangle \\ &= \frac{1}{V^2} \delta_{\sigma_1\sigma_3} \delta_{\sigma_2\sigma_4} \int d^3r_1 \int d^3r_2 \, e^{i(\mathbf{k}_3 - \mathbf{k}_1).\mathbf{r}_1} e^{i(\mathbf{k}_4 - \mathbf{k}_2).\mathbf{r}_2} v(\mathbf{r}_1 - \mathbf{r}_2). \end{aligned}$$

To proceed further, we replace $v(\mathbf{r}_1 - \mathbf{r}_2)$ by $(1/V) \sum_{\mathbf{q}} e^{i\mathbf{q}.(\mathbf{r}_1 - \mathbf{r}_2)} v_{\mathbf{q}}$,

$$\begin{aligned} M &= \frac{1}{V^3} \delta_{\sigma_1\sigma_3} \delta_{\sigma_2\sigma_4} \sum_{\mathbf{q}} v_{\mathbf{q}} \int d^3r_1 \, e^{i(\mathbf{k}_3 - \mathbf{k}_1 + \mathbf{q}).\mathbf{r}_1} \int d^3r_2 \, e^{i(\mathbf{k}_4 - \mathbf{k}_2 - \mathbf{q}).\mathbf{r}_2} \\ &= \frac{1}{V} \delta_{\sigma_1\sigma_3} \delta_{\sigma_2\sigma_4} \sum_{\mathbf{q}} v_{\mathbf{q}} \delta_{\mathbf{q},\mathbf{k}_1 - \mathbf{k}_3} \delta_{\mathbf{q},\mathbf{k}_4 - \mathbf{k}_2}. \end{aligned}$$

Therefore, in the expression for V_{int}, the sum vanishes unless $\mathbf{k}_1 = \mathbf{k}_3 + \mathbf{q}$, $\mathbf{k}_2 = \mathbf{k}_4 - \mathbf{q}$, $\sigma_1 = \sigma_3$, and $\sigma_2 = \sigma_4$; hence

$$V_{\text{int}} = \frac{1}{2V} \sum_{\mathbf{q}} \sum_{\mathbf{k}_3\sigma_3} \sum_{\mathbf{k}_4\sigma_4} v_{\mathbf{q}} c^\dagger_{\mathbf{k}_3+\mathbf{q}\sigma_3} c^\dagger_{\mathbf{k}_4-\mathbf{q}\sigma_4} c_{\mathbf{k}_4\sigma_4} c_{\mathbf{k}_3\sigma_3}.$$

Finally, relabeling indices: $\mathbf{k}_3\sigma_3 \to \mathbf{k}\sigma$, $\mathbf{k}_4\sigma_4 \to \mathbf{k}'\sigma'$, we obtain

$$V_{\text{int}} = \frac{1}{2V} \sum_{\mathbf{q}} \sum_{\mathbf{k}\sigma} \sum_{\mathbf{k}'\sigma'} v_{\mathbf{q}} c^\dagger_{\mathbf{k}+\mathbf{q}\sigma} c^\dagger_{\mathbf{k}'-\mathbf{q}\sigma'} c_{\mathbf{k}'\sigma'} c_{\mathbf{k}\sigma}. \tag{3.28}$$

Each term in the summation represents a scattering process in which two particles in states $|\mathbf{k}\sigma\rangle$ and $|\mathbf{k}'\sigma'\rangle$ are annihilated, and two particles are created in states

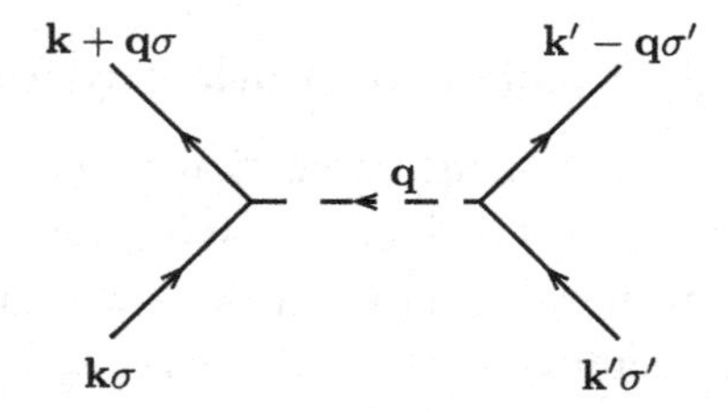

Figure 3.5 Schematic representation of the interaction of two particles. The two particles initially have wave vectors $\mathbf{k}$ and $\mathbf{k}'$. The interaction is viewed as a collision in which one particle transfers momentum $\hbar\mathbf{q}$ to the other.

$|\mathbf{k}+\mathbf{q}\sigma\rangle$ and $|\mathbf{k}'-\mathbf{q}\sigma'\rangle$. The scattering process may be represented pictorially, as shown in Figure 3.5.

3.9 Example: Coulomb interaction

In a system of N interacting electrons, the electron–electron interaction is

$$V_C = \frac{1}{2}\sum_{i\neq j} v(i,j) = \frac{1}{2}\sum_{i\neq j}\frac{e^2}{|\mathbf{r}_i - \mathbf{r}_j|} \qquad \text{(cgs)}.$$

In SI units, e^2 is replaced by $e^2/4\pi\varepsilon_0$. The system is translationally invariant. In order to express the Coulomb interaction V_C in second quantized form, we need to determine $v_{\mathbf{q}}$, the Fourier transform of $v(\mathbf{r}) = e^2/r$,

$$v_{\mathbf{q}} = e^2\int \frac{1}{r}e^{-i\mathbf{q}.\mathbf{r}}d^3r = e^2\int_0^\infty r^2dr\int_{-1}^{1}d(\cos\theta)\int_0^{2\pi}d\phi\,\frac{1}{r}\,e^{-iqr\cos\theta}.$$

Integration over ϕ gives 2π. Integrating over $\cos\theta$, we find

$$v_{\mathbf{q}} = 2\pi e^2\int_0^\infty \frac{e^{iqr}-e^{-iqr}}{iq}dr = \frac{4\pi e^2}{q}\int_0^\infty \sin(qr)dr.$$

The oscillatory behavior at infinity complicates the evaluation of the integral. We also note that the integral diverges at $q=0$ because in the limit $q\to 0$, $\sin(qr)/q \to r$, and $\int_0^\infty r dr = \infty$. What is to be done?

Rather than Fourier transforming the Coulomb potential, let us Fourier transform the Yukawa potential $v_Y = (e^2/r)e^{-\mu r}$. At the end, we replace μ by 0. Integration over ϕ and θ proceeds as before; we find

$$\begin{aligned} v_{\mathbf{q}} &= \lim_{\mu\to 0^+} e^2\int\frac{e^{-\mu r}}{r}e^{-i\mathbf{q}.\mathbf{r}}d^3r = \lim_{\mu\to 0^+}\frac{2\pi e^2}{iq}\int_0^\infty\left[e^{(iq-\mu)r}-e^{(-iq-\mu)r}\right]dr \\ &= \lim_{\mu\to 0^+}\frac{2\pi e^2}{iq}\left[\frac{-1}{iq-\mu}-\frac{1}{iq+\mu}\right] = \lim_{\mu\to 0^+}\frac{4\pi e^2}{q^2+\mu^2} = \frac{4\pi e^2}{q^2}. \end{aligned}$$

The expression for the Coulomb interaction thus becomes

$$V_C = \lim_{\mu\to 0} \frac{1}{2V} \sum_{\mathbf{q}} \sum_{\mathbf{k}\sigma} \sum_{\mathbf{k}'\sigma'} \frac{4\pi e^2}{q^2+\mu^2} c^\dagger_{\mathbf{k}+\mathbf{q}\sigma} c^\dagger_{\mathbf{k}'-\mathbf{q}\sigma'} c_{\mathbf{k}'\sigma'} c_{\mathbf{k}\sigma}$$

$$= \frac{1}{2V} \sum_{\mathbf{q}} \sum_{\mathbf{k}\sigma} \sum_{\mathbf{k}'\sigma'} \frac{4\pi e^2}{q^2} c^\dagger_{\mathbf{k}+\mathbf{q}\sigma} c^\dagger_{\mathbf{k}'-\mathbf{q}\sigma'} c_{\mathbf{k}'\sigma'} c_{\mathbf{k}\sigma}. \tag{3.29}$$

Here, V is the system's volume. The Coulomb energy diverges because of the $\mathbf{q} = 0$ term. We shall see in the next chapter that the $\mathbf{q} = 0$ term drops out of the sum as a result of charge neutrality: in a crystal, there are both electrons and positive ions.

3.10 Electrons in a periodic potential

We have discussed the second quantized formulation of the Hamiltonian for a system of N electrons using a basis set of plane waves. This is the convenient set to use in the study of the jellium model of a metal, where a uniform positive background replaces the lattice of positive ions. In the jellium model, the uniform positive background produces a constant potential, and the eigenstates of the single-particle Hamiltonian are plane waves. This is why an orthonormal basis of plane waves is most convenient. However, when the discrete nature of the lattice is taken into account, the eigenstates of the single-particle Hamiltonian are the Bloch states, and these form a more adequate basis in which to express the Hamiltonian.

Another important orthonormal basis is the set of Wannier states. Even though these are not eigenstates of the single-particle Hamiltonian, the formulation of the Hamiltonian in terms of Wannier states is at the heart of the tight binding methods that play an important role in the theoretical analysis of the electronic properties of crystals. In this section, we discuss the second quantized form of the Hamiltonian in terms of Bloch and Wannier states.

3.10.1 Bloch representation

A Bloch state $|n\mathbf{k}\sigma\rangle$ is characterized by a band index n, a wave vector $\mathbf{k}$, and a spin projection σ. The electronic Hamiltonian is

$$H = H_0 + V_C \tag{3.30}$$

$$H_0 = \sum_i \left[p_i^2/2m + v(\mathbf{r}_i)\right], \quad V_C = \frac{1}{2}\sum_{i\neq j} \frac{e^2}{|\mathbf{r}_i - \mathbf{r}_j|}. \tag{3.31}$$

$v(\mathbf{r})$ is the potential produced by the static periodic lattice of ions (the effects of ionic vibrations are studied in Chapter 11), and V_C is the Coulomb interaction between electrons. Using the Bloch states as a basis,

$$H_0 = \sum_{n\mathbf{k}\sigma} \sum_{n'\mathbf{k}'\sigma'} \langle n'\mathbf{k}'\sigma'| - \frac{\hbar^2}{2m}\nabla^2 + v(\mathbf{r})|n\mathbf{k}\sigma\rangle c^{\dagger}_{n'\mathbf{k}'\sigma'} c_{n\mathbf{k}\sigma}.$$

Since the Bloch states are eigenstates of the single-particle Hamiltonian,

$$\left[p^2/2m + v(\mathbf{r})\right] |n\mathbf{k}\sigma\rangle = \epsilon_{n\mathbf{k}}|n\mathbf{k}\sigma\rangle, \tag{3.32}$$

the above expression for H_0 reduces to

$$H_0 = \sum_{n\mathbf{k}\sigma} \epsilon_{n\mathbf{k}}\, c^{\dagger}_{n\mathbf{k}\sigma} c_{n\mathbf{k}\sigma}. \tag{3.33}$$

Setting $|\mathbf{r}_1 - \mathbf{r}_2| = r_{12}$, the second quantized form of V_C is

$$V_C = \frac{1}{2} \sum_{n_1\mathbf{k}_1\sigma_1} \sum_{n_2\mathbf{k}_2\sigma_2} \sum_{n'_1\mathbf{k}'_1\sigma'_1} \sum_{n'_2\mathbf{k}'_2\sigma'_2} \langle n'_1\mathbf{k}'_1\sigma'_1 n'_2\mathbf{k}'_2\sigma'_2|\frac{e^2}{r_{12}}|n_1\mathbf{k}_1\sigma_1 n_2\mathbf{k}_2\sigma_2\rangle$$
$$\times\, c^{\dagger}_{n'_1\mathbf{k}'_1\sigma'_1} c^{\dagger}_{n'_2\mathbf{k}'_2\sigma'_2} c_{n_2\mathbf{k}_2\sigma_2} c_{n_1\mathbf{k}_1\sigma_1}.$$

The matrix element $M = \langle n'_1\mathbf{k}'_1\sigma'_1 n'_2\mathbf{k}'_2\sigma'_2|\frac{e^2}{r_{12}}|n_1\mathbf{k}_1\sigma_1 n_2\mathbf{k}_2\sigma_2\rangle$ is given by

$$M = \delta_{\sigma_1\sigma'_1}\delta_{\sigma_2\sigma'_2} \int \psi^*_{n'_1\mathbf{k}'_1}(\mathbf{r}_1)\psi^*_{n'_2\mathbf{k}'_2}(\mathbf{r}_2)\frac{e^2}{|\mathbf{r}_1 - \mathbf{r}_2|}\psi_{n_1\mathbf{k}_1}(\mathbf{r}_1)\psi_{n_2\mathbf{k}_2}(\mathbf{r}_2)d^3r_1\,d^3r_2.$$

The Fourier transform of the Coulomb potential is $v_{\mathbf{q}} = 4\pi e^2/q^2$; hence,

$$\frac{e^2}{|\mathbf{r}_1 - \mathbf{r}_2|} = \frac{1}{V}\sum_{\mathbf{q}} \frac{4\pi e^2}{q^2} e^{i\mathbf{q}.(\mathbf{r}_1 - \mathbf{r}_2)} \tag{3.34}$$

where V is the volume of the system. The expression for M becomes

$$M = \delta_{\sigma_1\sigma'_1}\,\delta_{\sigma_2\sigma'_2} \sum_{\mathbf{q}} \frac{4\pi e^2}{Vq^2} \int \psi^*_{n'_1\mathbf{k}'_1}(\mathbf{r}_1)\, e^{i\mathbf{q}.\mathbf{r}_1}\, \psi_{n_1\mathbf{k}_1}(\mathbf{r}_1)\, d^3r_1$$
$$\times \int \psi^*_{n'_2\mathbf{k}'_2}(\mathbf{r}_2)\, e^{-i\mathbf{q}.\mathbf{r}_2}\, \psi_{n_2\mathbf{k}_2}(\mathbf{r}_2)\, d^3r_2.$$

Recall that, for any lattice vector $\mathbf{R}$, the Bloch function satisfies the relation

$$\psi_{n\mathbf{k}}(\mathbf{r} + \mathbf{R}) = e^{i\mathbf{k}.\mathbf{R}}\psi_{n\mathbf{k}}(\mathbf{r}). \tag{3.35}$$

This is the first form of Bloch's theorem (Section 2.3). If we now make a change of variable: $\mathbf{r}_1 \to \mathbf{r}_1 + \mathbf{R}$, the integral does not change, but the integrand gets multiplied by the factor $e^{i(\mathbf{k}_1 - \mathbf{k}'_1 + \mathbf{q}).\mathbf{R}}$; this factor must be equal to unity. Since this is true for every lattice vector $\mathbf{R}$, it follows that $\mathbf{k}'_1 = \mathbf{k}_1 + \mathbf{q} + \mathbf{G}$, where $\mathbf{G}$ is a reciprocal lattice vector. However, since $\mathbf{k}_1, \mathbf{k}'_1 \in$ FBZ, $\mathbf{G}$ will vanish if

$\mathbf{k}_1 + \mathbf{q} \in$ FBZ, but if $\mathbf{k}_1 + \mathbf{q} \notin$ FBZ, then $\mathbf{G}$ is the reciprocal lattice vector that brings $\mathbf{k}_1 + \mathbf{q}$ back into the first Brillouin zone. We thus require that $\mathbf{k}_1' = \mathbf{k}_1 + \mathbf{q}$, with the understanding that $\mathbf{k}_1, \mathbf{k}_1 + \mathbf{q} \in$ FBZ. A similar argument shows that $\mathbf{k}_2' = \mathbf{k}_2 - \mathbf{q}$. Hence,

$$M = \delta_{\sigma_1\sigma_1'}\delta_{\sigma_2\sigma_2'} \sum_{\mathbf{q}} \frac{4\pi e^2}{Vq^2} \delta_{\mathbf{k}_1',\mathbf{k}_1+\mathbf{q}} \delta_{\mathbf{k}_2',\mathbf{k}_2-\mathbf{q}} \int \psi^*_{n_1'\mathbf{k}_1'}(\mathbf{r}_1) e^{i\mathbf{q}.\mathbf{r}_1} \psi_{n_1\mathbf{k}_1}(\mathbf{r}_1) d^3r_1$$
$$\times \int \psi^*_{n_2'\mathbf{k}_2'}(\mathbf{r}_2) e^{-i\mathbf{q}.\mathbf{r}_2} \psi_{n_2\mathbf{k}_2}(\mathbf{r}_2) d^3r_2.$$

Therefore, the second quantized form of V_C, in the Bloch representation, is

$$V_C = \frac{1}{2V} \sum_{n_1n_1'\mathbf{k}_1\sigma_1} \sum_{n_2n_2'\mathbf{k}_2\sigma_2} \sum_{\mathbf{q}} \frac{4\pi e^2}{q^2} F^{n_1n_1'}_{\mathbf{k}_1,\mathbf{k}_1+\mathbf{q}} F^{n_2n_2'}_{\mathbf{k}_2,\mathbf{k}_2-\mathbf{q}}$$
$$\times\, c^\dagger_{n_1'\mathbf{k}_1+\mathbf{q}\sigma_1} c^\dagger_{n_2'\mathbf{k}_2-\mathbf{q}\sigma_2} c_{n_2\mathbf{k}_2\sigma_2} c_{n_1\mathbf{k}_1\sigma_1}. \tag{3.36}$$

The matrix elements in the above expression are defined as follows:

$$F^{nn'}_{\mathbf{k},\mathbf{k}\pm\mathbf{q}} = \langle n'\mathbf{k} \pm \mathbf{q}\sigma | e^{\pm i\mathbf{q}.\mathbf{r}} | n\mathbf{k}\sigma \rangle. \tag{3.37}$$

The plane wave, or momentum, representation is recovered from the Bloch representation by removing the sum over the band indices and setting $\psi_{n\mathbf{k}\sigma} = 1/\sqrt{V} e^{i\mathbf{k}.\mathbf{r}} |\sigma\rangle$, in which case $F^{n_1n_1'}_{\mathbf{k}_1,\mathbf{k}_1+\mathbf{q}} = F^{n_2n_2'}_{\mathbf{k}_2,\mathbf{k}_2-\mathbf{q}} = 1$.

3.10.2 Wannier representation

Let us consider a metal with one partially filled band. In terms of the Bloch states $|n\mathbf{k}\sigma\rangle$, the Wannier states $|ni\sigma\rangle$ are expressed as (see Eq. [2.31])

$$|ni\sigma\rangle = \frac{1}{\sqrt{N}} \sum_{\mathbf{k}} e^{-i\mathbf{k}.\mathbf{R}_i} |n\mathbf{k}\sigma\rangle. \tag{3.38}$$

Here, n is a band index, i is a lattice site index, $\mathbf{R}_i$ is the lattice vector from the origin (chosen as some lattice point) to the lattice site i, and the sum is over all $\mathbf{k}$-points in the first Brillouin zone (FBZ). Since our interest is only in the electrons in one band, we may drop the band index and write the Wannier state as $|i\sigma\rangle$ and the Bloch state as $|\mathbf{k}\sigma\rangle$. We define the operator $c^\dagger_{i\sigma}$ that creates an electron in the state $|i\sigma\rangle$,

$$c^\dagger_{i\sigma}|0\rangle = |i\sigma\rangle = \frac{1}{\sqrt{N}} \sum_{\mathbf{k}} e^{-i\mathbf{k}.\mathbf{R}_i} |\mathbf{k}\sigma\rangle = \frac{1}{\sqrt{N}} \sum_{\mathbf{k}} e^{-i\mathbf{k}.\mathbf{R}_i} c^\dagger_{\mathbf{k}\sigma}|0\rangle \tag{3.39}$$

where $|0\rangle$ is the vacuum state. The Wannier and Bloch operators are thus related according to

$$c_{i\sigma}^{\dagger} = \frac{1}{\sqrt{N}} \sum_{\mathbf{k}} e^{-i\mathbf{k}.\mathbf{R}_i} c_{\mathbf{k}\sigma}^{\dagger}, \quad c_{i\sigma} = \frac{1}{\sqrt{N}} \sum_{\mathbf{k}} e^{i\mathbf{k}.\mathbf{R}_i} c_{\mathbf{k}\sigma}. \tag{3.40}$$

The first equation, connecting the creation operators, is obtained directly from Eq. (3.39), while the second equation, connecting the annihilation operators, is obtained from the first equation by taking the adjoints on both sides of the equal sign. These equations can be inverted,

$$c_{\mathbf{k}\sigma}^{\dagger} = \frac{1}{\sqrt{N}} \sum_{i} e^{i\mathbf{k}.\mathbf{R}_i} c_{i\sigma}^{\dagger}, \quad c_{\mathbf{k}\sigma} = \frac{1}{\sqrt{N}} \sum_{i} e^{-i\mathbf{k}.\mathbf{R}_i} c_{i\sigma}. \tag{3.41}$$

Using the Wannier states as a basis, the Hamiltonian is represented as follows:

$$H = \sum_{i\sigma} \sum_{j\sigma'} \langle j\sigma' | h | i\sigma \rangle c_{j\sigma'}^{\dagger} c_{i\sigma}$$

$$+ \frac{1}{2} \sum_{i\sigma_1} \sum_{j\sigma_2} \sum_{i'\sigma_1'} \sum_{j'\sigma_2'} \left\langle i'\sigma_1' j'\sigma_2' \left| \frac{e^2}{r_{12}} \right| i\sigma_1 j\sigma_2 \right\rangle c_{i'\sigma_1'}^{\dagger} c_{j'\sigma_2'}^{\dagger} c_{j\sigma_2} c_{i\sigma_1}.$$

Since H_0 and V_C are spin-independent,

$$\langle j\sigma' | h | i\sigma \rangle = \delta_{\sigma\sigma'} \langle j | h | i \rangle \equiv \delta_{\sigma\sigma'} t_{ij}$$

$$\langle i'\sigma_1' j'\sigma_2' | v(1,2) | i\sigma_1 j\sigma_2 \rangle = \delta_{\sigma_1\sigma_1'} \delta_{\sigma_2\sigma_2'} \left\langle i'j' \left| \frac{e^2}{r_{12}} \right| ij \right\rangle \equiv \delta_{\sigma_1\sigma_1'} \delta_{\sigma_2\sigma_2'} U_{ij\, i'j'}.$$

The Hamiltonian in the Wannier representation takes the form:

$$H = \sum_{ij\sigma} t_{ij}\, c_{j\sigma}^{\dagger} c_{i\sigma} + \frac{1}{2} \sum_{\sigma\sigma'} \sum_{iji'j'} U_{ij\, i'j'}\, c_{i'\sigma}^{\dagger} c_{j'\sigma'}^{\dagger} c_{j\sigma'} c_{i\sigma}. \tag{3.42}$$

The matrix element $U_{ij\, i'j'}$ depends on the degree of overlap of the Wannier functions. When the overlap is very weak, the onsite Coulomb repulsion dominates the interaction. In this case, we ignore $U_{ij\, i'j'}$ except when $i = j = i' = j'$, and set $U_{ii\, ii} = U$. Keeping only nearest-neighbor contribution to the hopping matrix element t_{ij}, the Hamiltonian reduces to

$$H = \sum_{<ij>\sigma} t_{ij}\, c_{j\sigma}^{\dagger} c_{i\sigma} + U \sum_{i} n_{i\downarrow} n_{i\uparrow} \tag{3.43}$$

where $n_{i\sigma} = c_{i\sigma}^{\dagger} c_{i\sigma}$ is the operator that represents the number of electrons at site i, and $<ij>$ indicates that i and j are nearest-neighboring sites. In writing the interaction term, we have used the fact that $c_{i\sigma}^2 = 0$ and that $c_{i\sigma'}^{\dagger} c_{i\sigma} = -c_{i\sigma} c_{i\sigma'}^{\dagger}$ for $\sigma \neq \sigma'$. The model described by Eq. (3.43) is the Hubbard model (Hubbard, 1963).

It describes a situation where electrons are essentially localized on atomic sites, but they can tunnel to neighboring sites, with t_{ij} being the tunneling amplitude. Double occupancy of a site, however, is penalized through a rise in energy equal to the amount U.

3.11 Field operators

3.11.1 Definition

Thus far, we have expressed various operators mainly in the **k**-representation (momentum representation): a complete set of single-particle states $|\mathbf{k}\sigma\rangle$, that are eigenstates of the momentum operator, has been used. The position kets $|\mathbf{r}\sigma\rangle$ form another important set of single-particle states. $|\mathbf{r}\sigma\rangle$ is the state of a particle with a definite position $\mathbf{r}$ and spin projection σ. Given a complete set of orthonormal states $|\phi_1\rangle, |\phi_2\rangle, \ldots$, we write

$$|\mathbf{r}\sigma\rangle = \sum_\nu |\phi_\nu\rangle\langle\phi_\nu|\mathbf{r}\sigma\rangle,$$

where ν is a collective index that includes spin. It is advantageous to display the spin index explicitly; thus, we write $|\phi_\nu\rangle = |\phi_{n\lambda}\rangle$, where λ is the spin quantum number and n stands for orbital (spatial) quantum numbers. Then

$$|\mathbf{r}\sigma\rangle = \sum_{n\lambda} |\phi_{n\lambda}\rangle\langle\phi_{n\lambda}|\mathbf{r}\sigma\rangle = \sum_{n\lambda} |\phi_{n\lambda}\rangle\langle\phi_n|\mathbf{r}\rangle\langle\lambda|\sigma\rangle = \sum_{n\lambda} |\phi_{n\lambda}\rangle\phi_n^*(\mathbf{r})\delta_{\lambda\sigma}$$
$$= \sum_n |\phi_{n\sigma}\rangle\phi_n^*(\mathbf{r}) = \sum_n \phi_n^*(\mathbf{r})c_{n\sigma}^\dagger|0\rangle.$$

The field operator $\Psi_\sigma^\dagger(\mathbf{r})$ is defined as the operator that creates a particle with spin projection σ ($\uparrow$ or $\downarrow$ for an electron, for example) at position $\mathbf{r}$,

$$\Psi_\sigma^\dagger(\mathbf{r})|0\rangle = |\mathbf{r}\sigma\rangle.$$

A comparison with the previous expression for $|\mathbf{r}\sigma\rangle$ gives

$$\Psi_\sigma^\dagger(\mathbf{r}) = \sum_n \phi_n^*(\mathbf{r})c_{n\sigma}^\dagger. \tag{3.44}$$

The field operator that annihilates a particle with spin projection σ, located at $\mathbf{r}$, is the adjoint of $\Psi_\sigma^\dagger(\mathbf{r})$,

$$\Psi_\sigma(\mathbf{r}) = \sum_n \phi_n(\mathbf{r})c_{n\sigma}. \tag{3.45}$$

For example, if $|\phi_\nu\rangle = |\mathbf{k}\sigma\rangle$, then $\phi_\mathbf{k}(\mathbf{r}) = (1/\sqrt{V})e^{i\mathbf{k}.\mathbf{r}}$, where V is the volume of the system and $k_x, k_y, k_z = 0, \pm 2\pi/L, \pm 4\pi/L \ldots$; in this case,

$$\Psi_\sigma^\dagger(\mathbf{r}) = \frac{1}{\sqrt{V}}\sum_\mathbf{k} e^{-i\mathbf{k}.\mathbf{r}}c_{\mathbf{k}\sigma}^\dagger, \quad \Psi_\sigma(\mathbf{r}) = \frac{1}{\sqrt{V}}\sum_\mathbf{k} e^{i\mathbf{k}.\mathbf{r}}c_{\mathbf{k}\sigma}. \tag{3.46}$$

3.11.2 Commutation relations

The commutation relations of field operators can be deduced from the corresponding relations for creation and annihilation operators. For fermions,

$$\{\Psi_\sigma(\mathbf{r}), \Psi^\dagger_{\sigma'}(\mathbf{r}')\} = \sum_{nn'} \phi_n(\mathbf{r})\phi^*_{n'}(\mathbf{r}')\{c_{n\sigma}, c^\dagger_{n'\sigma'}\} = \sum_{nn'} \phi_n(\mathbf{r})\phi^*_{n'}(\mathbf{r}')\delta_{nn'}\delta_{\sigma\sigma'}$$
$$= \delta_{\sigma\sigma'} \sum_n \phi_n(\mathbf{r})\phi^*_n(\mathbf{r}').$$

Using the completeness property of single-particle states (see Eq. [1.20]),

$$\{\Psi_\sigma(\mathbf{r}), \Psi^\dagger_{\sigma'}(\mathbf{r}')\} = \delta_{\sigma\sigma'}\delta(\mathbf{r} - \mathbf{r}'). \tag{3.47}$$

Since $\{c_\nu, c_{\nu'}\} = \{c^\dagger_\nu, c^\dagger_{\nu'}\} = 0$, it immediately follows that

$$\{\Psi_\sigma(\mathbf{r}), \Psi_{\sigma'}(\mathbf{r}')\} = \{\Psi^\dagger_\sigma(\mathbf{r}), \Psi^\dagger_{\sigma'}(\mathbf{r}')\} = 0. \tag{3.48}$$

For bosons, the commutators of field operators are given by

$$[\Psi_\sigma(\mathbf{r}), \Psi_{\sigma'}(\mathbf{r}')] = [\Psi^\dagger_\sigma(\mathbf{r}), \Psi^\dagger_{\sigma'}(\mathbf{r}')] = 0 \tag{3.49}$$

$$[\Psi_\sigma(\mathbf{r}), \Psi^\dagger_{\sigma'}(\mathbf{r}')] = \delta_{\sigma\sigma'}\delta(\mathbf{r} - \mathbf{r}'). \tag{3.50}$$

3.11.3 One-body operators

We can express the one-body operator $H_0 = \sum_{i=1}^N h(i)$ in terms of field operators as follows:

$$H_0 = \sum_{n\sigma}\sum_{n'\sigma'} \langle\phi_{n'\sigma'}|h|\phi_{n\sigma}\rangle c^\dagger_{n'\sigma'}c_{n\sigma} = \sum_{n\sigma}\sum_{n'\sigma'} \int d^3r \phi^*_{n'}(\mathbf{r})\langle\sigma'|h|\sigma\rangle\phi_n(\mathbf{r})c^\dagger_{n'\sigma'}c_{n\sigma}$$
$$= \sum_\sigma \sum_{\sigma'} \int \Psi^\dagger_{\sigma'}(\mathbf{r})h_{\sigma'\sigma}(\mathbf{r})\Psi_\sigma(\mathbf{r})d^3r. \tag{3.51}$$

Here, $h_{\sigma'\sigma}(\mathbf{r}) = \langle\sigma'|h|\sigma\rangle$, and use is made of Eqs (3.44) and (3.45).

For the case of spin-1/2 particles, such as electrons, H_0 may be written in a more compact form. We define the two-component field operators

$$\Psi(\mathbf{r}) = \begin{pmatrix} \Psi_\uparrow(\mathbf{r}) \\ \Psi_\downarrow(\mathbf{r}) \end{pmatrix} \qquad \Psi^\dagger(\mathbf{r}) = \begin{pmatrix} \Psi^\dagger_\uparrow & \Psi^\dagger_\downarrow \end{pmatrix}.$$

We also define the matrix $h(\mathbf{r})$ by

$$h(\mathbf{r}) = \begin{bmatrix} h_{\uparrow\uparrow}(\mathbf{r}) & h_{\uparrow\downarrow}(\mathbf{r}) \\ h_{\downarrow\uparrow}(\mathbf{r}) & h_{\downarrow\downarrow}(\mathbf{r}) \end{bmatrix}.$$

Then it is straightforward to check that

$$H_0 = \int \Psi^\dagger(\mathbf{r})h(\mathbf{r})\Psi(\mathbf{r})d^3r. \tag{3.52}$$

Although the above equation looks like an expectation value formula, it is certainly not; $\Psi^\dagger(\mathbf{r})$ and $\Psi(\mathbf{r})$ are field operators, not wave functions. If $h(i)$ is spin-independent, the expression for H_0 may be simplified,

$$H_0 = \sum_{\sigma\sigma'} \int \Psi_{\sigma'}^\dagger(\mathbf{r})h(\mathbf{r})\delta_{\sigma\sigma'}\Psi_\sigma(\mathbf{r})d^3r = \sum_{\sigma} \int \Psi_{\sigma}^\dagger(\mathbf{r})h(\mathbf{r})\Psi_\sigma(\mathbf{r})d^3r.$$

3.11.4 Two-body operators

In terms of field operators, the two-body operator $V_{\text{int}} = (1/2)\sum_{i\neq j} v(i, j)$ is expressed as follows:

$$\begin{aligned} V_{\text{int}} &= \frac{1}{2}\sum_{\{n\sigma\}}\int d^3r_1 \int d^3r_2\, \phi_{n_1}^*(\mathbf{r}_1)\phi_{n_2}^*(\mathbf{r}_2)\langle\sigma_1\sigma_2|v(1,2)|\sigma_3\sigma_4\rangle\phi_{n_3}(\mathbf{r}_1)\phi_{n_4}(\mathbf{r}_2) \\ &\qquad \times\, c_{n_1\sigma_1}^\dagger c_{n_2\sigma_2}^\dagger c_{n_4\sigma_4} c_{n_3\sigma_3} \\ &= \frac{1}{2}\sum_{\sigma_1\sigma_2\sigma_3\sigma_4}\int d^3r_1 \int d^3r_2 \Psi_{\sigma_1}^\dagger(\mathbf{r}_1)\Psi_{\sigma_2}^\dagger(\mathbf{r}_2)v_{\sigma_1\sigma_2\sigma_3\sigma_4}\Psi_{\sigma_4}(\mathbf{r}_2)\Psi_{\sigma_3}(\mathbf{r}_1). \end{aligned} \tag{3.53}$$

Here, $\{n\sigma\} = n_1\sigma_1 n_2\sigma_2 n_3\sigma_3 n_4\sigma_4$. For spin-1/2 particles, each of σ_1, σ_2, σ_3, and σ_4 is either $\uparrow$ or $\downarrow$, and V_{int} is the sum of 16 terms. We can recast the above expression into a more compact form. We define

$$\Psi(\mathbf{r}_2)\Psi(\mathbf{r}_1) = \begin{pmatrix}\Psi_\uparrow(\mathbf{r}_2)\\ \Psi_\downarrow(\mathbf{r}_2)\end{pmatrix} \otimes \begin{pmatrix}\Psi_\uparrow(\mathbf{r}_1)\\ \Psi_\downarrow(\mathbf{r}_1)\end{pmatrix} = \begin{pmatrix}\Psi_\uparrow(\mathbf{r}_2)\Psi_\uparrow(\mathbf{r}_1)\\ \Psi_\uparrow(\mathbf{r}_2)\Psi_\downarrow(\mathbf{r}_1)\\ \Psi_\downarrow(\mathbf{r}_2)\Psi_\uparrow(\mathbf{r}_1)\\ \Psi_\downarrow(\mathbf{r}_2)\Psi_\downarrow(\mathbf{r}_1)\end{pmatrix}$$

$$\begin{aligned} \Psi^\dagger(\mathbf{r}_1)\Psi^\dagger(\mathbf{r}_2) &= \begin{pmatrix}\Psi_\uparrow^\dagger(\mathbf{r}_1) & \Psi_\downarrow^\dagger(\mathbf{r}_1)\end{pmatrix} \otimes \begin{pmatrix}\Psi_\uparrow^\dagger(\mathbf{r}_2) & \Psi_\downarrow^\dagger(\mathbf{r}_2)\end{pmatrix} \\ &= \begin{pmatrix}\Psi_\uparrow^\dagger(\mathbf{r}_1)\Psi_\uparrow^\dagger(\mathbf{r}_2) & \Psi_\uparrow^\dagger(\mathbf{r}_1)\Psi_\downarrow^\dagger(\mathbf{r}_2) & \Psi_\downarrow^\dagger(\mathbf{r}_1)\Psi_\uparrow^\dagger(\mathbf{r}_2) & \Psi_\downarrow^\dagger(\mathbf{r}_1)\Psi_\downarrow^\dagger(\mathbf{r}_2)\end{pmatrix}. \end{aligned}$$

We also define the 4×4 matrix,

$$v(\mathbf{r}_1, \mathbf{r}_2) = \begin{bmatrix} v_{\uparrow\uparrow\uparrow\uparrow} & v_{\uparrow\uparrow\downarrow\uparrow} & v_{\uparrow\uparrow\uparrow\downarrow} & v_{\uparrow\uparrow\downarrow\downarrow} \\ v_{\uparrow\downarrow\uparrow\uparrow} & v_{\uparrow\downarrow\downarrow\uparrow} & v_{\uparrow\downarrow\uparrow\downarrow} & v_{\uparrow\downarrow\downarrow\downarrow} \\ v_{\downarrow\uparrow\uparrow\uparrow} & v_{\downarrow\uparrow\downarrow\uparrow} & v_{\downarrow\uparrow\uparrow\downarrow} & v_{\downarrow\uparrow\downarrow\downarrow} \\ v_{\downarrow\downarrow\uparrow\uparrow} & v_{\downarrow\downarrow\downarrow\uparrow} & v_{\downarrow\downarrow\uparrow\downarrow} & v_{\downarrow\downarrow\downarrow\downarrow} \end{bmatrix}.$$

Although not shown explicitly, each of the 16 matrix elements in the above matrix is a function of $\mathbf{r}_1$ and $\mathbf{r}_2$. It is left as an exercise for the reader to show that V_{int} may be written as

$$V_{\text{int}} = \int \Psi^\dagger(\mathbf{r}_1)\Psi^\dagger(\mathbf{r}_2)v(\mathbf{r}_1, \mathbf{r}_2)\Psi(\mathbf{r}_2)\Psi(\mathbf{r}_1)\, d^3r_1\, d^3r_2. \tag{3.54}$$

If $v(i, j)$ is spin-independent, i.e., $v(1, 2) = v(\mathbf{r}_1, \mathbf{r}_2)$, then

$$\langle\sigma_1\sigma_2|v(1, 2)|\sigma_3\sigma_4\rangle = v(\mathbf{r}_1, \mathbf{r}_2)\delta_{\sigma_1\sigma_3}\delta_{\sigma_2\sigma_4},$$

and we can write

$$V_{\text{int}} = \frac{1}{2}\sum_{\sigma_1\sigma_2}\int \Psi^\dagger_{\sigma_1}(\mathbf{r}_1)\Psi^\dagger_{\sigma_2}(\mathbf{r}_2)v(\mathbf{r}_1, \mathbf{r}_2)\Psi_{\sigma_2}(\mathbf{r}_2)\Psi_{\sigma_1}(\mathbf{r}_1)\, d^3r_1\, d^3r_2.$$

3.11.5 Examples

3.11.5.1 Particle-number density

The particle-number density operator $n(\mathbf{r}) = \sum_i \delta(\mathbf{r} - \mathbf{r}'_i)$ is given in terms of field operators, by

$$n(\mathbf{r}) = \sum_\sigma \int \Psi^\dagger_\sigma(\mathbf{r}')\delta(\mathbf{r} - \mathbf{r}')\Psi_\sigma(\mathbf{r}')d^3r' = \sum_\sigma \Psi^\dagger_\sigma(\mathbf{r})\Psi_\sigma(\mathbf{r}). \tag{3.55}$$

3.11.5.2 Kinetic energy

The kinetic energy operator for a system of N particles is $\sum_{i=1}^N(-\hbar^2/2m)\nabla_i^2$. In terms of field operators, it is

$$T = -\frac{\hbar^2}{2m}\sum_\sigma \int \Psi^\dagger_\sigma(\mathbf{r})\nabla^2\Psi_\sigma(\mathbf{r})d^3r.$$

From the following equality

$$\nabla \cdot \left(\Psi^\dagger_\sigma(\mathbf{r})\nabla\Psi_\sigma(\mathbf{r})\right) = \nabla\Psi^\dagger_\sigma(\mathbf{r}) \cdot \nabla\Psi_\sigma(\mathbf{r}) + \Psi^\dagger_\sigma(\mathbf{r})\nabla^2\Psi_\sigma(\mathbf{r}),$$

it follows that

$$T = \frac{\hbar^2}{2m}\sum_\sigma \int \nabla\Psi^\dagger_\sigma(\mathbf{r}) \cdot \nabla\Psi_\sigma(\mathbf{r})d^3r - \frac{\hbar^2}{2m}\sum_\sigma \int \nabla \cdot \left(\Psi^\dagger_\sigma(\mathbf{r})\nabla\Psi_\sigma(\mathbf{r})\right) d^3r.$$

By the divergence theorem, the volume integral in the last term is converted into a surface integral,

$$\int \nabla \cdot \left(\Psi^\dagger_\sigma(\mathbf{r})\nabla\Psi_\sigma(\mathbf{r})\right) d^3r = \int \left(\Psi^\dagger_\sigma(\mathbf{r})\nabla\Psi_\sigma(\mathbf{r})\right) \cdot \hat{n}\, da$$

where $\hat{n}$ is an outward unit vector normal to the surface. If the surface is at infinity, the field operator $\Psi_\sigma^\dagger(\mathbf{r}) = \sum_n \phi_n^*(\mathbf{r})c_{n\sigma}^\dagger$ vanishes at the surface because $\phi_n^*(\mathbf{r})$ vanishes at infinity. On the other hand, if the particles are enclosed within a box of volume $V = L^3$ and periodic boundary conditions are employed, $\Psi_\sigma^\dagger(\mathbf{r})\nabla\Psi_\sigma(\mathbf{r})$ will be the same on opposite faces of the box, but the unit vector normal to one face will be opposite to the unit vector normal to the opposite face. In either case, the surface integral vanishes, and

$$T = \frac{\hbar^2}{2m}\sum_\sigma \int \nabla\Psi_\sigma^\dagger(\mathbf{r}).\nabla\Psi_\sigma(\mathbf{r})d^3r. \tag{3.56}$$

Further reading

Fetter, A.L. and Walecka, J.D. (1971). *Quantum Theory of Many-Particle Systems*. New York: McGraw-Hill.

Schwabl, F. (2008). *Advamced Quantum Mechanics*, 4th edn. Berlin: Springer.

Taylor, P.L. and Heinonen, O. (2002). *A Quantum Approach to Condensed Matter Physics*. Cambridge: Cambridge University Press.

Problems

3.1 *Noninteracting electrons on a square lattice.* Identical atoms sit at the lattice sites of a square lattice with lattice constant a. Assume that there is one Wannier orbital on each site, so that one band is formed from these orbitals. Neglecting electron–electron interaction, and assuming that an electron can hop from one site to only one of the nearest-neighboring sites, the hopping matrix element being $-t$, the Hamiltonian is

$$H = -t\sum_{<ij>\sigma} c_{i\sigma}^\dagger c_{j\sigma}.$$

Calculate the dispersion of the energy band.

3.2 *Graphene revisited.* In Problem 2.4, the energy bands in graphene are calculated using the tight binding method. Here, the calculation is repeated using the second quantized form of the Hamiltonian. Graphene consists of two sublattices, one of type A and one of type B. Sublattice A consists of all the sites of type A, and sublattice B consists of all the sites of type B. Assume that there is only one orbital centered on each site (p_z orbital). Neglect overlap between orbitals on different sites, and assume that an electron on one site can hop to only one of the three neighboring sites. With these assumptions, the tight binding Hamiltonian is

$$H = -t \sum_{i\sigma} \sum_{\delta=1}^{3} a_{i\sigma}^{\dagger} b_{i+\delta,\sigma} - t \sum_{i\sigma} \sum_{\delta=1}^{3} b_{i+\delta,\sigma}^{\dagger} a_{i\sigma}$$

where $a_{i\sigma}^{\dagger}(a_{i\sigma})$ creates (annihilates) an electron with spin projection σ on site i of type A, $b_{i\sigma}^{\dagger}(b_{i\sigma})$ creates (annihilates) an electron with spin projection σ on site i of type B, $-t$ is the hopping matrix element, and the sites $i+\delta$ are the nearest neighbors of site i.

For a given σ, there are N operators $a_{i\sigma}$, where N is the total number of primitive cells in the crystal. Define N new operators $a_{\mathbf{k}\sigma}$, $\mathbf{k} \in$ FBZ,

$$a_{\mathbf{k}\sigma} = \frac{1}{\sqrt{N}} \sum_{i} e^{-i\mathbf{k}.\mathbf{R}_i} a_{i\sigma}, \quad a_{\mathbf{k}\sigma}^{\dagger} = \frac{1}{\sqrt{N}} \sum_{i} e^{i\mathbf{k}.\mathbf{R}_i} a_{i\sigma}^{\dagger}.$$

Similar definitions are made for $b_{\mathbf{k}\sigma}$ and $b_{\mathbf{k}\sigma}^{\dagger}$. Show that

$$H = -t \sum_{\mathbf{k}\sigma} \begin{pmatrix} a_{\mathbf{k}\sigma}^{\dagger} & b_{\mathbf{k}\sigma}^{\dagger} \end{pmatrix} \begin{pmatrix} 0 & g_{\mathbf{k}} \\ g_{\mathbf{k}}^{*} & 0 \end{pmatrix} \begin{pmatrix} a_{\mathbf{k}\sigma} \\ b_{\mathbf{k}\sigma} \end{pmatrix}$$

where $g_{\mathbf{k}} = \sum_{\delta} e^{i\mathbf{k}.\delta} = \exp\left(i\frac{k_x a}{\sqrt{3}}\right) + \exp\left[i\left(\frac{-k_x a}{2\sqrt{3}} + \frac{k_y a}{2}\right)\right] + \exp\left[i\left(\frac{-k_x a}{2\sqrt{3}} - \frac{k_y a}{2}\right)\right]$. Reduce H to the form:

$$H = \sum_{n=1}^{2} \sum_{\mathbf{k}\sigma} E_{n\mathbf{k}} c_{n\mathbf{k}\sigma}^{\dagger} c_{n\mathbf{k}\sigma}$$

where $E_{1\mathbf{k}} = -t|g_{\mathbf{k}}|$, $E_{2\mathbf{k}} = t|g_{\mathbf{k}}|$, and $c_{1\mathbf{k}\sigma}$ and $c_{2\mathbf{k}\sigma}$ are electron operators that are linear combinations of $a_{\mathbf{k}\sigma}$ and $b_{\mathbf{k}\sigma}$.

3.3 *Commutators.* Calculate $[c_{\mathbf{k}\sigma}, \sum_{\mathbf{k}\sigma} \epsilon_k c_{\mathbf{k}\sigma}^{\dagger} c_{\mathbf{k}\sigma}]$ and $[c_{\mathbf{k}\sigma}^{\dagger}, \sum_{\mathbf{k}\sigma} \epsilon_k c_{\mathbf{k}\sigma}^{\dagger} c_{\mathbf{k}\sigma}]$.

3.4 *Field and number operators.* Show that, for bosons and fermions, the field operators and the total number of particles operator satisfy the following:

$$[N, \Psi_{\sigma}] = -\Psi_{\sigma}, \quad [N, \Psi_{\sigma}^{\dagger}] = \Psi_{\sigma}^{\dagger}.$$

Define $\tilde{\Psi}_{\sigma}(\theta) = e^{iN\theta}\Psi_{\sigma}e^{-iN\theta}$. Show that $\tilde{\Psi}_{\sigma}(\theta) = e^{-i\theta}\Psi_{\sigma}$ and $\tilde{\Psi}_{\sigma}^{\dagger}(\theta) = e^{i\theta}\Psi_{\sigma}^{\dagger}$. Since operators representing observables, when expressed in second quantized form, contain an equal number of creation and annihilation operators (to be Hermitian), they are invariant under the transformation $\Psi_{\sigma} \to \tilde{\Psi}_{\sigma}(\theta)$, $\Psi_{\sigma}^{\dagger} \to \tilde{\Psi}_{\sigma}^{\dagger}(\theta)$. Hence, in a many-particle system, any operator A that represents an observable satisfies the equation $e^{iN\theta}Ae^{-iN\theta} = A \Rightarrow [A, N] = 0$.

3.5 *Spin*. For a system of N electrons, $\mathbf{S} = \sum_{i=1}^{N} \mathbf{S}_i$. Let $\mathbf{S} = (S_x, S_y, S_z)$. Using the states $|\mathbf{k}\sigma\rangle$ as a basis, show that

$$\mathbf{S} = \frac{\hbar}{2} \sum_{\mathbf{k}} \left(\left[c^\dagger_{\mathbf{k}\uparrow} c_{\mathbf{k}\downarrow} + c^\dagger_{\mathbf{k}\downarrow} c_{\mathbf{k}\uparrow} \right], i \left[c^\dagger_{\mathbf{k}\downarrow} c_{\mathbf{k}\uparrow} - c^\dagger_{\mathbf{k}\uparrow} c_{\mathbf{k}\downarrow} \right], \left[c^\dagger_{\mathbf{k}\uparrow} c_{\mathbf{k}\uparrow} - c^\dagger_{\mathbf{k}\downarrow} c_{\mathbf{k}\downarrow} \right] \right).$$

3.6 *Number-density operator*. In a crystal, single-particle states are characterized by the quantum numbers n, $\mathbf{k}$, and σ, where n is the band index, $\mathbf{k} \in$ FBZ, and σ is the spin projection. Show that the Fourier transform of the electron number-density operator is given by

$$n_{\mathbf{q}} = \sum_{\mathbf{k}\sigma} \sum_{nn'} \langle n\mathbf{k}\sigma | e^{-i\mathbf{q}.\mathbf{r}} | n'\mathbf{k} + \mathbf{q}\sigma \rangle c^\dagger_{n\mathbf{k}\sigma} c_{n'\mathbf{k}+\mathbf{q}\sigma}.$$

3.7 *Electron current density*. In a course on electricity and magnetism, the electron current density is written as $\mathbf{j} = -en\mathbf{v}$, where $-e$ is the charge of the electron, n is the number of electrons per unit volume, and $\mathbf{v}$ is the average velocity of the electrons. The contribution of electron i to the current density is $-e\delta(\mathbf{r} - \mathbf{r}_i)\mathbf{v}_i = -e\delta(\mathbf{r} - \mathbf{r}_i)\mathbf{p}_i/m$. The quantum mechanical expression for the current density is thus given by

$$\mathbf{j}(\mathbf{r}) = -\frac{e}{2m} \sum_i [\mathbf{p}_i \delta(\mathbf{r} - \mathbf{r}_i) + \delta(\mathbf{r} - \mathbf{r}_i)\mathbf{p}_i].$$

We write it this way to ensure that $\mathbf{j}(\mathbf{r})$ is Hermitian ($\mathbf{r}_i$ and $\mathbf{p}_i$ do not commute). In the presence of a magnetic field, $\mathbf{p}_i \to \mathbf{p}_i + e\mathbf{A}(\mathbf{r}_i)/c$, where $\mathbf{A}$ is the vector potential. Show that the current-density operator is given by

$$\mathbf{j}(\mathbf{r}) = \mathbf{j}^P(\mathbf{r}) + \mathbf{j}^D(\mathbf{r})$$

where $\mathbf{j}^P$, the paramagnetic current density, and $\mathbf{j}^D$, the diamagnetic current density, are given by

$$\mathbf{j}^P(\mathbf{r}) = \frac{ie\hbar}{2m} \sum_\sigma \left[\Psi^\dagger_\sigma(\mathbf{r}) \nabla \Psi_\sigma(\mathbf{r}) - \left(\nabla \Psi^\dagger_\sigma(\mathbf{r}) \right) \Psi_\sigma(\mathbf{r}) \right]$$

$$\mathbf{j}^D(\mathbf{r}) = -\frac{e^2}{mc} \mathbf{A}(\mathbf{r}) n(\mathbf{r}), \quad n(\mathbf{r}) = \sum_\sigma \Psi^\dagger_\sigma(\mathbf{r}) \Psi_\sigma(\mathbf{r}).$$

Show that the Fourier transform of $\mathbf{j}^P(\mathbf{r})$ is given by

$$\mathbf{j}^P_{\mathbf{q}} = -\frac{e\hbar}{m} \sum_{\mathbf{k}\sigma} (\mathbf{k} + \frac{1}{2}\mathbf{q}) c^\dagger_{\mathbf{k}\sigma} c_{\mathbf{k}+\mathbf{q}\sigma}.$$

3.8 *Contact potential.* Consider a system where particles interact with each other via the contact potential

$$V = \frac{g}{2} \sum_{i \neq j} \delta(\mathbf{r}_i - \mathbf{r}_j).$$

Express V in second quantized form.

3.9 *Spin waves.* Consider a system of N particles, each of spin s, localized at the N lattice sites of a crystal. We assume that there exists an interaction between the particles that tends to align their spins. The Hamiltonian is assumed to be

$$H = -(J/2) \sum_{<ij>} \mathbf{S}_i . \mathbf{S}_j$$

where $J > 0$ and the summation is over nearest-neighboring sites. Here, J has units of energy, so the spin operators are dimensionless. The spin operators satisfy the usual commutation relations; e.g.,

$$[S_i^x, S_j^y] = i\delta_{ij} S_i^z.$$

Define the operators S_i^+ and S_i^- by

$$S_i^+ = S_i^x + iS_i^y, \quad S_i^- = S_i^x - iS_i^y.$$

We now transform to two *bosonic* operators, a_i and $a_i^\dagger$,

$$S_i^+ = (2s)^{1/2}[1 - a_i^\dagger a_i/2s]^{1/2} a_i, \quad S_i^- = (2s)^{1/2} a_i^\dagger [1 - a_i^\dagger a_i/2s]^{1/2}.$$

This is known as the Holstein–Primakoff transformation (Holstein and Primakoff, 1940).

(a) Using $(S_i^z)^2 = s(s+1) - (S_i^x)^2 - (S_i^y)^2$, show that $S_i^z = s - a_i^\dagger a_i$

(b) Define $a_\mathbf{k}$ and $a_\mathbf{k}^\dagger$ by

$$a_\mathbf{k} = \frac{1}{\sqrt{N}} \sum_i e^{i\mathbf{k}.\mathbf{R}_i} a_i, \quad a_\mathbf{k}^\dagger = \frac{1}{\sqrt{N}} \sum_i e^{-i\mathbf{k}.\mathbf{R}_i} a_i^\dagger.$$

Show that, to second order in the a-operators,

$$H = \text{const} + \sum_\mathbf{k} \hbar\omega_\mathbf{k} a_\mathbf{k}^\dagger a_\mathbf{k}.$$

What is the value of $\omega_\mathbf{k}$? The excitations of energy $\hbar\omega_\mathbf{k}$ describe the spin-wave excitations of the ferromagnet.

4

The electron gas

All exact science is dominated by the idea of approximation.

–Bertrand Russell

A metallic crystal has a large number of mobile electrons, of the order of Avogadro's number, and a correspondingly large number of ions. If our interest is in the bulk properties of a crystal, we may take the volume V of the crystal to be infinite, and the number of electrons N to be infinite, while keeping N/V, the number density of electrons, finite; this is called the thermodynamic limit. The ions incessantly vibrate about their equilibrium positions, but due to their large mass, they move very slowly in comparison with the electrons, so that the electrons quickly adjust their state to reflect whatever positions the ions occupy at any given time. Consequently, to a good approximation, one may solve the Schrödinger equation for electrons by assuming that the ions are fixed; this is the Born–Oppenheimer approximation. The influence of the ionic vibrations on the electronic states, described through the electron–phonon interaction, may be treated by perturbation theory; this is discussed in Chapter 11.

A more drastic approximation in the description of a metal is to replace the mesh of positive ions with a uniform positive background, which results in the so-called jellium model. In a model such as this, any results obtained are necessarily qualitative in nature. In this chapter, we study the jellium model. One of our goals in this study is to show that the divergent term in the Coulomb interaction, corresponding to $\mathbf{q} = \mathbf{0}$ (see Eq. [3.29]), is cancelled by contributions to the total energy from the positive background. This cancellation is a consequence of the charge neutrality of the crystal, and it holds true even if the approximation of a uniform positive background is relaxed. Another goal of this chapter is to show the necessity of performing perturbation expansions to higher, generally infinite, orders. We will later proceed to study Green's functions, whereby such a program may be carried out more easily.

4.1 The Hamiltonian in the jellium model

Let us consider the jellium model in thermodynamic limit: $N \to \infty$, $V \to \infty$, while N/V remains constant. The Hamiltonian consists of three terms,

$$H = H_e + H_b + H_{e-b}. \tag{4.1}$$

The first term is the sum of the kinetic energies of the electrons and their Coulomb interactions. From Eqs (3.19) and (3.29),

$$H_e = \sum_{\mathbf{k}\sigma} \frac{\hbar^2 k^2}{2m} c^\dagger_{\mathbf{k}\sigma} c_{\mathbf{k}\sigma} + \lim_{\mu \to 0} \frac{1}{2V} \sum_{\mathbf{q}} \sum_{\mathbf{k}\sigma} \sum_{\mathbf{k}'\sigma'} \frac{4\pi e^2}{q^2 + \mu^2} c^\dagger_{\mathbf{k}+\mathbf{q}\sigma} c^\dagger_{\mathbf{k}'-\mathbf{q}\sigma'} c_{\mathbf{k}'\sigma'} c_{\mathbf{k}\sigma}. \tag{4.2}$$

The second term, H_b, represents the Coulomb energy of the uniform positive background. To find the correct expression for H_b, consider a collection of point charges $q_1, q_2, \ldots$ at positions $\mathbf{r}_1, \mathbf{r}_2, \ldots$. Their Coulomb energy is

$$E_{\text{Coul}} = \frac{1}{2} \sum_{i \neq j} \frac{q_i q_j}{|\mathbf{r}_i - \mathbf{r}_j|} \qquad \text{(cgs units)}.$$

The factor $1/2$ ensures that pairs of point charges are counted only once. For a continuous charge distribution, q_i is replaced by $\rho(\mathbf{r}_i) d^3 r_i$, where $\rho(\mathbf{r})$ is the charge density, and the summation is replaced by integration. Therefore,

$$H_b = \lim_{\mu \to 0^+} \frac{1}{2} \int \frac{\rho(\mathbf{r})\rho(\mathbf{r}') e^{-\mu|\mathbf{r}-\mathbf{r}'|}}{|\mathbf{r}-\mathbf{r}'|} d^3 r d^3 r' = \lim_{\mu \to 0^+} \frac{N^2 e^2}{2V^2} \int \frac{e^{-\mu|\mathbf{r}-\mathbf{r}'|}}{|\mathbf{r}-\mathbf{r}'|} d^3 r d^3 r'. \tag{4.3}$$

For the uniform positive background, $\rho(\mathbf{r}) = Ne/V$. We have introduced an exponential term, as in the case of the Coulomb interaction between electrons (see Section 3.9). Why? Since we replaced the Coulomb potential in H_e with a Yukawa potential, we need to do the same for H_b and H_{e-b}, for the sake of consistency. More importantly, H_e, H_b, and H_{e-b} diverge. While H_e and H_b are positive, H_{e-b} is negative. We have to add and subtract infinities; to obtain meaningful results, we consider the infinities to arise in some limit.

Returning to H_b, we evaluate the integral in the limit $V \to \infty$, keeping μ fixed. As $V \to \infty$, we may shift the variables of integration without worrying about the limits of the integral. Defining $\mathbf{x} = \mathbf{r} - \mathbf{r}'$ and $x = |\mathbf{x}|$, we find

$$H_b = \lim_{\mu \to 0^+} \lim_{V \to \infty} \frac{N^2 e^2}{2V^2} \int d^3 r' \int d^3 x \frac{e^{-\mu x}}{x} = \lim_{\mu \to 0^+} \lim_{V \to \infty} \frac{N^2 e^2}{2V} \int d^3 x \frac{e^{-\mu x}}{x}. \tag{4.4}$$

Note how the limits are taken: first $V \to \infty$, then $\mu \to 0^+$. The above integral is easily evaluated,

$$\int d^3x \, \frac{e^{-\mu x}}{x} = 4\pi \int_0^\infty x e^{-\mu x} dx = -4\pi \frac{\partial}{\partial \mu} \int_0^\infty e^{-\mu x} dx = -4\pi \frac{\partial}{\partial \mu} \left(\frac{1}{\mu} \right) = \frac{4\pi}{\mu^2}$$

$$\Longrightarrow H_B = \lim_{\mu \to 0} \lim_{V \to \infty} \frac{2\pi N^2 e^2}{V \mu^2}. \tag{4.5}$$

The last term in the Hamiltonian is H_{eb}, the electron–background interaction. Denoting the positions of the electrons by $\mathbf{r}_1, \mathbf{r}_2, \ldots, \mathbf{r}_N$, we find

$$H_{eb} = -e \lim_{\mu \to 0^+} \sum_{i=1}^{N} \int d^3r \, \frac{\rho(\mathbf{r})}{|\mathbf{r} - \mathbf{r}_i|} e^{-\mu|\mathbf{r}-\mathbf{r}_i|} = - \lim_{\mu \to 0^+} \frac{Ne^2}{V} \sum_{i=1}^{N} \int d^3r \, \frac{e^{-\mu|\mathbf{r}-\mathbf{r}_i|}}{|\mathbf{r} - \mathbf{r}_i|}. \tag{4.6}$$

Replacing $\mathbf{r} - \mathbf{r}_i$ by $\mathbf{x}$, the integral is evaluated in the limit $V \to \infty$

$$H_{e-b} = - \lim_{\mu \to 0^+} \lim_{V \to \infty} \frac{Ne^2}{V} \sum_{i=1}^{N} \int d^3x \frac{e^{-\mu x}}{x} = - \lim_{\mu \to 0} \lim_{V \to \infty} \frac{Ne^2}{V} \sum_{i=1}^{N} \frac{4\pi}{\mu^2}$$

$$= \lim_{\mu \to 0} \lim_{V \to \infty} \frac{-4\pi N^2 e^2}{V \mu^2}. \tag{4.7}$$

We thus find

$$H_b + H_{e-b} = \lim_{\mu \to 0} \lim_{V \to \infty} \frac{-2\pi N^2 e^2}{V \mu^2}. \tag{4.8}$$

The above expression approaches $-\infty$ because $N \to \infty$ while N/V is finite.

The Coulomb interaction part of H_e diverges because of the $\mathbf{q} = \mathbf{0}$ term,

$$V_{C,\mathbf{q}=\mathbf{0}} = \lim_{\mu \to 0} \lim_{V \to \infty} \frac{2\pi e^2}{V \mu^2} \sum_{\mathbf{k}\sigma} \sum_{\mathbf{k}'\sigma'} c^\dagger_{\mathbf{k}\sigma} c^\dagger_{\mathbf{k}'\sigma'} c_{\mathbf{k}'\sigma'} c_{\mathbf{k}\sigma}. \tag{4.9}$$

The term comprising the product of operators may be rewritten as follows:

$$c^\dagger_{\mathbf{k}\sigma} c^\dagger_{\mathbf{k}'\sigma'} c_{\mathbf{k}'\sigma'} c_{\mathbf{k}\sigma} = -c^\dagger_{\mathbf{k}\sigma} c^\dagger_{\mathbf{k}'\sigma'} c_{\mathbf{k}\sigma} c_{\mathbf{k}'\sigma'} = -c^\dagger_{\mathbf{k}\sigma} (\delta_{\mathbf{k}\mathbf{k}'} \delta_{\sigma\sigma'} - c_{\mathbf{k}\sigma} c^\dagger_{\mathbf{k}'\sigma'}) c_{\mathbf{k}'\sigma'}$$

$$= c^\dagger_{\mathbf{k}\sigma} c_{\mathbf{k}\sigma} c^\dagger_{\mathbf{k}'\sigma'} c_{\mathbf{k}'\sigma'} - \delta_{\mathbf{k}\mathbf{k}'} \delta_{\sigma\sigma'} c^\dagger_{\mathbf{k}\sigma} c_{\mathbf{k}'\sigma'}$$

$$\Rightarrow V_{C,\mathbf{q}=\mathbf{0}} = \lim_{\mu \to 0} \lim_{V \to \infty} \frac{2\pi e^2}{V \mu^2} \left[\sum_{\mathbf{k}\sigma} c^\dagger_{\mathbf{k}\sigma} c_{\mathbf{k}\sigma} \sum_{\mathbf{k}'\sigma'} c^\dagger_{\mathbf{k}'\sigma'} c_{\mathbf{k}'\sigma'} - \sum_{\mathbf{k}\sigma} c^\dagger_{\mathbf{k}\sigma} c_{\mathbf{k}\sigma} \right]$$

$$= \lim_{\mu \to 0} \lim_{V \to \infty} \frac{2\pi e^2}{V \mu^2} \left(\hat{N}^2 - \hat{N} \right).$$

The number of electrons operator, $\hat{N}$, is given by $\hat{N} = \sum_{\mathbf{k}\sigma} c^\dagger_{\mathbf{k}\sigma} c_{\mathbf{k}\sigma}$. Since $\hat{N}$ commutes with the Hamiltonian H, the eigenstates of H are also eigenstates of $\hat{N}$ with eigenvalue N, the number of electrons, and we may replace the operator $\hat{N}$ by the number N. Hence,

$$E'/N \equiv \left(V_{C,\mathbf{q}=\mathbf{0}} + H_b + H_{e-b}\right)/N = \lim_{\mu\to 0} \lim_{V\to\infty} \frac{-2\pi e^2}{V\mu^2} = 0. \tag{4.10}$$

On the other hand, the average kinetic energy per electron is $3E_F/5$, where E_F is the Fermi energy (see Section 2.2); we are thus totally justified in ignoring E'. The effect of the positive background is thus to remove the $\mathbf{q} = \mathbf{0}$ term in the electron–electron interaction Hamiltonian. This is reasonable; after all, crystals are stable, so the energy per electron should be finite.

The reader may feel uneasy about the results we obtained; they rely on the mathematical artifact of introducing an exponential damping term, and on the sequence in which the limits are taken. The reader may rest assured that the results are correct. In fact, the same results are obtained without introducing the exponential term. Setting $\mu = 0$ while keeping the thermodynamic limit: $V \to \infty$, $N \to \infty$, $N/V =$ constant, Eqs (4.4) and (4.6) yield

$$H_b = \frac{N^2 e^2}{2V} \int \frac{d^3 r}{r}, \qquad H_{e-b} = -\frac{N^2 e^2}{V} \int \frac{d^3 r}{r}.$$

The $\mathbf{q} = \mathbf{0}$ term in V_C is

$$\begin{aligned} V_{C,\mathbf{q}=\mathbf{0}} &= \frac{1}{2V} v_{\mathbf{q}=\mathbf{0}} \sum_{\mathbf{k}'\sigma'} \sum_{\mathbf{k}\sigma} c^\dagger_{\mathbf{k}\sigma} c^\dagger_{\mathbf{k}'\sigma'} c_{\mathbf{k}'\sigma'} c_{\mathbf{k}\sigma} = \frac{1}{2V} v_{\mathbf{q}=\mathbf{0}} (N^2 - N) \\ &= \frac{(N^2 - N)e^2}{2V} \int \frac{d^3 r}{r}. \end{aligned}$$

Therefore,

$$\frac{E'}{N} = \frac{1}{N}\left(V_{C,\mathbf{q}=\mathbf{0}} + H_b + H_{e-b}\right) = -\frac{e^2}{2V} \int \frac{d^3 r}{r}.$$

If the linear dimension of the crystal is L, the integral $\int d^3 r/r$ is of the order L^2, whereas $V = L^3$; hence $E'/N = O(L^{-1})$, and $E'/N \to 0$ as $L \to \infty$. We arrive at the same conclusion as before: the effect of H_b and H_{e-b} is to cancel the $\mathbf{q} = \mathbf{0}$ term in the electron–electron interaction. With this in mind, the Hamiltonian for the electron gas can be written as

$$H = \sum_{\mathbf{k}\sigma} \frac{\hbar^2 k^2}{2m} c^\dagger_{\mathbf{k}\sigma} c_{\mathbf{k}\sigma} + \frac{1}{2V} \sum_{\mathbf{q}}{}' \sum_{\mathbf{k}\sigma} \sum_{\mathbf{k}'\sigma'} \frac{4\pi e^2}{q^2} c^\dagger_{\mathbf{k}+\mathbf{q}\sigma} c^\dagger_{\mathbf{k}'-\mathbf{q}\sigma'} c_{\mathbf{k}'\sigma'} c_{\mathbf{k}\sigma}. \tag{4.11}$$

The prime over the **q**-summation means that $\mathbf{q} = \mathbf{0}$ is excluded. H is given above in cgs units; to obtain H in SI units, simply replace e^2 with $e^2/4\pi\varepsilon_0$.

4.2 High density limit

Under what conditions could the Coulomb interaction between electrons be treated as a small perturbation? To answer this question, we define a dimensionless parameter r_s by

$$(4\pi/3)r_s^3 a_0^3 = V/N\,, \tag{4.12}$$

where $a_0 = \hbar^2/(me^2)$ is the Bohr radius (in SI units, $e^2 \to e^2/4\pi\varepsilon_0$). $r_s a_0$ is the radius of a sphere whose volume is equal to the average volume occupied by one electron. Defining the dimensionless quantities

$$V' = V/(r_s a_0)^3\,, \qquad \mathbf{K} = r_s a_0 \mathbf{k}\,, \qquad \mathbf{Q} = r_s a_0 \mathbf{q}\,,$$

we may recast the Hamiltonian in Eq. (4.11) into the following form:

$$H = \frac{e^2}{r_s^2 a_0}\left[\sum_{\mathbf{K}\sigma} K^2 c^\dagger_{\mathbf{k}\sigma} c_{\mathbf{k}\sigma} + \frac{r_s}{V'}\sum_{\mathbf{Q}}{}' \sum_{\mathbf{k}\sigma}\sum_{\mathbf{k}'\sigma'} \frac{2\pi}{Q^2} c^\dagger_{\mathbf{k}+\mathbf{q}\sigma} c^\dagger_{\mathbf{k}'-\mathbf{q}\sigma'} c_{\mathbf{k}'\sigma'} c_{\mathbf{k}\sigma}\right]. \tag{4.13}$$

This expression for H is very telling: compared to the kinetic energy of electrons, the Coulomb interaction is negligible in the high density limit, $r_s \to 0$. This conclusion appears to be counterintuitive, but a moment's reflection reveals its validity. Coulomb repulsion scales as $1/r_s$, and from Heisenberg's uncertainty principle, the electron's momentum also scales as $1/r_s$. Therefore, the kinetic energy scales as $1/r_s^2$. Thus, as $r_s \to 0$, even though the Coulomb energy grows larger, the kinetic energy of the electrons grows larger at a faster rate. We conclude that in the high-density limit, the Coulomb repulsion is weak in comparison with the kinetic energy, and it is permissible to treat it within the framework of perturbation theory. In real metals, $r_s = 2 - 6$, which is neither too small nor too large. Nevertheless, in most metals, the single-particle approximation explains many of their low energy properties. This is because the Coulomb interaction, even when it is strong, is not very effective at changing the momentum distribution of the electrons; most of the states into which they could scatter are already occupied.

This ineffectiveness of the Coulomb interaction, due to phase space limitations, lies at the heart of Landau's Fermi liquid theory (Landau, 1957a, 1957b, 1959). Consider an electron with wave vector **k** outside the Fermi sphere, $k > k_F$. At low temperatures, where almost all the states within the Fermi sphere are occupied, the electron can only decay, through Coulomb interaction, into states within a shell of width $k - k_F$ just above the Fermi sphere. The number of states in this shell is

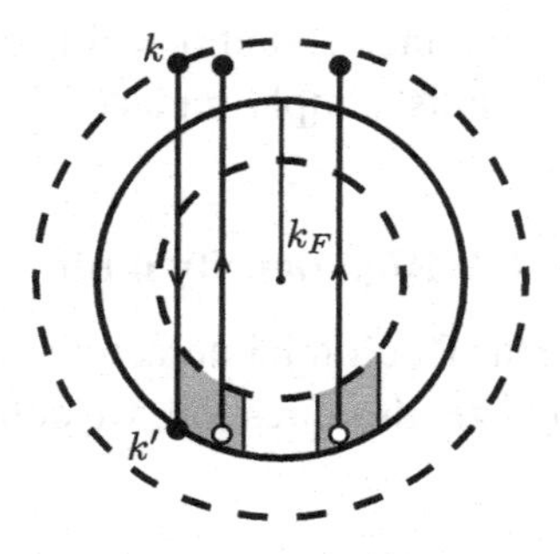

Figure 4.1 When an electron in a state of wave vector **k** above the Fermi surface is scattered out of this state, an electron-hole pair is created. At low temperatures, energy and wave vector conservation restricts the final state into which the electron can scatter to a shell of width $k - k_F$ just above the Fermi sphere. It also restricts the states of the holes created to the shaded regions, which lie within a shell of width $k - k_F$ just below the Fermi surface.

proportional to $k - k_F$. The decay process is accompanied by the creation of an electron-hole pair: an electron from within the Fermi sphere makes a transition to a state outside the Fermi sphere, leaving a hole behind. The requirement of energy and wave vector conservation restricts the state of the hole to a shell of width $k - k_F$ (see Figure 4.1). The number of states in the shaded regions of **k**-space is also proportional to $k - k_F$. The probability of decay is thus proportional to $(k - k_F)^2$, and it vanishes as $k \to k_F$. That is, low energy excited states (for which $k \gtrsim k_F$) are long-lived, since the probability of scattering out of these states due to Coulomb interaction is small.

4.3 Ground state energy

An interacting electron gas is described by the Hamiltonian $H = H_0 + V_C$. In the high density limit, the Coulomb term, V_C, is treated as a perturbation. In the absence of V_C, the Hamiltonian is simply the sum of the kinetic energies of the electrons, and the ground state of the noninteracting system at zero temperature, denoted by $|F\rangle$, is obtained by filling all states within the Fermi sphere. In Chapter 2, we found that the average energy per electron in this case is $E_0/N = 3E_F/5$, where $E_F = \hbar^2 k_F^2/2m$ is the Fermi energy, and $k_F = (3\pi^2 N/V)^{1/3}$ is the Fermi wave vector. It is easily shown that

$$E_0/N \simeq 2.21/r_s^2 \; Ry$$

where one Rydberg (Ry) is equal to $e^2/2a_0$, which is about 13.6 eV.

4.3.1 First order perturbation

Treating V_C as a perturbation, the energy per electron in the ground state is written as a perturbation series

$$E/N = E_0/N + E_1/N + E_2/N + \cdots . \tag{4.14}$$

E_1 is given by

$$E_1 = \frac{1}{2V} \sum_{\mathbf{q}}{}' \sum_{\mathbf{k}\sigma} \sum_{\mathbf{k}'\sigma'} \frac{4\pi e^2}{q^2} \langle F | c^\dagger_{\mathbf{k}+\mathbf{q}\sigma} c^\dagger_{\mathbf{k}'-\mathbf{q}\sigma'} c_{\mathbf{k}'\sigma'} c_{\mathbf{k}\sigma} | F \rangle. \tag{4.15}$$

The action of $c_{\mathbf{k}'\sigma'} c_{\mathbf{k}\sigma}$ on $|F\rangle$, for $k, k' < k_F$, removes two electrons in states $|\mathbf{k}\sigma\rangle$ and $|\mathbf{k}'\sigma'\rangle$; the action of $c^\dagger_{\mathbf{k}+\mathbf{q}\sigma} c^\dagger_{\mathbf{k}'-\mathbf{q}\sigma'}$ must restore the two electrons into these states if the matrix element is not to vanish. There are only two possibilities: (1) $\mathbf{k}+\mathbf{q} = \mathbf{k},\ \mathbf{k}'-\mathbf{q} = \mathbf{k}'$ and (2) $\mathbf{k}+\mathbf{q} = \mathbf{k}',\ \sigma = \sigma',\ \mathbf{k}'-\mathbf{q} = \mathbf{k}$. The first case holds if $\mathbf{q} = \mathbf{0}$, but the term $\mathbf{q} = \mathbf{0}$ is excluded, so we are left with only the second possibility. Therefore,

$$E_1 = \frac{1}{2V} \sum_{\mathbf{q}}{}' \sum_{\mathbf{k}\sigma} \frac{4\pi e^2}{q^2} \langle F | c^\dagger_{\mathbf{k}+\mathbf{q}\sigma} c^\dagger_{\mathbf{k}\sigma} c_{\mathbf{k}+\mathbf{q}\sigma} c_{\mathbf{k}\sigma} | F \rangle.$$

Since $\mathbf{q} \neq \mathbf{0}$, it follows that $c^\dagger_{\mathbf{k}\sigma} c_{\mathbf{k}+\mathbf{q}\sigma} = -c_{\mathbf{k}+\mathbf{q}\sigma} c^\dagger_{\mathbf{k}\sigma}$, and

$$\langle F | c^\dagger_{\mathbf{k}+\mathbf{q}\sigma} c^\dagger_{\mathbf{k}\sigma} c_{\mathbf{k}+\mathbf{q}\sigma} c_{\mathbf{k}\sigma} | F \rangle = -\langle F | c^\dagger_{\mathbf{k}+\mathbf{q}\sigma} c_{\mathbf{k}+\mathbf{q}\sigma} c^\dagger_{\mathbf{k}\sigma} c_{\mathbf{k}\sigma} | F \rangle.$$

The operator $c^\dagger_{\mathbf{k}\sigma} c_{\mathbf{k}\sigma}$ represents the number of electrons in state $|\mathbf{k}\sigma\rangle$, which is occupied by one electron if $k < k_F$ and vacant if $k > k_F$. Hence

$$c^\dagger_{\mathbf{k}\sigma} c_{\mathbf{k}\sigma} |F\rangle = \theta(k_F - k)|F\rangle,$$

where $\theta(k_F - k)$ is the step function,

$$\theta(k_F - k) = \begin{cases} 0 & k_F < k \\ 1 & k_F > k \end{cases}. \tag{4.16}$$

The first-order correction to the energy is now given by

$$E_1 = -\sum_{\mathbf{q}}{}' \sum_{\mathbf{k}\sigma} \frac{2\pi e^2}{V q^2}\, \theta(k_f - k)\, \theta(k_F - |\mathbf{k}+\mathbf{q}|).$$

Summation over σ gives a factor of 2. The sums over $\mathbf{k}$ and $\mathbf{q}$ are replaced by integrals,

$$E_1 = -\frac{4\pi e^2}{V} \frac{V^2}{(2\pi)^6} \int d^3q\, \frac{1}{q^2} \int d^3k\, \theta(k_f - k)\, \theta(k_F - |\mathbf{k}+\mathbf{q}|). \tag{4.17}$$

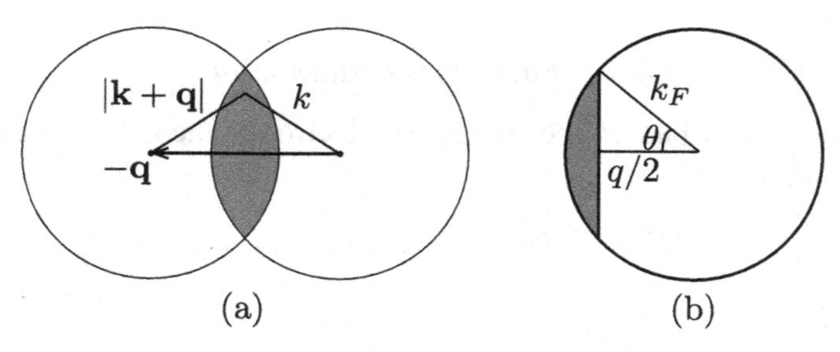

Figure 4.2 (a) Two spheres in **k**-space, one centered at $\mathbf{k} = \mathbf{0}$ and the other at $\mathbf{k} = -\mathbf{q}$. Each sphere's radius is k_F. The volume of the intersection region is V_Γ. (b) The shaded region has a volume of $V_{\Gamma/2} = V_\Gamma/2$.

The integral over **k** is simply the volume V_Γ of the region in **k**-space defined by $k < k_F$ and $|\mathbf{k} + \mathbf{q}| < k_f$. The expression for E_1 reduces to

$$E_1 = -4\pi e^2 \frac{V}{(2\pi)^6} \int d^3q \, \frac{1}{q^2} \, V_\Gamma. \tag{4.18}$$

Γ is the region of overlap of two spheres, each of radius k_F, one centered at $\mathbf{k} = \mathbf{0}$ and the other at $\mathbf{k} = -\mathbf{q}$ (see Figure 4.2a). Note from the figure that only in the overlap region are the conditions $k < k_F$ and $|\mathbf{k} + \mathbf{q}| < k_f$ satisfied. Since the two spheres have equal radii, $V_\Gamma = 2V_{\Gamma/2}$, where $V_{\Gamma/2}$ is the volume of the shaded region in Figure 4.2b. By inspection,

$$V_{\Gamma/2} = \frac{4\pi}{3} k_F^3 \frac{\Omega}{4\pi} - V_{\text{cone}}.$$

Here, Ω is the solid angle subtended at the center by the shaded region in Figure 4.2b, and V_{cone} is the volume of a cone with a half angle $\theta = \cos^{-1}(q/2k_F)$ at the vertex and a height $h = q/2$. They are given by

$$\Omega = \int_0^\theta \sin\theta' d\theta' \int_0^{2\pi} d\phi' = 2\pi(1 - q/2k_F), \quad V_{\text{cone}} = \frac{\pi q}{6}\,(k_F^2 - q^2/4).$$

Assembling the pieces together, and setting $q/2k_F = x$, we find

$$V_\Gamma = \frac{4\pi k_F^3}{3}\left[1 - \frac{3x}{2} + \frac{x^3}{2}\right]. \tag{4.19}$$

The integration over q in Eq. (4.18) is now carried out, noting that q varies from 0 to $2k_F$; for $q > 2k_F$ the spheres do not intersect. We find

$$E_1 = -4\pi e^2 \frac{V}{(2\pi)^6} \frac{4\pi k_F^3}{3} (4\pi)(2k_F) \int_0^1 (1 - 3x/2 + x^3/2) dx = -e^2 \left(V k_F^4 / 4\pi^3\right).$$

Since $k_F = (3\pi^2 N/V)^{1/3}$ (see Eq. [2.10]), it follows that

$$E_1/N = -\frac{3e^2}{4\pi} k_F.$$

Using the definition of r_s, and that $e^2/2a_0 = 1\ Ry$, we obtain

$$E_1/N = -\frac{3}{2\pi}\left(\frac{9\pi}{4}\right)^{1/3}\frac{1}{r_s}\ Ry \simeq -\frac{0.916}{r_s}\ Ry.$$

Hence, to first order in the perturbation

$$E/N \simeq \frac{2.21}{r_s^2} - \frac{0.916}{r_s}\ Ry. \tag{4.20}$$

The first term is the kinetic energy per electron, while the second term is known as the exchange energy per electron. The name is acquired because this term arises in the evaluation of the matrix element by having the creation operator $c^\dagger_{\mathbf{k}'-\mathbf{q}\sigma'}$ restore the electron annihilated by $c_{\mathbf{k}\sigma}$, while $c^\dagger_{\mathbf{k}+\mathbf{q}\sigma}$ restores the electron annihilated by $c_{\mathbf{k}'\sigma'}$. The exchange term is attractive, which may seem odd since it arises from the Coulomb interaction. The explanation for this situation is that the term arises when $\sigma = \sigma'$. Electrons in the same spin state cannot be located at the same point in space (Pauli exclusion principle), and they tend to stay away from each other, hence reducing the repulsive Coulomb interaction. This effect, which is quantum mechanical in nature, is not taken into account in the classical expression of the Coulomb interaction. In a way, the exchange term represents a quantum correction to an otherwise overestimated classical Coulomb repulsion.

4.3.2 Second order perturbation

The second order shift in energy, per one electron, is given by

$$E_2/N = \frac{1}{N}\sum_m{}' \frac{\langle F|V_C|m\rangle\langle m|V_C|F\rangle}{E_{|F\rangle} - E_{|m\rangle}}. \tag{4.21}$$

The prime indicates that the sum is over all intermediate states $|m\rangle \neq |F\rangle$. Consider $\langle m|V_C|F\rangle$. V_C annihilates two electrons in states $|\mathbf{k}\sigma\rangle$ and $|\mathbf{k}'\sigma'\rangle$ below the Fermi surface and creates two electrons in states $|\mathbf{k}+\mathbf{q}\sigma\rangle$ and $|\mathbf{k}'-\mathbf{q}\sigma'\rangle$; hence, if $|m\rangle \neq |F\rangle$, the two created electrons must lie outside the Fermi sphere, as shown in Figure 4.3. If the two electrons simply interchange their states, so that $|\mathbf{k}+\mathbf{q}\sigma\rangle = |\mathbf{k}'\sigma'\rangle$ and $|\mathbf{k}'-\mathbf{q}\sigma'\rangle = |\mathbf{k}\sigma\rangle$, then $|m\rangle$ and $|F\rangle$ will be the same. Therefore, $\langle m|V_C|F\rangle$ is nonzero if $k_F > k,\ k_F > k',\ k_F < |\mathbf{k}+\mathbf{q}|$, and $k_F < |\mathbf{k}'-\mathbf{q}|$. Considering the other matrix element, $\langle F|V_C|m\rangle$, we see that V_C must restore $|m\rangle$ to $|F\rangle$. This can be done by either a direct or an exchange process (see Figure 4.4). The contribution

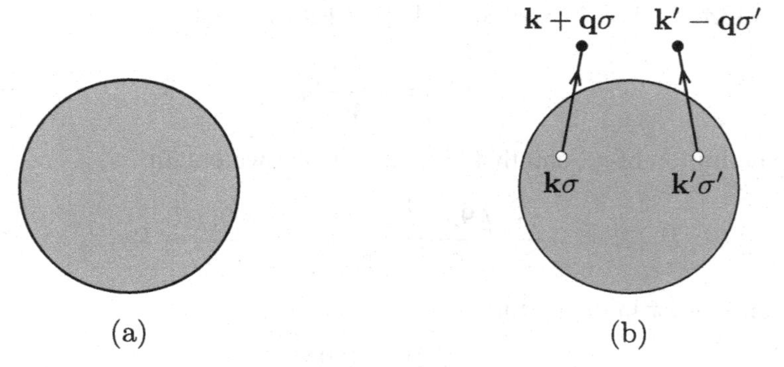

Figure 4.3 (a) Ground state $|F\rangle$ corresponding to a filled Fermi sphere F_S. (b) The action of V_C on $|F\rangle$: two electrons are annihilated from inside F_S, and two electrons are created outside F_S.

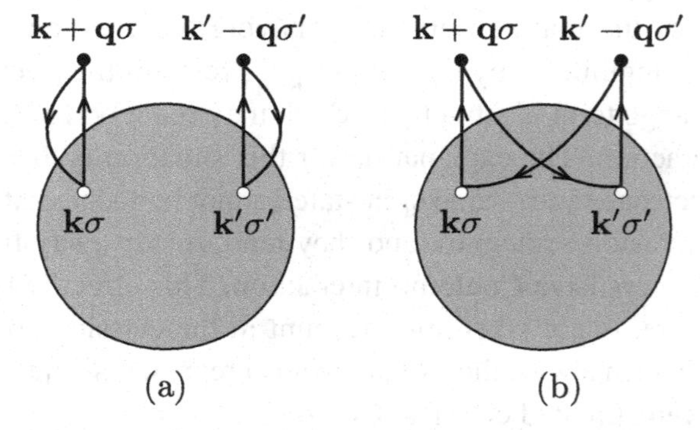

Figure 4.4 (a) A direct process: each electron recombines with the hole it left behind. (b) An exchange process: each electron recombines with the hole left behind by the other electron.

of the direct process to E_2/N is

$$\frac{E_{2,D}}{N} = \frac{1}{N}\sum_{\mathbf{q}}{}' \sum_{\mathbf{k}\sigma}\sum_{\mathbf{k}'\sigma'} \left(\frac{2\pi e^2}{Vq^2}\right)^2 \left(\frac{1}{E_{|F\rangle} - E_{|m\rangle}}\right)$$
$$\times\, \theta(k_F - k)\theta(k_F - k')\theta(|\mathbf{k}+\mathbf{q}| - k_F)\theta(|\mathbf{k}' - \mathbf{q}| - k_F). \qquad (4.22)$$

Evaluation of the RHS of Eq. (4.22) is not easy, but we can see from the following argument that it is divergent.

$$E_{|F\rangle} - E_{|m\rangle} = \frac{\hbar^2}{2m}\left[k^2 + k'^2 - (\mathbf{k}+\mathbf{q})^2 - (\mathbf{k}'-\mathbf{q})^2\right]$$
$$= \frac{\hbar^2}{2m}\left[2(\mathbf{k}'-\mathbf{k}).\mathbf{q} - 2q^2\right] = O(q) \text{ as } q \to 0.$$

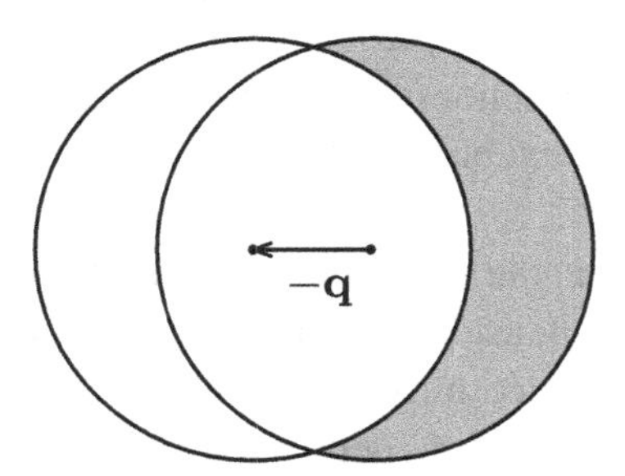

Figure 4.5 The region of integration over **k** in Eq. (4.22) is the shaded region. The centers of the spheres are separated by $-\mathbf{q}$ and each has radius k_F.

Replacing summation over **k** in Eq. (4.22) by integration, the presence of the product $\theta(k_F - k)\,\theta(|\mathbf{k}+\mathbf{q}| - k_F)$ in the integrand restricts the region of integration to the shaded space in Figure 4.5, whose volume $V_{\Gamma_\mathbf{k}}$ is equal to the volume of one sphere minus the volume of the overlap region of the two spheres (this was evaluated in the previous subsection). Therefore,

$$V_{\Gamma_\mathbf{k}} = \frac{4\pi k_F^3}{3}\left[\frac{3}{2}(q/2k_F) - \frac{1}{2}(q/2k_F)^3\right] = O(q) \text{ as } q \to 0.$$

Similarly, the volume of the region of integration over $\mathbf{k}'$ can be calculated; again we find

$$V_{\Gamma_{\mathbf{k}'}} = O(q) \text{ as } q \to 0.$$

The integral over **k** and $\mathbf{k}'$ in Eq. (4.22) can be written as

$$I = \int_{\Gamma_\mathbf{k}} d^3k \int_{\Gamma_{\mathbf{k}'}} d^3k' \frac{1}{E_{|F\rangle} - E_{|m\rangle}} = \frac{m}{\hbar^2}\int_{\Gamma_\mathbf{k}} d^3k \int_{\Gamma_{\mathbf{k}'}} d^3k' \, \frac{1}{q\left[k_z' - k_z - q\right]}.$$

In evaluating $E_{|F\rangle} - E_{|m\rangle}$, we have taken the z-direction as that of **q**. In the limit $q \to 0$, the above integral gives

$$I = \frac{m}{\hbar^2 q}\langle (k_z' - k_z)^{-1}\rangle V_{\Gamma_\mathbf{k}} V_{\Gamma_{\mathbf{k}'}} = O(q) \text{ as } q \to 0.$$

Here, $\langle (k_z' - k_z)^{-1}\rangle$ is the average of $(k_z' - k_z)^{-1}$ over the regions of integration $\Gamma_\mathbf{k}$ and $\Gamma_{\mathbf{k}'}$. Considering now the integration over **q**, we obtain

$$\frac{E_{2,D}}{N} \xrightarrow{q\to 0} \frac{V^3}{NV^2}\int d^3q \frac{O(q)}{O(q^4)} = \frac{V}{N}\int \frac{dq}{O(q)}.$$

The second-order correction to the energy, which arises from the direct processes, diverges logarithmically. Even though $\mathbf{q} = \mathbf{0}$ is excluded from the sum, in the thermodynamic limit, as the crystal's volume $V \to \infty$, q may get arbitrarily close to zero, leading to the logarithmic divergence. In the above expression, V^3 in the numerator arises from replacement of the summations over $\mathbf{k}, \mathbf{k}'$, and $\mathbf{q}$ by integration, while V^2 in the denominator results from V_C squared. We note, for completeness' sake, that the exchange term, following a similar analysis, turns out to be finite.

What we have found is that the second-order correction, rather than being smaller than the first-order correction, as one might expect, is actually divergent. Since the crystal is stable, this divergence must be eliminated somehow. Our only hope is to consider higher order terms in the perturbation. However, these also turn out to be divergent, so a sum of terms to infinite order needs to be carried out. Green's function formalism will provide a suitable framework within which to carry this out.

Finally, we note that the need to sum perturbation terms to infinite order arises as a mathematical necessity due to their divergence, which in turn results from the small values of q, or, equivalently, from the long-range nature of the Coulomb interaction. Physically, one needs to sum perturbation terms to infinite order to capture the effect of screening (this is discussed in detail in Chapter 10), which renders the Coulomb interaction short-ranged. In empty space, two electrons i and j would interact by exchanging momentum, but in the presence of a medium, the interaction can proceed in an infinite number of ways. Electron i may scatter, but the momentum it transfers could be picked up by an electron below the Fermi surface, which would then make a transition to a state above the Fermi surface, leaving behind a hole. The electron and hole could then recombine, transferring momentum to electron j. Alternatively, the electron–hole recombination could lead to the creation of another electron–hole pair which, upon recombination, would transfer momentum to electron j. This argument could be carried on at length, showing that there are infinite ways in which the interaction between two electrons could proceed. The net effect is that the Coulomb interaction becomes screened.

Further reading

Bruus, H. and Flensberg, K. (2004). *Many-Body Quantum Theory in Condensed Matter Physics*. Oxford: Oxford University Press.

Fetter, A.L. and Walecka, J.D. (1971). *Quantum Theory of Many-Particle Systems*. New York: McGraw-Hill.

Kittel, C. (1963). *Quantum Theory of Solids*. New York: Wiley.

Problems

4.1 *Constrained ground state*. In the ground state of a noninteracting electron gas at $T = 0$, there are $N/2$ spin-up electrons and $N/2$ spin-down electrons; the ground state is unpolarized ($S_z = 0$). A spin polarized state has $N_\uparrow$ spin-up electrons and $N_\downarrow$ spin-down electrons: $N_\uparrow = (1+p)N/2$, $N_\downarrow = (1-p)N/2$, where p is the fractional spin polarization: $-1 < p < 1$. The state of minimum energy, for a given value of p, is obtained by filling two Fermi spheres in **k**-space, one of radius $k_{F\uparrow}$ and one of radius $k_{F\downarrow}$. Show that, in three dimensions, the energy per electron in this constrained ground state is

$$\frac{E}{N} = \frac{E_0}{N} \frac{(1+p)^{5/3} + (1-p)^{5/3}}{2}$$

where $E_0/N = 3E_F/5$ is the energy per electron in the unpolarized ground state.

4.2 *Correlation function.* For a system of noninteracting electrons at $T = 0$, define the following correlation function

$$G_\sigma(\mathbf{r}, \mathbf{r}') = \langle \Phi_0 | \Psi_\sigma^\dagger(\mathbf{r}) \Psi_\sigma(\mathbf{r}') | \Phi_0 \rangle.$$

This is the probability amplitude for the state $|\alpha\rangle = \Psi_\sigma(\mathbf{r}')|\Phi_0\rangle$, in which a particle at $\mathbf{r}'$ is removed from the system in the ground state, to be found in state $|\beta\rangle = \Psi_\sigma(\mathbf{r})|\Phi_0\rangle$. $|\beta\rangle$ is the state obtained by removing a particle, with coordinates $(\mathbf{r}\sigma)$, from the ground state. Obtain an expression for $G_\sigma(\mathbf{r}, \mathbf{r}')$ in terms of $x = |\mathbf{r} - \mathbf{r}'|$.

4.3 *Pair correlation function.* For a noninteracting electron gas in the ground state $|\Phi_0\rangle$, evaluate $\langle \Phi_0 | \Psi_\sigma^\dagger(\mathbf{r}) \Psi_{\sigma'}^\dagger(\mathbf{r}') \Psi_{\sigma'}(\mathbf{r}') \Psi_\sigma(\mathbf{r}) | \Phi_0 \rangle$ in terms of $x = |\mathbf{r} - \mathbf{r}'|$ for $\sigma \neq \sigma'$ and for $\sigma = \sigma'$. What physical conclusion can you draw from your answer?

4.4 *Coulomb interaction in two dimensions.* Show that, in two dimensions, the Fourier transform of the Coulomb potential is $2\pi e^2/q$.
Hint: $J_0(x) = \frac{1}{2\pi} \int_0^{2\pi} e^{-ix\cos\theta} d\theta$.

4.5 *Exchange energy in two dimensions.* In three dimensions we found that the exchange energy per electron is $-3e^2 k_F/4\pi$. Show that, in two dimensions, the exchange energy per electron is $-4e^2 k_F/3\pi$.

5

A brief review of statistical mechanics

> Hence the importance of the role that is played in the physical sciences by the law of probability. We must thoroughly examine the principles on which it is based.
>
> –*Henri Poincare,* Science and Hypothesis

Since we will be dealing with systems at finite temperatures, we will need concepts that have been developed in the context of statistical mechanics. We devote this chapter to a brief review of the basic elements of statistical mechanics.

5.1 The fundamental postulate of statistical mechanics

Consider an isolated system of N noninteracting, identical particles confined to a region of volume V. The Hamiltonian is $H = \sum_{i=1}^{N} h(i)$, where $h(i)$ is the operator that represents the energy of particle i. The single-particle states are obtained by solving the Schrödinger equation $h|\phi_\nu\rangle = \epsilon_\nu|\phi_\nu\rangle$, where ν stands for all the quantum numbers that characterize the state. The energy ϵ_ν depends on V. For example, for a system of noninteracting particles confined to a cube of side length L, $\epsilon_{\mathbf{k}\sigma} = \hbar^2k^2/2m$, and if periodic boundary conditions are adopted, then $k_x, k_y, k_z = 0, \pm 2\pi/L, \pm 4\pi/L, \ldots$. The total energy of the system is $E = \sum_\nu n_\nu \epsilon_\nu$, where n_ν is the number of particles in state $|\phi_\nu\rangle$, and the total number of particles is $N = \sum_\nu n_\nu$. A *macrostate* of the system is defined by specifying the values of N, V, and E.

At the microscopic level, there are many different ways to distribute the energy E among the N particles that comprise the system. Each of these different ways defines a particular *microstate* that is consistent with the given macrostate. A microstate is a quantum state of the system described by a wave function $\psi(1, 2, \ldots, N)$. The number of microstates that are consistent with a given macrostate is a function of N, V, and E, and it is denoted by $\Omega(N, V, E)$. For a macroscopic system consisting of a large number of particles, of the order of Avogadro's number (6.22×10^{23}), $\Omega(N, V, E)$ will be, in general, a

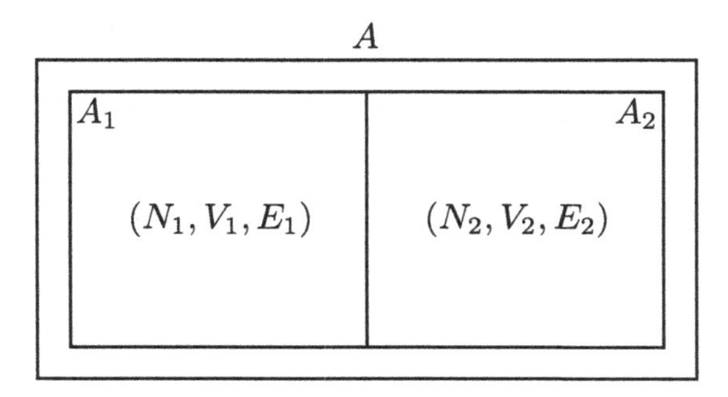

Figure 5.1 An isolated system A consists of two subsystems, A_1 and A_2, that are in thermal contact. Neither E_1 (the energy of A_1) nor E_2 (the energy of A_2) is constant, but $E_1 + E_2$ is constant.

fantastically large number. The fundamental postulate of statistical mechanics asserts the following: *an isolated system in equilibrium, in a given macrostate, is equally likely to be in any of the microstates that are consistent with the given macrostate.*

5.2 Contact between statistics and thermodynamics

Consider an isolated system A which consists of two subsystems, A_1 and A_2, that are in thermal contact and can exchange energy (see Figure 5.1). A_1, in macrostate (N_1, V_1, E_1), has $\Omega_1(N_1, V_1, E_1)$ microstates, while A_2 has $\Omega_2(N_2, V_2, E_2)$ microstates. Due to energy exchange, E_1 and E_2 are not constants, but the combined system A, being isolated, has a constant energy $E = E_1 + E_2$. Since A_1 is equally likely to be in any of its microstates, as is A_2, the total number of microstates of A is $\Omega_1(E_1)\Omega_2(E_2) = \Omega_1(E_1)\Omega_2(E - E_1) = \Omega(E, E_1)$, where the dependence on N_1, V_1, N_2, and V_2 is suppressed (these latter quantities are constants). Energy exchange between A_1 and A_2 persists until equilibrium is attained. Let the values of E_1 and E_2 at equilibrium be $\bar{E}_1$ and $\bar{E}_2$, respectively. We assert that at equilibrium $\Omega(E, E_1)$ is maximum. The idea is that a system, left to its own devices, settles into a macrostate that affords the largest possible number of microstates (e.g., a gas confined by a partition to the left half of a box will fill the box uniformly upon removal of the partition). Hence, at equilibrium

$$
\begin{aligned}
0 = \left.\frac{\partial \Omega}{\partial E_1}\right|_{\bar{E}_1} &= \left.\frac{\partial \Omega_1(E_1)}{\partial E_1}\right|_{\bar{E}_1} \Omega_2(\bar{E}_2) + \Omega_1(\bar{E}_1)\left.\frac{\partial \Omega_2(E_2)}{\partial E_1}\right|_{\bar{E}_1} \\
&= \left.\frac{\partial \Omega_1(E_1)}{\partial E_1}\right|_{\bar{E}_1} \Omega_2(\bar{E}_2) + \Omega_1(\bar{E}_1)\left.\frac{\partial \Omega_2(E_2)}{\partial E_2}\right|_{\bar{E}_2} \frac{\partial E_2}{\partial E_1}.
\end{aligned}
$$

Since $E_1 + E_2$ is constant, $\partial E_2/\partial E_1 = -1$, and the above relation yields

$$\left.\frac{\partial \ln \Omega_1(E_1)}{\partial E_1}\right|_{\bar{E}_1} = \left.\frac{\partial \ln \Omega_2(E_2)}{\partial E_2}\right|_{\bar{E}_2}.$$

The condition for equilibrium of A_1 and A_2 thus reduces to

$$\beta_1 = \beta_2 \tag{5.1}$$

where

$$\beta_i = \left.\frac{\partial \ln \Omega_i(N_i, V_i, E_i)}{\partial E_i}\right|_{N_i, V_i} \qquad i = 1, 2. \tag{5.2}$$

The parameters N_i and V_i, being held constant, are now written explicitly.

From thermodynamics, we know that at equilibrium the temperatures of both subsystems are equal

$$T_1 = T_2, \tag{5.3}$$

where T is given in terms of the entropy S by the thermodynamic relation

$$\frac{1}{T} = \left.\frac{\partial S}{\partial E}\right|_{N,V}. \tag{5.4}$$

Comparing Eqs (5.1) and (5.3), and (5.2) and (5.4), we are tempted to identify β with $1/T$. However, β has units of 1/Energy; hence, we write

$$\beta = 1/kT, \quad S = k \ln \Omega \tag{5.5}$$

where $k = 1.38 \times 10^{-23}$ J/K (Joules/degrees Kelvin) is Boltzmann's constant. Since, for a given system, the number of microstates Ω depends on N, V, and E, we can write

$$d(\ln \Omega) = \left.\frac{\partial \ln \Omega}{\partial E}\right|_{N,V} dE + \left.\frac{\partial \ln \Omega}{\partial V}\right|_{N,E} dV + \left.\frac{\partial \ln \Omega}{\partial N}\right|_{V,E} dN. \tag{5.6}$$

Using Eqs (5.4) and (5.5), the above relation is rewritten as

$$dS = \frac{1}{T} dE + \left.\frac{\partial S}{\partial V}\right|_{N,E} dV + \left.\frac{\partial S}{\partial N}\right|_{V,E} dN$$

$$\Longrightarrow dE = T dS - T\left.\frac{\partial S}{\partial V}\right|_{N,E} dV - T\left.\frac{\partial S}{\partial N}\right|_{V,E} dN. \tag{5.7}$$

Comparing this with the fundamental formula of thermodynamics

$$dE = TdS - PdV + \mu dN, \tag{5.8}$$

where P is the pressure and μ is the chemical potential, we find

$$P = kT \left. \frac{\partial \ln \Omega}{\partial V} \right|_{N,E}, \qquad \mu = -kT \left. \frac{\partial \ln \Omega}{\partial N} \right|_{V,E}. \tag{5.9}$$

The expressions for the thermodynamic quantities T, S, P, and μ, in terms of Ω, establish the connection between thermodynamics and statistics.

5.3 Ensembles

Given a system in a specific macrostate, what is the probability of finding the system in a particular microstate $|\psi_i\rangle$? Imagine a large collection or ensemble of systems, N_{ens}, with all systems in the same macrostate. We perform measurements on each system in the ensemble to determine its microstate. If N_i systems were found to be in state $|\psi_i\rangle$, we would say that the probability of the system being in state $|\psi_i\rangle$ is N_i/N_{ens}.

The information available about a system determines its macrostate. For example, a system may be isolated, in which case N, V, and E are fixed. A system may be in contact with a heat reservoir, in which case its temperature T is fixed, but only its mean energy is fixed. Whatever the constraints are, an ensemble is constructed such that all of its members are under the same physical conditions as the system of interest. Since one may imagine systems under various constraints, we are led to consider different kinds of ensembles. Below we discuss three important ensembles.

5.3.1 The microcanonical ensemble

Such an ensemble is representative of an isolated system with fixed energy. In fact, the exact value of a system's energy cannot be obtained by any measurement performed in a finite amount of time ($\Delta E \Delta t \sim \hbar$); all we can tell is that the energy of an isolated system is in a range $(E, E + \delta E)$, where $\delta E \ll E$. The fundamental postulate of statistical mechanics asserts that the probability of finding the system in state $|\psi_n\rangle$ is given by

$$p_n = \begin{cases} c & \text{if } E \leq E_n \leq E + \delta E \\ 0 & \text{otherwise} \end{cases} \tag{5.10}$$

where c is a constant equal to 1 divided by the total number of states accessible to the system.

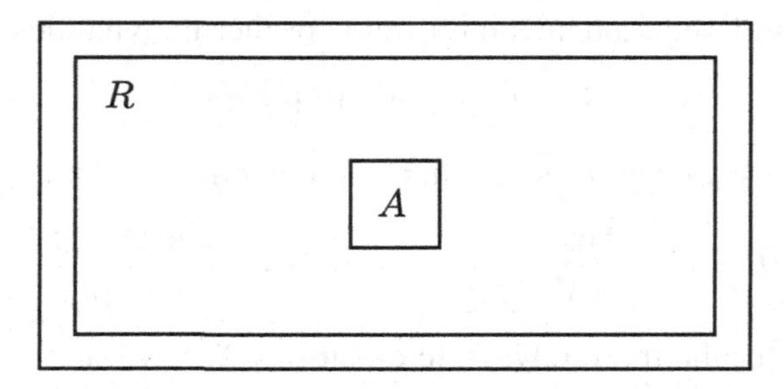

Figure 5.2 An isolated system consisting of a system A in thermal contact with a much larger heat reservoir R at temperature T. A and R exchange energy, but the total energy $E_A + E_R$ remains constant.

5.3.2 The canonical ensemble

A canonical ensemble is representative of a system at a fixed temperature T. We consider a small system A in contact with a heat reservoir R at temperature T (see Figure 5.2). The combined system, $A + R$, is isolated, and its total energy is E_0, which is a constant. System A is small in the sense that its degrees of freedom are far fewer than those of R. What is the probability p_n of finding A in state $|\psi_n\rangle$ with energy E_n, once equilibrium has been attained? If A is in state $|\psi_n\rangle$, then the number of states of the combined system is simply $\Omega_R(E_0 - E_n)$, which is the number of states accessible to R. Therefore, p_n is proportional to $\Omega_R(E_0 - E_n)$,

$$p_n = C\Omega_R(E_0 - E_n) = Ce^{\ln \Omega_R(E_0 - E_n)}.$$

We can expand $\ln \Omega_R(E_0 - E_n)$,

$$\begin{aligned}
\ln \Omega_R(E_0 - E_n) &= \ln \Omega_R(E_0) - \left.\frac{\partial \ln \Omega_R}{\partial E}\right|_{E_0} E_n + O(E_n^2) \\
&= \ln \Omega_R(E_0) - \left.\frac{\partial \ln \Omega_R}{\partial E}\right|_{E_R + E_n} E_n + O(E_n^2) \\
&= \ln \Omega_R(E_0) - \left.\frac{\partial \ln \Omega_R}{\partial E}\right|_{E_R} E_n + O(E_n^2).
\end{aligned}$$

Since $E_n \ll E_0$, we neglect terms of order higher than E_n. Using Eq. (5.2), the above equation may be written as

$$\ln \Omega_R(E_0 - E_n) = \ln \Omega_R(E_0) - \beta E_n$$

where $\beta = 1/kT$ and T is the temperature of the reservoir. Therefore,

$$p_n = Ce^{\ln \Omega_R(E_0) - \beta E_n} = C\Omega_R(E_0)e^{-\beta E_n} = C'e^{-\beta E_n}.$$

The constant C' is determined by the normalization condition

$$\sum_n p_n = 1 \Rightarrow p_n = \frac{e^{-\beta E_n}}{Z} \tag{5.11}$$

where Z, known as the partition function, is given by

$$Z = \sum_n e^{-\beta E_n}. \tag{5.12}$$

The Helmholtz free energy, F, of a system is defined by

$$F = E - TS \tag{5.13}$$

where E is the average energy of the system and S is its entropy. Since its derivation is too lengthy for a brief review such as this, we shall state without proof the following result, which establishes the connection between F and Z,

$$F = -kT \ln Z. \tag{5.14}$$

5.3.3 The grand canonical ensemble

A grand canonical ensemble is representative of a system at fixed temperature but consisting of a variable number of particles. Consider a system A in contact with a heat reservoir R at temperature T. The systems A and R exchange energy and particles (the particles in A and R are the same). Neither E_A nor N_A is fixed, but $E_A + E_R = E_0$ and $N_A + N_R = N_0$ are fixed. For any given number of particles N_s of system A, let the quantum states of A be labeled by r; the microstates of A are then labeled by (r, s). Following the same argument as in the previous subsection, the probability of finding system A in a particular state $|\psi_{rs}\rangle$, with energy E_r and number of particles N_s, is

$$p_{rs} = C\Omega_R(E_0 - E_r, N_0 - N_s).$$

Expanding $\ln \Omega_R$, and using Eqs (5.2) and (5.9), we find

$$\begin{aligned}\ln \Omega_R(E_0 - E_r, N_0 - N_s) &= \ln \Omega_R(E_0, N_0) - \left.\frac{\partial \ln \Omega_R}{\partial E}\right|_{E_0} E_r - \left.\frac{\partial \ln \Omega_R}{\partial N}\right|_{N_0} N_s \\ &= \ln \Omega_R(E_0, N_0) - \beta E_r + \beta\mu N_s.\end{aligned}$$

Higher orders in the expansion are neglected since $E_r \ll E_0$ and $N_s \ll N_0$. Following the same steps as we did in the case of the canonical ensemble, we find the probability that the system is in state $|\psi_{rs}\rangle \equiv |r, s\rangle$ to be

$$p_{rs} = \frac{e^{-\beta(E_r - \mu N_s)}}{Z_G}, \tag{5.15}$$

where

$$Z_G = \sum_{rs} e^{-\beta(E_r - \mu N_s)} \tag{5.16}$$

is the grand partition function. Note that Z_G may be written as

$$Z_G = \sum_{rs} \langle r, s | e^{-\beta(H-\mu N)} | r, s \rangle = \mathrm{Tr}\left[e^{-\beta(H-\mu N)} \right], \tag{5.17}$$

where H is the Hamiltonian, N is the number of particles operator for system A, and $\mathrm{Tr}[B]$ stands for the trace of operator B, the sum of the diagonal elements of the matrix that represents B.

Given an operator A acting on a system with a variable number of particles, its mean value in the state $|r, s\rangle$ is $\bar{A} = \langle r, s|A|r, s\rangle$. Since the system is in state $|r, s\rangle$ with probability p_{rs}, the ensemble average $\langle A \rangle$ is given by

$$\langle A \rangle = \sum_{rs} p_{rs} \bar{A} = \sum_{rs} \frac{e^{-\beta(E_r - \mu N_s)} \langle r, s|A|r, s\rangle}{Z_G} = \sum_{rs} \frac{\langle r, s| e^{-\beta(H-\mu N)} A |r, s\rangle}{Z_G}.$$

Defining the statistical operator for the grand canonical ensemble by

$$\rho_G = \frac{e^{-\beta(H-\mu N)}}{\mathrm{Tr}\left[e^{-\beta(H-\mu N)} \right]} \tag{5.18}$$

we can write

$$\langle A \rangle = \frac{\mathrm{Tr}\left[e^{-\beta(H-\mu N)} A \right]}{\mathrm{Tr}\left[e^{-\beta(H-\mu N)} \right]} = \mathrm{Tr}(\rho_G A). \tag{5.19}$$

Finally, we note that when a macroscopic system is in contact with a reservoir with which it can exchange energy and particles, the fluctuations about the mean energy and the mean number of particles are exceedingly small. Because this is so, physical properties of the system do not change in any appreciable way if it is removed from contact with the reservoir, resulting in it having fixed energy and a fixed number of particles. Therefore, when calculating mean values of various quantities, it makes no difference whether the system is isolated, in contact with a heat reservoir, or in contact with a reservoir with which it can exchange energy and particles. In other words, it makes no difference whether we are calculating mean values within a microcanonical, canonical, or grand canonical ensemble; mathematical convenience usually dictates the most appropriate choice.

5.4 The statistical operator for a general ensemble

5.4.1 Definition

Let us consider a system of identical particles under certain physical conditions, i.e., in a certain macrostate. For example, the system may be isolated, in which case the number of particles N, the volume V, and the energy E, are fixed. Alternatively, our system may be in contact with a heat reservoir, in which case N, V, and its temperature T are fixed. Regardless, we proceed to construct an ensemble of systems having exactly the same physical conditions as our original system. Let us assume that any particular microstate, characterized by the state vector $|\psi_i\rangle$, occurs in the ensemble with probability p_i. Clearly $\sum_i p_i = 1$. The ensemble average of an observable, represented by the hermitian operator A, is given by

$$\langle A\rangle = \sum_i p_i \langle\psi_i|A|\psi_i\rangle. \tag{5.20}$$

Notice that two types of averages are involved in writing the ensemble average. $\langle\psi_i|A|\psi_i\rangle$ is the usual quantum mechanical expectation value of A in state $|\psi_i\rangle$. The above equation also tells us that the quantum mechanical averages must be further weighted by the corresponding fractional occupation of the state $|\psi_i\rangle$; this second averaging is classical in nature.

Given a complete set $|1\rangle, |2\rangle, \ldots$ of single-particle states, we introduce two resolutions of identity ($\sum_n |n\rangle\langle n| = 1$) into Eq. (5.20),

$$\langle A\rangle = \sum_{nm}\sum_i p_i \langle\psi_i|n\rangle\langle n|A|m\rangle\langle m|\psi_i\rangle = \sum_{nm}\sum_i p_i \langle m|\psi_i\rangle\langle\psi_i|n\rangle\langle n|A|m\rangle.$$

In this form, the dependence of the average on the ensemble is factored out. We define the statistical, or density, operator as follows:

$$\rho = \sum_i p_i |\psi_i\rangle\langle\psi_i|. \tag{5.21}$$

The ensemble average of A is now written as

$$\langle A\rangle = \sum_{nm} \langle m|\rho|n\rangle\langle n|A|m\rangle = \sum_m \langle m|\rho A|m\rangle = \mathrm{Tr}[\rho A]. \tag{5.22}$$

Since the trace is independent of the basis set of single-particle states, $\mathrm{Tr}[\rho A]$ may be evaluated in any convenient basis set.

5.4.2 General properties

There are two important properties of the statistical operator:

(1) The statistical operator ρ is Hermitian; this follows from its definition:

$$\rho = \sum_i p_i |\psi_i\rangle\langle\psi_i| \Rightarrow \rho^\dagger = \sum_i p_i (|\psi_i\rangle\langle\psi_i|)^\dagger = \sum_i p_i |\psi_i\rangle\langle\psi_i| = \rho.$$

(2) The trace of ρ is unity,

$$\begin{aligned}\mathrm{Tr}[\rho] &= \sum_n \sum_i p_i \langle n|\psi_i\rangle\langle\psi_i|n\rangle = \sum_i p_i \sum_n \langle\psi_i|n\rangle\langle n|\psi_i\rangle = \sum_i p_i \langle\psi_i|\psi_i\rangle \\ &= \sum_i p_i = 1.\end{aligned}$$

5.4.3 Time evolution

At time t_0 the statistical operator is given by

$$\rho(t_0) = \sum_i p_i |\psi_i(t_0)\rangle\langle\psi_i(t_0)|.$$

How does ρ change with time? p_i is the probability of finding the microstate $|\psi_i\rangle$ in the ensemble. If the ensemble is left undisturbed, p_i cannot change with time. The evolution of ρ with time is thus completely governed by the time evolution of the states $|\psi_i\rangle$,

$$\rho(t) = \sum_i p_i |\psi_i(t)\rangle\langle\psi_i(t)|. \tag{5.23}$$

The equation of motion of the statistical operator is

$$i\hbar\frac{\partial\rho}{\partial t} = \sum_i p_i \left(i\hbar\frac{\partial}{\partial t}|\psi_i(t)\rangle\right)\langle\psi_i(t)| + \sum_i p_i |\psi_i(t)\rangle i\hbar\frac{\partial}{\partial t}\langle\psi_i(t)|.$$

From the Schrödinger equation and its complex conjugate:

$$i\hbar\frac{\partial}{\partial t}|\psi_i(t)\rangle = H|\psi_i(t)\rangle, \quad -i\hbar\frac{\partial}{\partial t}\langle\psi_i(t)| = \langle\psi_i(t)|H, \tag{5.24}$$

we obtain

$$i\hbar\frac{\partial\rho}{\partial t} = \sum_i p_i H|\psi_i(t)\rangle\langle\psi_i(t)| - \sum_i p_i |\psi_i(t)\rangle\langle\psi_i(t)|H.$$

Using the definition of ρ, we can write

$$i\hbar\frac{\partial\rho}{\partial t} = H\rho - \rho H = [H, \rho]. \tag{5.25}$$

We will discuss the time evolution of ρ again in Chapter 13, when we discuss the nonequilibrium Green's function.

5.5 Quantum distribution functions

Consider a system of noninteracting, identical particles. Let us denote the single-particle states by $\phi_1, \phi_2, \ldots$, with corresponding energies $\epsilon_1, \epsilon_2, \ldots$. The quantum states of the system are given in the number-representation by $|n_1\ n_2\ \ldots\rangle$, where n_i is the number of particles in single-particle state ϕ_i. The energy of the state $|n_1\ n_2\ \ldots\rangle$ is $\sum_{i=1}^{\infty} n_i\epsilon_i$, and the number of particles in the state is $\sum_{i=1}^{\infty} n_i$. We take the system to be a member of a grand canonical ensemble. The grand partition function is

$$\begin{aligned}
Z_G &= \mathrm{Tr}\left[e^{-\beta(H-\mu N)}\right] = \sum_{n_1 n_2 \ldots} \langle n_1\ n_2\ \ldots | e^{-\beta(H-\mu N)} | n_1\ n_2\ \ldots\rangle \\
&= \sum_{n_1 n_2 \ldots} \exp\left[-\beta \sum_{i=1}^{\infty} (n_i\epsilon_i - \mu n_i)\right] = \sum_{n_1} e^{-\beta(\epsilon_1-\mu)n_1} \sum_{n_2} e^{-\beta(\epsilon_2-\mu)n_2} \ldots \\
&= \prod_{i=1}^{\infty} \sum_{n_i} e^{-\beta(\epsilon_i-\mu)n_i}.
\end{aligned} \tag{5.26}$$

The ensemble average of the total number of particles is

$$\begin{aligned}
\langle N\rangle &= Z_G^{-1}\mathrm{Tr}\left[e^{-\beta(H-\mu N)}N\right] = Z_G^{-1}\sum_{n_1 n_2 \ldots} \langle n_1 n_2 \cdots | e^{-\beta(H-\mu N)} N | n_1 n_2 \cdots\rangle \\
&= Z_G^{-1} \sum_{n_1 n_2 \ldots} (n_1 + n_2 + \cdots) e^{-\beta[(n_1\epsilon_1+n_2\epsilon_2+\cdots)-\mu(n_1+n_2+\cdots)]} = \beta^{-1}\frac{\partial}{\partial\mu}\ln Z_G.
\end{aligned} \tag{5.27}$$

At this point, distinction is made between bosons and fermions. For bosons, n_i is unrestricted; it can vary from 0 to ∞. Equation (5.26) gives

$$Z_G^B = \prod_{i=1}^{\infty} \frac{1}{1-e^{-\beta(\epsilon_i-\mu)}} \Rightarrow \ln Z_G^B = -\sum_{i=1}^{\infty} \ln\left[1 - e^{-\beta(\epsilon_i-\mu)}\right]. \tag{5.28}$$

Using Eqs (5.27) and (5.28), we find

$$\langle N\rangle = \sum_i \frac{e^{-\beta(\epsilon_i-\mu)}}{1-e^{-\beta(\epsilon_i-\mu)}} = \sum_i \frac{1}{e^{\beta(\epsilon_i-\mu)}-1} = \sum_i n_i^{BE},$$

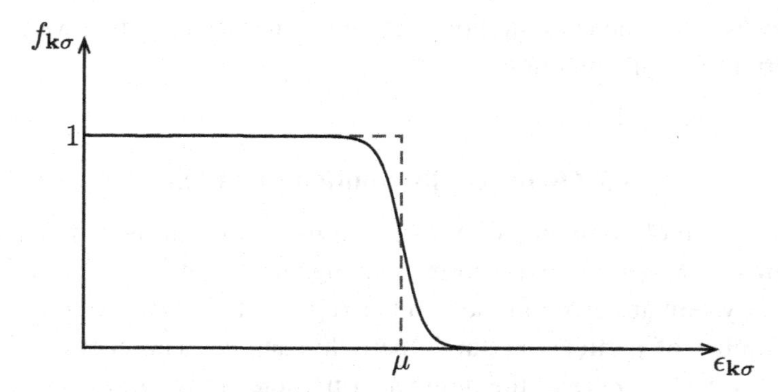

Figure 5.3 The Fermi–Dirac distribution function. The dashed line corresponds to the case of zero temperature.

where n_i^{BE}, the Bose–Einstein quantum distribution function, is the average number of particles in the single-particle state $|\phi_i\rangle$. It is given by

$$n_i^{BE} = \frac{1}{e^{\beta(\epsilon_i-\mu)} - 1} \qquad \text{bosons.} \tag{5.29}$$

For fermions, n_i in Eq. (5.26) is either 0 or 1; hence

$$Z_G^F = \prod_{i=1}^{\infty} \left[1 + e^{-\beta(\epsilon_i-\mu)}\right] \Rightarrow \ln Z_G^F = \sum_i \ln\left[1 + e^{-\beta(\epsilon_i-\mu)}\right]. \tag{5.30}$$

The ensemble average of the number of particles is

$$\langle N\rangle = \sum_i \frac{e^{-\beta(\epsilon_i-\mu)}}{1+e^{-\beta(\epsilon_i-\mu)}} = \sum_i \frac{1}{e^{\beta(\epsilon_i-\mu)}+1} = \sum_i f_i^{FD},$$

where f_i^{FD}, the Fermi–Dirac distribution function, is the average occupation number of single-particle state ϕ_i. It is given by

$$f_i^{FD} = \frac{1}{e^{\beta(\epsilon_i-\mu)} + 1} \qquad \text{fermions.} \tag{5.31}$$

Unless confusion may arise, we write f_i^{FD} simply as f_i. The Fermi–Dirac distribution function is depicted in Figure 5.3. In particular, for a system of noninteracting particles whose Hamiltonian is $H = \sum_{\mathbf{k}\sigma} \epsilon_{\mathbf{k}\sigma} c^\dagger_{\mathbf{k}\sigma} c_{\mathbf{k}\sigma}$, the occupation number of the single-particle state $|\mathbf{k}\sigma\rangle$ is given by

$$\langle n_{\mathbf{k}\sigma}\rangle = \langle c^\dagger_{\mathbf{k}\sigma} c_{\mathbf{k}\sigma}\rangle = \begin{cases} n^{BE}_{\mathbf{k}\sigma} = \left[e^{\beta(\epsilon_{\mathbf{k}\sigma}-\mu)} - 1\right]^{-1} & \text{bosons} \\ f_{\mathbf{k}\sigma} = \left[e^{\beta(\epsilon_{\mathbf{k}\sigma}-\mu)} + 1\right]^{-1} & \text{fermions.} \end{cases} \tag{5.32}$$

For a system with a fixed number of particles N, the chemical potential μ is obtained from the relation

$$N = \sum_i \frac{1}{e^{\beta(\epsilon_i - \mu)} \mp 1} \tag{5.33}$$

where the lower (upper) sign refers to fermions (bosons).

Further reading

Huang, K. (2001). *Introduction to Statistical Physics*. London: Taylor and Francis.
Pathria, R.K. (1996). *Statistical Mechanics*, 2nd edn. Oxford: Butterworth-Heinemann.
Reif, F. (1965). *Fundamentals of Statistical and Thermal Physics*. New York: McGraw-Hill.

Problems

5.1 *Stirling's formula for $N!$*. Starting from

$$N! = \int_0^\infty e^{-t} t^N dt,$$

replace t with $N + \sqrt{N}x$. Show that

$$N! = \sqrt{N} N^N e^{-N} \int_{-\sqrt{N}}^\infty f(x) dx, \qquad f(x) = e^{-\sqrt{N}x}\left(1 + \frac{x}{\sqrt{N}}\right)^N.$$

$f(x)$ is maximum at $x = 0$ and falls to zero on both sides of the maximum. Write $f(x) = e^{\ln f(x)}$, and expand $\ln f(x)$ around $x = 0$. Show that

$$N! = \sqrt{N} N^N e^{-N} \int_{-\sqrt{N}}^\infty \exp\left(-\frac{x^2}{2} + \frac{x^3}{3\sqrt{N}} - \dots\right).$$

For $N \gg 1$, keep only the leading order in the expansion. For $x < -\sqrt{N}$, $e^{-x^2/2}$ is exceedingly small, and the lower limit of integration may be pushed to $-\infty$. Show that

$$N! \simeq \sqrt{2\pi N} N^N e^{-N}.$$

Hence, prove Stirling's formula, valid for $N \gg 1$,

$$\ln N! \simeq N \ln - N.$$

5.2 *Vacancies and interstitials in graphene*. In graphene, assume that it costs energy ϵ to form a vacancy–interstitial pair. The pair is obtained by removing a carbon atom from a lattice site (the site becomes a vacancy) and placing it at the center of a hexagon (an interstitial site). The total number of carbon atoms is N, and the total energy is $E = M\epsilon$, where M is the number of vacancies (M

is also the number of interstitials). Also assume that N and M are much, much greater than 1. Determine (a) the system's entropy and (b) E as a function of temperature T.

5.3 *Magnetic susceptibility.* Consider a crystal where each atom has spin 1/2 and magnetic moment μ. The number of atoms per unit volume is n. A magnetic field **B** is applied. Choose one atom as the small system and treat the rest of the crystal as a heat reservoir at temperature T.
(a) Show that the magnetization (the mean magnetic moment per unit volume) of the crystal is $M = n\mu \tan h(\mu B/kT)$.
(b) For the case $\mu B/kT \ll 1$, show that the magnetic susceptibility, defined as $\chi = \partial M/\partial B$, varies as $1/T$ (Curie's law).

5.4 *Entropy and probabilty.* For a given macrostate of a system of n identical particles, let p_n be the probability that the system is in state $|\psi_n\rangle$. Using the relation $F = E - TS = -kT \ln Z$, show that the system's entropy is given by $S = -k\sum_n p_n \ln p_n$.

5.5 *Statistical operator.* Show that $\text{Tr}(\rho^2) \leq 1$.

5.6 *Ising model in one dimension.* The one-dimensional Ising model describes localized particles on a line, where each particle carries a spin s. The Hamiltonian is

$$H = -J\sum_{i=1}^{N} s_i s_{i+1} - h\sum_{i=1}^{N} s_i$$

where $-J$ represents the strength of the interaction between neighboring particles ($J > 0$) and h is proportional to a constant applied magnetic field. Assume periodic boundary conditions: $s_{N+1} = s_1$. The model also assumes that $s_i = 1$ or -1.
(a) Show that the partition function is given by $Z = \text{Tr}[T^N]$, where T is a real, symmetric matrix, given by

$$T = \begin{pmatrix} e^{\beta(J+h)} & e^{-\beta J} \\ e^{-\beta J} & e^{\beta(J-h)} \end{pmatrix}.$$

(b) By diagonalizing T, show that, as $N \to \infty$,

$$Z = -NJ - NkT \ln\left[\cosh(\beta h) + \sqrt{\sinh^2(\beta h) + e^{-4\beta J}}\right].$$

(c) Show that the mean magnetic moment, $\langle\sum_i s_i\rangle/N$, vanishes as $h \to 0$.

6

Real-time Green's and correlation functions

> Facts do not 'speak for themselves', they are read in the light of theory.
>
> –*Stephen Jay Gould*

A many-particle system is intrinsically quite complex. Its energy level spectrum is almost continuous, and the eigenfunctions that correspond to those energy levels are complicated functions of the particles' coordinates. The detailed form of its energy spectrum and wave functions is neither exactly calculable nor measurable; hence, we shall not be concerned with it.

In a typical experimental measurement that involves a many-particle system, a system in equilibrium is weakly perturbed in one or more ways: a particle may be added or removed, a weak electromagnetic field may be applied, a beam of electrons or neutrons may strike the system, a thermal gradient may be established across the system, and so on. Rather than attempting to calculate the full spectrum of a many-particle system, it is more useful to concentrate on understanding how a system responds to such external perturbations. The method of Green's function serves this purpose well. In this chapter, we focus on real-time functions for systems in equilibrium. Imaginary-time functions will be introduced in Chapter 8. For systems out of equilibrium, such as those featuring a metallic island between two metal electrodes and an applied bias voltage that causes current to flow through the island, another formalism, that of the nonequililbrium Green's function, is needed; it will be discussed in Chapter 13.

There are various types of real-time Green's and correlation functions. In this chapter, we develop the theory of these functions, with particular emphasis on retarded functions (the most useful). As we will see later, retarded correlation functions determine the response of a system to external probes, such as electromagnetic fields, electrons, or neutrons, and thus are directly related to experimentally measured quantities.

6.1 A plethora of functions

Consider a system of interacting particles, with a time-independent Hamiltonian H, at temperature T. It is convenient to allow the number of particles to vary, i.e., the system can exchange energy and particles with a reservoir. The system is a member of a grand canonical ensemble (see Section 5.3). Consider an operator $A(c, c^\dagger)$, represented in terms of fermion or boson annihilation and creation operators c and $c^\dagger$, acting upon the state space of the system. We define a modified Heisenberg picture operator $A(t)$ as follows:

$$A(t) = e^{i\bar{H}t/\hbar} A e^{-i\bar{H}t/\hbar}, \tag{6.1}$$

where

$$\bar{H} = H - \mu N, \tag{6.2}$$

N is the number of particles operator, and μ is the chemical potential. The standard Heisenberg picture is obtained if $\bar{H} \to H$.

6.1.1 *Correlation functions*

Considering now any two operators A and B, we define the real-time causal, or time-ordered, correlation function by

$$C^T_{AB}(t, t') = -i\langle T A(t) B(t')\rangle \tag{6.3}$$

where $\langle \cdots \rangle$ stands for the grand canonical ensemble average,

$$\langle \cdots \rangle = Z_G^{-1} \mathrm{Tr}\left[e^{-\beta\bar{H}} \cdots\right].$$

Here, $\beta = 1/kT$ (k is Boltzmann's constant), $Z_G = \mathrm{Tr}[e^{-\beta\bar{H}}]$ is the grand canonical partition function, and Tr stands for the trace. In Eq. (6.3), T is the time-ordering operator, sometimes written as T_t in order to distinguish it from the temperature. Acting on a product of operators, T orders them in increasing time order from right to left, and, in the process, introduces a minus sign every time two fermion operators are interchanged. Thus,

$$T A(t) B(t') = \begin{cases} A(t)B(t') & t > t' \\ \pm B(t')A(t) & t < t'. \end{cases} \tag{6.4}$$

The lower (upper) sign refers to the case when A and B are fermion (boson) operators. Note that the operator $A(c, c^\dagger)$ is bosonic if c and $c^\dagger$ are boson operators, or if it is of even order in c and $c^\dagger$; e.g., if c and $c^\dagger$ are fermion operators, an operator such as $c^\dagger c$ is considered a boson operator.

The retarded correlation function C^R is defined as follows. If A and B are fermion operators, then

$$C^R_{AB}(t,t') = -i\theta(t-t')\langle\{A(t), B(t')\}\rangle \tag{6.5}$$

where $\{A, B\} = AB + BA$ is the anticommutator of A and B, and $\theta(t-t')$ is the step function,

$$\theta(t-t') = \begin{cases} 0 & t < t' \\ 1 & t > t'. \end{cases} \tag{6.6}$$

$C^R_{AB}(t,t')$ is nonzero only if $t > t'$; hence the name "retarded." For the case when A and B are bosonic operators,

$$C^R_{AB}(t,t') = -i\theta(t-t')\langle[A(t), B(t')]\rangle \tag{6.7}$$

where $[A, B] = AB - BA$ is the commutator of A and B. We may combine both cases and write

$$C^R_{AB}(t,t') = -i\theta(t-t')\langle[A(t), B(t')]_{\mp}\rangle. \tag{6.8}$$

The lower (upper) sign refers to fermions (bosons). Similarly, we define the advanced correlation function, which is nonvanishing only if $t < t'$, as follows:

$$C^A_{AB}(t,t') = +i\theta(t'-t)\langle[A(t), B(t')]_{\mp}\rangle. \tag{6.9}$$

As we will see later, athough the different functions introduced above have different analytic properties, they are in fact closely related. Finally, we define one more correlation function, without a label,

$$C_{AB}(t,t') = \langle A(t)B(t')\rangle. \tag{6.10}$$

This function also turns out to be important in analyzing experimental data.

6.1.2 Time dependence

A general property of correlation functions is that, if the Hamiltonian is time-independent, they depend on $t - t'$ and not on t and t' independently. We prove this assertion for the retarded correlation function; similar proofs can be worked out for all other correlation functions. Consider

$$C^R_{AB} = -i\theta(t-t')Z_G^{-1}\left\{\mathrm{Tr}\left[e^{-\beta\bar{H}}e^{i\bar{H}t/\hbar}Ae^{-i\bar{H}(t-t')/\hbar}Be^{-i\bar{H}t'/\hbar}\right]\right.$$
$$\left.\mp\mathrm{Tr}\left[e^{-\beta\bar{H}}e^{i\bar{H}t'/\hbar}Be^{i\bar{H}(t-t')/\hbar}Ae^{-i\bar{H}t/\hbar}\right]\right\}.$$

In the second term on the RHS the lower (upper) sign refers to fermions (bosons). The time-independence of $\bar{H}$ makes possible the replacement of $e^{i\bar{H}t/\hbar}e^{-i\bar{H}t'/\hbar}$ with $e^{i\bar{H}(t-t')/\hbar}$. The trace is invariant under cyclic permutations: $\text{Tr}[AB\cdots CD] = \text{Tr}[DAB\cdots C]$; hence, we can move $e^{-i\bar{H}t'/\hbar}$ in the first term ($e^{-i\bar{H}t/\hbar}$ in the second term) to the leftmost position, and using the fact that $e^{-i\bar{H}t'/\hbar}$ (or $e^{-i\bar{H}t/\hbar}$) commutes with $e^{-\beta\bar{H}}$, we can write

$$C^R_{AB} = -i\theta(t-t')Z_G^{-1}\left\{\text{Tr}\left[e^{-\beta\bar{H}}e^{i\bar{H}(t-t')/\hbar}Ae^{-i\bar{H}(t-t')/\hbar}B\right]\right.$$
$$\left.\mp\text{Tr}\left[e^{-\beta\bar{H}}e^{-i\bar{H}(t-t')/\hbar}Be^{i\bar{H}(t-t')/\hbar}A\right]\right\}.$$

The above expression shows that C^R_{AB} is a function of $t-t'$; consequently, we may set $t'=0$ and consider C^R_{AB} to be a function of t. The same conclusion applies to C^T_{AB}, C^A_{AB}, and C_{AB}.

6.1.3 Single-particle Green's functions

An important special case of the correlation function is when $A = \Psi_\sigma(\mathbf{r})$ and $B = \Psi^\dagger_\sigma(\mathbf{r})$, where $\Psi_\sigma(\mathbf{r})$ ($\Psi^\dagger_\sigma(\mathbf{r})$) is the field operator that annihilates (creates) a particle with spin projection σ at position $\mathbf{r}$ (see Section 3.11). In this case, the causal, retarded, and advanced correlation functions are known as the single-particle real-time Green's functions, or simply real-time Green's functions. They are given by

$$G(\mathbf{r}\sigma t, \mathbf{r}'\sigma' t') = -i\langle T\Psi_\sigma(\mathbf{r}\,t)\Psi^\dagger_{\sigma'}(\mathbf{r}'\,t')\rangle \qquad \text{(causal)} \tag{6.11}$$

$$G^R(\mathbf{r}\sigma t, \mathbf{r}'\sigma' t') = -i\theta(t-t')\langle[\Psi_\sigma(\mathbf{r}\,t), \Psi^\dagger_{\sigma'}(\mathbf{r}'\,t')]_\mp\rangle \qquad \text{(retarded)} \tag{6.12}$$

$$G^A(\mathbf{r}\sigma t, \mathbf{r}'\sigma' t') = i\theta(t'-t)\langle[\Psi_\sigma(\mathbf{r}\,t), \Psi^\dagger_{\sigma'}(\mathbf{r}'\,t')]_\mp\rangle \qquad \text{(advanced).} \tag{6.13}$$

The lower (upper) sign refers to fermions (bosons). At this point we introduce two other single-particle functions that play an important role in the study of transport. The greater and lesser functions are defined by

$$G^>(\mathbf{r}\sigma t, \mathbf{r}'\sigma' t') = -i\langle\Psi_\sigma(\mathbf{r}\,t)\Psi^\dagger_{\sigma'}(\mathbf{r}'\,t')\rangle \qquad \text{(greater)} \tag{6.14}$$

$$G^<(\mathbf{r}\sigma t, \mathbf{r}'\sigma' t') = \mp i\langle\Psi^\dagger_{\sigma'}(\mathbf{r}'\,t')\Psi_\sigma(\mathbf{r}\,t)\rangle \qquad \text{(lesser).} \tag{6.15}$$

We note in passing that the ensemble average of the local particle number density can be expressed in terms of the lesser function,

$$\langle n_\sigma(\mathbf{r},t)\rangle = \langle\Psi^\dagger_\sigma(\mathbf{r}t)\Psi_\sigma(\mathbf{r}\,t)\rangle = \pm iG^<(\mathbf{r}\sigma t, \mathbf{r}\sigma t). \tag{6.16}$$

The single-particle correlation function is defined by

$$C(\mathbf{r}\sigma t, \mathbf{r}'\sigma' t') = \langle\Psi_\sigma(\mathbf{r}\,t)\Psi^\dagger_{\sigma'}(\mathbf{r}'\,t')\rangle; \tag{6.17}$$

it is simply $iG^>(\mathbf{r}\sigma t, \mathbf{r}'\sigma' t')$. The above definitions can be generalized: for any complete set $|\phi_1\rangle, |\phi_2\rangle, \ldots$ of single-particle states, Green's functions may be defined in terms of the corresponding annihilation and creation operators. The retarded Green's function, in the ϕ_ν-representation, is defined by

$$G^R(\nu t, \nu' t') = -i\theta(t - t')\langle[c_\nu(t), c^\dagger_{\nu'}(t')]_\mp\rangle.$$

Similar definitions can be made for the causal, advanced, greater, lesser, and correlation functions. For example, consider an atom with single-particle states $|\phi_{nlm\sigma}\rangle$, where n, l, m, and σ are, respectively, the principal, orbital, magnetic, and spin quantum numbers. We may define a retarded Green's function for electrons in this atom as

$$G^R(nlm\sigma t, n'l'm'\sigma' t') = -i\theta(t - t')\langle\{c_{nlm\sigma}(t), c^\dagger_{n'l'm'\sigma'}(t')\}\rangle.$$

6.2 Physical meaning of Green's functions

Let us consider the causal Green's function and assume that $t > t'$,

$$iG(\mathbf{r}\sigma t, \mathbf{r}'\sigma' t') = \langle\Psi_\sigma(\mathbf{r}\, t)\Psi^\dagger_{\sigma'}(\mathbf{r}'\, t')\rangle = Z_G^{-1}\mathrm{Tr}\left[e^{-\beta\bar{H}}\Psi_\sigma(\mathbf{r}\, t)\Psi^\dagger_{\sigma'}(\mathbf{r}'\, t')\right].$$

Given a complete set of states $|n\rangle$,

$$\mathrm{Tr}[\cdots] = \sum_n \langle n|\cdots|n\rangle.$$

Taking the states $|n\rangle$ to be eigenstates of $\bar{H}$ ($\bar{H}|n\rangle = \bar{E}_n|n\rangle$),

$$iG(\mathbf{r}\sigma t, \mathbf{r}'\sigma' t') = Z_G^{-1}\sum_n e^{-\beta\bar{E}_n}\langle n|e^{i\bar{H}t/\hbar}\Psi_\sigma(\mathbf{r})e^{-i\bar{H}(t-t')/\hbar}\Psi^\dagger_{\sigma'}(\mathbf{r}')e^{-i\bar{H}t'/\hbar}|n\rangle. \tag{6.18}$$

Defining the states $|\alpha\rangle$ and $|\beta\rangle$ by

$$|\alpha\rangle = e^{-i\bar{H}(t-t')/\hbar}\Psi^\dagger_{\sigma'}(\mathbf{r}')e^{-i\bar{H}t'/\hbar}|n\rangle, \qquad |\beta\rangle = \Psi^\dagger_\sigma(\mathbf{r})e^{-i\bar{H}t/\hbar}|n\rangle,$$

the matrix element in Eq. (6.18) may be written as

$$\langle n|\cdots|n\rangle = \langle\beta|\alpha\rangle;$$

it is the probability amplitude for a system in state $|\alpha\rangle$ to be found in state $|\beta\rangle$. Let us look more closely at state $|\alpha\rangle$. Starting from state $|n\rangle$ at $t = 0$, $e^{-i\bar{H}t'/\hbar}|n\rangle$ is the state after it has evolved to time t'. At this time, $\Psi^\dagger_{\sigma'}(\mathbf{r}')$ injects a particle with spin projection σ' into the system at position $\mathbf{r}'$, and the operator $e^{-i\bar{H}(t-t')/\hbar}$ carries the system to time t. Thus, $|\alpha\rangle$ is the state of the system at time t if a particle with coordinates ($\mathbf{r}'\sigma'$) was added to it at an earlier time t'. Similarly, $|\beta\rangle$ is the state of the system when it has an extra particle with coordinates ($\mathbf{r}\sigma$), added at time t. The

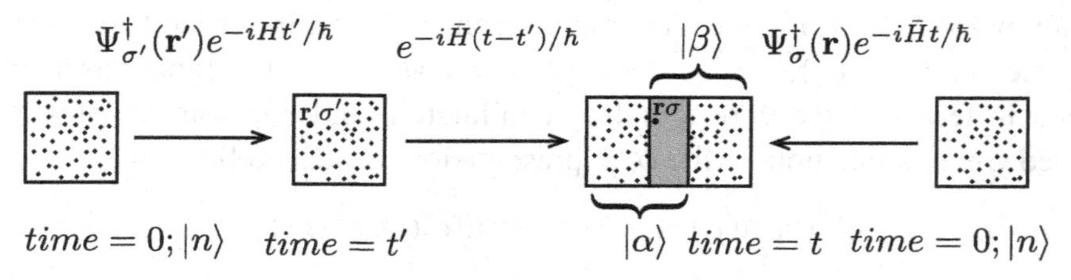

Figure 6.1 Definition of the causal Green's function in terms of an overlap of states $|\alpha\rangle$ and $|\beta\rangle$. State $|\alpha\rangle$ is obtained by letting state $|n\rangle$ evolve to time t', adding a particle with coordinates $(\mathbf{r}'\sigma')$, and allowing the system to evolve to time t. State $|\beta\rangle$ is obtained by letting $|n\rangle$ evolve and adding to it, at time t, a particle with spin projection σ at position $\mathbf{r}$.

matrix element $\langle n| \cdots |n\rangle$ is thus the probability amplitude of finding the system with an extra particle with coordinates $(\mathbf{r}\sigma)$ at time t if a particle with coordinates $(\mathbf{r}'\sigma')$ was injected at an earlier time t'. Loosely speaking, it is the probability amplitude for an added particle to propagate from $(\mathbf{r}'\sigma' t')$ to $(\mathbf{r}\sigma t)$, though we should be careful to note that, since the particles are indistinguishable, it is not meaningful to think of the particle with coordinates $(\mathbf{r}\sigma)$ as the same particle with spin projection σ' that was injected earlier at $\mathbf{r}'$. Since different states $|n\rangle$ occur with probabilities $Z_G^{-1}e^{-\beta\bar{E}_n}$, $iG(\mathbf{r}\sigma t, \mathbf{r}'\sigma' t')$ represents the ensemble average of the aforementioned propagation amplitude. For $t < t'$, $iG(\mathbf{r}\sigma t, \mathbf{r}'\sigma' t')$ represents the ensemble average of the propagation amplitude of a hole from $(\mathbf{r}\sigma t)$ to $(\mathbf{r}'\sigma' t')$. Similar meanings can be attached to the other Green's functions. The definition of $iG(\mathbf{r}\sigma t, \mathbf{r}'\sigma' t')$ is depicted pictorially in Figure 6.1.

The above discussion indicates that single-particle Green's functions are important tools in analyzing experiments where particles are added to or removed from a system. Examples include experiments that involve tunneling of electrons between two systems, at different chemical potentials, which are placed in contact with each other; and optical experiments whereby a photon removes an electron from a solid. Another example is the introduction of magnetic impurities into a host metal; tunneling takes place there between localized states on the impurity sites and the delocalized host states. We will explore some applications of Green's functions in the next chapter.

6.3 Spin-independent Hamiltonian, translational invariance

Let us assume that the Hamiltonian is spin-independent, i.e., that the system under consideration is nonmagnetic. Suppose that a particle with spin projection σ' is injected into the system at time t'. In the absence of any interactions that could flip the spin of a particle, we cannot expect that, at a later time t, the system will have an extra particle with a spin projection $\sigma \neq \sigma'$. The retarded Green's function thus vanishes unless $\sigma = \sigma'$,

$$G^R(\mathbf{r}\sigma t, \mathbf{r}'\sigma' t') = \delta_{\sigma\sigma'} G^R(\mathbf{r}\sigma t, \mathbf{r}'\sigma' t').$$

The same conclusion applies to all the correlation functions that we have introduced. We also note that, since $\bar{H}$ is assumed to be time-independent, G^R, being a special case of the more general retarded correlation functions, depends on $t - t'$, and not on t and t' separately, as was shown in the preceding section. This being the case, we may set $t' = 0$, and consider G^R to be a function of t.

Generally, we will be dealing with translationally invariant systems, where any function $f(\mathbf{r}, \mathbf{r}')$ of the positions of two particles (e.g., the interaction energy between two particles at positions $\mathbf{r}$ and $\mathbf{r}'$) does not change if both $\mathbf{r}$ and $\mathbf{r}'$ are shifted simultaneously by any vector $\mathbf{R}$. It follows that $f(\mathbf{r}, \mathbf{r}')$ depends only on $\mathbf{r} - \mathbf{r}'$, and not on $\mathbf{r}$ and $\mathbf{r}'$ separately. In particular, the single-particle Green's functions are functions of $\mathbf{r} - \mathbf{r}'$. The proof of this statement is the subject of Problem 6.2. Hence, in a translationally invariant system, with a time- and spin-independent Hamiltonian, we write the retarded Green's function as

$$G^R(\mathbf{r} - \mathbf{r}'\sigma, t) = -i\theta(t)\langle[\Psi_\sigma(\mathbf{r}t), \Psi^\dagger_\sigma(\mathbf{r}'0)]_\mp\rangle, \tag{6.19}$$

and do so similarly for the other Green's and correlation functions.

Considering a complete set of single-particle momentum states $|\mathbf{k}\sigma\rangle$ for which $\phi_{\mathbf{k}\sigma} = V^{-1/2}e^{i\mathbf{k}.\mathbf{r}}|\sigma\rangle$, V being the system's volume, the field operators are given by

$$\Psi_\sigma(\mathbf{r}t) = \frac{1}{\sqrt{V}}\sum_{\mathbf{k}} e^{i\mathbf{k}.\mathbf{r}} c_{\mathbf{k}\sigma}(t), \quad \Psi^\dagger_\sigma(\mathbf{r}, t) = \frac{1}{\sqrt{V}}\sum_{\mathbf{k}} e^{-i\mathbf{k}.\mathbf{r}} c^\dagger_{\mathbf{k}\sigma}(t) \tag{6.20}$$

(see Section 3.11). The operator $c^\dagger_{\mathbf{k}\sigma}(c_{\mathbf{k}\sigma})$ creates (annihilates) a particle in state $|\mathbf{k}\sigma\rangle$. Inserting the above relations into Eq. (6.19), we obtain

$$\begin{aligned} G^R(\mathbf{r} - \mathbf{r}'\sigma, t) &= -i\theta(t)\frac{1}{V}\sum_{\mathbf{k}\mathbf{k}'} e^{i\mathbf{k}.\mathbf{r}} e^{-i\mathbf{k}'.\mathbf{r}'}\langle[c_{\mathbf{k}\sigma}(t), c^\dagger_{\mathbf{k}'\sigma}(0)]_\mp\rangle \\ &= -i\theta(t)\frac{1}{V}\sum_{\mathbf{k}\mathbf{k}'} e^{i\mathbf{k}.(\mathbf{r}-\mathbf{r}')} e^{i(\mathbf{k}-\mathbf{k}').\mathbf{r}'}\langle[c_{\mathbf{k}\sigma}(t), c^\dagger_{\mathbf{k}'\sigma}(0)]_\mp\rangle. \end{aligned}$$

Since G^R depends only on $\mathbf{r} - \mathbf{r}'$, it does not change if $\mathbf{r}$ and $\mathbf{r}'$ are shifted simultaneously to $\mathbf{r} + \mathbf{R}$ and $\mathbf{r}' + \mathbf{R}$, for any vector $\mathbf{R}$; hence,

$$G^R(\mathbf{r} - \mathbf{r}'\sigma, t) = -i\theta(t)\frac{1}{V}\sum_{\mathbf{k}\mathbf{k}'} e^{i\mathbf{k}.(\mathbf{r}+\mathbf{R}-\mathbf{r}'-\mathbf{R})} e^{i(\mathbf{k}-\mathbf{k}').(\mathbf{r}'+\mathbf{R})}\langle[c_{\mathbf{k}\sigma}(t), c^\dagger_{\mathbf{k}'\sigma}(0)]_\mp\rangle.$$

Comparing the above two expressions for G^R, we find

$$e^{i(\mathbf{k}-\mathbf{k}').\mathbf{R}} = 1, \quad \forall \mathbf{R} \in \mathbb{R}^3.$$

This is satisfied only if $\mathbf{k}' = \mathbf{k}$. Therefore,

$$\begin{aligned} G^R(\mathbf{r} - \mathbf{r}'\sigma, t) &= -i\theta(t)\frac{1}{V}\sum_{\mathbf{k}} e^{i\mathbf{k}.(\mathbf{r}-\mathbf{r}')}\langle[c_{\mathbf{k}\sigma}(t), c^\dagger_{\mathbf{k}\sigma}(0)]_\mp\rangle \\ &= \frac{1}{V}\sum_{\mathbf{k}} e^{i\mathbf{k}.(\mathbf{r}-\mathbf{r}')}G^R(\mathbf{k}\sigma, t). \end{aligned} \tag{6.21}$$

We have introduced $G^R(\mathbf{k}\sigma, t)$, the spatial Fourier transform of $G^R(\mathbf{r} - \mathbf{r}'\sigma, t)$,

$$G^R(\mathbf{k}\sigma, t) = -i\theta(t)\langle[c_{\mathbf{k}\sigma}(t), c^\dagger_{\mathbf{k}\sigma}(0)]_\mp\rangle. \tag{6.22}$$

Similarly,

$$G^A(\mathbf{k}\sigma, t) = i\theta(-t)\langle[c_{\mathbf{k}\sigma}(t), c^\dagger_{\mathbf{k}\sigma}(0)]_\mp\rangle \tag{6.23}$$

$$G(\mathbf{k}\sigma, t) = -i\langle T c_{\mathbf{k}\sigma}(t)c^\dagger_{\mathbf{k}\sigma}(0)\rangle \tag{6.24}$$

$$G^>(\mathbf{k}\sigma, t) = -i\langle c_{\mathbf{k}\sigma}(t)c^\dagger_{\mathbf{k}\sigma}(0)\rangle \tag{6.25}$$

$$G^<(\mathbf{k}\sigma, t) = \mp i\langle c^\dagger_{\mathbf{k}\sigma}(0)c_{\mathbf{k}\sigma}(t)\rangle \tag{6.26}$$

and

$$C(\mathbf{k}\sigma, t) = \langle c_{\mathbf{k}\sigma}(t)c^\dagger_{\mathbf{k}\sigma}(0)\rangle. \tag{6.27}$$

6.4 Spectral representation

What does a spectral representation mean? And why is it useful? To get an idea of what a spectral representation is, consider a system with Hamiltonian H and orthonormal eigenkets $|n\rangle$: $H|n\rangle = E_n|n\rangle$. Introducing two resolutions of identity ($1 = \sum |n\rangle\langle n|$), we write H as

$$H = \sum_{nm} |n\rangle\langle n|H|m\rangle\langle m| = \sum_{nm} E_n\delta_{nm}|n\rangle\langle m| = \sum_n E_n|n\rangle\langle n|.$$

The expression on the RHS forms a spectral representation of H, in the sense that H is written in terms of its spectrum of energy levels and eigenstates. We will follow a similar procedure in deriving the spectral representation of Green's functions: we shall introduce resolutions of identity and express the functions in terms of the exact energy spectrum and eigenstates of the system. An answer to the second question will unfold in later chapters. For now, it suffices to note that merely expressing Green's function in terms of the exact eigenstates of the system is not, in and of itself, a worthwhile goal. After all, the exact eigenstates are not known; if they were, the problem would be completely solved. At finite temperature, real-time Green's (or correlation) functions are not amenable to a treatment of interacting systems by means of perturbation theory. The burden will fall upon the imaginary-time Green's function, which will be discussed in Chapter 8. Nevertheless, experiments are carried out in real time, and their interpretation requires knowledge of real-time Green's functions. How do we find the real-time Green's function if we know the imaginary-time function? By studying the spectral representations of these functions, we will arrive at a simple method for obtaining the real-time Green's function from its imaginary-time counterpart. With this in mind, let us next proceed to determine the spectral representations of $G^R(\mathbf{k}\sigma, t)$, $C(\mathbf{k}\sigma, t)$, and, more generally, $C^R_{AB}(t)$ and $C_{AB}(t)$.

6.4.1 Retarded and advanced Green's functions

The retarded Green's function is given by

$$G^R(\mathbf{k}\sigma, t) = -i\theta(t)\langle c_{\mathbf{k}\sigma}(t)c^\dagger_{\mathbf{k}\sigma}(0)\rangle \pm i\theta(t)\langle c^\dagger_{\mathbf{k}\sigma}(0)c_{\mathbf{k}\sigma}(t)\rangle = \mathcal{A} \mp \mathcal{B}. \tag{6.28}$$

The lower (upper) sign refers to fermions (bosons). Consider the first term,

$$\langle c_{\mathbf{k}\sigma}(t)c^\dagger_{\mathbf{k}\sigma}(0)\rangle = Z_G^{-1}\mathrm{Tr}\left[e^{-\beta\bar{H}}e^{i\bar{H}t/\hbar}c_{\mathbf{k}\sigma}e^{-i\bar{H}t/\hbar}c^\dagger_{\mathbf{k}\sigma}\right] \tag{6.29}$$

where $c_{\mathbf{k}\sigma} = c_{\mathbf{k}\sigma}(t=0)$ and $c^\dagger_{\mathbf{k}\sigma} = c^\dagger_{\mathbf{k}\sigma}(t=0)$. The trace of an operator is the sum of its diagonal elements,

$$\begin{aligned}
\langle c_{\mathbf{k}\sigma}(t)c^\dagger_{\mathbf{k}\sigma}(0)\rangle &= Z_G^{-1}\sum_n \langle n|e^{-\beta\bar{H}}e^{i\bar{H}t/\hbar}c_{\mathbf{k}\sigma}e^{-i\bar{H}t/\hbar}c^\dagger_{\mathbf{k}\sigma}|n\rangle \\
&= Z_G^{-1}\sum_{nm} \langle n|e^{-\beta\bar{H}}e^{i\bar{H}t/\hbar}c_{\mathbf{k}\sigma}|m\rangle\langle m|e^{-i\bar{H}t/\hbar}c^\dagger_{\mathbf{k}\sigma}|n\rangle \\
&= Z_G^{-1}\sum_{nm} e^{-\beta\bar{E}_n}e^{-i(\bar{E}_m-\bar{E}_n)t/\hbar}\langle n|c_{\mathbf{k}\sigma}|m\rangle\langle m|c^\dagger_{\mathbf{k}\sigma}|n\rangle \\
&= -\int_{-\infty}^{\infty} P(\mathbf{k}\sigma, \epsilon)e^{-i\epsilon t}\frac{d\epsilon}{2\pi}.
\end{aligned} \tag{6.30}$$

Here $\bar{E}_i = E_i - \mu N_i$, where N_i is the number of particles in state $|i\rangle$, and

$$P(\mathbf{k}\sigma, \epsilon) = -2\pi Z_G^{-1}\sum_{nm} e^{-\beta\bar{E}_n}\left|\langle m|c^\dagger_{\mathbf{k}\sigma}(0)|n\rangle\right|^2\delta\left(\epsilon - (\bar{E}_m - \bar{E}_n)/\hbar\right) \tag{6.31}$$

is a spectral function (we will call it the P-spectral function). Therefore,

$$\mathcal{A} = i\theta(t)\int_{-\infty}^{\infty} P(\mathbf{k}\sigma, \epsilon)e^{-i\epsilon t}\frac{d\epsilon}{2\pi}.$$

Next, we consider the second term in Eq. (6.28),

$$\langle c_{\mathbf{k}\sigma}^{\dagger}(0)c_{\mathbf{k}\sigma}(t)\rangle = Z_G^{-1}\mathrm{Tr}\left[e^{-\beta\bar{H}}c_{\mathbf{k}\sigma}^{\dagger}e^{i\bar{H}t/\hbar}c_{\mathbf{k}\sigma}e^{-i\bar{H}t/\hbar}\right].$$

Using the invariance property of the trace under cyclic permutations, we first move $e^{-i\bar{H}t/\hbar}$ to the leftmost position, then move $c_{\mathbf{k}\sigma}(0)$ to the leftmost position, and finally move $e^{i\bar{H}t/\hbar}$ to the leftmost position; the result is

$$\langle c_{\mathbf{k}\sigma}^{\dagger}(0)c_{\mathbf{k}\sigma}(t)\rangle = Z_G^{-1}\mathrm{Tr}\left[e^{i\bar{H}t/\hbar}c_{\mathbf{k}\sigma}e^{-i\bar{H}(t-i\beta\hbar)/\hbar}c_{\mathbf{k}\sigma}^{\dagger}\right].$$

Introducing $1 = e^{-\beta\bar{H}}e^{\beta\bar{H}}$ at the leftmost position, we obtain

$$\langle c_{\mathbf{k}\sigma}^{\dagger}(0)c_{\mathbf{k}\sigma}(t)\rangle = Z_G^{-1}\mathrm{Tr}\left[e^{-\beta\bar{H}}e^{i\bar{H}(t-i\beta\hbar)/\hbar}c_{\mathbf{k}\sigma}e^{-i\bar{H}(t-i\beta\hbar)/\hbar}c_{\mathbf{k}\sigma}^{\dagger}\right].$$

The RHS of the above equation is the same as the RHS of Eq. (6.29) with $t \to t - i\beta\hbar$; hence,

$$\langle c_{\mathbf{k}\sigma}^{\dagger}(0)c_{\mathbf{k}\sigma}(t)\rangle = -\int_{-\infty}^{\infty} P(\mathbf{k}\sigma, \epsilon)e^{-i\epsilon(t-i\beta\hbar)}\frac{d\epsilon}{2\pi} = -\int_{-\infty}^{\infty} e^{-\beta\hbar\epsilon}P(\mathbf{k}\sigma, \epsilon)e^{-i\epsilon t}\frac{d\epsilon}{2\pi}. \tag{6.32}$$

The expression for $\mathcal{B}$ in Eq. (6.28) is thus obtained. The Fourier transform of $G^R(\mathbf{k}\sigma, t)$ is given by

$$G^R(\mathbf{k}\sigma, \omega) = \int_{-\infty}^{\infty} e^{i\omega t}G^R(\mathbf{k}\sigma, t)dt. \tag{6.33}$$

Noting that $G^R(\mathbf{k}\sigma, t)$ vanishes for $t < 0$, we can write

$$\begin{aligned} G^R(\mathbf{k}\sigma, \omega) &= \int_0^{\infty} e^{i\omega t}G^R(\mathbf{k}\sigma, t)dt = \int_0^{\infty} e^{i\omega t}(\mathcal{A} \mp \mathcal{B})dt \\ &= i\int_{-\infty}^{\infty} P(\mathbf{k}\sigma, \epsilon)(1 \mp e^{-\beta\hbar\epsilon})\frac{d\epsilon}{2\pi}\int_0^{\infty} e^{i(\omega-\epsilon)t}dt. \end{aligned}$$

The integral over t is oscillatory at infinity; we evaluate it as follows:

$$\begin{aligned} \int_0^{\infty} e^{i(\omega-\epsilon)t}dt &= \lim_{\eta\to 0^+}\int_0^{\infty} e^{i(\omega-\epsilon+i\eta)t}dt = \lim_{\eta\to 0^+}\left.\frac{e^{i(\omega-\epsilon+i\eta)t}}{i(\omega-\epsilon+i\eta)}\right|_0^{\infty} \\ &= \lim_{\eta\to 0^+}\frac{-1}{i(\omega-\epsilon+i\eta)}. \end{aligned} \tag{6.34}$$

Introducing the spectral density function $A(\mathbf{k}\sigma, \epsilon)$, defined by

$$A(\mathbf{k}\sigma, \epsilon) = -P(\mathbf{k}\sigma, \epsilon)(1 \mp e^{-\beta\hbar\epsilon})$$
$$= 2\pi Z_G^{-1} \sum_{nm} e^{-\beta\bar{E}_n} \left|\langle m|c^\dagger_{\mathbf{k}\sigma}|n\rangle\right|^2 (1 \mp e^{-\beta\hbar\epsilon})\delta\left(\epsilon - (\bar{E}_m - \bar{E}_n)/\hbar\right), \quad (6.35)$$

the spectral representation of $G^R(\mathbf{k}\sigma, \omega)$ reduces to

$$G^R(\mathbf{k}\sigma, \omega) = \lim_{\eta\to 0^+} \int_{-\infty}^{\infty} \frac{A(\mathbf{k}\sigma, \epsilon)}{\omega - \epsilon + i\eta} \frac{d\epsilon}{2\pi}. \quad (6.36)$$

A similar derivation yields the spectral representation of the advanced Green's function $G^A(\mathbf{k}\sigma, \omega)$,

$$G^A(\mathbf{k}\sigma, \omega) = \lim_{\eta\to 0^+} \int_{-\infty}^{\infty} \frac{A(\mathbf{k}\sigma, \epsilon)}{\omega - \epsilon - i\eta} \frac{d\epsilon}{2\pi}. \quad (6.37)$$

The derivation of the above result, as well as the spectral representation of the causal Green's function, is relegated to the Problems section.

6.4.2 Single-particle correlation function

Turning now to the correlation function, the same initial steps as above yield

$$C(\mathbf{k}\sigma, t) = -\int_{-\infty}^{\infty} P(\mathbf{k}\sigma, \epsilon) e^{-i\epsilon t} \frac{d\epsilon}{2\pi}.$$

The Fourier transform is given by

$$C(\mathbf{k}\sigma, \omega) = \int_{-\infty}^{\infty} e^{i\omega t} C(\mathbf{k}\sigma, t)dt = -\int_{-\infty}^{\infty} P(\mathbf{k}\sigma, \epsilon) \frac{d\epsilon}{2\pi} \int_{-\infty}^{\infty} e^{i(\omega-\epsilon)t} dt. \quad (6.38)$$

The integral over t is straightforward (see Eq. [1.15])

$$\int_{-\infty}^{\infty} e^{i(\omega-\epsilon)t} dt = 2\pi\delta(\omega - \epsilon). \quad (6.39)$$

Substituting this into Eq. (6.38), we find

$$C(\mathbf{k}\sigma, \omega) = -P(\mathbf{k}\sigma, \omega). \quad (6.40)$$

We can establish a relationship between $G^R(\mathbf{k}\sigma, \omega)$ and $C(\mathbf{k}\sigma, \omega)$. Using

$$\frac{1}{x \pm i0^+} = P\left(\frac{1}{x}\right) \mp i\pi\delta(x) \quad (6.41)$$

where $P(1/x)$ is the principal value of $1/x$, and noting that $A(\mathbf{k}\sigma, \epsilon)$ is real, Eq. (6.36) gives

$$G^R(\mathbf{k}\sigma, \omega) = \frac{1}{2\pi}\int_{-\infty}^{\infty} A(\mathbf{k}\sigma, \epsilon)\left[P\left(\frac{1}{\omega - \epsilon}\right) - i\pi\delta(\omega - \epsilon)\right] d\epsilon$$

$$\Longrightarrow A(\mathbf{k}\sigma, \omega) = -2\, Im\, G^R(\mathbf{k}\sigma, \omega). \tag{6.42}$$

On the other hand, Eqs (6.35) and (6.40) give

$$C(\mathbf{k}\sigma, \omega) = (1 \mp e^{-\beta\hbar\omega})^{-1} A(\mathbf{k}\sigma, \omega) = A(\mathbf{k}\sigma, \omega)\begin{cases}(1 + n_\omega) & \text{bosons}\\ (1 - f_\omega) & \text{fermions}\end{cases} \tag{6.43}$$

where f_ω and n_ω are the Fermi–Dirac and Bose–Einstein distribution functions, respectively, for the case when energy is measured from the chemical potential, i.e., when μ is set equal to zero,

$$f_\omega = \frac{1}{e^{\beta\hbar\omega} + 1}, \quad n_\omega = \frac{1}{e^{\beta\hbar\omega} - 1}. \tag{6.44}$$

The above expressions for $G^R(\mathbf{k}\sigma, \omega)$ and $C(\mathbf{k}\sigma, \omega)$ imply that

$$C(\mathbf{k}\sigma, \omega) = -2\, Im\, G^R(\mathbf{k}\sigma, \omega)\begin{cases}(1 + n_\omega) & \text{bosons}\\ (1 - f_\omega) & \text{fermions.}\end{cases} \tag{6.45}$$

This relation is one form of the fluctuation–dissipation theorem. The correlation function measures the mean square fluctuation in the operator. However, energy dissipation in the system is proportional to the imaginary part of some retarded function. Further discussion of the fluctuation–dissipation theorem will occur at the end of this section.

We end this subsection by deriving a relationship between the number of particles in state $|\mathbf{k}\sigma\rangle$ and G^R. Setting $t = 0$ in Eq. (6.32), we can write

$$\langle c^\dagger_{\mathbf{k}\sigma} c_{\mathbf{k}\sigma}\rangle = -\int_{-\infty}^{\infty} P(\mathbf{k}\sigma, \epsilon) e^{-\beta\hbar\epsilon}\frac{d\epsilon}{2\pi} = \int_{-\infty}^{\infty}\frac{A(\mathbf{k}\sigma, \epsilon)e^{-\beta\hbar\epsilon}}{1 \mp e^{-\beta\hbar\epsilon}}\frac{d\epsilon}{2\pi}$$

$$= \int_{-\infty}^{\infty}\frac{A(\mathbf{k}\sigma, \epsilon)}{e^{\beta\hbar\epsilon} \mp 1}\frac{d\epsilon}{2\pi} = \int_{-\infty}^{\infty}\frac{d\epsilon}{2\pi} A(\mathbf{k}\sigma, \epsilon)\begin{cases}n_\epsilon & \text{bosons}\\ f_\epsilon & \text{fermions.}\end{cases}$$

With the help of Eq. (6.42), the above relation is rewritten as

$$\langle c^\dagger_{\mathbf{k}\sigma} c_{\mathbf{k}\sigma}\rangle = \int_{-\infty}^{\infty} d\epsilon\left(\frac{-1}{\pi}\right) Im\, G^R(\mathbf{k}\sigma, \epsilon)\begin{cases}n_\epsilon & \text{bosons}\\ f_\epsilon & \text{fermions.}\end{cases} \tag{6.46}$$

This equation provides a method for calculating the number of particles in a given state once the retarded Green's function has been found.

6.4.3 Retarded correlation function

Consider the retarded correlation function generated by operators A and B,

$$C^R_{AB}(t) = -i\theta(t)\langle[A(t), B(0)]_{\mp}\rangle.$$

The lower (upper) sign refers to the case where A and B are fermion (boson) operators. Expanding the anticommutator/commutator, we can write

$$\begin{aligned}
C^R_{AB}(t) &= -i\theta(t)\left[\langle A(t)B(0)\rangle \mp \langle B(0)A(t)\rangle\right] \\
&= -i\theta(t)Z_G^{-1}\mathrm{Tr}\left[e^{-\beta\bar{H}}e^{i\bar{H}t/\hbar}Ae^{-i\bar{H}t/\hbar}B \mp e^{-\beta\bar{H}}Be^{i\bar{H}t/\hbar}Ae^{-i\bar{H}t/\hbar}\right] \\
&= -i\theta(t)Z_G^{-1}\sum_n e^{-\beta\bar{E}_n}\left[e^{i\bar{E}_nt/\hbar}\langle n|Ae^{-i\bar{H}t/\hbar}B|n\rangle \mp e^{-i\bar{E}_nt/\hbar}\langle n|Be^{i\bar{H}t/\hbar}A|n\rangle\right] \\
&= -i\theta(t)Z_G^{-1}\sum_{nm}\left[e^{-\beta\bar{E}_n}e^{i(\bar{E}_n-\bar{E}_m)t/\hbar}\langle n|A|m\rangle\langle m|B|n\rangle\right. \\
&\quad \left.\mp e^{-\beta\bar{E}_n}e^{-i(\bar{E}_n-\bar{E}_m)t/\hbar}\langle n|B|m\rangle\langle m|A|n\rangle\right].
\end{aligned}$$

Relabeling indices in the second term: $n \to m,\ m \to n$, we obtain

$$C^R_{AB}(t) = -i\theta(t)Z_G^{-1}\sum_{nm} e^{i(\bar{E}_n-\bar{E}_m)t/\hbar}\langle n|A|m\rangle\langle m|B|n\rangle\left(e^{-\beta\bar{E}_n} \mp e^{-\beta\bar{E}_m}\right).$$

We now take the Fourier transform,

$$\begin{aligned}
C^R_{AB}(\omega) &= \int_{-\infty}^{\infty} e^{i\omega t}C^R_{AB}(t)dt \\
&= -iZ_G^{-1}\sum_{nm}\langle n|A|m\rangle\langle m|B|n\rangle\left(e^{-\beta\bar{E}_n} \mp e^{-\beta\bar{E}_m}\right)\int_0^{\infty} e^{i(\hbar\omega+\bar{E}_n-\bar{E}_m)t/\hbar}dt \\
&= Z_G^{-1}\sum_{nm}\frac{\langle n|A|m\rangle\langle m|B|n\rangle\left(e^{-\beta\bar{E}_n} \mp e^{-\beta\bar{E}_m}\right)}{\omega - (\bar{E}_m - \bar{E}_n)/\hbar + i0^+}. \qquad (6.47)
\end{aligned}$$

This is the spectral representation of the retarded correlation function. Notice that all the poles lie below the real ω-axis; the retarded function is analytic in the upper half of the complex ω-plane. Of course, the same conclusion applies to the retarded single-particle Green's function $G^R(\mathbf{k}\sigma, \omega)$, since it is a special case of the more general retarded correlation function.

We can go a step further and express the spectral representation in a form similar to Eq. (6.36). Define the spectral density function by

$$S(\epsilon) = 2\pi Z_G^{-1}\sum_{nm} e^{-\beta\bar{E}_n}\langle n|A|m\rangle\langle m|B|n\rangle\left(1 \mp e^{-\beta\hbar\epsilon}\right)\delta\left(\epsilon - \left(\bar{E}_m - \bar{E}_n\right)/\hbar\right).$$

The spectral representation of the retarded correlation is now given by

$$C^R_{AB}(\omega) = \int_{-\infty}^{\infty} \frac{S(\epsilon)}{\omega - \epsilon + i0^+} \frac{d\epsilon}{2\pi}.$$

6.4.4 Correlation function

Consider two operators A and B, and the correlation function

$$\begin{aligned} C_{AB}(t) = \langle A(t)B(0)\rangle &= Z_G^{-1} \sum_n e^{-\beta \bar{E}_n} e^{i\bar{E}_n t/\hbar} \langle n|Ae^{-i\bar{H}t/\hbar} B|n\rangle \\ &= Z_G^{-1} \sum_{nm} e^{-\beta \bar{E}_n} e^{i(\bar{E}_n - \bar{E}_m)t/\hbar} \langle n|A|m\rangle\langle m|B|n\rangle. \end{aligned} \tag{6.48}$$

Taking the Fourier transform,

$$C_{AB}(\omega) = Z_G^{-1} \sum_{nm} e^{-\beta \bar{E}_n} \langle n|A|m\rangle\langle m|B|n\rangle \int_{-\infty}^{\infty} e^{i(\hbar\omega + \bar{E}_n - \bar{E}_m)t/\hbar} dt.$$

The integral over t gives $2\pi\delta\left(\omega - \left(\bar{E}_m - \bar{E}_n\right)/\hbar\right)$; hence,

$$C_{AB}(\omega) = 2\pi Z_G^{-1} \sum_{nm} e^{-\beta \bar{E}_n} \langle n|A|m\rangle\langle m|B|n\rangle \delta\left(\omega - \left(\bar{E}_m - \bar{E}_n\right)/\hbar\right).$$

The function $C^R_{AB}(\omega)$ is given in Eq. (6.47). Using Eq. (6.41), we find

$$\begin{aligned} Im\, C^R_{AB}(\omega) &= -\pi Z_G^{-1} \sum_{nm} e^{-\beta \bar{E}_n} \langle n|A|m\rangle\langle m|B|n\rangle \left(1 \mp e^{-\beta(\bar{E}_m - \bar{E}_n)}\right) \\ &\quad \times \delta\left(\omega - \left(\bar{E}_m - \bar{E}_n\right)/\hbar\right) \\ &= -\pi Z_G^{-1} \sum_{nm} e^{-\beta \bar{E}_n} \langle n|A|m\rangle\langle m|B|n\rangle \left(1 \mp e^{-\beta\hbar\omega}\right)\delta\left(\omega - \left(\bar{E}_m - \bar{E}_n\right)/\hbar\right). \end{aligned}$$

In the last step, we replaced the exponent $(\bar{E}_m - \bar{E}_n)/\hbar$ with ω (this is made possible by the presence of the delta function). Comparing $C_{AB}(\omega)$ with $Im\, C^R_{AB}(\omega)$, we obtain

$$\begin{aligned} C_{AB}(\omega) &= -2\, Im\, C^R_{AB}(\omega) \left(1 \mp e^{-\beta\hbar\omega}\right)^{-1} \Longrightarrow \\ C_{AB}(\omega) &= -2\, Im\, C^R_{AB}(\omega) \begin{cases} (1 + n_\omega) & \text{bosons} \\ (1 - f_\omega) & \text{fermions.} \end{cases} \end{aligned} \tag{6.49}$$

This is the fluctuation–dissipation theorem (Nyquist, 1928; Callen and Welton, 1951). To better understand the content of this theorem, we assume that $A = B$, and let $\tilde{A}(t) = A(t) - \langle A\rangle$; i.e., $\tilde{A}$ is the deviation of A from its thermal average value. $\langle A \rangle$ does not depend on time, since the Hamiltonian is assumed to be time-independent. The correlation function $C_{\tilde{A}\tilde{A}}(t) = \langle \tilde{A}(t)\tilde{A}(0)\rangle = \langle A(t)A(0)\rangle - \langle A \rangle^2$ describes the quantum thermal fluctuations in the operator A. On the other hand, as we shall see later in this chapter, $C^R_{\tilde{A}\tilde{A}}(t)$ describes the response of the system to an external field; its imaginary part is usually related to energy dissipation. For example, an external electromagnetic field couples to the current density $\mathbf{j}$. In this case $A = \mathbf{j}$. Whereas $C_{\tilde{A}\tilde{A}}(t)$ describes the quantum thermal fluctuations in the current density, the imaginary part of $C^R_{\tilde{A}\tilde{A}}(t)$ turns out to be related to the resistance in the system, and hence to the mode of dissipation of energy supplied to the system by the external field. Thus, it is usually the case that the LHS of Eq. (6.49) represents fluctutations, while the RHS describes dissipation.

To see more explicitly that the imaginary part of the retarded correlation function describes dissipation, consider an applied external field that couples to some observable of the system. We take the perturbation to be

$$H' = fA^\dagger e^{-i\omega t} + f^* A e^{i\omega t} \tag{6.50}$$

where f is proportional to the strength of the applied field, and A is the operator that represents the observable of the system (such as the current density) to which the field is coupled. Since A is hermitian ($A^\dagger = A$), it must have an equal number of creation and annihilation operators when it is expressed in second quantized form; hence, it is a bosonic operator. The transition rate (transition probability per unit time) from stationary state $|n\rangle$ to stationary state $|m\rangle$ (eigenstates of the unperturbed Hamiltonian) is

$$w_{n\to m} = \frac{2\pi}{\hbar}|f|^2|\langle m|A|n\rangle|^2 \left[\delta(\bar{E}_m - \bar{E}_n - \hbar\omega) + \delta(\bar{E}_m - \bar{E}_n + \hbar\omega)\right] \tag{6.51}$$

(see Problem 1.13). Assuming that the system is a member of a grand canonical ensemble, the energy absorbed by the system per unit time (the power delivered by the field to the system) is given by

$$\begin{aligned} P &= Z_G^{-1} \sum_{nm} e^{-\beta \bar{E}_n}(\bar{E}_m - \bar{E}_n) w_{n\to m} \\ &= \frac{2\pi}{\hbar}|f|^2 \hbar\omega Z_G^{-1} \sum_{nm} e^{-\beta \bar{E}_n}|A_{mn}|^2 \left[\delta(\bar{E}_m - \bar{E}_n - \hbar\omega) - \delta(\bar{E}_m - \bar{E}_n + \hbar\omega)\right]. \end{aligned}$$

where $A_{mn} = \langle m|A|n\rangle$. Interchanging n and m in the second summation, and noting that $\delta(-ax) = \delta(ax) = (1/|a|)\delta(x)$, we obtain

$$P = \frac{2\pi}{\hbar}|f|^2 \omega Z_G^{-1} \sum_{nm} |\langle m|A|n\rangle|^2 (e^{-\beta \bar{E}_n} - e^{-\beta \bar{E}_m})\delta\left[\omega - (\bar{E}_m - \bar{E}_n)/\hbar\right]. \tag{6.52}$$

From the spectral representation of the retarded correlation function, as given in Eq. (6.47), we find

$$Im\, C^R_{AA}(\omega) = -\pi Z_G^{-1} \sum_{nm} |\langle m|A|n\rangle|^2 (e^{-\beta \bar{E}_n} - e^{-\beta \bar{E}_m})\delta\left[\omega - (\bar{E}_m - \bar{E}_n)/\hbar\right].$$

The power (energy per unit time) dissipated in the system is thus given by

$$P = \frac{2}{\hbar}\omega |f|^2 \left[-Im\, C^R_{AA}(\omega)\right]. \tag{6.53}$$

We note that it is indeed proportional to the imaginary part of the retarded correlation function.

6.5 Example: Green's function of a noninteracting system

As an example, we shall calculate the retarded Green's function $G^{R,0}(\mathbf{k}\sigma, \omega)$ of a system of noninteracting particles. The Hamiltonian is given by

$$\bar{H} = \sum_{\mathbf{k}\sigma} (\epsilon_{\mathbf{k}\sigma} - \mu) c^\dagger_{\mathbf{k}\sigma} c_{\mathbf{k}\sigma} = \sum_{\mathbf{k}\sigma} \bar{\epsilon}_{\mathbf{k}\sigma} c^\dagger_{\mathbf{k}\sigma} c_{\mathbf{k}\sigma}$$

where $\bar{\epsilon}_{\mathbf{k}\sigma}$ is the single-particle state energy relative to the chemical potential. Below, we calculate $G^{R,0}(\mathbf{k}\sigma, \omega)$ using two different methods.

6.5.1 Derivation from the spectral density function

The spectral density function is given by Eq. (6.35),

$$A(\mathbf{k}\sigma, \epsilon) = 2\pi Z_G^{-1} \sum_{nm} e^{-\beta \bar{E}_n} \left|\langle m|c^\dagger_{\mathbf{k}\sigma}|n\rangle\right|^2 (1 \mp e^{-\beta\hbar\epsilon})\delta\left(\epsilon - (\bar{E}_m - \bar{E}_n)/\hbar\right).$$

For $\langle m|c^\dagger_{\mathbf{k}\sigma}|n\rangle$ to be nonzero, $|m\rangle$ must differ from $|n\rangle$ by an extra particle in state $|\mathbf{k}\sigma\rangle$. Since the system is noninteracting, $\bar{E}_m - \bar{E}_n = \bar{\epsilon}_{\mathbf{k}\sigma}$, and

$$A(\mathbf{k}\sigma, \epsilon) = 2\pi Z_G^{-1} \delta(\epsilon - \bar{\epsilon}_{\mathbf{k}\sigma}/\hbar)(1 \mp e^{-\beta\hbar\epsilon}) \sum_{nm} e^{-\beta \bar{E}_n} \langle n|c_{\mathbf{k}\sigma}|m\rangle\langle m|c^\dagger_{\mathbf{k}\sigma}|n\rangle.$$

The sum over m gives 1 ($\sum |m\rangle\langle m| = 1$). Therefore,

$$\begin{aligned} A(\mathbf{k}\sigma, \epsilon) &= 2\pi Z_G^{-1} \delta(\epsilon - \bar{\epsilon}_{\mathbf{k}\sigma}/\hbar)(1 \mp e^{-\beta\hbar\epsilon}) \sum_{n} e^{-\beta \bar{E}_n} \langle n|c_{\mathbf{k}\sigma} c^\dagger_{\mathbf{k}\sigma}|n\rangle \\ &= 2\pi \delta(\epsilon - \bar{\epsilon}_{\mathbf{k}\sigma}/\hbar)(1 \mp e^{-\beta\hbar\epsilon})\langle c_{\mathbf{k}\sigma} c^\dagger_{\mathbf{k}\sigma}\rangle. \end{aligned} \tag{6.54}$$

The definition of the grand canonical ensemble average is used in the last step. From the commutation property of the c-operators, we can write

$$\langle c_{\mathbf{k}\sigma} c^{\dagger}_{\mathbf{k}\sigma} \rangle = \langle 1 \pm c^{\dagger}_{\mathbf{k}\sigma} c_{\mathbf{k}\sigma} \rangle = \begin{cases} 1 + n_{\mathbf{k}\sigma} & \text{bosons} \\ 1 - f_{\mathbf{k}\sigma} & \text{fermions.} \end{cases}$$

$n_{\mathbf{k}\sigma} = (e^{\beta \bar{\epsilon}_{\mathbf{k}\sigma}} - 1)^{-1}$ and $f_{\mathbf{k}\sigma} = (e^{\beta \bar{\epsilon}_{\mathbf{k}\sigma}} + 1)^{-1}$ are the Bose–Einstein and Fermi–Dirac distribution functions, respectively. The spectral density function reduces to

$$A(\mathbf{k}\sigma, \epsilon) = 2\pi \delta(\epsilon - \bar{\epsilon}_{\mathbf{k}\sigma}/\hbar)(1 \mp e^{-\beta\hbar\epsilon}) \begin{cases} 1 + n_{\mathbf{k}\sigma} & \text{bosons} \\ 1 - f_{\mathbf{k}\sigma} & \text{fermions.} \end{cases}$$

The Dirac-delta function has the property: $\delta(x - a) f(x) = \delta(x - a) f(a)$ for any function $f(x)$. The factor $(1 \mp e^{-\beta\hbar\epsilon})$ in the above expression is thus replaced with $(1 \mp e^{-\beta \bar{\epsilon}_{\mathbf{k}\sigma}})$. It is then straightforward to show that

$$A(\mathbf{k}\sigma, \omega) = 2\pi \delta(\omega - \bar{\epsilon}_{\mathbf{k}\sigma}/\hbar). \tag{6.55}$$

This is the spectral density function for noninteracting particles. Inserting this into Eq. (6.36) gives the noninteracting retarded Green's function

$$G^R(\mathbf{k}\sigma, \omega) = \frac{1}{\omega - \bar{\epsilon}_{\mathbf{k}\sigma}/\hbar + i0^+}. \tag{6.56}$$

We note that the poles of $G^R(\mathbf{k}\sigma, \omega)$ occur at the excitation energies of the system. In the presence of interactions, the spectral density function will no longer be a delta function; instead, the sharp peak representing the delta function will broaden, yielding information about the energies of the excited states and their lifetimes.

6.5.2 An alternative derivation

For any modified Heisenberg picture operator $A(t)$,

$$\begin{aligned} \frac{dA(t)}{dt} &= \frac{d}{dt}\left(e^{i\bar{H}t/\hbar} A e^{-i\bar{H}t/\hbar}\right) = \frac{i}{\hbar}\bar{H}A(t) + e^{i\bar{H}t/\hbar}\,\frac{\partial A}{\partial t}\,e^{-i\bar{H}t/\hbar} - \frac{i}{\hbar}A(t)\bar{H} \\ &= \frac{i}{\hbar}\left[\bar{H}, A(t)\right] + \frac{\partial A}{\partial t}(t). \end{aligned}$$

We have assumed that $\bar{H}$ is time-independent, and used the fact that $\bar{H}$ commutes with $e^{-i\bar{H}t/\hbar}$. The last term in the above equation is a Heisenberg operator. If $A = c_{\mathbf{k}\sigma}$, we find

$$\frac{d}{dt} c_{\mathbf{k}\sigma}(t) = \frac{i}{\hbar}\left[\bar{H}, c_{\mathbf{k}\sigma}(t)\right].$$

For a system of noninteracting particles,

$$\bar{H} = \sum_{\mathbf{k}\sigma} \bar{\epsilon}_{\mathbf{k}\sigma} c_{\mathbf{k}\sigma}^{\dagger} c_{\mathbf{k}\sigma}.$$

It follows that

$$\frac{d}{dt} c_{\mathbf{k}\sigma}(t) = \frac{i}{\hbar}\left[\bar{H}, e^{i\bar{H}t/\hbar} c_{\mathbf{k}\sigma} e^{-i\bar{H}t/\hbar}\right] = \frac{i}{\hbar} e^{i\bar{H}t/\hbar}\left[\bar{H}, c_{\mathbf{k}\sigma}\right] e^{-i\bar{H}t/\hbar}. \tag{6.57}$$

The commutator is given by

$$\left[\bar{H}, c_{\mathbf{k}\sigma}\right] = \sum_{\mathbf{k}'\sigma'} \bar{\epsilon}_{\mathbf{k}'\sigma'} [c_{\mathbf{k}'\sigma'}^{\dagger} c_{\mathbf{k}'\sigma'}, c_{\mathbf{k}\sigma}].$$

Using the relation $[AB, C] = A\{B, C\} - \{A, C\}B$, or, $[AB, C] = A[B, C] + [A, C]B$, the above commutator gives, for fermions,

$$[c_{\mathbf{k}'\sigma'}^{\dagger} c_{\mathbf{k}'\sigma'}, c_{\mathbf{k}\sigma}] = c_{\mathbf{k}'\sigma'}^{\dagger}\{c_{\mathbf{k}'\sigma'}, c_{\mathbf{k}\sigma}\} - \{c_{\mathbf{k}'\sigma'}^{\dagger}, c_{\mathbf{k}\sigma}\} c_{\mathbf{k}'\sigma'} = 0 - \delta_{\mathbf{k}\mathbf{k}'}\delta_{\sigma\sigma'} c_{\mathbf{k}'\sigma'}$$

and, for bosons,

$$[c_{\mathbf{k}'\sigma'}^{\dagger} c_{\mathbf{k}'\sigma'}, c_{\mathbf{k}\sigma}] = c_{\mathbf{k}'\sigma'}^{\dagger}[c_{\mathbf{k}'\sigma'}, c_{\mathbf{k}\sigma}] + [c_{\mathbf{k}'\sigma'}^{\dagger}, c_{\mathbf{k}\sigma}] c_{\mathbf{k}'\sigma'} = 0 - \delta_{\mathbf{k}\mathbf{k}'}\delta_{\sigma\sigma'} c_{\mathbf{k}'\sigma'}.$$

It follows that

$$[\bar{H}, c_{\mathbf{k}\sigma}] = -\sum_{\mathbf{k}'\sigma'} \bar{\epsilon}_{\mathbf{k}'\sigma'} \delta_{\mathbf{k}\mathbf{k}'}\delta_{\sigma\sigma'} c_{\mathbf{k}'\sigma'} = -\bar{\epsilon}_{\mathbf{k}\sigma} c_{\mathbf{k}\sigma}. \tag{6.58}$$

Putting this into Eq. (6.57), we find

$$\frac{d}{dt} c_{\mathbf{k}\sigma}(t) = (-i\bar{\epsilon}_{\mathbf{k}\sigma}/\hbar) c_{\mathbf{k}\sigma}(t). \tag{6.59}$$

This is easily solved,

$$c_{\mathbf{k}\sigma}(t) = e^{-i\bar{\epsilon}_{\mathbf{k}\sigma}t/\hbar} c_{\mathbf{k}\sigma}(0). \tag{6.60a}$$

Taking the adjoint on both sides, we obtain

$$c_{\mathbf{k}\sigma}^{\dagger}(t) = e^{i\bar{\epsilon}_{\mathbf{k}\sigma}t/\hbar} c_{\mathbf{k}\sigma}^{\dagger}(0). \tag{6.60b}$$

The retarded Green's function is given by

$$\begin{aligned} G^R(\mathbf{k}\sigma, t) &= -i\theta(t)\langle [c_{\mathbf{k}\sigma}(t), c_{\mathbf{k}\sigma}^{\dagger}(0)]_{\mp}\rangle = -i\theta(t) e^{-i\bar{\epsilon}_{\mathbf{k}\sigma}t/\hbar}\langle [c_{\mathbf{k}\sigma}(0), c_{\mathbf{k}\sigma}^{\dagger}(0)]_{\mp}\rangle \\ &= -i\theta(t) e^{-i\bar{\epsilon}_{\mathbf{k}\sigma}t/\hbar}\langle 1\rangle = -i\theta(t) e^{-i\bar{\epsilon}_{\mathbf{k}\sigma}t/\hbar}. \end{aligned} \tag{6.61}$$

Its Fourier transform is

$$G^R(\mathbf{k}\sigma, \omega) = \int_{-\infty}^{\infty} G^R(\mathbf{k}\sigma, t) e^{i\omega t} dt = -i \int_0^{\infty} e^{i(\omega - \bar{\epsilon}_{\mathbf{k}\sigma}/\hbar)t} dt$$

$$= -i \lim_{\eta \to 0^+} \int_0^{\infty} e^{i(\omega - \bar{\epsilon}_{\mathbf{k}\sigma}/\hbar + i\eta)t} dt = \frac{1}{\omega - \bar{\epsilon}_{\mathbf{k}\sigma}/\hbar + i0^+}.$$

This is the same expression obtained earlier from the spectral density function.

6.6 Linear response theory

A typical measurement on a system is carried out by perturbing the system in the vicinity of a point $\mathbf{r}'$, at time t', by a probe such as an electromagnetic field, electrons, or neutrons, and measuring the response of the system near a point $\mathbf{r}$ at a later time t. For example, if a weak electromagnetic field impinges on a metal, the scalar potential $\phi(\mathbf{r}, t)$ couples to the local electronic charge density $\rho(\mathbf{r}) = -en(\mathbf{r})$, where $n(\mathbf{r})$ is the electron number density, causing a disturbance that propagates to other parts of the system. Similarly, the vector potential $\mathbf{A}(\mathbf{r})$ couples to the local current density $\mathbf{j}(\mathbf{r})$. On the other hand, neutrons couple to the local spin density; neutron scattering is used to characterize the state of a magnetic system. If the interaction of the probe with the system is weak, which is usually the case (if the interaction was strong, the probe would modify the properties of the system, and we would be studying a system different from the original one), a calculation of the system's response to first order (linear) in the external perturbation provides a good approximation.

The external field couples locally to a system's operator A. In general, the perturbation produced by the external field is given by the Hamiltonian

$$H^{\text{ext}}(t) = \int d^3r\, F(\mathbf{r}, t) A(\mathbf{r}). \tag{6.62}$$

$F(\mathbf{r}, t)$ is a "generalized force." For example, the scalar potential $\phi(\mathbf{r}, t)$ of the electromagnetic field couples to the number density of electrons $n(\mathbf{r})$,

$$H^{\text{ext}}(t) = -e \int d^3r\, \phi(\mathbf{r}, t) n(\mathbf{r}).$$

The generalized force in this case is $-e\phi(\mathbf{r}, t)$ and $A(\mathbf{r}) = n(\mathbf{r})$. The external perturbation drives the system out of its unperturbed state, leading to a measurable effect: the ensemble average $\langle A \rangle$ shifts to a new value $\langle A \rangle_{\text{ext}}$. For example, in the absence of the scalar potential, $\langle n(\mathbf{r}, t) \rangle$ is constant, but it will no longer be constant once $\phi(\mathbf{r}, t)$ is turned on, or a current (nonexistent in an isolated metal) begins to flow upon the application of an external voltage. The experiment measures $\delta\langle A \rangle = \langle A \rangle_{\text{ext}} - \langle A \rangle$ as a function of the generalized force F. From a theoretical perspective, we can say that, if F is weak, the response of the system, $\delta\langle A\rangle$, will be, to a good approximation, linear in F,

$$\delta\langle A\rangle(\mathbf{r},t) = \int d^3r' \int dt' \chi(\mathbf{r}t,\mathbf{r}'t')F(\mathbf{r}',t'). \tag{6.63}$$

In an experiment, F is varied at will (input) and $\delta\langle A\rangle$ is measured (output). On the other hand, $\chi(\mathbf{r}t,\mathbf{r}'t')$, the generalized susceptibility, is an intrinsic property of the system, and it determines how the system responds to external perturbations; its calculation is one of the goals of the theory. As we will see later, the generalized susceptibility is expressed as a retarded correlation function.

Consider a system of identical particles with a time-independent Hamiltonian H, subjected to a time-dependent external perturbation $H^{\text{ext}}(t)$ (the effect of the probe). We assume that $H^{\text{ext}}(t)$ is turned on at time t_0. For $t < t_0$, the state of the system evolves according to H,

$$|\Psi(t)\rangle = e^{-iHt/\hbar}|\Psi(0)\rangle \qquad t < t_0.$$

For $t > t_0$, the state evolves according to the Schrödinger equation

$$i\hbar\frac{\partial}{\partial t}|\Psi(t)\rangle = \left(H + H^{\text{ext}}(t)\right)|\Psi(t)\rangle. \tag{6.64}$$

We consider a solution of the form:

$$|\Psi(t)\rangle = e^{-iHt/\hbar}U(t)|\Psi(0)\rangle \tag{6.65}$$

where $U(t)$ is an operator to be determined. We note that

$$U(t) = 1 \qquad t \le t_0. \tag{6.66}$$

Inserting this solution into the Schrödinger equation, we find

$$\begin{aligned}
i\hbar\,\frac{\partial}{\partial t}\left[e^{-iHt/\hbar}\,U(t)\right]|\Psi(0)\rangle &= \left[H + H^{\text{ext}}(t)\right]e^{-iHt/\hbar}\,U(t)|\Psi(0)\rangle \\
&\Rightarrow \left[He^{-iHt/\hbar}\,U(t) + i\hbar\,e^{-iHt/\hbar}\,\partial U/\partial t\right]|\Psi(0)\rangle \\
&= \left[He^{-iHt/\hbar}\,U(t) + H^{\text{ext}}(t)\,e^{-iHt/\hbar}\,U(t)\right]|\Psi(0)\rangle \\
&\Rightarrow i\hbar\,e^{-iHt/\hbar}\,\partial U/\partial t = H^{\text{ext}}(t)\,e^{-iHt/\hbar}\,U(t).
\end{aligned}$$

The last equality is obtained since $|\Psi(0)\rangle$ is arbitrary. Multiplying both sides by $e^{iHt/\hbar}$ on the left, we obtain

$$i\hbar\,\partial U/\partial t = H_H^{\text{ext}}(t)U(t) \tag{6.67}$$

where $H_H^{\text{ext}}(t)$ is $H^{\text{ext}}(t)$ in the Heisenberg picture,

$$H_H^{\text{ext}}(t) = e^{iHt/\hbar} H^{\text{ext}}(t) e^{-iHt/\hbar}. \tag{6.68}$$

The operator $U(t)$ is determined by Eq. (6.67), a differential equation, along with the boundary condition, Eq. (6.66). We integrate both sides of this equation from t_0 to t; since $U(t_0) = 1$, we obtain

$$U(t) = 1 - \frac{i}{\hbar} \int_{t_0}^{t} H_H^{\text{ext}}(t') U(t') dt'. \tag{6.69}$$

This is an integral equation for $U(t)$; it can be solved by iteration

$$\begin{aligned} U(t) &= 1 - \frac{i}{\hbar} \int_{t_0}^{t} dt' H_H^{\text{ext}}(t') \left[1 - \frac{i}{\hbar} \int_{t_0}^{t'} H_H^{\text{ext}}(t'') U(t'') dt'' \right] \\ &= 1 - \frac{i}{\hbar} \int_{t_0}^{t} dt' H_H^{\text{ext}}(t') + \left(\frac{-i}{\hbar} \right)^2 \int_{t_0}^{t} dt' \int_{t_0}^{t'} dt'' H_H^{\text{ext}}(t') H_H^{\text{ext}}(t'') U(t''). \end{aligned}$$

We can continue to iterate; we find

$$U(t) = 1 - \frac{i}{\hbar} \int_{t_0}^{t} dt_1 H_H^{\text{ext}}(t_1) + \left(\frac{-i}{\hbar} \right)^2 \int_{t_0}^{t} dt_1 \int_{t_0}^{t_1} dt_2 H_H^{\text{ext}}(t_1) H_H^{\text{ext}}(t_2) + \cdots . \tag{6.70}$$

We now consider the response of the system to the external perturbation. In particular, we are interested in the effect of H^{ext} on the expectation value of an operator A that represents an observable of the system, such as its charge or current density. Let the eigenstates of H (interacting, but unperturbed Hamiltonian) and the number operator N be denoted by $|n\rangle$,

$$H|n\rangle = E_n|n\rangle \ , \quad N|n\rangle = N|n\rangle.$$

The states $|n\rangle$ are time-independent; they may be considered the stationary states at $t = 0$, and they evolve in time, in the absence of H^{ext}, as

$$|n, t\rangle = e^{-iHt/\hbar}|n\rangle.$$

In the absence of H^{ext}, the expectation value of A, in state $|n\rangle$, at time t, is

$$\langle n, t|A|n, t\rangle|_{H^{\text{ext}}=0} = \langle n|e^{iHt/\hbar} A e^{-iHt/\hbar}|n\rangle = \langle n|A_H(t)|n\rangle$$

where $A_H(t)$ is operator A in the Heisenberg picture.

In the presence of H^{ext}, on the other hand, the state evolves according to Eq. (6.65); the expectation value of A at time $t > t_0$ is

$$\begin{aligned}
\langle n, t|A|n, t\rangle &= \langle n|U^\dagger(t)e^{iHt/\hbar}Ae^{-iHt/\hbar}U(t)|n\rangle = \langle n|U^\dagger(t)A_H(t)U(t)|n\rangle \\
&= \langle n|\left[1 + \frac{i}{\hbar}\int_{t_0}^{t} dt' H_H^{\text{ext}}(t') + \cdots\right] A_H(t) \left[1 - \frac{i}{\hbar}\int_{t_0}^{t} dt' H_H^{\text{ext}}(t') + \cdots\right]|n\rangle \\
&= \langle n|A_H(t)|n\rangle + \frac{i}{\hbar}\int_{t_0}^{t} dt' \langle n|\left[H_H^{\text{ext}}(t')A_H(t) - A_H(t)H_H^{\text{ext}}(t')\right]|n\rangle + \cdots \\
&= \langle n|A_H(t)|n\rangle - \frac{i}{\hbar}\int_{t_0}^{t} dt' \langle n|\left[A_H(t), H_H^{\text{ext}}(t')\right]|n\rangle + \cdots . \qquad (6.71)
\end{aligned}$$

If H^{ext} is weak, we are justified in ignoring higher-order terms in H^{ext} and keeping only the first-order term. The first term on the RHS is just the expectation value of A in the absence of H^{ext}; hence, the change in the expectation value of A, brought about by H^{ext}, is given by

$$\delta\langle n, t|A|n, t\rangle = -\frac{i}{\hbar}\int_{t_0}^{t} dt' \langle n|\left[A_H(t), H_H^{\text{ext}}(t')\right]|n\rangle. \qquad (6.72)$$

For $t < t_0$, the system is in equilibrium and state $|n\rangle$ is occupied with probability $p_n = Z_G^{-1}e^{-\beta(E_n-\mu N)}$. We thus consider the change $\delta\langle A\rangle$ in the ensemble average of A, caused by H^{ext}. Taking the ensemble average on both sides of Eq. (6.72), we obtain, for $t > t_0$,

$$\delta\langle A\rangle(\mathbf{r}, t) = -\frac{i}{\hbar}\int_{t_0}^{t} dt' \langle\left[A_H(\mathbf{r}, t), H_H^{\text{ext}}(t')\right]\rangle. \qquad (6.73)$$

On the RHS of the above equation, the ensemble average is taken over the interacting, but unperturbed, system. One may object to this because p_n may change as a result of the perturbation. However, the above expression for $\delta\langle A\rangle$ is already first order in H^{ext}; any modification brought about by considering a change in p_n will be of higher order. Thus, within a linear response theory, where $\delta\langle A\rangle$ is calculated only to first order in H^{ext}, we are justified in taking the ensemble average over the unperturbed system. Another way to arrive at this conclusion is as follows. Before the external perturbation is turned on at t_0, the system has been in contact with a reservoir, with which it exchanges energy and particles, for a sufficiently long time for equilibrium to be established. If we assume that the time $t - t_0$, during which the system is observed, is too short in comparison with the equilibration time, the probabilities of occupation of the states $|n\rangle$ will remain unchanged. Stated differently, the process of measuring of the system's response is finished long before the reservoir is able to cause a repopulation of the states of the system through exchange of energy and particles.

For H^{ext} of the form given in Eq. (6.62), the response of the system is

$$\delta\langle A\rangle(\mathbf{r},t) = -\frac{i}{\hbar}\int_{t_0}^{t} dt' \int d^3r' \langle [A_H(\mathbf{r},t), A_H(\mathbf{r}',t')]\rangle F(\mathbf{r}',t')$$
$$= \frac{1}{\hbar}\int_{t_0}^{t} dt' \int d^3r' D^R(\mathbf{r}t,\mathbf{r}'t')F(\mathbf{r}',t') \tag{6.74}$$

where

$$D^R(\mathbf{r}t,\mathbf{r}'t') = -i\theta(t-t')\langle [A_H(\mathbf{r},t), A_H(\mathbf{r}',t')]\rangle. \tag{6.75}$$

Since $t > t'$, the integration over t' being from t_0 to $t > t_0$, the step function $\theta(t-t')$ is equal to 1, and its introduction into Eq. (6.74) is a totally innocuous step. Equation (6.74) is Kubo's formula for the linear response of a system in equilibrium to an external perturbation (Kubo, 1957). Since the operator A represents an observable, it commutes with the number operator N (see Problem 3.4). Because H also commutes with N, we can write

$$A_H(\mathbf{r},t) = e^{iHt/\hbar}A(\mathbf{r})e^{-iHt/\hbar} = e^{i(H-\mu N)t/\hbar}A(\mathbf{r})e^{-i(H-\mu N)t/\hbar}$$
$$= e^{i\bar{H}t/\hbar}A(\mathbf{r})e^{-i\bar{H}t/\hbar} = A_{\bar{H}}(\mathbf{r},t)$$
$$\Rightarrow D^R(\mathbf{r}t,\mathbf{r}'t') = -i\theta(t-t')\langle [A_{\bar{H}}(\mathbf{r},t), A_{\bar{H}}(\mathbf{r}',t')]\rangle. \tag{6.76}$$

D^R is thus a retarded correlation function. The generalized susceptibility $\chi(\mathbf{r}t,\mathbf{r}'t')$ is given by

$$\chi(\mathbf{r}t,\mathbf{r}'t') = \frac{1}{\hbar}D^R(\mathbf{r}t,\mathbf{r}'t'). \tag{6.77}$$

Since $\bar{H}$ is time-independent, the retarded correlation function, and hence χ, depend on $t-t'$, and not on t and t' separately. Furthermore, if the system is translationally invariant, χ depends on $\mathbf{r}-\mathbf{r}'$. Thus,

$$\delta\langle A\rangle(\mathbf{r},t) = \int_{t_0}^{t} dt' \int d^3r' \chi(\mathbf{r}-\mathbf{r}',t-t')F(\mathbf{r}',t')$$
$$= \int_{-\infty}^{\infty} dt' \int d^3r' \chi(\mathbf{r}-\mathbf{r}',t-t')F(\mathbf{r}',t'). \tag{6.78}$$

Changing the limits of integration over t' is justifiable: for $t' < t_0$, $F(\mathbf{r}',t')$ vanishes, and the value of the integral is unchanged by extending the integration range to $-\infty$; similarly, for $t' > t$, $\chi(\mathbf{r}-\mathbf{r}',t-t')$ vanishes due to the factor $\theta(t-t')$ contained in $D^R(\mathbf{r}t,\mathbf{r}'t')$. Taking the Fourier transform with respect to time, we find

$$\delta\langle A\rangle(\mathbf{r},\omega)=\int_{-\infty}^{\infty}dte^{i\omega t}\delta\langle A\rangle(\mathbf{r},t)$$
$$=\int_{-\infty}^{\infty}dt\int_{-\infty}^{\infty}dt'\int d^3r'e^{i\omega(t-t')}\chi(\mathbf{r}-\mathbf{r}',t-t')e^{i\omega t'}F(\mathbf{r}',t').$$

Noting that

$$\int_{-\infty}^{\infty}dt\int_{-\infty}^{\infty}dt'\cdots=\int_{-\infty}^{\infty}dt'\int_{-\infty}^{\infty}d(t-t')\ldots,$$

we obtain

$$\delta\langle A\rangle(\mathbf{r},\omega)=\int d^3r'\chi(\mathbf{r}-\mathbf{r}',\omega)F(\mathbf{r}',\omega). \tag{6.79}$$

Similarly, we can Fourier transform with respect to spatial coordinates,

$$\delta\langle A\rangle(\mathbf{q},\omega)=\int d^3re^{-i\mathbf{q}.\mathbf{r}}\delta\langle A\rangle(\mathbf{r},\omega)$$
$$=\int d^3r\int d^3r'e^{-i\mathbf{q}.(\mathbf{r}-\mathbf{r}')}\chi(\mathbf{r}-\mathbf{r}',\omega)e^{-i\mathbf{q}.\mathbf{r}'}F(\mathbf{r}',\omega).$$

In the thermodynamic limit, where the volume $V\to\infty$,

$$\int d^3r\int d^3r'\cdots=\int d^3r'\int d^3x\ldots$$

where $\mathbf{x}=\mathbf{r}-\mathbf{r}'$. Although less transparent, the above replacement is valid if V is finite and periodic boundary conditions are adopted. The reader should convince himself/herself of this. From the above, it follows that

$$\delta\langle A\rangle(\mathbf{q},\omega)=\chi(\mathbf{q},\omega)F(\mathbf{q},\omega). \tag{6.80}$$

Thus, the system responds at the wave vector and frequency of the external field; if these match the wave vector and frequency of an intrinsic excitation of the system, a resonance effect occurs and a peak in $\delta\langle A\rangle$ is registered.

6.7 Noninteracting electron gas in an external potential

As an example, let us consider the response of a noninteracting electron gas to an external electric potential $\phi(\mathbf{r},t)$. In this case

$$H^{\text{ext}}(t)=-e\int d^3r\phi(\mathbf{r},t)n(\mathbf{r}). \tag{6.81}$$

Within linear response theory, the change in the ensemble average of n is

$$\delta\langle n\rangle(\mathbf{r},t) = (-e/\hbar)\int_{t_0}^{t} dt' \int d^3r' D^R(\mathbf{r}t,\mathbf{r}'t')\phi(\mathbf{r}',t') \tag{6.82}$$

where D^R is the retarded density–density correlation function of the noninteracting system,

$$D^R(\mathbf{r}t,\mathbf{r}'t') = -i\theta(t-t')\langle[n_{\bar{H}}(\mathbf{r},t), n_{\bar{H}}(\mathbf{r}',t')]\rangle. \tag{6.83}$$

Since $\bar{H}$ is time-independent and the system is translationally invariant, D^R depends on $\mathbf{r}-\mathbf{r}'$ and $t-t'$: $D^R(\mathbf{r}t,\mathbf{r}'t') = D^R(\mathbf{r}-\mathbf{r}',t-t')$. Hence

$$D^R(\mathbf{r}t,\mathbf{r}'t') = \frac{1}{V}\sum_{\mathbf{q}} e^{i\mathbf{q}.(\mathbf{r}-\mathbf{r}')} D^R(\mathbf{q},t-t'). \tag{6.84}$$

Similarly, decomposing $n_{\bar{H}}(\mathbf{r},t)$ and $n_{\bar{H}}(\mathbf{r}',t')$ into Fourier components, we find

$$\begin{aligned} D^R(\mathbf{r}t,\mathbf{r}'t') &= -i\theta(t-t')\frac{1}{V^2}\sum_{\mathbf{q}\mathbf{q}'} e^{i\mathbf{q}.\mathbf{r}} e^{i\mathbf{q}'.\mathbf{r}'}\langle[n_{\bar{H}}(\mathbf{q},t), n_{\bar{H}}(\mathbf{q}',t')]\rangle \\ &= -i\theta(t-t')\frac{1}{V^2}\sum_{\mathbf{q}\mathbf{q}'} e^{i\mathbf{q}.(\mathbf{r}-\mathbf{r}')} e^{i(\mathbf{q}+\mathbf{q}').\mathbf{r}'}\langle[n_{\bar{H}}(\mathbf{q},t), n_{\bar{H}}(\mathbf{q}',t')]\rangle. \end{aligned} \tag{6.85}$$

Since the RHS must depend on $\mathbf{r}-\mathbf{r}'$ and not independently on $\mathbf{r}'$, it follows that $\mathbf{q}'=-\mathbf{q}$. Alternatively, we may argue that if $\mathbf{r}$ and $\mathbf{r}'$ are shifted simultaneously by any vector $\mathbf{R}$, the RHS must remain unchanged since it depends only on $\mathbf{r}-\mathbf{r}'$. However, such a shift brings a factor of $e^{i(\mathbf{q}+\mathbf{q}').\mathbf{R}}$ into Eq. (6.85); this factor should be equal to 1 for any vector $\mathbf{R}$, and we conclude that $\mathbf{q}'=-\mathbf{q}$. Removing the summation over $\mathbf{q}'$, replacing $\mathbf{q}'$ with $-\mathbf{q}$, and comparing Eq. (6.84) with Eq. (6.85), we obtain

$$D^R(\mathbf{q},t-t') = -i\theta(t-t')\frac{1}{V}\langle[n_{\bar{H}}(\mathbf{q},t), n_{\bar{H}}(-\mathbf{q},t')]\rangle. \tag{6.86}$$

Using Eq. (3.25), we can write

$$\begin{aligned} n_{\bar{H}}(\mathbf{q},t) &= e^{i\bar{H}t/\hbar}\, n_{\mathbf{q}}\, e^{-i\bar{H}t/\hbar} = \sum_{\mathbf{k}\sigma} e^{i\bar{H}t/\hbar}\, c^{\dagger}_{\mathbf{k}\sigma}\, c_{\mathbf{k}+\mathbf{q}\sigma}\, e^{-i\bar{H}t/\hbar} \\ &= \sum_{\mathbf{k}\sigma} e^{i\bar{H}t/\hbar}\, c^{\dagger}_{\mathbf{k}\sigma}\, e^{-i\bar{H}t/\hbar}\, e^{i\bar{H}t/\hbar}\, c_{\mathbf{k}+\mathbf{q}\sigma}\, e^{-i\bar{H}t/\hbar} = \sum_{\mathbf{k}\sigma} c^{\dagger}_{\mathbf{k}\sigma}(t)\, c_{\mathbf{k}+\mathbf{q}\sigma}(t). \end{aligned} \tag{6.87}$$

Up to this point, our treatment applies to an interacting electron gas. In the simpler case of a noninteracting electron gas, $c_{\mathbf{k}\sigma}(t)$ and $c^{\dagger}_{\mathbf{k}\sigma}(t)$ are given by Eq. (6.60), and

$$n_{\bar{H}}(\mathbf{q},t) = \sum_{\mathbf{k}\sigma} c^{\dagger}_{\mathbf{k}\sigma}\, c_{\mathbf{k}+\mathbf{q}\sigma}\, e^{i(\bar{\epsilon}_{\mathbf{k}\sigma}-\bar{\epsilon}_{\mathbf{k}+\mathbf{q}\sigma})t/\hbar}. \tag{6.88}$$

The retarded function is now expressed as follows:

$$D^{R,0}(\mathbf{q}, t-t') = -i\theta(t-t')\frac{1}{V}\sum_{\mathbf{k}\sigma}\sum_{\mathbf{k}'\sigma'} e^{i(\bar{\epsilon}_{\mathbf{k}\sigma}-\bar{\epsilon}_{\mathbf{k}+\mathbf{q}\sigma})t/\hbar}\, e^{i(\bar{\epsilon}_{\mathbf{k}'\sigma'}-\bar{\epsilon}_{\mathbf{k}'-\mathbf{q}\sigma'})t'/\hbar}$$
$$\times \langle [c^{\dagger}_{\mathbf{k}\sigma}\, c_{\mathbf{k}+\mathbf{q}\sigma}\,,\, c^{\dagger}_{\mathbf{k}'\sigma'}\, c_{\mathbf{k}'-\mathbf{q}\sigma'}]\rangle .$$

The commutator is evaluated using the general formula

$$[AB, CD] = A\{B, C\}D - AC\{B, D\} + \{A, C\}DB - C\{A, D\}B \tag{6.89}$$

which can be easily verified; we find

$$[c^{\dagger}_{\mathbf{k}\sigma}\, c_{\mathbf{k}+\mathbf{q}\sigma}\,,\, c^{\dagger}_{\mathbf{k}'\sigma'}\, c_{\mathbf{k}'-\mathbf{q}\sigma'}] = \left(c^{\dagger}_{\mathbf{k}\sigma}\, c_{\mathbf{k}'-\mathbf{q}\sigma'} - c^{\dagger}_{\mathbf{k}'\sigma'}\, c_{\mathbf{k}+\mathbf{q}\sigma}\right)\delta_{\sigma\sigma'}\,\delta_{\mathbf{k}+\mathbf{q},\mathbf{k}'}.$$

The retarded correlation function can now be written as

$$D^{R,0}(\mathbf{q}, t-t') = -i\theta(t-t')\frac{1}{V}\sum_{\mathbf{k}\sigma} e^{i(\bar{\epsilon}_{\mathbf{k}\sigma}-\bar{\epsilon}_{\mathbf{k}+\mathbf{q}\sigma})(t-t')/\hbar}\langle c^{\dagger}_{\mathbf{k}\sigma} c_{\mathbf{k}\sigma} - c^{\dagger}_{\mathbf{k}+\mathbf{q}\sigma} c_{\mathbf{k}+\mathbf{q}\sigma}\rangle$$
$$= -i\theta(t-t')\frac{1}{V}\sum_{\mathbf{k}\sigma} e^{i(\bar{\epsilon}_{\mathbf{k}\sigma}-\bar{\epsilon}_{\mathbf{k}+\mathbf{q}\sigma})(t-t')/\hbar}\left(f_{\mathbf{k}\sigma} - f_{\mathbf{k}+\mathbf{q}\sigma}\right).$$

where $f_{\mathbf{k}\sigma}$ is the Fermi–Dirac distribution function. Taking the Fourier transform with respect to time,

$$D^{R,0}(\mathbf{q}, \omega) = \int_{-\infty}^{\infty} dt\, e^{i\omega t} D^{R,0}(\mathbf{q}, t) = \frac{-i}{V}\sum_{\mathbf{k}\sigma}\left(f_{\mathbf{k}\sigma} - f_{\mathbf{k}+\mathbf{q}\sigma}\right)\int_0^{\infty} dt\, e^{[i\omega + (\bar{\epsilon}_{\mathbf{k}\sigma}-\bar{\epsilon}_{\mathbf{k}+\mathbf{q}\sigma})/\hbar]t}$$
$$= \frac{-i}{V}\sum_{\mathbf{k}\sigma}\left(f_{\mathbf{k}\sigma} - f_{\mathbf{k}+\mathbf{q}\sigma}\right)\lim_{\eta\to 0^+}\int_0^{\infty} dt\, e^{i[\omega + (\bar{\epsilon}_{\mathbf{k}\sigma}-\bar{\epsilon}_{\mathbf{k}+\mathbf{q}\sigma})/\hbar + i\eta]t}$$
$$= \frac{1}{V}\sum_{\mathbf{k}\sigma}\frac{f_{\mathbf{k}\sigma} - f_{\mathbf{k}+\mathbf{q}\sigma}}{\omega + (\bar{\epsilon}_{\mathbf{k}\sigma} - \bar{\epsilon}_{\mathbf{k}+\mathbf{q}\sigma})/\hbar + i0^+}. \tag{6.90}$$

According to our general result, Eq. (6.80), the response of the system, $\delta\langle n\rangle(\mathbf{q}, \omega)$, is given by

$$\delta\langle n\rangle(\mathbf{q}, \omega) = \chi^0(\mathbf{q}, \omega)F(\mathbf{q}, \omega) = -\frac{e}{\hbar}D^{R,0}(\mathbf{q}, \omega)\phi(\mathbf{q}, \omega) \tag{6.91}$$

where $F(\mathbf{q}, \omega) = -e\phi(\mathbf{q}, \omega)$ is the generalized force, and

$$\chi^0(\mathbf{q}, \omega) = \frac{1}{\hbar}D^{R,0}(\mathbf{q}, \omega) = \frac{1}{\hbar V}\sum_{\mathbf{k}\sigma}\frac{f_{\mathbf{k}\sigma} - f_{\mathbf{k}+\mathbf{q}\sigma}}{\omega + (\bar{\epsilon}_{\mathbf{k}\sigma} - \bar{\epsilon}_{\mathbf{k}+\mathbf{q}\sigma})/\hbar + i0^+} \tag{6.92}$$

is the polarizability of the noninteracting electron gas. The function on the RHS of Eq. (6.92) is known as the Lindhard function (Lindhard, 1954). Equation (6.91) is also valid for an interacting electron gas if $D^{R,0} \to D^R$.

6.8 Dielectric function of a noninteracting electron gas

The dielectric function $\varepsilon(\mathbf{q}, \omega)$ is defined by the relation

$$\phi_{\text{tot}}(\mathbf{q}, \omega) = \phi_{\text{ext}}(\mathbf{q}, \omega)/\varepsilon(\mathbf{q}, \omega)$$

The total potential ϕ_{tot} is the sum of the external and induced potentials,

$$\begin{aligned}\phi_{\text{tot}}(\mathbf{q}, \omega) &= \phi_{\text{ext}}(\mathbf{q}, \omega) + \phi_{\text{ind}}(\mathbf{q}, \omega)\\ &\Longrightarrow \varepsilon(\mathbf{q}, \omega) = [1 + \phi_{\text{ind}}(\mathbf{q}, \omega)/\phi_{\text{ext}}(\mathbf{q}, \omega)]^{-1}. \qquad (6.93)\end{aligned}$$

The induced potential results from the induced charge density $\rho_{\text{ind}} = -e\delta\langle n\rangle$; it is given by

$$\phi_{\text{ind}}(\mathbf{r}, t) = \int d^3r' \, \frac{\rho_{\text{ind}}(\mathbf{r}', t)}{|\mathbf{r} - \mathbf{r}'|} \qquad \text{(cgs)}.$$

Multiplying both sides by e^2, and noting that $e^2/|\mathbf{r} - \mathbf{r}'|$ is the Coulomb interaction, which we can expand in a Fourier series, we obtain,

$$e^2\phi_{\text{ind}}(\mathbf{r}, t) = \int d^3r' \rho_{\text{ind}}(\mathbf{r}', t)\frac{1}{V}\sum_{\mathbf{q}} v_{\mathbf{q}} e^{i\mathbf{q}.(\mathbf{r}-\mathbf{r}')} = \frac{1}{V}\sum_{\mathbf{q}} v_{\mathbf{q}} e^{i\mathbf{q}.\mathbf{r}} \rho_{\text{ind}}(\mathbf{q}, t)$$

where $v_{\mathbf{q}} = 4\pi e^2/q^2$ is the Fourier transform of the Coulomb potential. The above expression implies that

$$e^2\phi_{\text{ind}}(\mathbf{q}, \omega) = v_{\mathbf{q}}\rho_{\text{ind}}(\mathbf{q}, \omega) = -ev_{\mathbf{q}}\delta\langle n\rangle(\mathbf{q}, \omega) = \frac{e^2}{\hbar}v_{\mathbf{q}}D^R(\mathbf{q}, \omega)\phi_{\text{ext}}(\mathbf{q}, \omega).$$

In the last step, Eq. (6.91) was used. The total potential is thus given by

$$\phi_{\text{tot}}(\mathbf{q}, \omega) = \left[1 + (1/\hbar)v_{\mathbf{q}}D^R(\mathbf{q}, \omega)\right]\phi_{\text{ext}}(\mathbf{q}, \omega).$$

Thus, for a noninteracting electron gas,

$$\varepsilon(\mathbf{q}, \omega) = \left[1 + (1/\hbar)v_{\mathbf{q}}D^{R,0}(\mathbf{q}, \omega)\right]^{-1} = \left[1 + v_{\mathbf{q}}\chi^0(\mathbf{q}, \omega)\right]^{-1}. \qquad (6.94)$$

The polarizability $\chi^0(\mathbf{q}, \omega)$ of the noninteracting electron gas is given in Eq. (6.92).

6.9 Paramagnetic susceptibility of a noninteracting electron gas

Let us consider another example of the application of linear response theory, the spin response of a noninteracting electron gas to a magnetic field $\mathbf{B}(\mathbf{r}, t)$ applied at $t = t_0$. The effect of the magnetic field on the orbital motion of the electrons complicates the situation considerably. Here, we ignore the orbital response of the electrons and consider only the interaction of the magnetic field with the electrons' spins.

The magnetic moment of an electron is $\boldsymbol{\mu} = -(ge/2mc)\mathbf{S} \simeq -(e/mc)\mathbf{S}$, where $g \simeq 2$ is the gyromagnetic factor, $-e$ is the charge of the electron, m is its mass, $\mathbf{S}$ is its spin, and c is the speed of light (in SI units, $\boldsymbol{\mu} \simeq -(e/m)\mathbf{S}$). Since the energy of a magnetic dipole in a magnetic field is $-\boldsymbol{\mu}.\mathbf{B}$, the external perturbation due to the applied field is written as

$$H^{\text{ext}}(t) = -\int \mathbf{m}(\mathbf{r}).\mathbf{B}(\mathbf{r}, t) d^3 r \qquad t > t_0. \tag{6.95}$$

Here, $\mathbf{m}(\mathbf{r})$ is the magnetic moment density operator; it is given by

$$\mathbf{m}(\mathbf{r}) = -\frac{e}{mc}\mathbf{s}(\mathbf{r}) \tag{6.96}$$

where $\mathbf{s}(\mathbf{r})$ is the spin-density operator. $H^{\text{ext}}(t)$ has the standard form:

$$H^{\text{ext}}(t) = \int \mathbf{A}(\mathbf{r}).\mathbf{F}(\mathbf{r}, t) d^3 r \tag{6.97}$$

where $\mathbf{A}(\mathbf{r}) = -\mathbf{m}(\mathbf{r})$ and the generalized force $\mathbf{F}(\mathbf{r}, t) = \mathbf{B}(\mathbf{r}, t)$. Using Kubo's formula (see Eq. [6.74]), derived within linear response theory,

$$\delta\langle -m_i\rangle(\mathbf{r}, t) = \frac{-i}{\hbar}\int_{t_0}^{t} dt' \int d^3 r' \sum_j \langle[-m_i(\mathbf{r}, t), -m_j(\mathbf{r}', t')]\rangle B_j(\mathbf{r}', t')$$

where $i, j = x, y, z$. Since $t > t'$, we can introduce $\theta(t - t')$ on the RHS,

$$\delta\langle m_i\rangle(\mathbf{r}, t) = \frac{-e^2}{\hbar m^2 c^2}\int_{t_0}^{t} dt' \int d^3 r' \sum_j D_{ij}^R(\mathbf{r}t, \mathbf{r}'t') B_j(\mathbf{r}', t'). \tag{6.98}$$

D_{ij}^R is the retarded spin-density correlation function

$$D_{ij}^R(\mathbf{r}t, \mathbf{r}'t') = -i\theta(t - t')\langle[s_i(\mathbf{r}, t), s_j(\mathbf{r}', t')]\rangle. \tag{6.99}$$

For N electrons at positions $\mathbf{r}_1, \mathbf{r}_2, \ldots, \mathbf{r}_N$, the spin-density operator is

$$\mathbf{s}(\mathbf{r}) = \sum_{i=1}^{N} \delta(\mathbf{r} - \mathbf{r}_i)\mathbf{S}_i \tag{6.100}$$

where $\mathbf{S}_i$ is the spin operator for electron i. Employing a basis set of plane waves $|\mathbf{k}\sigma\rangle$, we cast the spin-density operator into second quantized form:

$$\begin{aligned}
\mathbf{s}(\mathbf{r}) &= \sum_{\mathbf{k}\sigma_1}\sum_{\mathbf{k}'\sigma_1'}\langle\mathbf{k}\sigma_1|\delta(\mathbf{r}-\mathbf{r}')\mathbf{S}|\mathbf{k}'\sigma_1'\rangle c^\dagger_{\mathbf{k}\sigma_1}c_{\mathbf{k}'\sigma_1'} \\
&= \frac{\hbar}{2V}\sum_{\sigma_1\sigma_1'}\langle\sigma_1|\boldsymbol{\sigma}|\sigma_1'\rangle\sum_{\mathbf{k}\mathbf{k}'}\int d^3r' e^{-i\mathbf{k}.\mathbf{r}'}\delta(\mathbf{r}-\mathbf{r}')e^{i\mathbf{k}'.\mathbf{r}'}c^\dagger_{\mathbf{k}\sigma_1}c_{\mathbf{k}'\sigma_1'} \\
&= \frac{\hbar}{2V}\sum_{\sigma_1\sigma_1'}\sum_{\mathbf{k}\mathbf{q}}e^{i\mathbf{q}.\mathbf{r}}\langle\sigma_1|\boldsymbol{\sigma}|\sigma_1'\rangle c^\dagger_{\mathbf{k}\sigma_1}c_{\mathbf{k}+\mathbf{q}\sigma_1'} = \frac{1}{V}\sum_{\mathbf{q}}e^{i\mathbf{q}.\mathbf{r}}\mathbf{s}(\mathbf{q}) \qquad (6.101)
\end{aligned}$$

where $\mathbf{s}(\mathbf{q})$ is the Fourier transform of $\mathbf{s}(\mathbf{r})$,

$$\mathbf{s}(\mathbf{q}) = \frac{\hbar}{2}\sum_{\mathbf{k}\sigma_1\sigma_1'}\langle\sigma_1|\boldsymbol{\sigma}|\sigma_1'\rangle c^\dagger_{\mathbf{k}\sigma_1}c_{\mathbf{k}+\mathbf{q}\sigma_1'}. \qquad (6.102)$$

In Eq. (6.101), we have replaced the spin operator $\mathbf{S}$ with $\hbar\boldsymbol{\sigma}/2$, where σ_x, σ_y, and σ_z are the Pauli spin matrices. For the noninteracting electron gas, $c^\dagger_{\mathbf{k}\sigma}(t) = c^\dagger_{\mathbf{k}\sigma}e^{i\bar{\epsilon}_{\mathbf{k}}t/\hbar}$ and $c_{\mathbf{k}\sigma}(t) = c_{\mathbf{k}\sigma}e^{-i\bar{\epsilon}_{\mathbf{k}}t/\hbar}$. It follows that

$$\mathbf{s}(\mathbf{q},t) = \frac{\hbar}{2}\sum_{\mathbf{k}\sigma_1\sigma_1'}\langle\sigma_1|\boldsymbol{\sigma}|\sigma_1'\rangle c^\dagger_{\mathbf{k}\sigma_1}c_{\mathbf{k}+\mathbf{q}\sigma_1'}e^{i(\bar{\epsilon}_{\mathbf{k}}-\bar{\epsilon}_{\mathbf{k}+\mathbf{q}})t/\hbar}. \qquad (6.103)$$

Owing to the time-independence of the unperturbed Hamiltonian and the translational invariance of the unperturbed system, an analysis similar to the one carried out in Section 6.7 shows that

$$\begin{aligned}
D^{R,0}_{ij}(\mathbf{q},t) &= -i\theta(t)\langle[s_i(\mathbf{q},t), s_j(-\mathbf{q},0)]\rangle = -i\theta(t)\frac{\hbar^2}{4} \\
&\times\left\langle\left[\sum_{\mathbf{k}_1\sigma_1\sigma_1'}\langle\sigma_1|\sigma_i|\sigma_1'\rangle c^\dagger_{\mathbf{k}_1\sigma_1}c_{\mathbf{k}_1+\mathbf{q}\sigma_1'}e^{i(\bar{\epsilon}_{\mathbf{k}_1}-\bar{\epsilon}_{\mathbf{k}_1+\mathbf{q}})t/\hbar}, \sum_{\mathbf{k}_2\sigma_2\sigma_2'}\langle\sigma_2|\sigma_j|\sigma_2'\rangle c^\dagger_{\mathbf{k}_2\sigma_2}c_{\mathbf{k}_2-\mathbf{q}\sigma_2'}\right]\right\rangle \\
&= -i\theta(t)\frac{\hbar^2}{4}\sum_{\sigma_1\sigma_1'\sigma_2\sigma_2'}\langle\sigma_1|\sigma_i|\sigma_1'\rangle\langle\sigma_2|\sigma_j|\sigma_2'\rangle \\
&\times\sum_{\mathbf{k}_1\mathbf{k}_2}e^{i(\bar{\epsilon}_{\mathbf{k}_1}-\bar{\epsilon}_{\mathbf{k}_1+\mathbf{q}})t/\hbar}\left\langle[c^\dagger_{\mathbf{k}_1\sigma_1}c_{\mathbf{k}_1+\mathbf{q}\sigma_1'}, c^\dagger_{\mathbf{k}_2\sigma_2}c_{\mathbf{k}_2-\mathbf{q}\sigma_2'}]\right\rangle.
\end{aligned}$$

The commutator is evaluated using Eq. (6.89),

$$\begin{aligned}
&\left\langle[c^\dagger_{\mathbf{k}_1\sigma_1}c_{\mathbf{k}_1+\mathbf{q}\sigma_1'}, c^\dagger_{\mathbf{k}_2\sigma_2}c_{\mathbf{k}_2-\mathbf{q}\sigma_2'}]\right\rangle = \delta_{\mathbf{k}_2,\mathbf{k}_1+\mathbf{q}}\delta_{\sigma_1'\sigma_2}\langle c^\dagger_{\mathbf{k}_1\sigma_1}c_{\mathbf{k}_1\sigma_2'}\rangle \\
&\quad - \delta_{\mathbf{k}_2,\mathbf{k}_1+\mathbf{q}}\delta_{\sigma_1\sigma_2'}\langle c^\dagger_{\mathbf{k}_1+\mathbf{q}\sigma_2}c_{\mathbf{k}_1+\mathbf{q}\sigma_1'}\rangle = \delta_{\mathbf{k}_2,\mathbf{k}_1+\mathbf{q}}\delta_{\sigma_1'\sigma_2}\delta_{\sigma_1\sigma_2'}(f_{\mathbf{k}_1}-f_{\mathbf{k}_1+\mathbf{q}})
\end{aligned}$$

where $f_{\mathbf{k}_1}$ and $f_{\mathbf{k}_1+\mathbf{q}}$ are Fermi–Dirac distribution functions. Therefore,

$$D_{ij}^{R,0}(\mathbf{q},t) = -i\theta(t)\frac{\hbar^2}{4}\sum_{\sigma_1\sigma_2}\langle\sigma_1|\sigma_i|\sigma_2\rangle\langle\sigma_2|\sigma_j|\sigma_1\rangle\sum_{\mathbf{k}}(f_{\mathbf{k}} - f_{\mathbf{k}+\mathbf{q}})e^{i(\bar{\epsilon}_{\mathbf{k}}-\bar{\epsilon}_{\mathbf{k}+\mathbf{q}})t/\hbar}. \tag{6.104}$$

Using the completeness property of the spin states ($\sum_{\sigma_2}|\sigma_2\rangle\langle\sigma_2| = 1$),

$$\sum_{\sigma_1\sigma_2}\langle\sigma_1|\sigma_i|\sigma_2\rangle\langle\sigma_2|\sigma_j|\sigma_1\rangle = \sum_{\sigma_1}\langle\sigma_1|\sigma_i\sigma_j|\sigma_1\rangle = \mathrm{Tr}(\sigma_i\sigma_j)$$
$$= \frac{1}{2}\left[\mathrm{Tr}(\sigma_i\sigma_j) + \mathrm{Tr}(\sigma_j\sigma_i)\right] = \frac{1}{2}\mathrm{Tr}\{\sigma_i,\sigma_j\} = \frac{1}{2}\mathrm{Tr}[2\delta_{ij}I] = 2\delta_{ij}. \tag{6.105}$$

We have made use of the invariance of the trace under cyclic permutations: $\mathrm{Tr}[\sigma_i\sigma_j] = \mathrm{Tr}[\sigma_j\sigma_i]$, and the fact that $\{\sigma_i,\sigma_j\} = 2\delta_{ij}I$, where I is the 2×2 identity matrix. The expression for the retarded function thus reduces to

$$D_{ij}^{R,0}(\mathbf{q},t) = -i\theta(t)\delta_{ij}\frac{\hbar^2}{2}\sum_{\mathbf{k}}(f_{\mathbf{k}} - f_{\mathbf{k}+\mathbf{q}})e^{i(\bar{\epsilon}_{\mathbf{k}}-\bar{\epsilon}_{\mathbf{k}+\mathbf{q}})t/\hbar}.$$

Its Fourier transform is

$$D_{ij}^{R,0}(\mathbf{q},\omega) = \delta_{ij}\frac{\hbar^2}{2V}\sum_{\mathbf{k}}\frac{f_{\mathbf{k}} - f_{\mathbf{k}+\mathbf{q}}}{\omega + (\bar{\epsilon}_{\mathbf{k}} - \bar{\epsilon}_{\mathbf{k}+\mathbf{q}})/\hbar + i0^+}. \tag{6.106}$$

Hence,

$$\delta\langle m_i\rangle(\mathbf{q},\omega) = -\frac{\hbar^2e^2B_i(\mathbf{q},\omega)}{2m^2c^2V}\sum_{\mathbf{k}}\frac{f_{\mathbf{k}} - f_{\mathbf{k}+\mathbf{q}}}{\hbar\omega + \bar{\epsilon}_{\mathbf{k}} - \bar{\epsilon}_{\mathbf{k}+\mathbf{q}} + i0^+}. \tag{6.107}$$

The paramagnetic susceptibility is

$$\chi_{ij}^P(\mathbf{q},\omega) = \frac{\partial}{\partial B_j(\mathbf{q},\omega)}\delta\langle m_i\rangle(\mathbf{q},\omega) = \chi^P(\mathbf{q},\omega)\delta_{ij}. \tag{6.108}$$

We thus find

$$\chi^P(\mathbf{q},\omega) = -\mu_B^2\frac{1}{V}\sum_{\mathbf{k}\sigma}\frac{f_{\mathbf{k}} - f_{\mathbf{k}+\mathbf{q}}}{\hbar\omega + \bar{\epsilon}_{\mathbf{k}} - \bar{\epsilon}_{\mathbf{k}+\mathbf{q}} + i0^+} \tag{6.109}$$

where $\mu_B = \hbar e/(2mc)$ is the Bohr magneton, and we have used $\sum_{\mathbf{k}} F(\mathbf{k}) = (1/2)\sum_{\mathbf{k}\sigma} F(\mathbf{k})$. The Lindhard function has made another appearance, as it often does in the theory of the electron gas.

The paramagnetic susceptibility can be evaluated at zero temperature for the case of a static field ($\omega = 0$) in the long wave length limit ($\mathbf{q}\to 0$); χ^P is then real (it is easy to see that $Im\,\chi^P = 0$ when $\omega = 0$) and is given by

$$\chi^P(\mathbf{q} \to 0, \omega = 0) = -\mu_B^2 \frac{1}{V} \sum_{\mathbf{k}\sigma} \lim_{\mathbf{q}\to 0} \frac{f_{\mathbf{k}} - f_{\mathbf{k}+\mathbf{q}}}{\bar{\epsilon}_{\mathbf{k}} - \bar{\epsilon}_{\mathbf{k}+\mathbf{q}}} = \mu_B^2 \frac{1}{V} \sum_{\mathbf{k}\sigma} \left(-\frac{\partial f_{\mathbf{k}}}{\partial \bar{\epsilon}_{\mathbf{k}}} \right).$$

At $T = 0$, $f_{\mathbf{k}} = 1$ for $\bar{\epsilon}_{\mathbf{k}} < 0$ ($\epsilon_{\mathbf{k}} < \epsilon_F$), and $f_{\mathbf{k}} = 0$ for $\bar{\epsilon}_{\mathbf{k}} > 0$ ($\epsilon_{\mathbf{k}} > \epsilon_F$); hence $-\partial f_{\mathbf{k}} / \partial \bar{\epsilon}_{\mathbf{k}} = \delta(\epsilon_{\mathbf{k}} - \epsilon_F)$. Therefore, at $T = 0$,

$$\chi(\mathbf{q} \to 0, \omega = 0) = \mu_B^2 \frac{1}{V} \sum_{\mathbf{k}\sigma} \delta(\epsilon_{\mathbf{k}} - \epsilon_F) = \mu_B^2\, d(\epsilon_F) \tag{6.110}$$

where $d(\epsilon_F)$ is the density of states, per unit volume, at the Fermi energy. The above expression for χ is the well-known Pauli paramagnetic susceptibility of noninteracting, or independent, electrons.

6.10 Equation of motion

Next, we shall develop an equation of motion satisfied by the retarded Green's function. This approach allows us to calculate G^R for an interacting system provided that we adopt some approximations. We focus here on fermionic systems; the bosonic case is considered in the Problems section.

Consider a system of interacting fermions whose time-independent Hamiltonian is

$$\bar{H} = \bar{H}_0 + V = \sum_{\mathbf{k}\sigma} \bar{\epsilon}_{\mathbf{k}\sigma} c_{\mathbf{k}\sigma}^\dagger c_{\mathbf{k}\sigma} + V$$

where V represents the interaction between the particles, or the interaction of the particles with an external field. The retarded Green's function is

$$G^R(\mathbf{k}\sigma, t) = -i\theta(t) \left\langle \left\{ c_{\mathbf{k}\sigma}(t),\, c_{\mathbf{k}\sigma}^\dagger(0) \right\} \right\rangle.$$

Recalling that the derivative of the step function is the Dirac-delta function, we can write

$$i \frac{\partial}{\partial t} G^R(\mathbf{k}\sigma, t) = \delta(t) \left\langle \left\{ c_{\mathbf{k}\sigma}(t),\, c_{\mathbf{k}\sigma}^\dagger(0) \right\} \right\rangle + \theta(t) \left\langle \left\{ \frac{\partial}{\partial t} c_{\mathbf{k}\sigma}(t),\, c_{\mathbf{k}\sigma}^\dagger(0) \right\} \right\rangle.$$

Since $\delta(x) f(x) = \delta(x) f(0)$, the first term on the RHS is written as

$$\delta(t) \left\langle \left\{ c_{\mathbf{k}\sigma}(t),\, c_{\mathbf{k}\sigma}^\dagger(0) \right\} \right\rangle = \delta(t) \left\langle \left\{ c_{\mathbf{k}\sigma}(0),\, c_{\mathbf{k}\sigma}^\dagger(0) \right\} \right\rangle = \delta(t) \langle 1 \rangle = \delta(t).$$

As for the second term,

$$\frac{\partial}{\partial t} c_{\mathbf{k}\sigma}(t) = \frac{i}{\hbar} \left[\bar{H}, c_{\mathbf{k}\sigma}(t) \right] = \frac{i}{\hbar} \left[\bar{H}_0(t), c_{\mathbf{k}\sigma}(t) \right] + \frac{i}{\hbar} \left[V(t), c_{\mathbf{k}\sigma}(t) \right].$$

We have made use of the fact that since $\bar{H}$ is time-independent, $\bar{H} = \bar{H}(t)$. In the equation above, the first commutator $[\bar{H}_0(t), c_{\mathbf{k}\sigma}(t)]$ is equal to $-\bar{\epsilon}_{\mathbf{k}\sigma} c_{\mathbf{k}\sigma}(t)$. It follows that

$$\frac{\partial}{\partial t} c_{\mathbf{k}\sigma}(t) = -\frac{i}{\hbar} \bar{\epsilon}_{\mathbf{k}\sigma} c_{\mathbf{k}\sigma}(t) + \frac{i}{\hbar} \left[V(t), c_{\mathbf{k}\sigma}(t) \right].$$

The equation of motion for $G^R(\mathbf{k}\sigma, t)$ becomes

$$\left(i\hbar \frac{\partial}{\partial t} - \bar{\epsilon}_{\mathbf{k}\sigma} \right) G^R(\mathbf{k}\sigma, t) = \hbar\delta(t) + F^R(\mathbf{k}\sigma, t) \tag{6.111}$$

where

$$F^R(\mathbf{k}\sigma, t) = -i\theta(t) \langle\{- \left[V(t), c_{\mathbf{k}\sigma}(t) \right], c_{\mathbf{k}\sigma}(0)\}\rangle \tag{6.112}$$

is a retarded correlation function that describes the effect of the interactions in the system. To proceed further, we would need to evaluate $F^R(\mathbf{k}\sigma, t)$. In general, an exact solution is not possible; an approximation scheme must be used.

6.11 Example: noninteracting electron gas

We use the equation of motion to evaluate G^R for a system of noninteracting electrons: $V = 0$. In this case, Eq. (6.111) simplifies to

$$\left(i\frac{\partial}{\partial t} - \bar{\epsilon}_{\mathbf{k}\sigma}/\hbar \right) G^R(\mathbf{k}\sigma, t) = \delta(t).$$

This equation can be solved by Fourier decomposition

$$\left(i\frac{\partial}{\partial t} - \bar{\epsilon}_{\mathbf{k}\sigma}/\hbar \right) \frac{1}{2\pi} \int_{-\infty}^{\infty} e^{-i\omega t} G^R(\mathbf{k}\sigma, \omega) d\omega = \frac{1}{2\pi} \int_{-\infty}^{\infty} e^{-i\omega t} d\omega.$$

The integral on the RHS is one of the representations of the Dirac-delta function. Hence,

$$\int_{-\infty}^{\infty} (\omega - \bar{\epsilon}_{\mathbf{k}\sigma}/\hbar) e^{-i\omega t} G^R(\mathbf{k}\sigma, \omega) d\omega = \int_{-\infty}^{\infty} e^{-i\omega t} d\omega$$

$$\Rightarrow G^R(\mathbf{k}\sigma, \omega) = \frac{1}{\omega - \bar{\epsilon}_{\mathbf{k}\sigma}/\hbar}.$$

There is one problem with this expression for $G^R(\mathbf{k}\sigma, \omega)$. Suppose that we try to calculate $G^R(\mathbf{k}\sigma, t)$ from $G^R(\mathbf{k}\sigma, \omega)$,

$$G^R(\mathbf{k}\sigma, t) = \frac{1}{2\pi} \int_{-\infty}^{\infty} \frac{e^{-i\omega t} d\omega}{\omega - \bar{\epsilon}_{\mathbf{k}\sigma}/\hbar}.$$

The integral is problematic because of the pole at $\bar{\epsilon}_{\mathbf{k}\sigma}/\hbar$. To circumvent this problem, we can shift the pole slightly, either above or below the real ω-axis. For $t < 0$, the integral along the semicircle at infinity in the upper half of the complex ω-plane vanishes because of the $e^{-i\omega t}$ factor in the integrand. In this case, $G^R(\mathbf{k}\sigma, t)$ is equal to the contour integral along the closed contour consisting of the real ω-axis and the semicircle at infinity in the upper half ω-plane. Since $G^R(\mathbf{k}\sigma, t) = 0$ for $t < 0$, the pole should not lie above the real axis, because the residue theorem would then yield a nonvanishing value for $G^R(\mathbf{k}\sigma, t)$. Causality thus dictates that the pole needs to be shifted slightly below the real axis to $\bar{\epsilon}_{\mathbf{k}\sigma}/\hbar - i0^+$. Hence, the correct expression for $G^R(\mathbf{k}\sigma, \omega)$ is

$$G^R(\mathbf{k}\sigma, \omega) = \frac{1}{\omega - \bar{\epsilon}_{\mathbf{k}\sigma}/\hbar + i0^+},$$

in conformity with the expression obtained earlier.

6.12 Example: an atom adsorbed on graphene

In the Problems sections of Chapters 2 and 3, the dispersion of the two energy π-bands of graphene was calculated. Let us now consider a system consisting of one atom adsorbed on graphene, with a model Hamiltonian

$$H = \sum_{n\mathbf{k}\sigma} \epsilon_{n\mathbf{k}} c^\dagger_{n\mathbf{k}\sigma} c_{n\mathbf{k}\sigma} + \sum_{\sigma} \epsilon_d d^\dagger_\sigma d_\sigma + \sum_{n\mathbf{k}\sigma} V_{n\mathbf{k}d} c^\dagger_{n\mathbf{k}\sigma} d_\sigma + \sum_{n\mathbf{k}\sigma} V^*_{n\mathbf{k}d} d^\dagger_\sigma c_{n\mathbf{k}\sigma} + U n_{d\uparrow} n_{d\downarrow}.$$

In the first term, n is a band index which can take two values, 1 and 2, $\mathbf{k} = (k_x, k_y)$ is a vector in the first Brillouin zone of graphene, and $\sigma = \uparrow$ or $\downarrow$. We assume that the adsorbed atom has one orbital of energy ϵ_d; because of Coulomb repulsion, the energy increases by U if this orbital is doubly occupied, as indicated by the last term in the Hamiltonian. The operator $d^\dagger_\sigma$ (d_σ) creates (annihilates) an electron with spin projection σ in the atomic orbital. The third and fourth terms in H describe the hybridization between the orbital on the adsorbed atom and the π-states of graphene: electrons can hop from the adsorbed atom onto graphene, and vice versa. In this example, we assume that $U = 0$, and we calculate the spectral density function of the adsorbed atom. For the isolated atom, this is a Dirac-delta function, peaked at $\hbar\omega = \epsilon_d$; we will see that interactions broaden the peak into a Lorentzian.

The retarded Green's function of the adsorbed atom is given by

$$G^R(dd\sigma, t) = -i\theta(t)\left\langle\left\{d_\sigma(t), d^\dagger_\sigma(0)\right\}\right\rangle.$$

Taking the derivative with respect to t, we obtain

$$i\frac{\partial}{\partial t}G^R(dd\sigma, t) = \delta(t) + \frac{i}{\hbar}\theta(t)\left\langle\left\{[H, d_\sigma(t)], d_\sigma^\dagger(0)\right\}\right\rangle.$$

Note that $H(t) = e^{iHt/\hbar}He^{-iHt/\hbar} = H$. The commutator is given by

$$\begin{aligned}[H, d_\sigma] &= \sum_{\sigma'}\epsilon_d\left[d_{\sigma'}^\dagger d_{\sigma'}, d_\sigma\right] + \sum_{n\mathbf{k}\sigma'}V_{n\mathbf{k}d}^*\left[d_{\sigma'}^\dagger c_{n\mathbf{k}\sigma'}, d_\sigma\right] \\ &= -\sum_{\sigma'}\epsilon_d\left\{d_{\sigma'}^\dagger, d_\sigma\right\}d_{\sigma'} - \sum_{n\mathbf{k}\sigma'}V_{n\mathbf{k}d}^*\left\{d_{\sigma'}^\dagger, d_\sigma\right\}c_{n\mathbf{k}\sigma'} = -\epsilon_d d_\sigma - \sum_{n\mathbf{k}}V_{n\mathbf{k}d}^* c_{n\mathbf{k}\sigma}.\end{aligned}$$

We have used the relation $[AB, C] = A\{B, C\} - \{A, C\}B$ and the commutation relations of fermion annihilation and creation operators. Thus,

$$i\frac{\partial}{\partial t}G^R(dd\sigma, t) = \delta(t) + \frac{\epsilon_d}{\hbar}G^R(dd\sigma, t) + \frac{1}{\hbar}\sum_{n\mathbf{k}}V_{n\mathbf{k}d}^*G^R(n\mathbf{k}d\sigma, t). \quad (6.113)$$

We have introduced the graphene-adsorbed-atom retarded Green's function

$$G^R(n\mathbf{k}d\sigma, t) = -i\theta(t)\left\langle\left\{c_{n\mathbf{k}\sigma}(t), d_\sigma^\dagger(0)\right\}\right\rangle. \quad (6.114)$$

Its equation of motion is

$$i\frac{\partial}{\partial t}G^R(n\mathbf{k}d\sigma, t) = \frac{\epsilon_{n\mathbf{k}}}{\hbar}\,G^R(n\mathbf{k}d\sigma, t) + \frac{1}{\hbar}\,V_{n\mathbf{k}d}\,G^R(dd\sigma, t). \quad (6.115)$$

Upon Fourier decomposition:

$$G^R(dd\sigma, t) = \frac{1}{2\pi}\int_{-\infty}^{\infty}e^{-i\omega t}G^R(dd\sigma, \omega)d\omega, \quad \delta(t) = \frac{1}{2\pi}\int_{-\infty}^{\infty}e^{-i\omega t}d\omega,$$

and similar decomposition for $G^R(n\mathbf{k}d\sigma, t)$, we obtain

$$(\omega - \epsilon_d/\hbar)\,G^R(dd\sigma, \omega) = 1 + (1/\hbar)\sum_{n\mathbf{k}}V_{n\mathbf{k}d}^*G^R(n\mathbf{k}d\sigma, \omega) \quad (6.116)$$

and

$$(\omega - \epsilon_{n\mathbf{k}}/\hbar)\,G^R(n\mathbf{k}d\sigma, \omega) = (1/\hbar)V_{n\mathbf{k}d}G^R(dd\sigma, \omega). \quad (6.117)$$

These equations can be solved for $G^R(dd\sigma, \omega)$; we find

$$G^R(dd\sigma, \omega) = \frac{\hbar}{\hbar\omega - \epsilon_d - \sum_{n\mathbf{k}}\dfrac{|V_{n\mathbf{k}d}|^2}{\hbar\omega - \epsilon_{n\mathbf{k}}}}.$$

At this point, an argument is made similar to the one outlined in the previous section: causality dictates that $\omega \to \omega + i0^+$; hence,

$$G^R(dd\sigma,\omega) = \frac{\hbar}{\hbar\omega + i0^+ - \epsilon_d - \sum_{n\mathbf{k}} \frac{|V_{n\mathbf{k}d}|^2}{\hbar\omega - \epsilon_{n\mathbf{k}} + i0^+}}$$

$$= \frac{\hbar}{\hbar\omega - \epsilon_d - \sum_{n\mathbf{k}} P\left(\frac{|V_{n\mathbf{k}d}|^2}{\hbar\omega - \epsilon_{n\mathbf{k}}}\right) + i\pi \sum_{n\mathbf{k}} |V_{n\mathbf{k}d}|^2\, \delta(\hbar\omega - \epsilon_{n\mathbf{k}})}$$

where P stands for the principal value. To proceed further, we assume that $V_{n\mathbf{k}d}$ is small except for **k**-points in the first Brillouin zone near K and K', where it takes the constant value $\bar{V}$. Under this assumption,

$$\sum_{n\mathbf{k}} |V_{n\mathbf{k}d}|^2\, \delta(\hbar\omega - \epsilon_{n\mathbf{k}}) = \bar{V}^2 \sum_{n\mathbf{k}} \delta(\hbar\omega - \epsilon_{n\mathbf{k}}) = \bar{V}^2 D_\sigma(\hbar\omega).$$

$D_\sigma(\hbar\omega)$ is the density of states, per spin, in graphene (see Problem 2.4). The spectral density function, $A(dd\sigma,\omega) = -2\, Im\, G^R(dd\sigma,\omega)$, is thus given by

$$A(dd\sigma,\omega) = \frac{2\pi\hbar \bar{V}^2 D_\sigma(\hbar\omega)}{\left[\hbar\omega - \epsilon_d - \sum_{n\mathbf{k}} P\left(\frac{|\bar{V}|^2}{\hbar\omega - \epsilon_{n\mathbf{k}}}\right)\right]^2 + \left[\pi \bar{V}^2 D_\sigma(\hbar\omega)\right]^2}. \tag{6.118}$$

As expected, the presence of interactions causes a shift, and a broadening into a Lorentzian, of the Dirac-delta peak, which characterizes the spectral density of a noninteracting system. The shift is equal to the change in the energy of the atomic orbital, while the width of the Lorentzian determines the lifetime of the atomic state.

Further reading

Altland, A. and Simon, B. (2006). *Condensed Matter Field Theory*. Cambridge: Cambridge University Press.

Bruus, H. and Flensberg, K. (2004). *Many-Body Quantum Theory in Condensed Matter Physics*. Oxford: Oxford University Press.

Fetter, A.L. and Walecka, J.D. (1971). *Quantum Theory of Many-Particle Systems*. New York: McGraw-Hill.

Giuliani, G.F. and Vignale, G. (2005). *Quantum Theory of the Electron Liquid*. Cambridge: Cambridge University Press.

Mahan, G.D. (2000). *Many-Particle Physics*, 3rd edn. New York: Kluwer Academic/Plenum Publishers.

Problems

6.1 *Time independence.* If H is time-independent, show that the time-ordered correlation function $-i\langle TA(t)B(t')\rangle$ depends on $t-t'$ and not on t and t' separately.

6.2 *Translational invariance.* In a translationally invariant system, the Hamiltonian does not change if the positions of all the particles are shifted by the same vector $\mathbf{R}$: $H(\mathbf{r}_1,\ldots,\mathbf{r}_N) = H(\mathbf{r}_1+\mathbf{R},\ldots,\mathbf{r}_N+\mathbf{R})$. H thus commutes with the translation operator. Since the momentum operator $\mathbf{P}$ is the generator of translations, H and $\mathbf{P}$ commute. $\mathbf{P}$ is given by

$$\mathbf{P} = \sum_j(-i\hbar\nabla_j) = \sum_\sigma \int \Psi_\sigma^\dagger(\mathbf{r})(-i\hbar\nabla)\Psi_\sigma(\mathbf{r})d^3r.$$

(a) Show that $[\Psi_\sigma^\dagger(\mathbf{r}),\mathbf{P}] = -i\hbar\nabla\Psi_\sigma(\mathbf{r})$. Deduce that

$$\Psi_\sigma(\mathbf{r}) = e^{-i\mathbf{P}.\mathbf{r}/\hbar}\Psi_\sigma(0)e^{i\mathbf{P}.\mathbf{r}/\hbar} = T(\mathbf{r})\Psi_\sigma(0)T^{-1}(\mathbf{r})$$

where $T(\mathbf{r}) = e^{-i\mathbf{P}.\mathbf{r}/\hbar}$ is the translation operator that translates the positions of all particles by $\mathbf{r}$.

(b) Let $C(\mathbf{r}\sigma t,\mathbf{r}'\sigma' t') = \langle\Psi(\mathbf{r}\sigma t)\Psi^\dagger(\mathbf{r}'\sigma' t')\rangle$. Using the fact that $T(\mathbf{r})$ commutes with $e^{-\beta\bar{H}}$, along with the cyclic property of the trace, show that C is a function of $\mathbf{r}-\mathbf{r}'$. Deduce that in a translationally invariant system, all single-particle Green's functions are functions of $\mathbf{r}-\mathbf{r}'$.

6.3 *Spectral density function.* Show that the spectral density function $A(\mathbf{k}\sigma,\omega)$ satisfies the normalization condition

$$\int_{-\infty}^{\infty} A(\mathbf{k}\sigma,\omega)d\omega = 2\pi.$$

6.4 *Advanced Green's function.* Derive the spectral representation of $G^A(\mathbf{k}\sigma,\omega)$.

6.5 *Advanced correlation function.* Derive the spectral representation of the advanced correlation function $C^A_{AB}(\omega)$. Show that all the poles lie above the real ω-axis.

6.6 *Greater and lesser functions.* Show that, for fermions

$$iG^>(\mathbf{k}\sigma,\omega) = A(\mathbf{k}\sigma,\omega)[1-f_\omega], \quad iG^<(\mathbf{k}\sigma,\omega) = -A(\mathbf{k}\sigma,\omega)f_\omega$$

while for bosons

$$iG^{>}(\mathbf{k}\sigma,\omega) = A(\mathbf{k}\sigma,\omega)[1+n_\omega], \quad iG^{<}(\mathbf{k}\sigma,\omega) = A(\mathbf{k}\sigma,\omega)n_\omega.$$

6.7 *Causal Green's function.* Derive the spectral representation of the causal (time-ordered) Green's function.

6.8 *Relation among Green's functions.*
(a) Show that $Re\, G(\mathbf{k}\sigma,\omega) = Re\, G^R(\mathbf{k}\sigma,\omega) = Re\, G^A(\mathbf{k}\sigma,\omega)$.
(b) Show that for fermions

$$Im\, G^R(\mathbf{k}\sigma,\omega) = -Im\, G^A(\mathbf{k}\sigma,\omega) = [\tan h(\beta\hbar\omega/2)]^{-1} Im\, G(\mathbf{k}\sigma,\omega)$$

while for bosons

$$Im\, G^R(\mathbf{k}\sigma,\omega) = -Im\, G^A(\mathbf{k}\sigma,\omega) = \tan h(\beta\hbar\omega/2)\, Im\, G(\mathbf{k}\sigma,\omega).$$

6.9 *Greater and lesser correlation functions.* For two observables represented by operators A and B, $iC^{>}_{AB}(t) = \langle A(t)B(0)\rangle$, and $iC^{<}_{AB}(t) = \langle B(0)A(t)\rangle$. Assuming that H is time-independent, show that $C^{>}_{AB}(t - i\beta\hbar) = C^{<}_{AB}(t)$. Deduce that $C^{<}_{AB}(\omega) = e^{-\beta\hbar\omega}C^{>}_{AB}(\omega)$.

6.10 *Susceptibility.* Let $\hbar\chi_{AB}(t) = -i\theta(t)\langle[A(t), B(0)]\rangle$, where A and B are hermitian operators.
(a) Show that $\chi_{AB}(t)$ is real.
(b) Deduce that $[\chi_{AB}(\omega)]^* = \chi_{AB}(-\omega)$. This shows that $Re\,\chi_{AB}(\omega)$ is an even function of ω, while $Im\,\chi_{AB}(\omega)$ is an odd function of ω.

6.11 *Kramers–Kronig relations.* Assume that a function $\chi(\omega)$ satisfies the following:
(a) The poles of $\chi(\omega)$ are all below the real ω-axis.
(b) $\int d\omega\chi(\omega)/\omega = 0$ if the integration is around a semicircle at infinity in the upper half ω-plane.
(c) The real part of $\chi(\omega)$ is an even function of ω, while the imaginary part of $\chi(\omega)$ is an odd function of ω.
Show that

$$Re\,\chi(\omega) = \frac{2}{\pi}P\int_0^\infty \frac{\omega'\, Im\,\chi(\omega')d\omega'}{\omega'^2-\omega^2}, \quad Im\,\chi(\omega) = \frac{2\omega}{\pi}P\int_0^\infty \frac{Re\,\chi(\omega')d\omega'}{\omega'^2-\omega^2}$$

where P stands for the principal value. To prove this, consider the integral

$$I = \int_C \frac{\chi(\omega')d\omega'}{\omega'-\omega}$$

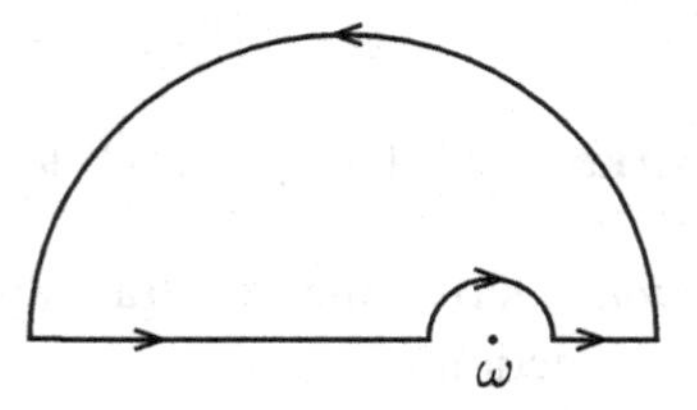

Figure 6.2 The contour C. The large semicircle is at infinity, while the radius of the small semicircle is infinitesimal.

where C is the contour shown in Figure 6.2. The large semicircle is at infinity, while the radius of the small semicircle is infinitesimal. Show that

$$\chi(\omega) = \frac{1}{i\pi} P \int_{-\infty}^{\infty} \frac{\chi(\omega')d\omega'}{\omega' - \omega}$$

and then equate the real and imaginary parts on both sides of the above equation.

6.12 *Polarizability.* Starting from the spectral representation of the retarded correlation function, derive the expression for the polarizability of a noninteracting electron gas.

6.13 *Equation of motion.* Derive the equation of motion for the retarded Green's function for an interacting system of bosons.

6.14 *Mixed retarded function.* Derive Eq. (6.115).

6.15 *Polarizability at zero temperature.* The polarizability of a noninteracting electron gas is given by Eq. (6.92). Show that, at $T = 0$

$$Re\, \chi^0(\mathbf{q}, \omega) = -d(\epsilon_F) \left[\frac{1}{2} + \frac{1 - z_-^2}{4q/k_F} \ln \left| \frac{1 - z_-}{1 + z_-} \right| - \frac{1 - z_+^2}{4q/k_F} \ln \left| \frac{1 - z_+}{1 + z_+} \right| \right]$$

$$Im\, \chi^0(\mathbf{q}, \omega) = -d(\epsilon_F) \frac{\pi}{4q/k_F} [(1 - z_-^2)\theta(1 - z_-^2) - (1 - z_+^2)\theta(1 - z_+^2)]$$

where $z_\pm = \omega/qv_F \pm q/2k_F$, k_F is the Fermi wave vector, $v_F = \hbar k_F/m$ is the Fermi velocity, and $d(\epsilon_F) = mk_F/\pi^2\hbar^2$ is the density of states, per unit volume, at the Fermi energy.

Hint: The expression for $\chi^0(\mathbf{q}, \omega)$ consists of two terms, each involving a sum over $\mathbf{k}$ and σ. In the second term, replace $\mathbf{k}$ with $-\mathbf{k} - \mathbf{q}$. The second term becomes the same as the first term, but with $\hbar\omega + i0^+ \to -\hbar\omega - i0^+$. Now replace the sum over $\mathbf{k}$ with integration. Also note that

$$\int x \ln|x + a| dx = \frac{x^2 - a^2}{2} \ln|x + a| - \frac{1}{4}(x - a)^2.$$

6.16 *Polarizability*. Show that, at zero temperature

$$\chi^0(\mathbf{q}, \omega = 0) = \begin{cases} -d(\epsilon_F)\left[\dfrac{1}{2} - \dfrac{4 - q'^2}{8q'} \ln\left|\dfrac{2 - q'}{2 + q'}\right|\right] & \text{3D} \\ -d(\epsilon_F)\left[1 - \theta(q' - 2)\dfrac{\sqrt{q'^2 - 4}}{q'}\right] & \text{2D} \\ -d(\epsilon_F)\left[\dfrac{1}{q'} \ln\left|\dfrac{2 + q'}{2 - q'}\right|\right] & \text{1D} \end{cases}$$

where $q' = q/k_F$.

7

Applications of real-time Green's functions

> Theory leads to application, and application brings to mind the Source of all theory and the theory itself. He who never applies a theory will find that he has lost it. He who does apply a theory will find he can't exhaust it.
>
> –*Niffari, tenth century mystic,* Mawaqif (Spiritual Stations)

We are now ready to apply the analytical methods developed in Chapter 6. We begin by studying a single-level quantum dot, a system which has one energy level that can accommodate up to two electrons. Next, we consider a system consisting of a single-level quantum dot in contact with a metal, where electrons can tunnel back and forth from the metal to the dot. Finally, we consider two metal electrodes separated by a thin insulating layer, and derive an expression for the tunneling current as a function of the bias voltage utilizing linear response theory. We will return to these model systems in Chapter 13, when we discuss transport in terms of the nonequilibrium Green's function method.

7.1 Single-level quantum dot

The model Hamiltonian for the single-level quantum dot is

$$H_D = \epsilon \sum_{\sigma} d_{\sigma}^{\dagger} d_{\sigma} + U n_{\uparrow} n_{\downarrow}. \tag{7.1}$$

Here, ϵ is the energy of the level, $d_{\sigma}^{\dagger}$ (d_{σ}) creates (annihilates) an electron of spin projection σ in that level, $n_{\uparrow}$ ($n_{\downarrow}$) is the number operator for spin-up (-down) electrons, and $U > 0$ is the onsite Coulomb repulsion. The second term in the Hamiltonian tends to prevent double occupancy of the energy level. If only one electron occupies the level, the energy of the dot is ϵ. If two electrons, one with

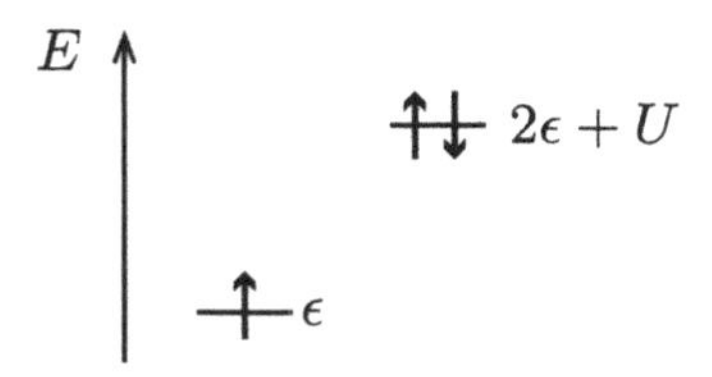

Figure 7.1 Energy of a quantum dot with a single level. If only one electron occupies the level, the energy of the dot is ϵ. If there are two electrons in the dot, the energy is $2\epsilon + U$.

spin up and the other with spin down occupy the level, in accordance with the Pauli exclusion principle, then the energy of the dot is $2\epsilon + U$ (see Figure 7.1).

The retarded Green's function of the system is

$$G_\sigma^R(t) = -i\theta(t)\langle\{d_\sigma(t), d_\sigma^\dagger(0)\}\rangle \tag{7.2}$$

where $\langle\cdots\rangle$ stands for thermal averaging, $\theta(t)$ is the step function, and $\{A, B\} = AB + BA$ is the anticommutator of operators A and B. We proceed to determine the retarded function by the equation of motion approach,

$$i\frac{\partial}{\partial t}G_\sigma^R(t) = \delta(t) + \theta(t)\langle\{\dot{d}_\sigma(t), d_\sigma^\dagger(0)\}\rangle \tag{7.3}$$

where $\dot{d}_\sigma(t) = \partial d_\sigma(t)/\partial t$. The Heisenberg equation of motion gives

$$\dot{d}_\sigma(t) = -\frac{i}{\hbar}\left[d_\sigma(t), H_D(t)\right]. \tag{7.4}$$

The commutator on the RHS of the above equation is easily determined:

$$[d_\sigma, \sum_{\sigma'} d_{\sigma'}^\dagger d_{\sigma'}] = d_\sigma$$

$$\begin{aligned}\left[d_\sigma, n_\uparrow n_\downarrow\right] &= n_\uparrow\left[d_\sigma, n_\downarrow\right] + \left[d_\sigma, n_\uparrow\right]n_\downarrow = n_\uparrow\left[d_\sigma, d_\downarrow^\dagger d_\downarrow\right] + \left[d_\sigma, d_\uparrow^\dagger d_\uparrow\right]n_\downarrow \\ &= n_\uparrow d_\downarrow \delta_{\sigma\downarrow} + d_\uparrow n_\downarrow \delta_{\sigma\uparrow} = n_{\bar{\sigma}} d_\sigma\end{aligned}$$

where $\bar{\sigma} = -\sigma$. In obtaining these results, we have used the relation $[A, BC] = B[A, C] + [A, B]C = \{A, B\}C - B\{A, C\}$, and the fact that d_σ commutes with $n_{\bar{\sigma}}$. Therefore,

$$\dot{d}_\sigma = \frac{-i}{\hbar}\left[\epsilon + U n_{\bar{\sigma}}(t)\right] d_\sigma(t). \tag{7.5}$$

Upon substituting the above result into Eq. (7.3), we obtain

$$i\frac{\partial}{\partial t}G_\sigma^R(t) = \delta(t) + \frac{\epsilon}{\hbar}G_\sigma^R(t) + \frac{U}{\hbar}\Gamma_\sigma^R(t). \tag{7.6}$$

The retarded correlation function $\Gamma_\sigma^R(t)$ is defined by

$$\Gamma_\sigma^R(t) = -i\theta(t)\langle\{n_{\bar{\sigma}}(t)d_\sigma(t), d_\sigma^\dagger(0)\}\rangle. \tag{7.7}$$

Next, we construct the equation of motion for $\Gamma_\sigma^R(t)$,

$$i\frac{\partial}{\partial t}\Gamma_\sigma^R(t) = \delta(t)\langle\{n_{\bar{\sigma}}(0)d_\sigma(0), d_\sigma^\dagger(0)\}\rangle + \theta(t)\langle\{\dot{n}_{\bar{\sigma}}(t)d_\sigma(t), d_\sigma^\dagger(0)\}\rangle$$
$$+ \theta(t)\langle\{n_{\bar{\sigma}}(t)\dot{d}_\sigma(t), d_\sigma^\dagger(0)\}\rangle.$$

Since $n_{\bar{\sigma}}$ commutes with d_σ and $d_\sigma^\dagger$, the first term is simply $\delta(t)\langle n_{\bar{\sigma}}\rangle$. The second term vanishes because $n_{\bar{\sigma}}$ commutes with the Hamiltonian H_D. Evaluating the third term, with the help of Eq. (7.5), we obtain a term containing the product $n_{\bar{\sigma}}(t)n_{\bar{\sigma}}(t)$. Noting that

$$n_{\bar{\sigma}}n_{\bar{\sigma}} = d_{\bar{\sigma}}^\dagger d_{\bar{\sigma}} d_{\bar{\sigma}}^\dagger d_{\bar{\sigma}} = d_{\bar{\sigma}}^\dagger\left(1 - d_{\bar{\sigma}}^\dagger d_{\bar{\sigma}}\right)d_{\bar{\sigma}} = d_{\bar{\sigma}}^\dagger d_{\bar{\sigma}} - d_{\bar{\sigma}}^\dagger d_{\bar{\sigma}}^\dagger d_{\bar{\sigma}} d_{\bar{\sigma}} = d_{\bar{\sigma}}^\dagger d_{\bar{\sigma}} = n_{\bar{\sigma}},$$

the equation of motion for the retarded correlation function reduces to

$$i\frac{\partial}{\partial t}\Gamma_\sigma^R(t) = \delta(t)\langle n_{\bar{\sigma}}\rangle + \frac{1}{\hbar}(\epsilon + U)\Gamma_\sigma^R(t). \tag{7.8}$$

Fourier transforming Eqs (7.6) and (7.8), we find

$$(\omega - \epsilon/\hbar + i0^+)G_\sigma^R(\omega) = 1 + (U/\hbar)\Gamma_\sigma^R(\omega) \tag{7.9}$$

$$(\omega - \epsilon/\hbar - U/\hbar + i0^+)\Gamma_\sigma^R(\omega) = \langle n_{\bar{\sigma}}\rangle. \tag{7.10}$$

These equations are readily solved,

$$G_\sigma^R(\omega) = \frac{1 - \langle n_{\bar{\sigma}}\rangle}{\omega - \epsilon/\hbar + i0^+} + \frac{\langle n_{\bar{\sigma}}\rangle}{\omega - (\epsilon + U)/\hbar + i0^+}. \tag{7.11}$$

This is the exact retarded Green's function for an isolated single-level quantum dot. The spin-resolved density of states D_σ, given by $-(1/\pi\hbar)\, Im\, G_\sigma^R(\omega)$, has two delta-function peaks, one at $\hbar\omega = \epsilon$ with a weight of $1 - \langle n_{\bar{\sigma}}\rangle$, which corresponds to the level being singly occupied, and another at $\hbar\omega = \epsilon + U$, with a weight of $\langle n_{\bar{\sigma}}\rangle$, which corresponds to double occupancy of the level.

The result (7.11) is plausible: if we add an electron, of spin projection σ, to an empty quantum dot ($\langle n_{\bar{\sigma}}\rangle = 0$), $G_\sigma^R(\omega)$ will be the first term in Eq. (7.11), with $\langle n_{\bar{\sigma}}\rangle = 0$, and it will have one pole at $\epsilon/\hbar$; hence, the energy of the added electron is ϵ. Now suppose that we add an electron, of spin projection σ, to a single-level quantum dot which is already occupied by one electron. For this to be possible, the electron initially present in the quantum dot must have spin projection $\bar{\sigma}$ and energy ϵ. In this case, $\langle n_{\bar{\sigma}}\rangle = 1$, and $G_\sigma^R(\omega)$ consists of only the second term in

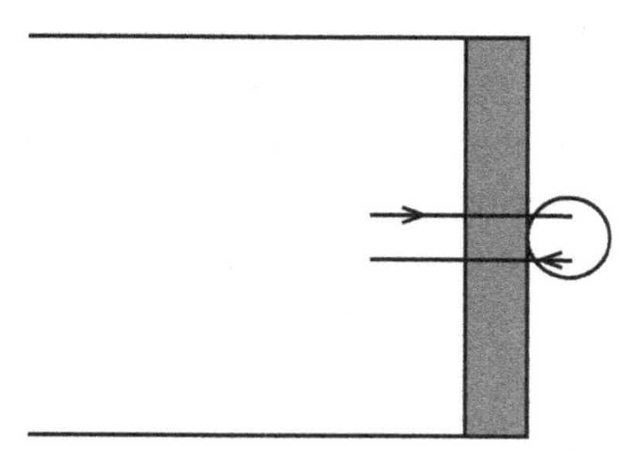

Figure 7.2 A thin insulating layer separates a metal and a quantum dot. Electrons tunnel back and forth from the metal to the dot.

Eq. (7.11), which has a pole at $(\epsilon + U)/\hbar$. The energy of the added electron is thus $\epsilon + U$, and the total energy of the two electrons is $2\epsilon + U$, as expected.

7.2 Quantum dot in contact with a metal: Anderson's model

We now consider a system which consists of a single-level quantum dot in contact with a metal surface (see Figure 7.2), paying particular attention to the effect of the interaction between the dot and the metal on the energy level in the dot. The system is described by the following model Hamiltonian:

$$H = H_e + H_D + H_T. \tag{7.12}$$

H_e is the Hamiltonian for the metal, H_D is that for the dot, and H_T describes the interaction between the metal and the dot. We assume that the electrons in the metal are noninteracting, or that each electron interacts with other electrons through a self-consistent average potential. Thus,

$$H_e = \sum_{\mathbf{k}\sigma} \epsilon_{\mathbf{k}} c^{\dagger}_{\mathbf{k}\sigma} c_{\mathbf{k}\sigma} \tag{7.13}$$

where $\epsilon_{\mathbf{k}} = \hbar^2 k^2/2m^* - \mu$, m^* is the effective electron mass, and μ is the chemical potential of the metal (the Fermi energy). As in the previous section, the Hamiltonian for the single-level quantum dot is

$$H_D = \epsilon \sum_{\sigma} d^{\dagger}_{\sigma} d_{\sigma} + U n_{\uparrow} n_{\downarrow}. \tag{7.14}$$

The interaction between the metal and the dot is described by a tunneling Hamiltonian:

$$H_T = \sum_{\mathbf{k}\sigma} \left(V_{\mathbf{k}} c^{\dagger}_{\mathbf{k}\sigma} d_{\sigma} + V^*_{\mathbf{k}} d^{\dagger}_{\sigma} c_{\mathbf{k}\sigma} \right). \tag{7.15}$$

The first term in H_T describes tunneling of an electron from the dot to the metal, a process whose matrix element $V_{\mathbf{k}}$ is assumed to be spin-independent. The second term describes tunneling from the metal to the dot. In writing H_T, we have assumed that no spin flipping occurs during tunneling. The model described above, summarized by Eqs (7.12–7.15), is known as Anderson's impurity model. It was first used to describe the localized states of a magnetic impurity inside a metal (Anderson, 1961).

We proceed as before; the retarded Green's function of the dot, now denoted by $G^R_{d\sigma}(t)$, satisfies the following equation of motion

$$i\hbar\frac{\partial}{\partial t}G^R_{d\sigma}(t) = \hbar\delta(t) + \epsilon G^R_{d\sigma}(t) + U\Gamma^R_{d\sigma}(t) + \sum_{\mathbf{k}} V^*_{\mathbf{k}} G^R_{\mathbf{k}d\sigma}(t) \tag{7.16}$$

where

$$G^R_{\mathbf{k}d\sigma}(t) = -i\theta(t)\left\langle\left\{c_{\mathbf{k}\sigma}(t), d^\dagger_\sigma(0)\right\}\right\rangle \tag{7.17}$$

is a mixed retarded Green's function whose equation of motion is

$$i\hbar\frac{\partial}{\partial t}G^R_{\mathbf{k}d\sigma}(t) = \epsilon_{\mathbf{k}} G^R_{\mathbf{k}d\sigma}(t) + V_{\mathbf{k}} G^R_{d\sigma}(t). \tag{7.18}$$

In Eq. (7.16), the term $\Gamma^R_{d\sigma}(t)$ is given by

$$\Gamma^R_{d\sigma}(t) = -i\theta(t)\langle\{n_{\bar\sigma}(t)d_\sigma(t), d^\dagger_\sigma(0)\}\rangle. \tag{7.19}$$

The equation of motion for this term is

$$i\hbar\frac{\partial}{\partial t}\Gamma^R_{d\sigma}(t) = \hbar\delta(t)\langle n_{\bar\sigma}\rangle + (\epsilon + U)\Gamma^R_{d\sigma}(t) + \sum_{\mathbf{k}} V^*_{\mathbf{k}} B^R_{\mathbf{k}d\sigma}(t) - \sum_{\mathbf{k}} V_{\mathbf{k}} C^R_{\mathbf{k}d\sigma}(t) - \sum_{\mathbf{k}} V^*_{\mathbf{k}} D^R_{\mathbf{k}d\sigma}(t) \tag{7.20}$$

where

$$B^R_{\mathbf{k}d\sigma}(t) = -i\theta(t)\langle\{n_{\bar\sigma}(t)c_{\mathbf{k}\sigma}(t), d^\dagger_\sigma(0)\}\rangle \tag{7.21}$$

$$C^R_{\mathbf{k}d\sigma}(t) = -i\theta(t)\langle\{c^\dagger_{\mathbf{k}\bar\sigma}(t)d_{\bar\sigma}(t)d_\sigma(t), d^\dagger_\sigma(0)\}\rangle \tag{7.22}$$

$$D^R_{\mathbf{k}d\sigma}(t) = -i\theta(t)\langle\{c_{\mathbf{k}\bar\sigma}(t)d^\dagger_{\bar\sigma}(t)d_\sigma(t), d^\dagger_\sigma(0)\}\rangle. \tag{7.23}$$

We have generated three new retarded functions. Equations of motion for these functions will clearly generate complicated functions whose equations of motion will produce even more complicated functions. It is a never-ending story, and we have to be satisfied with a truncated version of it if we want to obtain expressions in closed form. We use a mean field (Hartree–Fock) approximation and write

$$B^R_{\mathbf{k}d\sigma}(t) \simeq -i\theta(t)\langle n_{\bar{\sigma}}\rangle\langle\{c_{\mathbf{k}\sigma}(t), d^\dagger_\sigma(0)\}\rangle = \langle n_{\bar{\sigma}}\rangle G^R_{\mathbf{k}d\sigma}(t). \tag{7.24}$$

Within mean field approximation, $C^R_{\mathbf{k}d\sigma}$ and $D^R_{\mathbf{k}d\sigma}$ vanish, because quantities such as $\langle c^\dagger d\rangle$, $\langle d_{\bar{\sigma}} d_\sigma\rangle$, and $\langle d^\dagger_{\bar{\sigma}} d_\sigma\rangle$ are equal to zero, and Eq. (7.20) reduces to

$$i\hbar\frac{\partial}{\partial t}\Gamma^R_{d\sigma}(t) = \hbar\delta(t)\langle n_{\bar{\sigma}}\rangle + (\epsilon + U)\Gamma^R_{d\sigma}(t) + \langle n_{\bar{\sigma}}\rangle\sum_{\mathbf{k}} V^*_{\mathbf{k}} G^R_{\mathbf{k}d\sigma}(t). \tag{7.25}$$

In Fourier space, Eqs (7.16), (7.18), and (7.25) take the following forms:

$$(\hbar\omega - \epsilon + i0^+)G^R_{d\sigma}(\omega) = \hbar + U\Gamma^R_{d\sigma}(\omega) + \sum_{\mathbf{k}} V^*_{\mathbf{k}} G^R_{\mathbf{k}d\sigma}(\omega) \tag{7.26}$$

$$(\hbar\omega - \epsilon_{\mathbf{k}} + i0^+)G^R_{\mathbf{k}d\sigma}(\omega) = V_{\mathbf{k}} G^R_{d\sigma}(\omega) \tag{7.27}$$

$$(\hbar\omega - \epsilon - U + i0^+)\Gamma^R_{d\sigma}(\omega) = \langle n_{\bar{\sigma}}\rangle\left(\hbar + \sum_{\mathbf{k}} V^*_{\mathbf{k}} G^R_{\mathbf{k}d\sigma}(\omega)\right). \tag{7.28}$$

Solving these equations, we find the dot's retarded Green's function

$$G^R_{d\sigma}(\omega) = \frac{\hbar\omega - \epsilon - U + \langle n_{\bar{\sigma}}\rangle U}{(\omega - \epsilon/\hbar)(\hbar\omega - \epsilon - U) - \Sigma^R(\omega)(\hbar\omega - \epsilon - U + \langle n_{\bar{\sigma}}\rangle U)} \tag{7.29}$$

where $\Sigma^R(\omega)$ is a self energy term given by

$$\Sigma^R(\omega) = \frac{1}{\hbar}\sum_{\mathbf{k}}\frac{|V_{\mathbf{k}}|^2}{\hbar\omega - \epsilon_{\mathbf{k}} + i0^+}. \tag{7.30}$$

If we assume that Σ^R is independent of ω and that $\hbar|\Sigma^R| << U$, the poles of $G^R_{d\sigma}(\omega)$ will occur at $\omega \simeq \epsilon/\hbar + (1 - \langle n_{\bar{\sigma}}\rangle)\Sigma^R$ and $\omega = (\epsilon + U)/\hbar + \langle n_{\bar{\sigma}}\rangle\Sigma^R$. Under these assumptions, the retarded function is approximately given by

$$G^R_{d\sigma}(\omega) \simeq \frac{1 - \langle n_{\bar{\sigma}}\rangle}{\omega - \epsilon/\hbar - (1 - \langle n_{\bar{\sigma}}\rangle)\Sigma^R} + \frac{\langle n_{\bar{\sigma}}\rangle}{\omega - (\epsilon + U)/\hbar - \langle n_{\bar{\sigma}}\rangle\Sigma^R}. \tag{7.31}$$

We saw that the density of states in the isolated quantum dot consists of two delta-function peaks at ϵ and $\epsilon + U$ (see Eq. [7.11]). We now see that the effect of the quantum dot's interaction with the metal is to shift the two peaks and to broaden them: the amount of shift is proportional to the real part of Σ^R, and the amount of broadening is proportional to the imaginary part of Σ^R.

7.3 Tunneling in solids

As a final example, we calculate the tunneling current between two semi-infinite metal electrodes separated by a thin insulating layer. A current flows upon the

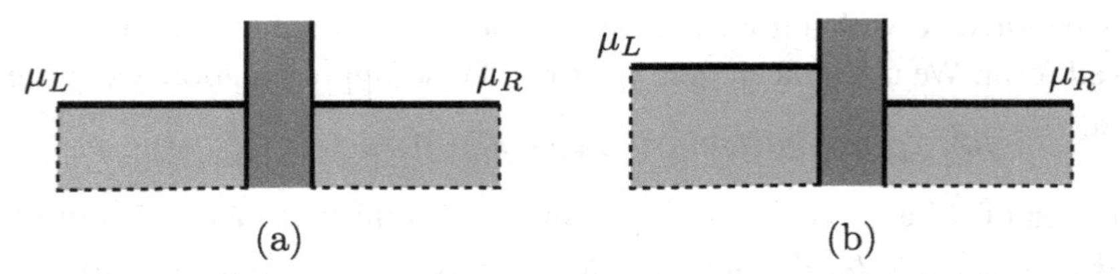

Figure 7.3 Two metallic electrodes separated by a thin insulating layer (usually metal oxide). In (a) no bias voltage is applied; the system is in equilibrium and the chemical potentials on the left and right sides of the tunnel junction are equal. In (b) a bias voltage V is applied; $eV = \mu_L - \mu_R$ and a tunneling current flows.

application of a bias voltage which raises the chemical potential of one electrode relative to the other (see Figure 7.3). We assume that the electrodes are metals in the normal state; at sufficiently low temperatures all states below the chemical potential are occupied. The treatment, however, could be generalized to allow either one or both electrodes to be superconductive. For a detailed treatment of tunneling in solids, the reader is referred to the book by Duke (Duke, 1969).

The model Hamiltonian for the tunnel junction is written as

$$H = H_L + H_R + H_T ,$$

where H_L (H_R) is the Hamiltonian for the left (right) electrode, and H_T is the tunneling Hamiltonian. In general, H_L may include the interparticle interactions in the left electrode, and H_R may include those in the right electrode. The tunneling Hamiltonian takes the form:

$$H_T = \sum_{\mathbf{kq}\sigma} \left(V_{\mathbf{kq}} b^\dagger_{\mathbf{q}\sigma} c_{\mathbf{k}\sigma} + V^*_{\mathbf{kq}} c^\dagger_{\mathbf{k}\sigma} b_{\mathbf{q}\sigma} \right) , \tag{7.32}$$

where $c^\dagger_{\mathbf{k}\sigma}$ ($c_{\mathbf{k}\sigma}$) creates (annihilates) an electron in the left electrode in the single-particle state $|\mathbf{k}\sigma\rangle$, and $b^\dagger_{\mathbf{q}\sigma}$ ($b_{\mathbf{q}\sigma}$) creates (annihilates) an electron in the right electrode in the single-particle state $|\mathbf{q}\sigma\rangle$. We assume that the creation and annihilation operators in the left electrode anticommute with those in the right electrode.

The first term in H_T describes tunneling of an electron from the state $|\mathbf{k}\sigma\rangle$ in the left electrode to the state $|\mathbf{q}\sigma\rangle$ in the right electrode; the amplitude for this process is the matrix element $V_{\mathbf{kq}}$, which is assumed to be spin-independent. The second term in H_T describes tunneling in the opposite direction. We assume that no spin flipping takes place during tunneling, which is generally true if the metal electrodes and the insulating layer are nonmagnetic. In general, a tunneling process is either elastic, in which case the energy of the electron does not change, or inelastic, whereby electron tunneling is accompanied by an excitation in the insulating layer.

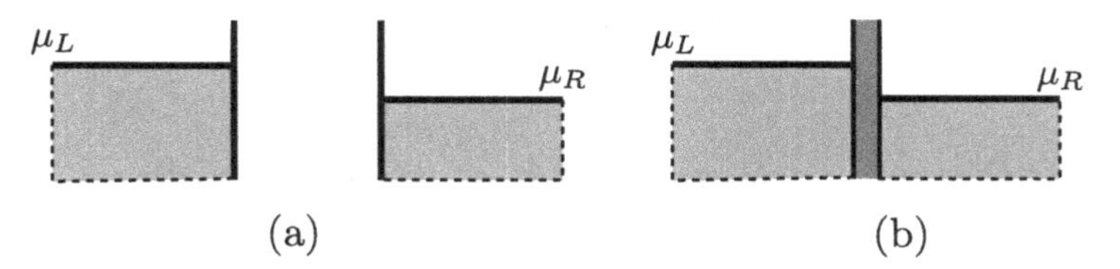

Figure 7.4 In (a) the left and right electrodes are separated: $H_T = 0$. In (b) the electrodes are brought into contact with an insulating layer: $H_T \neq 0$.

In an experimental setup, the system is initially in equilibrium, both chemical potentials μ_L and μ_R being the same. Current flows in response to a bias voltage which raises the chemical potential on one side relative to the other. The external perturbation thus corresponds to a constant upward shift in the energies of the electrons on only one side.

Our approach to the calculation of the current, however, will differ. We assume that the left and right electrodes are initially in equilibrium at their own chemical potentials, such that $\mu_L = \mu_R + eV$, but that no tunneling takes place. The initial Hamiltonian is $H_0 = H_L + H_R$; the left and right electrodes are separated. The two electrodes are then brought into contact with the insulating layer at $t = t_0$, and tunneling is switched on:

$$H = \begin{cases} H_0 = H_L + H_R & t < t_0 \\ H_L + H_R + H_T & t \geq t_0. \end{cases} \tag{7.33}$$

The external perturbation is H_T, and the current flows in response to this perturbation. This is illustrated in Figure 7.4.

The electron current is obtained from the rate of change in the number of electrons in one of the electrodes, say the left electrode,

$$I = -e\langle \dot{N}_L \rangle \tag{7.34}$$

where N_L is the number of electrons operator for the left electrode,

$$N_L = \sum_{\mathbf{k}\sigma} c^{\dagger}_{\mathbf{k}\sigma} c_{\mathbf{k}\sigma}. \tag{7.35}$$

The rate of change of N_L is given by the Heisenberg equation of motion,

$$\dot{N}_L = -\frac{i}{\hbar}[N_L, H_0 + H_T] = -\frac{i}{\hbar}[N_L, H_T]. \tag{7.36}$$

The last equality follows since N_L commutes with H_L and H_R. The commutator on the RHS is easily evaluated,

$$\dot{N}_L = \frac{i}{\hbar}\sum_{\mathbf{kq}\sigma}\left(V_{\mathbf{kq}} b^{\dagger}_{\mathbf{q}\sigma} c_{\mathbf{k}\sigma} - V^{*}_{\mathbf{kq}} c^{\dagger}_{\mathbf{k}\sigma} b_{\mathbf{q}\sigma}\right). \tag{7.37}$$

According to linear response theory (see Section 6.6), $\langle \dot{N}_L(t)\rangle_{\text{ext}}$ is given by Kubo's formula

$$\langle \dot{N}_L(t)\rangle_{\text{ext}} = \langle \dot{N}_L(t)\rangle_0 - \frac{i}{\hbar}\int_{t_0}^{t} dt' \langle [\dot{N}_L(t), H_T(t')]\rangle_0. \tag{7.38}$$

The zero subscript means that the operators evolve according to H_0,

$$\dot{N}_L(t) = e^{iH_0t/\hbar}\dot{N}_L e^{-iH_0t/\hbar}, \quad H_T(t) = e^{iH_0t/\hbar}H_T e^{-iH_0t/\hbar}. \tag{7.39}$$

Since $\langle c^{\dagger}_{\mathbf{k}\sigma} b_{\mathbf{q}\sigma}\rangle_0 = \langle b^{\dagger}_{\mathbf{q}\sigma} c_{\mathbf{k}\sigma}\rangle_0 = 0$, it follows that $\langle \dot{N}_L(t)\rangle_0 = 0$, which merely expresses the fact that, in the absence of tunneling, the current is zero. Hence,

$$\langle \dot{N}_L(t)\rangle_{\text{ext}} = -\frac{i}{\hbar}\int_{t_0}^{t} dt' \; \langle [\dot{N}_L(t), H_T(t')]\rangle_0. \tag{7.40}$$

To express the RHS of Eq. (7.40) as a retarded correlation function, we need to rewrite the Heisenberg operators $\dot{N}_L(t)$ and $H_T(t)$ as modified Heisenberg operators that evolve according to $\bar{H}_0 = H_0 - \mu_L N_L - \mu_R N_R$. Care must be exercised, since $\dot{N}_L$ and H_T do not commute with N_L and N_R.

Since both $\dot{N}_L$ and H_T are linear combinations of $b^{\dagger}_{\mathbf{q}\sigma} c_{\mathbf{k}\sigma}$ and $c^{\dagger}_{\mathbf{k}\sigma} b_{\mathbf{q}\sigma}$, we first rewrite these as modified Heisenberg operators. We note that

$$e^{iH_0t/\hbar} = e^{i\bar{H}_0t/\hbar}\, e^{i(\mu_L N_L + \mu_R N_R)t/\hbar}.$$

This equality is valid because $[\bar{H}_0, N_L] = [\bar{H}_0, N_R] = 0$. Now consider

$$\begin{aligned}
b^{\dagger}_{\mathbf{q}\sigma}(t) c_{\mathbf{k}\sigma}(t) &= e^{iH_0t/\hbar} b^{\dagger}_{\mathbf{q}\sigma} c_{\mathbf{k}\sigma} e^{-iH_0t/\hbar} \\
&= e^{i\bar{H}_0t/\hbar} e^{i(\mu_L N_L+\mu_R N_R)t/\hbar} b^{\dagger}_{\mathbf{q}\sigma} c_{\mathbf{k}\sigma} e^{-i(\mu_L N_L+\mu_R N_R)t/\hbar} e^{-i\bar{H}_0t/\hbar} \\
&= e^{i\bar{H}_0t/\hbar}\, X(t)\, e^{-i\bar{H}_0t/\hbar}
\end{aligned}$$

where we have defined the operator

$$X(t) = e^{i(\mu_L N_L+\mu_R N_R)t/\hbar}\, b^{\dagger}_{\mathbf{q}\sigma}\, c_{\mathbf{k}\sigma}\, e^{-i(\mu_L N_L+\mu_R N_R)t/\hbar}.$$

To determine $X(t)$, we take its derivative with respect to time,

$$\dot{X}(t) = \frac{i}{\hbar} e^{i(\mu_L N_L+\mu_R N_R)t/\hbar}\left[\mu_L N_L + \mu_R N_R,\, b^{\dagger}_{\mathbf{q}\sigma}\, c_{\mathbf{k}\sigma}\right] e^{-i(\mu_L N_L+\mu_R N_R)t/\hbar}.$$

The evaluation of the commutator on the RHS is straightforward; we find

$$\dot{X}(t) = -(ieV/\hbar)X(t) \Rightarrow X(t) = X(0)e^{-ieVt/\hbar} = b^{\dagger}_{\mathbf{q}\sigma}\, c_{\mathbf{k}\sigma}\, e^{-ieVt/\hbar}$$

where $eV = \mu_L - \mu_R$ and V is the bias voltage. Therefore,

$$b^{\dagger}_{\mathbf{q}\sigma}(t) c_{\mathbf{k}\sigma}(t) = e^{-ieVt/\hbar} e^{i\bar{H}_0t/\hbar}\, b^{\dagger}_{\mathbf{q}\sigma}\, c_{\mathbf{k}\sigma}\, e^{-i\bar{H}_0t/\hbar}.$$

Taking the adjoints on both sides of the above equation, we obtain

$$c_{\mathbf{k}\sigma}^{\dagger}(t)b_{\mathbf{q}\sigma}(t) = e^{ieVt/\hbar}e^{i\bar{H}_0 t/\hbar}c_{\mathbf{k}\sigma}^{\dagger}\,b_{\mathbf{q}\sigma}\,e^{-i\bar{H}_0 t/\hbar}.$$

The expression for the current now becomes

$$I(V,T) = -e\langle\dot{N}_L\rangle = \frac{e}{\hbar^2}\int_{t_0}^{t}dt'\Bigg\langle\Bigg[\sum_{\mathbf{kq}\sigma}\left(e^{-ieVt/\hbar}V_{\mathbf{kq}}b_{\mathbf{q}\sigma}^{\dagger}(t)c_{\mathbf{k}\sigma}(t) - H.C.\right),$$

$$\sum_{\mathbf{kq}\sigma}\left(e^{-ieVt'/\hbar}V_{\mathbf{kq}}b_{\mathbf{q}\sigma}^{\dagger}(t')c_{\mathbf{k}\sigma}(t') + H.C.\right)\Bigg]\Bigg\rangle_0$$

where $H.C.$ stands for hermitian conjugate. The current depends on the bias voltage V and temperature T (from thermal averaging). In the above expression, the time development of the creation and annihilation operators is governed by $\bar{H}_0 = H_L + H_R - \mu_L N_L - \mu_R N_R$. Defining the operator $A(t)$ by

$$A(t) = \sum_{\mathbf{kq}\sigma}V_{\mathbf{kq}}b_{\mathbf{q}\sigma}^{\dagger}(t)\,c_{\mathbf{k}\sigma}(t), \tag{7.41}$$

the expression for the current takes the form:

$$I(V,T) = \frac{e}{\hbar^2}\int_{t_0}^{t}dt'\Big\langle\Big[e^{-ieVt/\hbar}\,A(t) - e^{ieVt/\hbar}\,A^{\dagger}(t),$$

$$e^{-ieVt'/\hbar}\,A(t') + e^{ieVt'/\hbar}\,A^{\dagger}(t')\Big]\Big\rangle_0.$$

Since the electrodes are normal metals, $\langle A(t)A(t')\rangle_0 = \langle A^{\dagger}(t)A^{\dagger}(t')\rangle_0 = 0$, because quantities such as $\langle c_{\mathbf{k}\sigma}^{\dagger}(t)c_{\mathbf{k}\sigma}^{\dagger}(t')\rangle_0$ and $\langle c_{\mathbf{k}\sigma}(t)c_{\mathbf{k}\sigma}(t')\rangle_0$ vanish; however, this is not true if the electrodes are superconductors, as we will see in Chapter 12.

We are interested in evaluating the current long after the external perturbation has been turned on: $t \gg t_0$ (steady state); equivalently, we set $t_0 = -\infty$. Under this assumption, the current is given by

$$I = \frac{e}{\hbar^2}\int_{-\infty}^{t}dt'\left\{e^{-ieV(t-t')/\hbar}\Big\langle[A(t),A^{\dagger}(t')]\Big\rangle_0 - e^{ieV(t-t')/\hbar}\Big\langle[A^{\dagger}(t),A(t')]\Big\rangle_0\right\}.$$

A simplification of the above expression results from noting that

$$[A^{\dagger}(t),A(t')] = -[A(t),A^{\dagger}(t')]^{\dagger} \Rightarrow \Big\langle[A^{\dagger}(t),A(t')]\Big\rangle_0 = -\Big\langle[A(t),A^{\dagger}(t')]\Big\rangle_0^{*}.$$

The expression for the current thus reduces to

$$I(V,T) = \frac{2e}{\hbar^2}Re\int_{-\infty}^{t}dt'e^{-ieV(t-t')/\hbar}\Big\langle[A(t),A^{\dagger}(t')]\Big\rangle_0.$$

Since $\bar{H}_0$ is time-independent, the ensemble average on the RHS of the above equation depends on $t - t'$ and not on t and t' separately. Moreover,

$$\int_{-\infty}^{t} dt' \cdots = -\int_{\infty}^{0} d(t-t') \cdots = \int_{0}^{\infty} d(t-t') \cdots = \int_{-\infty}^{\infty} d(t-t')\, \theta(t-t') \cdots .$$

Relabeling $t - t'$ as t, we obtain

$$I(V,T) = \frac{2e}{\hbar^2}\, Re \int_{-\infty}^{\infty} dt\, \theta(t)\, e^{-ieVt/\hbar} \left\langle [A(t),\, A^{\dagger}(0)] \right\rangle_0 .$$

Finally, introducing the retarded correlation function

$$D^{R}_{AA}(t) = -i\,\theta(t) \left\langle [A(t),\, A^{\dagger}(0)] \right\rangle_0 \tag{7.42}$$

the following is obtained:

$$I(V,T) = -\frac{2e}{\hbar^2}\, Im\, D^{R}_{AA}(\omega)|_{\omega=-eV/\hbar}. \tag{7.43}$$

We have succeeded in expressing the tunneling current in terms of a retarded correlation function. We stop here, since the evaluation of this function is most easily carried out by evaluating the corresponding imaginary-time function, followed by analytic continuation. Imaginary-time Green's functions will be discussed in the next chapter. In one special case, however, the retarded function can be evaluated directly. If we assume that tunneling is elastic, and that the electrons in the left and right electrodes are noninteracting, the evaluation of the current is not difficult (see Problem 7.4).

Further reading

Bruus, H. and Flensberg, K. (2004). *Many-Body Quantum Theory in Condensed Matter Physics*. Oxford: Oxford University Press.

Mahan, G.D. (2000). *Many-Particle Physics*, 3rd edn. New York: Kluwer Academic/Plenum Publishers.

Problems

7.1 *Equation of motion for* $G^{R}_{\mathbf{k}d\sigma}$. Derive Eq. (7.18).

7.2 *Equation of motion for* $\Gamma^{R}_{d\sigma}$. Derive Eq. (7.20).

7.3 *The dot and metal.* For a single-level quantum dot in contact with a metal, assuming that Σ^R is independent of ω and that $|\Sigma^R| \ll U/\hbar$, derive Eq. (7.31). Calculate the spin resolved density of states and show that it consists of two

Lorentzians. Calculate the width of each Lorentzian. What is the physical interpretation of this result?

7.4 *Tunneling current at* $T = 0$. In Section 7.3, the tunneling current was obtained in terms of a retarded correlation function. Assume that

$$H_0 = \sum_{\mathbf{k}\sigma} \epsilon_{\mathbf{k}} c^\dagger_{\mathbf{k}\sigma} c_{\mathbf{k}\sigma} + \sum_{\mathbf{q}\sigma} \epsilon_{\mathbf{q}} b^\dagger_{\mathbf{q}\sigma} b_{\mathbf{q}\sigma}$$

where operators with **k**-subscript are for electrons in the left electrode, while those with **q**-subscript refer to electrons in the right electrode. Creation and annihilation operators with **k**-subscript anticommute with creation and annihilation operators with **q**-subscript.

(a) Show that the current is given by

$$I = \frac{4\pi e}{\hbar} \sum_{\mathbf{kq}} |V_{\mathbf{kq}}|^2 \left(f_{\mathbf{q}} - f_{\mathbf{k}}\right) \delta\left(\bar{\epsilon}_{\mathbf{q}} - \bar{\epsilon}_{\mathbf{k}} - eV\right)$$

where $\bar{\epsilon}_{\mathbf{k}(\mathbf{q})} = \epsilon_{\mathbf{k}(\mathbf{q})} - \mu_{L(R)}$, $f_{\mathbf{k}} = \left(e^{\beta\bar{\epsilon}_{\mathbf{k}}} + 1\right)^{-1}$, $f_{\mathbf{q}} = \left(e^{\beta\bar{\epsilon}_{\mathbf{q}}} + 1\right)^{-1}$.

(b) Now assume that $T = 0$, and that the bias voltage is small so that only electrons near the Fermi surface are involved in tunneling. Replace $\sum_{\mathbf{k}}$ with $D_L(0) \int d\bar{\epsilon}_{\mathbf{k}}$, and $\sum_{\mathbf{q}}$ with $D_R(0) \int d\bar{\epsilon}_{\mathbf{q}}$; $D_L(0)$ and $D_R(0)$ are the densities of states at the Fermi energy in the left and right electrodes, respectively. Assume that $V_{\mathbf{kq}}$ is independent of **k** and **q**, and is given by $\bar{V}$. Under these conditions, show that the current in the tunnel junction obeys Ohm's law: $I = V/R$, where

$$1/R = \frac{4\pi e^2}{\hbar} |\bar{V}|^2 D_L(0) D_R(0).$$

7.5 *Magnetic impurity in a metal host.* An impurity with a single level is embedded in a metal host with Fermi energy ϵ_F. The model Hamiltonian is given in Eq. (7.12). The equations of motion for $G^R_{d\sigma}(t)$ and $G^R_{\mathbf{k}d\sigma}(t)$ are given in Eqs (7.16) and (7.18). However, instead of writing an equation of motion for $\Gamma^R_{d\sigma}(t)$ and then resorting to mean field approximation, as we did in Section 7.2, let us apply the mean field approximation directly to $\Gamma^R_{d\sigma}(t)$,

$$\begin{aligned}\Gamma^R_{d\sigma}(t) &= -i\theta(t)\langle\{n_{\bar{\sigma}}(t)d_\sigma(t), d^\dagger_\sigma(0)\}\rangle \simeq -i\theta(t)\langle n_{\bar{\sigma}}\rangle\langle\{d_\sigma(t), d^\dagger_\sigma(0)\}\rangle \\ &= \langle n_{\bar{\sigma}}\rangle G^R_{d\sigma}(t).\end{aligned}$$

(a) Show that, with this approximation

$$G^R_{d\sigma}(\omega) = \left[\omega - \epsilon/\hbar - (U/\hbar)\langle n_{\bar{\sigma}}\rangle - \Sigma^R(\omega)\right]^{-1}.$$

(b) Ignoring $Re\,\Sigma^R(\omega)$, and assuming that $\Delta \equiv -\hbar\, Im\,\Sigma^R(\omega)$ is independent of ω, show that, at zero temperature

$$\langle n_\sigma \rangle = \frac{1}{\pi}\cot^{-1}\frac{\epsilon + U\langle n_{\bar{\sigma}}\rangle - \epsilon_F}{\Delta}.$$

(c) Let $m = \langle n_\uparrow \rangle - \langle n_\downarrow \rangle$. Find the transcendental equation satisfied by m. Under what conditions is $m \neq 0$?

8

Imaginary-time Green's and correlation functions

No white nor red was ever seen
So amorous as this lovely Green.
–*Andrew Marvel,* The Garden

In studying the properties of a many-particle system at finite temperature, one usually calculates the system's free energy, from which its equilibrium properties may be derived. If interactions are present, we need to use perturbation theory, since an exact solution is generally not possible. It often turns out that a perturbation expansion to lowest orders is insufficient, mainly due to the occurrence of divergent terms; we saw an example of this in Chapter 4. In general, we must carry out perturbation theory to infinite order. Clearly, however, a straightforward application of perturbation theory is not feasible (to appreciate the difficulty, try to write down the third and fourth order terms in the perturbation expansion). The Feynman diagram technique offers us a way out, since it allows for systematic study of the structure of perturbation terms of any order. At finite temperature, the diagram technique can be constructed for a particular quantity, the imaginary-time Green's function.

Real-time Green's or correlation functions, which were discussed in the previous two chapters, involve the ensemble average of the product of two operators at different times. Consider

$$\langle A(t)B(0)\rangle = Z_G^{-1}\,\mathrm{Tr}\,[e^{-\beta\bar{H}}e^{i\bar{H}t/\hbar}Ae^{-i\bar{H}t/\hbar}B],$$

where $Z_G = \mathrm{Tr}\,[e^{-\beta\bar{H}}]$ is the grand canonical partition function. The main problem arises from a mismatch in the exponents; whereas $-\beta\bar{H}$ is real, $\pm i\bar{H}t/\hbar$ is imaginary. This renders a perturbation expansion of the RHS a most difficult task. To circumvent this difficulty, we replace $it \to \tau$ (hence the name "imaginary time") and treat τ as a real quantity. As a result of this replacement, perturbation expansion becomes possible. Once the imaginary-time function is calculated, a simple recipe will yield the real-time functions that are more intimately connected to experiments. Not only that! We will find that the imaginary-time Green's function is not merely a means to an end, but that it stands on its own as a quantity of intrinsic significance: it yields the equilibrium thermodynamic properties of the system.

A perturbation expansion is also possible for the real-time Green's function at zero temperature. At $T = 0$, the ensemble average of $A(t)B(0)$ reduces to the expectation value of this quantity in the ground state; the factor $e^{-\beta \bar{H}}$ disappears, and, along with it, the problem noted earlier of the mismatch of exponents.

8.1 Imaginary-time correlation function

Given two operators $A(c, c^\dagger)$ and $B(c, c^\dagger)$ expressed through fermion or boson annihilation and creation operators, we define the imaginary-time, or finite-temperature, or Matsubara correlation function as follows

$$c^T_{AB}(\tau, \tau') = -\langle T A(\tau) B(\tau') \rangle \tag{8.1}$$

where

$$A(\tau) = e^{\bar{H}\tau/\hbar} A e^{-\bar{H}\tau/\hbar}, \quad B(\tau') = e^{\bar{H}\tau'/\hbar} B e^{-\bar{H}\tau'/\hbar} \tag{8.2}$$

are modified Heisenberg operators, $\bar{H} = H - \mu N$, μ is the chemical potential, N is the number of particles operator, and $\langle \cdots \rangle$ stands for a grand canonical ensemble average. In Eq. (8.1), T is the time-ordering operator introduced in Chapter 6,

$$T A(\tau) B(\tau') = \begin{cases} A(\tau) B(\tau') & \text{if } \tau > \tau' \\ \pm B(\tau') A(\tau) & \text{if } \tau < \tau'. \end{cases} \tag{8.3}$$

The lower (upper) sign refers to the case when A and B are fermion (boson) operators. As discussed in Chapter 6, $A(c, c^\dagger)$ is considered a boson operator if the annihilation and creation operators c and $c^\dagger$ are boson operators, or if A consists of an even number of creation and annihilation operators (e.g., if $A = c^\dagger c$, then A is a boson operator).

From the definition of the T-operator, we may rewrite Eq. (8.1) as

$$c^T_{AB}(\tau, \tau') = -\theta(\tau - \tau') \langle A(\tau) B(\tau') \rangle \mp \theta(\tau' - \tau) \langle B(\tau') A(\tau) \rangle, \tag{8.4}$$

where $\theta(\tau - \tau')$ is the step function,

$$\theta(\tau - \tau') = \begin{cases} 0 & \tau < \tau' \\ 1 & \tau > \tau'. \end{cases}$$

8.1.1 Time-dependence

If $\bar{H}$ does not depend on time, as is often the case, the imaginary-time correlation function depends on $\tau - \tau'$, not on τ and τ' separately. The proof of this statement is as follows:

$$
\begin{aligned}
c^T_{AB}(\tau, \tau') = &- \theta(\tau - \tau')\, Z_G^{-1}\, \mathrm{Tr}\, [e^{-\beta \bar{H}} e^{\bar{H}\tau/\hbar} A e^{-\bar{H}(\tau-\tau')/\hbar} B e^{-\bar{H}\tau'/\hbar}] \\
&\mp \theta(\tau' - \tau)\, Z_G^{-1}\, \mathrm{Tr}\, [e^{-\beta \bar{H}} e^{\bar{H}\tau'/\hbar} B e^{\bar{H}(\tau-\tau')/\hbar} A e^{-\bar{H}\tau/\hbar}].
\end{aligned}
$$

Using the cyclic property of the trace, we move $e^{-\bar{H}\tau'/\hbar}$ in the first term to the leftmost position, and then commute it through $e^{-\beta\bar{H}}$. In the second term we commute $e^{\bar{H}\tau'/\hbar}$ with $e^{-\beta\hbar}$, then move it to the rightmost position; we obtain

$$
\begin{aligned}
c^T_{AB}(\tau, \tau') = &- \theta(\tau - \tau')\, Z_G^{-1}\, \mathrm{Tr}\, [e^{-\beta \bar{H}} e^{\bar{H}(\tau-\tau')/\hbar} A e^{-\bar{H}(\tau-\tau')/\hbar} B] \\
&\mp \theta(\tau' - \tau)\, Z_G^{-1}\, \mathrm{Tr}\, [e^{-\beta \bar{H}} B e^{\bar{H}(\tau-\tau')/\hbar} A e^{-\bar{H}(\tau-\tau'/\hbar}] \\
= &- \theta(\tau - \tau')\langle A(\tau - \tau')B(0)\rangle \mp \theta(\tau' - \tau)\langle B(0)A(\tau - \tau')\rangle \\
= &- \langle T\, A(\tau - \tau')B(0)\rangle = c^T_{AB}(\tau - \tau').
\end{aligned}
$$

Thus, we may set $\tau' = 0$ and consider c^T_{AB} to depend only on τ:

$$
c^T_{AB}(\tau) = -\langle T\, A(\tau)B(0)\rangle. \tag{8.5}
$$

8.1.2 Periodicity

Now suppose that $\tau > 0$. Then

$$
c^T_{AB}(\tau > 0) = -\langle A(\tau)B(0)\rangle = -Z_G^{-1}\, \mathrm{Tr}\, [e^{-\beta\bar{H}} e^{\bar{H}\tau/\hbar} A e^{-\bar{H}\tau/\hbar} B]
$$

where $A = A(0)$ and $B = B(0)$. Now perform the following three steps in succession: (1) move B to the leftmost position, (2) introduce $1 = e^{+\beta\bar{H}} e^{-\beta\bar{H}}$ at the far right, and (3) move $e^{-\beta\bar{H}}$ from the rightmost to the leftmost position. We end up with

$$
\begin{aligned}
c^T_{AB}(\tau > 0) = &- Z_G^{-1}\, \mathrm{Tr}\, [e^{-\beta\bar{H}} B e^{\bar{H}(\tau-\beta\hbar)/\hbar} A e^{-\bar{H}(\tau-\beta\hbar)/\hbar}] \\
= &- Z_G^{-1}\, \mathrm{Tr}\, [e^{-\beta\bar{H}} B(0)A(\tau - \beta\hbar)] \\
= &- \langle B(0)A(\tau - \beta\hbar)\rangle.
\end{aligned} \tag{8.6}
$$

To make use of the above result, we assume that τ is restricted to vary between $-\beta\hbar$ and $\beta\hbar$: $\tau \in [-\beta\hbar, \beta\hbar]$. Then, if $\tau > 0$, $\tau - \beta\hbar$ will be negative, and the RHS of Eq. (8.6) will be $\pm c^T_{AB}(\tau - \beta\hbar)$. Hence, if $\tau > 0$,

$$
c^T_{AB}(\tau) = \pm c^T_{AB}(\tau - \beta\hbar). \tag{8.7}
$$

The lower (upper) sign refers to fermions (bosons). Since $-\beta\hbar \le \tau \le \beta\hbar$, we can decompose $c^T_{AB}(\tau)$ into a Fourier series

$$c^T_{AB}(\tau) = \frac{1}{\beta\hbar}\sum_n c^T_{AB}(\omega_n)e^{-i\omega_n\tau}. \tag{8.8}$$

The constraint imposed by Eq. (8.7) implies that

$$e^{-i\omega_n\tau} = \pm e^{-i\omega_n(\tau-\beta\hbar)} \Rightarrow e^{i\omega_n\beta\hbar} = \pm 1$$

$$\Rightarrow \omega_n = \begin{cases} 2n\pi/\beta\hbar & n \in \mathbb{Z} \quad \text{bosons} \\ (2n+1)\pi/\beta\hbar & n \in \mathbb{Z} \quad \text{fermions.} \end{cases} \tag{8.9}$$

We can obtain $c^T_{AB}(\omega_n)$ in terms of $c^T_{AB}(\tau)$: multiply Eq. (8.8) by $e^{i\omega_m\tau}$ and integrate over τ from $-\beta\hbar$ to $\beta\hbar$,

$$\int_{-\beta\hbar}^{\beta\hbar} e^{i\omega_m\tau} c^T_{AB}(\tau)d\tau = \frac{1}{\beta\hbar}\sum_n c^T_{AB}(\omega_n)\int_{-\beta\hbar}^{\beta\hbar} e^{i(\omega_m-\omega_n)\tau}d\tau.$$

Since $\omega_m - \omega_n = 2(m-n)\pi/\beta\hbar$, the integral on the RHS vanishes unless $n = m$, in which case it is equal to $2\beta\hbar$. Hence,

$$c^T_{AB}(\omega_n) = \frac{1}{2}\int_{-\beta\hbar}^{\beta\hbar} c^T_{AB}(\tau)e^{i\omega_n\tau}d\tau. \tag{8.10a}$$

This is true for both fermions and bosons. We proceed further,

$$c^T_{AB}(\omega_n) = \frac{1}{2}\left[\int_{-\beta\hbar}^{0} c^T_{AB}(\tau)e^{i\omega_n\tau}d\tau + \int_{0}^{\beta\hbar} c^T_{AB}(\tau)e^{i\omega_n\tau}d\tau\right].$$

Making use of $c^T_{AB}(\tau < 0) = \pm c^T_{AB}(\tau + \beta\hbar)$ and $e^{i\omega_n\beta\hbar} = \pm 1$, we can write

$$\int_{-\beta\hbar}^{0} c^T_{AB}(\tau)e^{i\omega_n\tau}d\tau = \int_{-\beta\hbar}^{0} c^T_{AB}(\tau+\beta\hbar)e^{i\omega_n(\tau+\beta\hbar)}d\tau = \int_{0}^{\beta\hbar} c^T_{AB}(\tau)e^{i\omega_n\tau}d\tau.$$

In the last step, we have made a change of variable: $\tau \to \tau + \beta\hbar$. Hence,

$$c^T_{AB}(\omega_n) = \int_{0}^{\beta\hbar} c^T_{AB}(\tau)e^{i\omega_n\tau}d\tau. \tag{8.10b}$$

8.2 Imaginary-time Green's function

The imaginary-time Green's function, also known as the finite-temperature Green's function, or Matsubara Green's function, is defined as

$$g(\mathbf{r}\sigma\tau, \mathbf{r}'\sigma'\tau') = -\langle T\,\Psi_\sigma(\mathbf{r}\tau)\Psi^\dagger_{\sigma'}(\mathbf{r}'\tau')\rangle \tag{8.11}$$

where the τ-dependent field operators are given by

$$\Psi_\sigma(\mathbf{r}\tau) = e^{\bar{H}\tau/\hbar}\Psi_\sigma(\mathbf{r})e^{-\bar{H}\tau/\hbar}, \quad \Psi_\sigma^\dagger(\mathbf{r}\tau) = e^{\bar{H}\tau/\hbar}\Psi_\sigma^\dagger(\mathbf{r})e^{-\bar{H}\tau/\hbar}.$$

The imaginary-time Green's function is a special case of the imaginary-time correlation function $c_{AB}^T(\tau, \tau')$, obtained by setting $A = \Psi_\sigma(\mathbf{r})$ and $B = \Psi_{\sigma'}^\dagger(\mathbf{r}')$. We note that $\Psi_\sigma^\dagger(\mathbf{r}\tau)$ is not the adjoint of $\Psi_\sigma(\mathbf{r}\tau)$.

From the definition of the T-operator, we can write

$$g(\mathbf{r}\sigma\tau, \mathbf{r}'\sigma'\tau') = \begin{cases} -\langle\Psi_\sigma(\mathbf{r}\tau)\Psi_{\sigma'}^\dagger(\mathbf{r}'\tau')\rangle & \tau > \tau' \\ \mp\langle\Psi_{\sigma'}^\dagger(\mathbf{r}'\tau')\Psi_\sigma(\mathbf{r}\tau)\rangle & \tau < \tau'. \end{cases} \tag{8.12}$$

The lower (upper) sign refers to fermions (bosons). For $\tau > \tau'$, $g(\mathbf{r}\sigma\tau, \mathbf{r}'\sigma'\tau')$ is the probability amplitude of finding the system with one extra particle of spin projection σ at position $\mathbf{r}$ and time τ if a particle with spin projection σ' was added to the system at position $\mathbf{r}'$ at an earlier time τ'. For $\tau < \tau'$, $g(\mathbf{r}\sigma\tau, \mathbf{r}'\sigma'\tau')$ is the probability amplitude of finding the system with one less particle of spin projection σ' at time τ' if one particle with spin projection σ was removed from position $\mathbf{r}$ at an earlier time τ. We note the following:

1. In the absence of spin-dependent interactions that could flip a particle's spin, σ and σ' must be the same.
2. Since $\bar{H}$ is time-independent, $g(\mathbf{r}\sigma\tau, \mathbf{r}'\sigma'\tau')$ depends on $\tau - \tau'$, not on τ and τ' independently.
3. For a translationally invariant system, $g(\mathbf{r}\sigma\tau, \mathbf{r}'\sigma'\tau')$ does not change if $\mathbf{r} \to \mathbf{r} + \mathbf{R}$, $\mathbf{r}' \to \mathbf{r}' + \mathbf{R}$; hence, $g(\mathbf{r}\sigma\tau, \mathbf{r}'\sigma'\tau')$ depends on $\mathbf{r} - \mathbf{r}'$, not on $\mathbf{r}$ and $\mathbf{r}'$ independently.

With these thoughts in mind, the imaginary-time Green's function is written $g(\mathbf{r} - \mathbf{r}'\sigma, \tau)$, and we consider its spatial Fourier transform $g(\mathbf{k}\sigma, \tau)$,

$$g(\mathbf{r} - \mathbf{r}'\sigma, \tau) = \frac{1}{V}\sum_{\mathbf{k}} e^{i\mathbf{k}\cdot(\mathbf{r}-\mathbf{r}')}g(\mathbf{k}\sigma, \tau) \tag{8.13}$$

where V is the system's volume, and

$$g(\mathbf{k}\sigma, \tau) = -\langle T c_{\mathbf{k}\sigma}(\tau)c_{\mathbf{k}\sigma}^\dagger(0)\rangle. \tag{8.14}$$

This expression for $g(\mathbf{k}\sigma, \tau)$ is obtained by expanding the field operators:

$$\Psi_\sigma(\mathbf{r}\tau) = \frac{1}{\sqrt{V}}\sum_{\mathbf{k}} e^{i\mathbf{k}\cdot\mathbf{r}}c_{\mathbf{k}\sigma}(\tau), \quad \Psi_\sigma^\dagger(\mathbf{r}\tau) = \frac{1}{\sqrt{V}}\sum_{\mathbf{k}} e^{-i\mathbf{k}\cdot\mathbf{r}}c_{\mathbf{k}\sigma}^\dagger(\tau) \tag{8.15}$$

and using the translational invariance property, exactly as we did in Chapter 6 when we found $G^R(\mathbf{k}\sigma, t)$. Equations (8.11) and (8.14) are the definitions of the imaginary-time Green's function in the position and momentum representations, respectively. We may consider a more general definition using the ν-representation,

$$g(\nu\tau, \nu'\tau') = -\langle T c_\nu(\tau) c^\dagger_{\nu'}(\tau')\rangle$$

where $\{|\phi_\nu\rangle\}$ is a complete set of single-particle states, ν stands for all the quantum numbers that characterize the states, and $c^\dagger_\nu$ (c_ν) creates (annihilates) a particle in the single-particle state $|\phi_\nu\rangle$.

8.3 Significance of the imaginary-time Green's function

Once Green's function is determined, the thermodynamic equilibrium properties of the system can be found. Let $\tau^+ = \tau + 0^+$, and consider

$$\begin{aligned}\sum_\sigma g(\mathbf{r}\sigma\tau, \mathbf{r}\sigma\tau^+) &= -\sum_\sigma \langle T\Psi_\sigma(\mathbf{r}\tau)\Psi^\dagger_\sigma(\mathbf{r}\tau^+)\rangle = \mp\sum_\sigma \langle \Psi^\dagger_\sigma(\mathbf{r}\tau^+)\Psi_\sigma(\mathbf{r}\tau)\rangle \\ &= \mp Z_G^{-1}\sum_\sigma \mathrm{Tr}\,[e^{-\beta\bar{H}}\, e^{\bar{H}\tau/\hbar}\,\Psi^\dagger_\sigma(\mathbf{r})\Psi_\sigma(\mathbf{r})e^{-\bar{H}\tau/\hbar}].\end{aligned}$$

The lower (upper) sign refers to fermions (bosons). Using the cyclic property of the trace, we move $e^{-\bar{H}\tau/\hbar}$ to the far left and commute it through $e^{-\beta\bar{H}}$,

$$\begin{aligned}\sum_\sigma g(\mathbf{r}\sigma\tau, \mathbf{r}\sigma\tau^+) &= \mp Z_G^{-1}\sum_\sigma \mathrm{Tr}\,[e^{-\beta\bar{H}}\,\Psi^\dagger_\sigma(\mathbf{r})\Psi_\sigma(\mathbf{r})] \\ &= \mp Z_G^{-1}\mathrm{Tr}\,[e^{-\beta\bar{H}}\sum_\sigma \Psi^\dagger_\sigma(\mathbf{r})\Psi_\sigma(\mathbf{r})] = \mp\langle n(\mathbf{r})\rangle \qquad (8.16)\end{aligned}$$

where $n(\mathbf{r})$ is the particle-number density operator. The ensemble average of the number of particles in a system of volume V is

$$N(V, T, \mu) = \int_V \langle n(\mathbf{r})\rangle d^3r = \mp\int_V \sum_\sigma g(\mathbf{r}\sigma\tau, \mathbf{r}\sigma\tau^+)d^3r. \qquad (8.17)$$

The dependence of N on T and μ results from $\beta\bar{H} = \beta(H - \mu N)$ in the ensemble average of $n(\mathbf{r})$. We can solve the above equation for $\mu(N, T, V)$ and determine the Helmholtz free energy F from the relation

$$\mu = \left.\frac{\partial F}{\partial N}\right|_{T,V}. \qquad (8.18)$$

Once F is found, the thermodynamic properties of the system can be derived. For a translationally invariant system, we can also write

$$N(V,T,\mu) = \sum_{\mathbf{k}\sigma} \langle c^\dagger_{\mathbf{k}\sigma} c_{\mathbf{k}\sigma} \rangle = \sum_{\mathbf{k}\sigma} \langle c^\dagger_{\mathbf{k}\sigma}(0) c_{\mathbf{k}\sigma}(0) \rangle = \pm \sum_{\mathbf{k}\sigma} \langle T\, c_{\mathbf{k}\sigma}(0) c^\dagger_{\mathbf{k}\sigma}(0^+) \rangle$$

$$\Rightarrow N(V,T,\mu) = \mp \sum_{\mathbf{k}\sigma} g(\mathbf{k}\sigma, \tau = 0^-) \tag{8.19}$$

where $g(\mathbf{k}\sigma, \tau = 0^-) = -\langle T\, c_{\mathbf{k}\sigma}(0) c^\dagger_{\mathbf{k}\sigma}(0^+) \rangle$. In this case, the dependence of N on V results from the replacement $\sum_{\mathbf{k}} \to V/(2\pi)^3 \int d^3k$.

In general, we can express the ensemble average of any one-body operator, such as the number density $n(\mathbf{r})$, in terms of Green's function. Consider a one-body operator $F = \sum_i f(i)$. Its second quantized form is

$$F = \sum_{\nu,\nu'} \langle \phi_{\nu'} | f | \phi_\nu \rangle c^\dagger_{\nu'} c_\nu \tag{8.20}$$

where ν stands for all the quantum numbers that characterize the single-particle state $|\phi_\nu\rangle$. Taking the ensemble average, we obtain

$$\langle F \rangle = \sum_{\nu,\nu'} \langle \phi_{\nu'} | f | \phi_\nu \rangle \langle c^\dagger_{\nu'} c_\nu \rangle. \tag{8.21}$$

The matrix element, a c-number, has been moved outside the ensemble average. Writing ϕ_ν as $\phi_n(\mathbf{r})|\sigma\rangle$, where n represents the orbital (spatial) quantum numbers, and denoting $\langle \sigma' | f | \sigma \rangle$ by $f_{\sigma'\sigma}(\mathbf{r})$, Eq. (8.21) becomes

$$\begin{aligned}\langle F \rangle &= \sum_{\sigma,\sigma'} \sum_{n,n'} \int d^3r\, \phi^*_{n'}(\mathbf{r}) f_{\sigma'\sigma}(\mathbf{r}) \phi_n(\mathbf{r}) \langle c^\dagger_{n'\sigma'} c_{n\sigma} \rangle \\ &= \sum_{\sigma,\sigma'} \int d^3r \lim_{\mathbf{r}' \to \mathbf{r}} f_{\sigma'\sigma}(\mathbf{r}) \langle \Psi^\dagger_{\sigma'}(\mathbf{r}') \Psi_\sigma(\mathbf{r}) \rangle.\end{aligned}$$

We have used the relations connecting the field operators to the creation and annihilation operators,

$$\Psi_\sigma(\mathbf{r}) = \sum_n \phi_n(\mathbf{r}) c_{n\sigma}, \quad \Psi^\dagger_\sigma(\mathbf{r}) = \sum_n \phi^*_n(\mathbf{r}) c^\dagger_{n\sigma}. \tag{8.22}$$

Note the necessity of introducing $\mathbf{r}'$ and taking the limit $\mathbf{r}' \to \mathbf{r}$: $f_{\sigma'\sigma}(\mathbf{r})$ is an operator that acts on $\phi_n(\mathbf{r})$; by introducing $\mathbf{r}'$, we make it possible for $f_{\sigma'\sigma}(\mathbf{r})$ to act on the product $\langle \psi^\dagger_{\sigma'}(\mathbf{r}') \psi_\sigma(\mathbf{r}) \rangle$. For any two operators A and B,

$$\begin{aligned}\langle A(\mathbf{r}'\tau) B(\mathbf{r}\tau) \rangle &= Z_G^{-1}\, \mathrm{Tr}[e^{-\beta \bar{H}} e^{\bar{H}\tau/\hbar} A(\mathbf{r}') B(\mathbf{r}) e^{-\bar{H}\tau/\hbar}] \\ &= Z_G^{-1}\, \mathrm{Tr}[e^{-\beta \bar{H}} A(\mathbf{r}') B(\mathbf{r})] = \langle A(\mathbf{r}') B(\mathbf{r}) \rangle.\end{aligned}$$

In the penultimate step, $e^{-\bar{H}\tau/\hbar}$ was moved to the left and commuted through $e^{-\beta\bar{H}}$. It follows that

$$\begin{aligned}\langle F\rangle &= \sum_{\sigma,\sigma'}\int d^3r \lim_{\mathbf{r}'\to\mathbf{r}} f_{\sigma'\sigma}(\mathbf{r})\langle\Psi^\dagger_{\sigma'}(\mathbf{r}',\tau)\Psi_\sigma(\mathbf{r},\tau)\rangle\\
&= \sum_{\sigma,\sigma'}\int d^3r \lim_{\mathbf{r}'\to\mathbf{r}}\lim_{\tau'\to\tau^+} f_{\sigma'\sigma}(\mathbf{r})\langle\Psi^\dagger_{\sigma'}(\mathbf{r}',\tau')\Psi_\sigma(\mathbf{r},\tau)\rangle\\
&= \pm\sum_{\sigma,\sigma'}\int d^3r \lim_{\mathbf{r}'\to\mathbf{r}}\lim_{\tau'\to\tau^+} f_{\sigma'\sigma}(\mathbf{r})\,\langle T\Psi_\sigma(\mathbf{r},\tau)\Psi^\dagger_{\sigma'}(\mathbf{r}',\tau')\rangle\\
&= \mp\int d^3r \lim_{\mathbf{r}'\to\mathbf{r}}\lim_{\tau'\to\tau^+}\sum_{\sigma\sigma'} f_{\sigma'\sigma}(\mathbf{r})g(\mathbf{r}\sigma\tau,\mathbf{r}'\sigma'\tau').\end{aligned} \tag{8.23}$$

As an example, the ensemble average of the kinetic energy is

$$\langle T\rangle = \mp\int d^3r \lim_{\mathbf{r}'\to\mathbf{r}}\sum_\sigma\left(-\frac{\hbar^2}{2m}\nabla^2\right)g(\mathbf{r}\sigma\tau,\mathbf{r}'\sigma\tau^+).$$

Likewise, we can express $\langle V\rangle$, the ensemble average of the potential energy (assumed to arise from pairwise interaction), in terms of Green's function; it is

$$\langle V\rangle = \mp\frac{1}{2}\int d^3r \lim_{\mathbf{r}'\to\mathbf{r}}\lim_{\tau'\to\tau^+}\left(-\hbar\frac{\partial}{\partial\tau}+\frac{\hbar^2}{2m}\nabla^2+\mu\right)\sum_\sigma g(\mathbf{r}\sigma\tau,\mathbf{r}'\sigma\tau'). \tag{8.24}$$

The internal energy $E(N,V,T)$ of a system of interacting particles, given by $\langle T\rangle+\langle V\rangle$, can be expressed as

$$E(N,V,T) = \mp\frac{1}{2}\int d^3r \lim_{\mathbf{r}'\to\mathbf{r}}\lim_{\tau'\to\tau^+}\left[-\hbar\frac{\partial}{\partial\tau}-\frac{\hbar^2}{2m}\nabla^2+\mu\right]\sum_\sigma g(\mathbf{r}\sigma\tau,\mathbf{r}'\sigma\tau'). \tag{8.25}$$

The thermodynamic potential, $\Omega(T,V,\mu)$, is given by

$$\begin{aligned}\Omega(T,V,\mu) = \Omega_0(T,V,\mu) &\mp \frac{1}{2}\int_0^1\frac{d\lambda}{\lambda}\int d^3r \lim_{\mathbf{r}'\to\mathbf{r}}\lim_{\tau'\to\tau^+}\left(-\hbar\frac{\partial}{\partial\tau}+\frac{\hbar^2}{2m}\nabla^2+\mu\right)\\
&\times\sum_\sigma g^\lambda(\mathbf{r}\sigma\tau,\mathbf{r}'\sigma\tau')\end{aligned} \tag{8.26}$$

where $g^\lambda(\mathbf{r}\sigma\tau,\mathbf{r}'\sigma\tau')$ is Green's function for a system with Hamiltonian $\bar{H}(\lambda) = \bar{H}_0+\lambda V$, and Ω_0 is the thermodynamic potential for a system of noninteracting particles. In Problems 8.1 and 8.2, the method used to derive expressions for $\langle V\rangle$, E, and Ω is outlined.

8.4 Spectral representation, relation to real-time functions

Our next task is to derive spectral representations of imaginary-time Green's and correlation functions. We shall obtain the real-time functions from their imaginary-time counterparts.

8.4.1 Imaginary-time Green's function

To evaluate $g(\mathbf{k}\sigma, \omega_n)$, we proceed as follows. From Eq. (8.10b),

$$g(\mathbf{k}\sigma, \omega_n) = \int_0^{\beta\hbar} g(\mathbf{k}\sigma, \tau)e^{i\omega_n \tau} d\tau = \int_0^{\beta\hbar} g^{>}(\mathbf{k}\sigma, \tau)e^{i\omega_n \tau} d\tau \tag{8.27}$$

where

$$\begin{aligned} g^{>}(\mathbf{k}\sigma, \tau) &= g(\mathbf{k}\sigma, \tau > 0) = -Z_G^{-1}\mathrm{Tr}[e^{-\beta\bar{H}} c_{\mathbf{k}\sigma}(\tau)c^{\dagger}_{\mathbf{k}\sigma}(0)] \\ &= -Z_G^{-1}\mathrm{Tr}[e^{-\beta\bar{H}} e^{\bar{H}\tau/\hbar} c_{\mathbf{k}\sigma} e^{-\bar{H}\tau/\hbar} c^{\dagger}_{\mathbf{k}\sigma}] \\ &= -Z_G^{-1}\sum_{n,m}\langle n|e^{-\beta\bar{H}} e^{\bar{H}\tau/\hbar} c_{\mathbf{k}\sigma}|m\rangle\langle m|e^{-\bar{H}\tau/\hbar} c^{\dagger}_{\mathbf{k}\sigma}|n\rangle \\ &= -Z_G^{-1}\sum_{n,m} e^{-\beta\bar{E}_n} e^{-(\bar{E}_m-\bar{E}_n)\tau/\hbar}\langle n|c_{\mathbf{k}\sigma}|m\rangle\langle m|c^{\dagger}_{\mathbf{k}\sigma}|n\rangle \\ &= \int_{-\infty}^{\infty} P^{>}(\mathbf{k}\sigma, \epsilon)e^{-\epsilon\tau}\frac{d\epsilon}{2\pi} \end{aligned} \tag{8.28}$$

where

$$P^{>}(\mathbf{k}\sigma, \epsilon) = -2\pi\; Z_G^{-1}\sum_{n,m} e^{-\beta\bar{E}_n}\; |\langle m|c^{\dagger}_{\mathbf{k}\sigma}|n\rangle|^2\delta\left(\epsilon - \frac{1}{\hbar}(\bar{E}_m - \bar{E}_n)\right). \tag{8.29}$$

This is exactly the same function which we obtained in Chapter 6 (see Eq. [6.31]) when we developed the spectral representation of the retarded Green's function. Equation (8.27) now becomes

$$\begin{aligned} g(\mathbf{k}\sigma, \omega_n) &= \int_{-\infty}^{\infty} P^{>}(\mathbf{k}\sigma, \epsilon)\frac{d\epsilon}{2\pi}\int_0^{\beta\hbar} e^{(i\omega_n-\epsilon)\tau}d\tau \\ &= \int_{-\infty}^{\infty} P^{>}(\mathbf{k}\sigma, \epsilon)\frac{d\epsilon}{2\pi}\left.\frac{e^{(i\omega_n-\epsilon)\tau}}{i\omega_n-\epsilon}\right|_0^{\beta\hbar} = -\int_{-\infty}^{\infty} P^{>}(\mathbf{k}\sigma, \epsilon)\frac{(1\mp e^{-\beta\hbar\epsilon})}{i\omega_n-\epsilon}\frac{d\epsilon}{2\pi} \\ &= \int_{-\infty}^{\infty}\frac{A(\mathbf{k}\sigma, \epsilon)}{i\omega_n-\epsilon}\frac{d\epsilon}{2\pi} \end{aligned} \tag{8.30}$$

where

$$A(\mathbf{k}\sigma, \epsilon) = -P^{>}(\mathbf{k}\sigma, \epsilon)(1\mp e^{-\beta\hbar\epsilon}) \tag{8.31}$$

is the spectral density function, and the lower (upper) sign refers to fermions (bosons). The P-greater function can be written as

$$P^{>}(\mathbf{k}\sigma,\epsilon) = \frac{-A(\mathbf{k}\sigma,\epsilon)}{1 \mp e^{-\beta\hbar\epsilon}} = \begin{cases} -(1+n_\epsilon)A(\mathbf{k}\sigma,\epsilon) & \text{bosons} \\ -(1-f_\epsilon)A(\mathbf{k}\sigma,\epsilon) & \text{fermions.} \end{cases} \tag{8.32}$$

Using the periodicity/antiperiodicity property of the boson/fermion Green's function, we can also write for $g^{<}(\mathbf{k}\sigma,\tau) = g(\mathbf{k}\sigma,\tau<0)$,

$$g^{<}(\mathbf{k}\sigma,\tau) = \int_{-\infty}^{\infty} P^{<}(\mathbf{k}\sigma,\epsilon)e^{-\epsilon\tau}\frac{d\epsilon}{2\pi} \tag{8.33}$$

where

$$P^{<}(\mathbf{k}\sigma,\epsilon) = \begin{cases} -n_\epsilon A(\mathbf{k}\sigma,\epsilon) & \text{bosons} \\ f_\epsilon A(\mathbf{k}\sigma,\epsilon) & \text{fermions.} \end{cases} \tag{8.34}$$

On the other hand, the retarded Green's function is given by Eq. (6.36),

$$G^R(\mathbf{k}\sigma,\omega) = \int_{-\infty}^{\infty} \frac{A(\mathbf{k}\sigma,\epsilon)}{\omega-\epsilon+i0^+}\frac{d\epsilon}{2\pi}.$$

Assuming that $g(\mathbf{k}\sigma,\omega_n)$ is found for all positive values of $i\omega_n$ (these form discrete points on the upper half of the imaginary axis in the complex ω-plane), how do we construct $G^R(\mathbf{k}\sigma,\omega)$? Consider the function $F(\mathbf{k}\sigma,z)$ of the complex variable z, defined by

$$F(\mathbf{k}\sigma,z) = \int_{-\infty}^{\infty} \frac{A(\mathbf{k}\sigma,\epsilon)}{z-\epsilon}\frac{d\epsilon}{2\pi}.$$

This function is analytic everywhere except on the real axis. Furthermore,

$$G^R(\mathbf{k}\sigma,\omega) = F(\mathbf{k}\sigma,z=\omega+i0^+), \quad g(\mathbf{k}\sigma,\omega_n) = F(\mathbf{k}\sigma,z=i\omega_n).$$

Therefore, both G^R and g can be found once F is known. From knowing $g(\mathbf{k}\sigma,\omega_n)$, we can know $F(\mathbf{k}\sigma,z)$ only on a discrete set of points along the imaginary axis. To obtain $F(\mathbf{k}\sigma,z)$ everywhere in the upper half-plane, we need to analytically continue $F(\mathbf{k}\sigma,i\omega_n)$ from the discrete set of points onto the entire upper half-plane. If we succeed in doing that, replacement of z in $F(\mathbf{k}\sigma,z)$ by $\omega+i0^+$ will produce $G^R(\mathbf{k}\sigma,\omega)$. In other words

$$G^R(\mathbf{k}\sigma,\omega) = g(\mathbf{k}\sigma,i\omega_n)|_{i\omega_n\to\omega+i0^+}\,. \tag{8.35}$$

This is the analytical continuation recipe for obtaining the real-time retarded Green's function from its imaginary-time counterpart. We note that the advanced real-time Green's function is obtained from the imaginary-time Green's function by a similar recipe: $i\omega_n \to \omega - i0^+$.

The construction of G^R from g hinges on the ability to analytically continue g, from a discrete set of points on the upper half of the imaginary ω-axis, onto the upper half ω-plane. Although there is no definite algorithm for doing so, in practice, we first calculate $g(\mathbf{k}\sigma, \omega_n)$, then replace $i\omega_n$ with z; if the resulting function is analytic in the upper half-plane, then we have found $F(\mathbf{k}\sigma, z)$, and $G^R(\mathbf{k}\sigma, \omega)$ is obtained by replacing z with $\omega + i0^+$. If this procedure fails, we can still obtain the retarded Green's function by analytically continuing the Feynman diagrams of the imaginary-time Green's function. This is discussed in Chapter 9.

8.4.2 Imaginary-time correlation function

The imaginary-time correlation function and its Fourier transform are given by Eqs (8.1) and (8.10b),

$$c^T_{AB}(\tau) = -\langle T A(\tau)B(0)\rangle, \quad c^T_{AB}(\omega_n) = -\int_0^{\beta\hbar} \langle A(\tau)B(0)\rangle e^{i\omega_n\tau}\,d\tau.$$

In writing $c^T_{AB}(\omega_n)$ we dropped the T-operator since $\tau > 0$, the integration over τ being from 0 to $\beta\hbar$. We rewrite the ensemble average, introducing a resolution of identity,

$$\begin{aligned}\langle A(\tau)B(0)\rangle &= Z_G^{-1}\,\mathrm{Tr}\,[e^{-\beta\bar{H}}e^{\bar{H}\tau/\hbar}Ae^{-\bar{H}\tau/\hbar}B] \\ &= Z_G^{-1}\sum_{n,m}\langle n|e^{-\beta\bar{H}}e^{\bar{H}\tau/\hbar}A|m\rangle\langle m|e^{-\bar{H}\tau/\hbar}B|n\rangle \\ &= Z_G^{-1}\sum_{n,m}e^{-\beta\bar{E}_n}e^{-(\bar{E}_m-\bar{E}_n)\tau/\hbar}\langle n|A|m\rangle\langle m|B|n\rangle.\end{aligned}$$

Therefore,

$$c^T_{AB}(\omega_n) = -Z_G^{-1}\sum_{n,m}e^{-\beta\bar{E}_n}\langle n|A|m\rangle\langle m|B|n\rangle\int_0^{\beta\hbar} e^{[i\omega_n-(\bar{E}_m-\bar{E}_n)/\hbar]\tau}\,d\tau.$$

Since $e^{i\omega_n\beta\hbar} = \pm 1$, the above expression reduces to

$$c^T_{AB}(\omega_n) = Z_G^{-1}\sum_{n,m}\frac{\langle n|A|m\rangle\langle m|B|n\rangle(e^{-\beta\bar{E}_n}\mp e^{-\beta\bar{E}_m})}{i\omega_n-(\bar{E}_m-\bar{E}_n)/\hbar}. \tag{8.36}$$

Comparing this expression with that for $C^R_{AB}(\omega)$, Eq. (6.47), and bearing in mind our discussion in the previous subsection regarding analytic continuation and its possible complications, we deduce that

$$C^R_{AB}(\omega) = c^T_{AB}(\omega_n)\big|_{i\omega_n=\omega+i0^+}. \tag{8.37}$$

Therefore, in order to calculate $C^R_{AB}(\omega)$, we first calculate $c^T_{AB}(\omega_n)$ and then replace $i\omega_n$ with $\omega + i0^+$. As a final cautionary remark, we note that Eq. (8.37) is valid only if $\langle A\rangle$ and/or $\langle B\rangle$ vanish (see Problem 8.8).

8.5 Example: Green's function for noninteracting particles

As an example, let us calculate the imaginary-time Green's function for a system of noninteracting particles. The Hamiltonian is given by

$$\bar{H}_0 = \sum_{\mathbf{k}\sigma}(\epsilon_{\mathbf{k}\sigma} - \mu)c^\dagger_{\mathbf{k}\sigma}c_{\mathbf{k}\sigma} = \sum_{\mathbf{k}} \bar{\epsilon}_{\mathbf{k}\sigma}c^\dagger_{\mathbf{k}\sigma}c_{\mathbf{k}\sigma}. \tag{8.38}$$

$\bar{\epsilon}_{\mathbf{k}\sigma}$ is the single-particle state energy relative to the chemical potential.

8.5.1 Derivation from the spectral density function

The spectral density function for noninteracting particles is given by

$$A^0(\mathbf{k}\sigma, \epsilon) = 2\pi\delta(\epsilon - \bar{\epsilon}_{\mathbf{k}\sigma}/\hbar) \tag{8.39}$$

(see Eq. [6.55]). Thus, the imaginary-time Green's function for a system of noninteracting particles (bosons or fermions) is given by

$$g^0(\mathbf{k}\sigma, \omega_n) = \int_{-\infty}^{\infty} \frac{A^0(\mathbf{k}\sigma, \epsilon)}{i\omega_n - \epsilon}\frac{d\epsilon}{2\pi} = \int_{-\infty}^{\infty} \frac{\delta(\epsilon - \bar{\epsilon}_{\mathbf{k}\sigma}/\hbar)}{i\omega_n - \epsilon}d\epsilon$$

$$\Rightarrow g^0(\mathbf{k}\sigma, \omega_n) = \frac{1}{i\omega_n - \bar{\epsilon}_{\mathbf{k}\sigma}/\hbar}. \tag{8.40}$$

The retarded Green's function, obtained from $g^0(\mathbf{k}\sigma, \omega_n)$ through the replacement: $i\omega_n \to \omega + i0^\dagger$, is

$$G^{R,0}(\mathbf{k}\sigma, \omega) = \frac{1}{\omega - \bar{\epsilon}_{\mathbf{k}\sigma}/\hbar + i0^\dagger}. \tag{8.41}$$

This is in agreement with the expression obtained in Chapter 6.

8.5.2 An alternative derivation

Starting from $c_{\mathbf{k}\sigma}(\tau) = e^{\bar{H}\tau/\hbar}c_{k\sigma}e^{-\bar{H}\tau/\hbar}$, we find

$$\frac{d}{d\tau}c_{\mathbf{k}\sigma}(\tau) = \frac{1}{\hbar}[\bar{H}, c_{\mathbf{k}\sigma}(\tau)].$$

For the noninteracting system, $\bar{H} = \bar{H}_0$. It is easily verified that

$$[\bar{H}_0, c_{\mathbf{k}\sigma}(\tau)] = -\bar{\epsilon}_{\mathbf{k}\sigma}c_{\mathbf{k}\sigma}(\tau). \tag{8.42}$$

Therefore,

$$\frac{d}{d\tau}c_{\mathbf{k}\sigma}(\tau) = -\frac{\bar{\epsilon}_{\mathbf{k}\sigma}}{\hbar}c_{\mathbf{k}\sigma}(\tau) \Rightarrow c_{\mathbf{k}\sigma}(\tau) = c_{\mathbf{k}\sigma}(0)e^{-\bar{\epsilon}_{\mathbf{k}\sigma}\tau/\hbar} \tag{8.43a}$$

Similarly,

$$c^{\dagger}_{\mathbf{k}\sigma}(\tau) = c^{\dagger}_{\mathbf{k}\sigma}(0)e^{\bar{\epsilon}_{\mathbf{k}\sigma}\tau/\hbar}. \tag{8.43b}$$

Note that $c^{\dagger}_{\mathbf{k}\sigma}(\tau)$ is not the adjoint of $c_{\mathbf{k}\sigma}(\tau)$. The imaginary-time Green's function is given by

$$\begin{aligned}
g^0(\mathbf{k}\sigma,\tau) &= -\langle T c_{\mathbf{k}\sigma}(\tau)c^{\dagger}_{\mathbf{k}\sigma}(0)\rangle = -\theta(\tau)\langle c_{\mathbf{k}\sigma}(\tau)c^{\dagger}_{\mathbf{k}\sigma}(0)\rangle \mp \theta(-\tau)\langle c^{\dagger}_{\mathbf{k}\sigma}(0)c_{\mathbf{k}\sigma}(\tau)\rangle \\
&= \left[-\theta(\tau)\langle c_{\mathbf{k}\sigma}(0)c^{\dagger}_{\mathbf{k}\sigma}(0)\rangle \mp \theta(-\tau)\langle c^{\dagger}_{\mathbf{k}\sigma}(0)c_{\mathbf{k}\sigma}(0)\rangle\right]e^{-\bar{\epsilon}_{\mathbf{k}\sigma}\tau/\hbar} \\
&= \left[-\theta(\tau)\begin{Bmatrix}1+n_{\mathbf{k}\sigma}\\ 1-f_{\mathbf{k}\sigma}\end{Bmatrix} \mp \theta(-\tau)\begin{Bmatrix}n_{\mathbf{k}\sigma}\\ f_{\mathbf{k}\sigma}\end{Bmatrix}\right]e^{-\bar{\epsilon}_{\mathbf{k}\sigma}\tau/\hbar}.
\end{aligned}$$

The Fourier transform $g^0(\mathbf{k}\sigma,\omega_n)$ is

$$\begin{aligned}
g^0(\mathbf{k}\sigma,\omega_n) &= \int_0^{\beta\hbar} g^0(\mathbf{k}\sigma,\tau)e^{i\omega_n\tau}d\tau = -\begin{Bmatrix}1+n_{\mathbf{k}\sigma}\\ 1-f_{\mathbf{k}\sigma}\end{Bmatrix}\int_0^{\beta\hbar} e^{(i\omega_n-\bar{\epsilon}_{\mathbf{k}\sigma}/\hbar)\tau}d\tau \\
&= \begin{Bmatrix}1+n_{\mathbf{k}\sigma}\\ 1-f_{\mathbf{k}\sigma}\end{Bmatrix}\frac{1\mp e^{-\beta\bar{\epsilon}_{\mathbf{k}\sigma}}}{i\omega_n-\bar{\epsilon}_{\mathbf{k}\sigma}/\hbar} \Rightarrow g^0(\mathbf{k}\sigma,\omega_n) = \frac{1}{i\omega_n-\bar{\epsilon}_{\mathbf{k}\sigma}/\hbar}.
\end{aligned}$$

8.6 Example: Green's function for 2-DEG in a magnetic field

In Chapter 2, we considered a two-dimensional electron gas confined in the x–y plane in the presence of a uniform static magnetic field **B** that is in the z-direction. We showed that the single-particle states are described by three quantum numbers: n, k, and σ. The spatial functions are given by

$$\phi_{nk}(x,y) = \frac{A_n}{\sqrt{L_x}}e^{ikx}H_n(a(y-y_o))e^{[-a^2(y-y_o)^2/2]}.$$

L_x is the sample length in the x-direction, $k = 0, \pm 2\pi/L_x, \pm 4\pi/L_x, \ldots$, $n = 0, 1, 2\ldots$, H_n is the Hermite polynomial of degree n, A_n is a normalization constant, $a = (m\omega/\hbar)^{1/2}$, m is the electron mass, $\omega = eB/mc$ is the cyclotron frequency, and $y_o = \hbar ck/eB$. The corresponding single-particle energies are $\epsilon_{nk\sigma} = (n+1/2)\hbar\omega + g\mu_B B\sigma$, where g is the gyromagnetic factor for the electron spin, μ_B is the Bohr magneton, and $\sigma = -1/2, +1/2$. The field operators are given by

$$\Psi_\sigma(\mathbf{r}) = \sum_{nk} \phi_{nk}(\mathbf{r}) c_{nk\sigma}, \quad \Psi_\sigma^\dagger(\mathbf{r}) = \sum_{nk} \phi_{nk}^*(\mathbf{r}) c_{nk\sigma}^\dagger.$$

Assuming that the electrons are noninteracting, the imaginary-time Green's function $g^0(\mathbf{r}\sigma\tau, \mathbf{r}'\sigma'0) = -\langle T\Psi_\sigma(\mathbf{r}\tau)\Psi_{\sigma'}^\dagger(\mathbf{r}'0)\rangle$ is given by

$$\begin{aligned} g^0(\mathbf{r}\sigma\tau, \mathbf{r}'\sigma'0) &= -\sum_{nk}\sum_{n'k'} \phi_{nk}(\mathbf{r})\phi_{n'k'}^*(\mathbf{r}')\langle T\, c_{nk\sigma}(\tau) c_{n'k'\sigma'}^\dagger(0)\rangle \\ &= \sum_{nk}\sum_{n'k'} \phi_{nk}(\mathbf{r})\, g(nk\sigma\tau, n'k'\sigma'0)\phi_{n'k'}^*(\mathbf{r}') \end{aligned}$$

where

$$\begin{aligned} g^0(nk\sigma\tau, n'k'\sigma'0) &= -\langle T c_{nk\sigma}(\tau) c_{n'k'\sigma'}^\dagger(0)\rangle = -e^{-\bar{\epsilon}_{nk\sigma}\tau/\hbar}[\theta(\tau)\langle c_{nk\sigma} c_{n'k'\sigma'}^\dagger\rangle \\ &- \theta(-\tau)\langle c_{n'k'\sigma'}^\dagger c_{nk\sigma}\rangle] = -e^{-\bar{\epsilon}_{nk\sigma}\tau/\hbar}[\theta(\tau)(1 - f_{nk\sigma}) - \theta(-\tau) f_{nk\sigma}]\delta_{nn'}\delta_{kk'}\delta_{\sigma\sigma'}. \end{aligned}$$

Hence,

$$g^0(\mathbf{r}\sigma\tau, \mathbf{r}'\sigma'0) = -\delta_{\sigma\sigma'} \sum_{nk} \phi_{nk}(\mathbf{r})\phi_{nk}^*(\mathbf{r}') e^{-\bar{\epsilon}_{nk\sigma}\tau/\hbar}[\theta(\tau)(1 - f_{nk\sigma}) - \theta(-\tau) f_{nk\sigma}].$$

The Fourier transform of the Green's function

$$g^0(\mathbf{r}\sigma, \mathbf{r}'\sigma', \omega_m) = \int_0^{\beta\hbar} g^0(\mathbf{r}\sigma\tau, \mathbf{r}'\sigma'0) e^{i\omega_m\tau} d\tau$$

is readily obtained; we find

$$g(\mathbf{r}\sigma\mathbf{r}'\sigma', \omega_m) = \delta_{\sigma\sigma'} \sum_{nk} \frac{\phi_{nk}(\mathbf{r})\phi_{nk}^*(\mathbf{r}')}{i\omega_m - \epsilon_{nk\sigma}/\hbar}.$$

The presence of a magnetic field breaks the translational invariance of the two-dimensional electron gas; Green's function in this case does not depend on $\mathbf{r} - \mathbf{r}'$, but rather on $\mathbf{r}$ and $\mathbf{r}'$ separately.

8.7 Green's function and the $\hat{U}$-operator

In Section 8.5 we calculated Green's function for a system of noninteracting particles. In the presence of interactions, it is generally true that the Schrödinger equation is not exactly soluble, and it would be too much to hope that Green's function would be exactly calculable; we must resort to perturbation theory. Before applying perturbation theory, however, Green's function must be recast into a different form. Towards that end we introduce the interaction picture and the $\hat{U}$-operator.

8.7.1 The Interaction picture

Consider an interacting system with a time-independent Hamiltonian,

$$\bar{H} = H_0 - \mu N + V = \bar{H}_0 + V \tag{8.44}$$

where V represents the interaction terms. We have already introduced the modified Heisenberg picture where an operator $A(\tau)$ is given by

$$A(\tau) = e^{\bar{H}\tau/\hbar} A e^{-\bar{H}\tau/\hbar}.$$

In the interaction picture, the operator $\hat{A}(\tau)$ is defined by

$$\hat{A}(\tau) = e^{\bar{H}_0\tau/\hbar} A e^{-\bar{H}_0\tau/\hbar}. \tag{8.45}$$

Note that a hat "$\wedge$" above an operator identifies it as an interaction picture operator. This definition differs from that of the standard interaction picture of quantum mechanics (see Problem 1.11) in that $it \to \tau$ and $H_0 \to \bar{H}_0$. In a way, $\hat{A}(\tau)$ is a modified interaction picture operator, but we will refer to it as an interaction picture operator. Since the imaginary-time Green's function is defined in terms of a product of two Heisenberg operators, we consider the product

$$\begin{aligned} A(\tau)B(\tau') &= e^{\bar{H}\tau/\hbar} A e^{-\bar{H}\tau/\hbar} e^{\bar{H}\tau'/\hbar} B e^{-\bar{H}\tau'/\hbar} \\ &= e^{\bar{H}\tau/\hbar} e^{-\bar{H}_0\tau/\hbar} \hat{A}(\tau) e^{\bar{H}_0\tau/\hbar} e^{-\bar{H}(\tau-\tau')/\hbar} e^{-\bar{H}_0\tau'/\hbar} \hat{B}(\tau') e^{\bar{H}_0\tau'/\hbar} e^{-\bar{H}\tau'/\hbar}. \end{aligned}$$

We have used Eq. (8.45) to express A and B in terms of $\hat{A}(\tau)$ and $\hat{B}(\tau')$.

8.7.2 The $\hat{U}$-operator

The above equation motivates the definition of the $\hat{U}$-operator,

$$\hat{U}(\tau, \tau') = e^{\bar{H}_0\tau/\hbar} e^{-\bar{H}(\tau-\tau')/\hbar} e^{-\bar{H}_0\tau'/\hbar}. \tag{8.46}$$

The product of the Heisenberg operators, then, reduces to

$$A(\tau)B(\tau') = \hat{U}(0, \tau)\hat{A}(\tau)\hat{U}(\tau, \tau')\hat{B}(\tau')\hat{U}(\tau', 0). \tag{8.47}$$

The following two properties of the $\hat{U}$-operator are easily verified

$$\hat{U}(\tau, \tau) = 1 \tag{8.48}$$

$$\hat{U}(\tau, \tau')\hat{U}(\tau', \tau'') = \hat{U}(\tau, \tau''). \tag{8.49}$$

We can express a Heisenberg operator in term of the interaction picture operator as follows:

$$A(\tau) = e^{\bar{H}\tau/\hbar} A e^{-\bar{H}\tau/\hbar} = e^{\bar{H}\tau/\hbar} e^{-\bar{H}_0\tau/\hbar} \hat{A}(\tau) e^{\bar{H}_0\tau/\hbar} e^{-\bar{H}\tau/\hbar}$$
$$= \hat{U}(0,\tau)\hat{A}(\tau)\hat{U}(\tau,0). \tag{8.50}$$

From the definition of $\hat{U}(\tau,\tau')$, Eq. (8.46), we find

$$\frac{\partial}{\partial\tau}\hat{U}(\tau,\tau') = \frac{1}{\hbar}\bar{H}_0\hat{U}(\tau,\tau') - \frac{1}{\hbar}e^{\bar{H}_0\tau/\hbar}\bar{H}e^{-\bar{H}(\tau-\tau')/\hbar}e^{-\bar{H}_0\tau'/\hbar}.$$

Writing $\bar{H}$ in the second term as $\bar{H}_0 + V$, and noting that $\bar{H}_0$ commutes with $e^{\bar{H}_0\tau/\hbar}$, the above equation reduces to

$$\frac{\partial}{\partial\tau}\hat{U}(\tau,\tau') = -\frac{1}{\hbar}e^{\bar{H}_0\tau/\hbar}Ve^{-\bar{H}(\tau-\tau')/\hbar}e^{-\bar{H}_0\tau'/\hbar}$$
$$= -\frac{1}{\hbar}e^{\bar{H}_0\tau/\hbar}Ve^{-\bar{H}_0\tau/\hbar}e^{\bar{H}_0\tau/\hbar}e^{-\bar{H}(\tau-\tau')/\hbar}e^{-\bar{H}_0\tau'/\hbar} = -\frac{1}{\hbar}\hat{V}(\tau)\hat{U}(\tau,\tau').$$

Integrating both sides from τ' to τ, we find

$$\int_{\tau'}^{\tau}\frac{\partial}{\partial\tau_1}\hat{U}(\tau_1,\tau')d\tau_1 = -\frac{1}{\hbar}\int_{\tau'}^{\tau}\hat{V}(\tau_1)\hat{U}(\tau_1,\tau')d\tau_1$$
$$\Rightarrow \hat{U}(\tau,\tau') - \hat{U}(\tau',\tau') = -\frac{1}{\hbar}\int_{\tau'}^{\tau}\hat{V}(\tau_1)\hat{U}(\tau_1,\tau')d\tau_1$$
$$\Rightarrow \hat{U}(\tau,\tau') = 1 - \frac{1}{\hbar}\int_{\tau'}^{\tau}\hat{V}(\tau_1)\hat{U}(\tau_1,\tau')d\tau_1. \tag{8.51}$$

This is an integral equation for $\hat{U}$; we solve it by iteration,

$$\hat{U}(\tau,\tau') = 1 - \frac{1}{\hbar}\int_{\tau'}^{\tau}d\tau_1\hat{V}(\tau_1)\left[1 - \frac{1}{\hbar}\int_{\tau'}^{\tau_1}\hat{V}(\tau_2)\hat{U}(\tau_2,\tau')d\tau_2\right]$$
$$= 1 - \frac{1}{\hbar}\int_{\tau'}^{\tau}d\tau_1\hat{V}(\tau_1) + \left(-\frac{1}{\hbar}\right)^2\int_{\tau'}^{\tau}d\tau_1\int_{\tau'}^{\tau_1}d\tau_2\hat{V}(\tau_1)\hat{V}(\tau_2)\hat{U}(\tau_2,\tau').$$

We continue to iterate in the same fashion; we find

$$\hat{U}(\tau,\tau') = 1 - \frac{1}{\hbar}\int_{\tau'}^{\tau}d\tau_1\hat{V}(\tau_1) + \left(-\frac{1}{\hbar}\right)^2\int_{\tau'}^{\tau}d\tau_1\int_{\tau'}^{\tau_1}d\tau_2\,\hat{V}(\tau_1)\hat{V}(\tau_2)$$
$$+ \left(-\frac{1}{\hbar}\right)^3\int_{\tau'}^{\tau}d\tau_1\int_{\tau'}^{\tau_1}d\tau_2\int_{\tau'}^{\tau_2}d\tau_3\,\hat{V}(\tau_1)\hat{V}(\tau_2)\hat{V}(\tau_3) + \cdots. \tag{8.52}$$

Let us consider the double integral on the RHS of the above equation,

$$I \equiv \int_{\tau'}^{\tau} d\tau_1 \int_{\tau'}^{\tau_1} d\tau_2 \, \hat{V}(\tau_1)\hat{V}(\tau_2) = \int_{\tau'}^{\tau} d\tau_1 \int_{\tau'}^{\tau} d\tau_2 \, \hat{V}(\tau_1)\hat{V}(\tau_2)\theta(\tau_1 - \tau_2)$$
$$= \int_{\tau'}^{\tau} d\tau_1 \int_{\tau'}^{\tau} d\tau_2 \, \hat{V}(\tau_2)\hat{V}(\tau_1)\theta(\tau_2 - \tau_1).$$

The step function $\theta(\tau_1 - \tau_2)$ ensures that $\tau_2 < \tau_1$. The last equality is obtained by interchanging the variables of integration τ_1 and τ_2. Therefore,

$$I = \frac{1}{2}\int_{\tau'}^{\tau} d\tau_1 \int_{\tau'}^{\tau} d\tau_2 \, [\hat{V}(\tau_1)\hat{V}(\tau_2)\theta(\tau_1 - \tau_2) + \hat{V}(\tau_2)\hat{V}(\tau_1)\theta(\tau_2 - \tau_1)]$$
$$= \frac{1}{2}\int_{\tau'}^{\tau} d\tau_1 \int_{\tau'}^{\tau} d\tau_2 \, T\,\hat{V}(\tau_1)\hat{V}(\tau_2).$$

where T is the time-ordering operator.

Encouraged by the above result, we consider the nth order term,

$$\int_{\tau'}^{\tau} d\tau_1 \int_{\tau'}^{\tau_1} d\tau_2 \cdots \int_{\tau'}^{\tau_{n-1}} d\tau_n \, \hat{V}(\tau_1)\hat{V}(\tau_2)\cdots\hat{V}(\tau_n)$$
$$= \int_{\tau'}^{\tau} d\tau_1 \int_{\tau'}^{\tau} d\tau_2 \cdots \int_{\tau'}^{\tau} d\tau_n \, \hat{V}(\tau_1)\cdots\hat{V}(\tau_n)\theta(\tau_1 - \tau_2)\theta(\tau_2 - \tau_3)\cdots\theta(\tau_{n-1} - \tau_n)$$
$$= \int_{\tau'}^{\tau} d\tau_1 \int_{\tau'}^{\tau} d\tau_2 \cdots \int_{\tau'}^{\tau} d\tau_n \, \hat{V}(\tau_{P(1)})\hat{V}(\tau_{P(2)})\cdots\hat{V}(\tau_{P(n)})$$
$$\times\, \theta(\tau_{P(1)} - \tau_{P(2)})\theta(\tau_{P(2)} - \tau_{P(3)})\cdots\theta(\tau_{P(n-1)} - \tau_{P(n)}).$$

$P(1), P(2), \ldots, P(n)$ in any permutation of $1, 2, \ldots, n$. The last equality holds because the last integral is obtained from the preceding one by relabeling the integration variables: $\tau_1, \tau_2, ..., \tau_n \to \tau_{P(1)}, \tau_{P(2)}, ..., \tau_{P(n)}$. Since there are $n!$ permutations of $1, 2, ..., n$, we can write

$$\int_{\tau'}^{\tau} d\tau_1 \int_{\tau'}^{\tau_1} d\tau_2 \cdots \int_{\tau'}^{\tau_{n-1}} d\tau_n \, \hat{V}(\tau_1)\hat{V}(\tau_2)\ldots\hat{V}(\tau_n) = \frac{1}{n!}\int_{\tau'}^{\tau} d\tau_1 \ldots \int_{\tau'}^{\tau} d\tau_n$$
$$\sum_P \hat{V}(\tau_{P(1)})\ldots\, V(\tau_{P(n)})\theta(\tau_{P(1)} - \tau_{P(2)})\ldots\theta(\tau_{P(n-1)} - \tau_{P(n)})$$
$$= \frac{1}{n!}\int_{\tau'}^{\tau} d\tau_1 \ldots \int_{\tau'}^{\tau} d\tau_n \, T[\hat{V}(\tau_1)\ldots\hat{V}(\tau_n)].$$

Note that when the time-ordering operator T rearranges $\hat{V}(\tau_1)$, ..., $\hat{V}(\tau_n)$ in ascending time order, from right to left, no minus sign is introduced whenever $\hat{V}(\tau_i)$ and $\hat{V}(\tau_j)$ are interchanged, even if $\hat{V}$ describes interactions among fermions. This is because $\hat{V}$, when expressed in second quantized form, consists of an even number of operators; i.e., $\hat{V}$ is a bosonic operator. The $\hat{U}$-operator now has the following perturbation expansion:

$$\hat{U}(\tau,\tau') = \sum_{n=0}^{\infty} \frac{1}{n!}\left(-\frac{1}{\hbar}\right)^n \int_{\tau'}^{\tau} d\tau_1 \ldots \int_{\tau'}^{\tau} d\tau_n\, T[\hat{V}(\tau_1)\ldots\hat{V}(\tau_n)]. \tag{8.53}$$

8.7.3 Green's function and the $\hat{U}$-operator

The $\hat{U}$-operator was defined by

$$\hat{U}(\tau,\tau') = e^{\bar{H}_0\tau/\hbar} e^{-\bar{H}(\tau-\tau')/\hbar} e^{-\bar{H}_0\tau'/\hbar}.$$

If we set $\tau = \beta\hbar$ ($\beta\hbar$ and τ have the same units) and $\tau' = 0$, we obtain

$$\hat{U}(\beta\hbar, 0) = e^{\beta\bar{H}_0} e^{-\beta\bar{H}} \Rightarrow e^{-\beta\bar{H}} = e^{-\beta\bar{H}_0}\hat{U}(\beta\hbar, 0). \tag{8.54}$$

We may thus write the imaginary-time Green's function as follows:

$$\begin{aligned} g(\mathbf{k}\sigma,\tau) &= -\langle T\, c_{\mathbf{k}\sigma}(\tau) c^{\dagger}_{\mathbf{k}\sigma}(0)\rangle = -Z_G^{-1}\,\mathrm{Tr}[e^{-\beta\bar{H}}\, T\, c_{\mathbf{k}\sigma}(\tau) c^{\dagger}_{\mathbf{k}\sigma}(0)] \\ &= -Z_G^{-1}\,\mathrm{Tr}[e^{-\beta\bar{H}_0}\hat{U}(\beta\hbar,0)\, T\, c_{\mathbf{k}\sigma}(\tau) c^{\dagger}_{\mathbf{k}\sigma}(0)]. \end{aligned}$$

Using Eq. (8.50), the Heisenberg operators $c_{\mathbf{k}\sigma}(\tau)$ and $c^{\dagger}_{\mathbf{k}\sigma}(0)$ are written as

$$c_{\mathbf{k}\sigma}(\tau) = \hat{U}(0,\tau)\hat{c}_{\mathbf{k}\sigma}(\tau)\hat{U}(\tau,0), \quad c^{\dagger}_{\mathbf{k}\sigma}(0) = \hat{c}^{\dagger}_{\mathbf{k}\sigma}(0). \tag{8.55}$$

First, consider the case $\tau > 0$,

$$\begin{aligned} g(\mathbf{k}\sigma,\tau>0) &= -Z_G^{-1}\mathrm{Tr}[e^{-\beta\bar{H}_0}\hat{U}(\beta\hbar,0)\hat{U}(0,\tau)\hat{c}_{\mathbf{k}\sigma}(\tau)\hat{U}(\tau,0)\hat{c}^{\dagger}_{\mathbf{k}\sigma}(0)] \\ &= -Z_G^{-1}\mathrm{Tr}[e^{-\beta\bar{H}_0}\hat{U}(\beta\hbar,\tau)\hat{c}_{\mathbf{k}\sigma}(\tau)\hat{U}(\tau,0)\hat{c}^{\dagger}_{\mathbf{k}\sigma}(0)] \end{aligned} \tag{8.56}$$

where Eq. (8.49) is used to establish the second equality. Consider the following expression:

$$T[\hat{U}(\beta\hbar,0)\hat{c}_{\mathbf{k}\sigma}(\tau)\hat{c}^{\dagger}_{\mathbf{k}\sigma}(0)] = T[\hat{U}(\beta\hbar,\tau)\hat{U}(\tau,0)\hat{c}_{\mathbf{k}\sigma}(\tau)\hat{c}^{\dagger}_{\mathbf{k}\sigma}(0)].$$

Equation (8.53) shows that all operators $\hat{V}$ in the expansion of $\hat{U}(\tau,0)$ occur at times between 0 and τ, and all operators $\hat{V}$ in the expansion of $\hat{U}(\beta\hbar,\tau)$ occur at times between τ and $\beta\hbar$ (recall that $\tau \le \beta\hbar$); hence,

$$T[\hat{U}(\beta\hbar,\tau)\hat{U}(\tau,0)\hat{c}_{\mathbf{k}\sigma}(\tau)\hat{c}^{\dagger}_{\mathbf{k}\sigma}(0)] = \hat{U}(\beta\hbar,\tau)\hat{c}_{\mathbf{k}\sigma}(\tau)\hat{U}(\tau,0)\hat{c}^{\dagger}_{\mathbf{k}\sigma}(0).$$

No minus signs are introduced in the above reordering because $\hat{V}$ consists of an even number of creation and annihilation operators. Thus, Eq. (8.56) may be written as

$$g(\mathbf{k}\sigma,\tau>0) = -Z_G^{-1}\mathrm{Tr}[e^{-\beta\bar{H}_0}\, T\,\hat{U}(\beta\hbar,0)\hat{c}_{\mathbf{k}\sigma}(\tau)\hat{c}^{\dagger}_{\mathbf{k}\sigma}(0)]. \tag{8.57}$$

Next, we consider the case when $\tau < 0$,

$$\begin{aligned} g(\mathbf{k}\sigma, \tau < 0) &= -\langle T c_{\mathbf{k}\sigma}(\tau) c^{\dagger}_{\mathbf{k}\sigma}(0) \rangle = \mp \langle c^{\dagger}_{\mathbf{k}\sigma}(0) c_{\mathbf{k}\sigma}(\tau) \rangle \\ &= \mp Z_G^{-1} \mathrm{Tr}[e^{-\beta \bar{H}_0} \hat{U}(\beta\hbar, 0) c^{\dagger}_{\mathbf{k}\sigma}(0) c_{\mathbf{k}\sigma}(\tau)] \\ &= \mp Z_G^{-1} \mathrm{Tr}[e^{-\beta \bar{H}_0} T\, \hat{U}(\beta\hbar, 0) c^{\dagger}_{\mathbf{k}\sigma}(0) c_{\mathbf{k}\sigma}(\tau)]. \end{aligned}$$

The introduction of the T operator is justified by the fact that in the product $\hat{U}(\beta\hbar, 0) c^{\dagger}_{\mathbf{k}\sigma}(0) c_{\mathbf{k}\sigma}(\tau)$, the operators are already ordered in increasing time order from right to left (recall that $\tau < 0$). Thus

$$g(\mathbf{k}\sigma, \tau < 0) = -Z_G^{-1} \mathrm{Tr}[e^{-\beta \bar{H}_0} T\, \hat{U}(\beta\hbar, 0) c_{\mathbf{k}\sigma}(\tau) c^{\dagger}_{\mathbf{k}\sigma}(0)]$$

where the interchange of the creation and annihilation operators brought about a minus sign in case they were fermion operators. Expressing the creation and annihilation operators in the interaction picture, we obtain

$$g(\mathbf{k}\sigma, \tau < 0) = -Z_G^{-1} \mathrm{Tr}[e^{-\beta \bar{H}_0} T\, \hat{U}(\beta\hbar, 0) \hat{U}(0, \tau) \hat{c}_{\mathbf{k}\sigma}(\tau) \hat{U}(\tau, 0) \hat{c}^{\dagger}_{\mathbf{k}\sigma}(0)].$$

Since the operators $\hat{V}$ that appears in $\hat{U}(0, \tau)$ are bosonic, we can interchange $\hat{U}(0, \tau)$ and $\hat{c}_{\mathbf{k}\sigma}(\tau)$ without incurring a minus sign. Using $\hat{U}(0, \tau)\hat{U}(\tau, 0) = \hat{U}(0, 0) = 1$, the above expression reduces to

$$g(\mathbf{k}\sigma, \tau < 0) = -Z_G^{-1} \mathrm{Tr}[e^{-\beta \bar{H}_0} T\, \hat{U}(\beta\hbar, 0) \hat{c}_{\mathbf{k}\sigma}(\tau) \hat{c}^{\dagger}_{\mathbf{k}\sigma}(0)],$$

which is the same expression as in Eq. (8.57).

In summary, we found that the imaginary-time Green's function can be expressed in terms of interaction picture operators as

$$g(\mathbf{k}\sigma, \tau) = -Z_G^{-1} \mathrm{Tr}[e^{-\beta \bar{H}_0} T\, \hat{U}(\beta\hbar, 0) \hat{c}_{\mathbf{k}\sigma}(\tau) \hat{c}^{\dagger}_{\mathbf{k}\sigma}(0)].$$

The grand partition function is given by

$$Z_G = \mathrm{Tr}[e^{-\beta \bar{H}}] = \mathrm{Tr}[e^{-\beta \bar{H}_0} e^{\beta \bar{H}_0} e^{-\beta \bar{H}}] = \mathrm{Tr}[e^{-\beta \bar{H}_0} \hat{U}(\beta\hbar, 0)],$$

where we used Eq. (8.54). Therefore, we can write

$$g(\mathbf{k}\sigma, \tau) = -\frac{\mathrm{Tr}[e^{-\beta \bar{H}_0} T\, \hat{U}(\beta\hbar, 0) \hat{c}_{\mathbf{k}\sigma}(\tau) \hat{c}^{\dagger}_{\mathbf{k}\sigma}(0)]}{\mathrm{Tr}[e^{-\beta \bar{H}_0} \hat{U}(\beta\hbar, 0)]}.$$

Dividing the numerator and denominator by $Z_{G,0} = \mathrm{Tr}[e^{-\beta H_0}]$, we arrive at the following formula, which is the main goal of this subsection,

$$g(\mathbf{k}\sigma, \tau) = -\frac{\langle T\, \hat{U}(\beta\hbar, 0) \hat{c}_{\mathbf{k}\sigma}(\tau) \hat{c}^{\dagger}_{\mathbf{k}\sigma}(0) \rangle_0}{\langle \hat{U}(\beta\hbar, 0) \rangle_0}. \tag{8.58}$$

The subscript "0" indicates that the ensemble average is over the noninteracting system,

$$\langle \cdots \rangle_0 = \frac{\mathrm{Tr}[e^{-\beta \bar{H}_0} \cdots]}{\mathrm{Tr}[e^{-\beta \bar{H}_0}]}.$$

Finally, we note that in Eq. (8.58) we can move $\hat{c}_{\mathbf{k}\sigma}(\tau)$ to the left, or move the product $\hat{c}_{\mathbf{k}\sigma}(\tau)\hat{c}^{\dagger}_{\mathbf{k}\sigma}(0)$ to the left, without incurring a minus sign because the $\hat{V}$ operators in $\hat{U}(\beta\hbar, 0)$ are bosonic. Hence, $g(\mathbf{k}\sigma, \tau)$ is also given by

$$g(\mathbf{k}\sigma, \tau) = -\frac{\langle T\, \hat{c}_{\mathbf{k}\sigma}(\tau)\hat{c}^{\dagger}_{\mathbf{k}\sigma}(0)\,\hat{U}(\beta\hbar, 0)\rangle_0}{\langle \hat{U}(\beta\hbar, 0)\rangle_0}. \tag{8.59}$$

8.7.4 *Perturbation expansion of the imaginary-time Green's function*

Using the perturbation series for the $\hat{U}$-operator, Eq. (8.53), we can write

$$g(\mathbf{k}\sigma, \tau) = -\frac{\sum_{n=0}^{\infty} \frac{1}{n!}\left(-\frac{1}{\hbar}\right)^n \langle \int_0^{\beta\hbar} d\tau_1 \cdots \int_0^{\beta\hbar} d\tau_n T\, \hat{c}_{\mathbf{k}\sigma}(\tau)\hat{c}^{\dagger}_{\mathbf{k}\sigma}(0)\hat{V}(\tau_1)\ldots\hat{V}(\tau_n)\rangle_0}{\sum_{n=0}^{\infty} \frac{1}{n!}\left(-\frac{1}{\hbar}\right)^n \langle \int_0^{\beta\hbar} d\tau_1 \ldots \int_0^{\beta\hbar} d\tau_n T\, \hat{V}(\tau_1)\ldots\hat{V}(\tau_n)\rangle_0}. \tag{8.60}$$

Although the denominator appears to make the above expression unwieldy, it actually makes matters simpler. As we will discuss later, it cancels out the disconnected Feynman diagrams in the numerator.

8.8 Wick's theorem

In order to evaluate $g(\mathbf{k}\sigma, \tau)$, Eq. (8.60), we need a method to determine the ensemble average, over the noninteracting system, of the time-ordered product of interaction picture operators. This is provided by Wick's theorem (Wick, 1950). In the following discussion, we assume that the operators in the time-ordered product are fermion operators. The theorem is valid for both fermions and bosons (the proof in the case of bosonic operators is left as an exercise for the reader). In the following, we proceed in a series of steps leading to the proof of this important theorem.

8.8.1 *Contractions*

Given two interaction picture operators $\hat{A}$ and $\hat{B}$, we define a contraction of $\hat{A}$ and $\hat{B}$ by

$$\overbracket{\hat{A}\;\hat{B}} = \langle T\,\hat{A}\hat{B}\rangle_0 = \mathrm{Tr}[e^{-\beta \bar{H}_0} T\,\hat{A}\hat{B}]/\mathrm{Tr}[e^{-\beta \bar{H}_0}].$$

For example,

$$\overbracket{\hat{c}_{\mathbf{k}\sigma}(\tau)\;\hat{c}^{\dagger}_{\mathbf{k}\sigma}}(0) = \langle T\; \hat{c}_{\mathbf{k}\sigma}(\tau)\hat{c}^{\dagger}_{\mathbf{k}\sigma}(0)\rangle_0 = -g^0(\mathbf{k}\sigma, \tau).$$

We note that $g^0(\mathbf{k}\sigma, \tau)$ is the single-particle imaginary-time Green's function for a noninteracting system. For such a system, the Heisenberg and interaction pictures coincide; hence, the above replacement of the time-ordered product of interaction picture operators by $-g^0$ is certainly valid. Since the expectation value, in any eigenstate of $\bar{H}_0$, of two annihilation or two creation operators, is zero, the ensemble average of two annihilation or two creation operators is zero. Hence,

$$\overbracket{\hat{c}_{\mathbf{k}\sigma}(\tau_1)\,\hat{c}_{\mathbf{k}'\sigma'}(\tau_2)} = \overbracket{\hat{c}^{\dagger}_{\mathbf{k}\sigma}(\tau_1)\,\hat{c}^{\dagger}_{\mathbf{k}'\sigma'}(\tau_2)} = 0.$$

8.8.2 Statement of Wick's theorem

Wick's theorem states that the ensemble average over a noninteracting system of the time-ordered product of interaction picture operators is equal to the sum over all possible contracted pairs,

$$\langle T[\hat{A}\hat{B}\hat{C}\hat{D}\ldots]\rangle_o = \overbracket{\hat{A}\hat{B}}\,\overbracket{\hat{C}\hat{D}}\cdots + \hat{A}^{\bullet}\hat{B}^{\circ}\hat{C}^{\bullet}\hat{D}^{\circ}\cdots + \overbracket{\hat{A}\underbracket{\hat{B}\hat{C}}\hat{D}}\cdots + \cdots. \tag{8.61}$$

If $\hat{A}, \hat{B}, \hat{C}, \hat{D}\cdots$ are fermion operators, a term such as $\hat{A}^{\bullet}\hat{B}^{\circ}\hat{C}^{\bullet}\hat{D}^{\circ}\cdots$ is to be interpreted as $-\overbracket{\hat{A}\hat{C}}\,\overbracket{\hat{B}\hat{D}}\cdots$, since we need to interchange $\hat{B}$ and $\hat{C}$.

We can write Wick's theorem in a more compact way. Clearly, for the ensemble average of the time-ordered product of operators to be nonzero, there must be an equal number of creation and annihilation operators. Assuming that the total number of operators is $2n$, Wick's theorem states that

$$\langle T\prod_{i=1}^{2n}\hat{a}_i\rangle_o = \sum(-1)^P\prod\langle T\ \hat{a}_j\hat{a}_k\rangle_o \tag{8.62}$$

where $\hat{a}_i$ is a creation or an annihilation operator, and the sum is over all possible ways of picking n pairs from among the $2n$ operators. On the RHS of the above equation, $P = 1$ if the permutation of fermion operators required to arrange the pairs as they appear on the RHS, starting from the arrangement on the LHS, is odd; otherwise $P = 0$. Each summand on the RHS consists of a product of n contracted pairs.

8.8.3 An example

Let us use Wick's theorem to evaluate

$$A = \langle T\,\hat{c}_{\mathbf{k}\sigma}(\tau)\hat{c}^{\dagger}_{\mathbf{k}\sigma}(0)\hat{c}^{\dagger}_{\mathbf{k}'\sigma'}(\tau_1)\hat{c}_{\mathbf{k}'\sigma'}(\tau_1)\rangle_0$$

where $0 \leq \tau_1 \leq \tau$, and the operators are fermion operators. Excluding pairs of annihilation or creation operators (their ensemble average in zero), we are left with two possible ways to pick pairs from among the four operators,

$$A = \langle T\,\hat{c}_{\mathbf{k}\sigma}(\tau)\hat{c}^{\dagger}_{\mathbf{k}\sigma}(0)\rangle_0 \langle T\,\hat{c}^{\dagger}_{\mathbf{k}'\sigma'}(\tau_1)\hat{c}_{\mathbf{k}'\sigma'}(\tau_1)\rangle_0$$
$$- \langle T\,\hat{c}_{\mathbf{k}\sigma}(\tau)\hat{c}^{\dagger}_{\mathbf{k}'\sigma'}(\tau_1)\rangle_0 \langle T\,\hat{c}^{\dagger}_{\mathbf{k}\sigma}(0)\hat{c}_{\mathbf{k}'\sigma'}(\tau_1)\rangle_0.$$

The minus sign in the second term arises from the interchange of the second and third operators in the original time-ordered product. The first term on the RHS is somewhat problematic: it involves a time-ordered product of two operators with equal time arguments, a case for which time ordering is ill-defined. How should we interpret $\langle T\,\hat{c}^{\dagger}_{\mathbf{k}'\sigma'}(\tau_1)\hat{c}_{\mathbf{k}'\sigma'}(\tau_1)\rangle_0$? We note that when an operator of a many-particle system is written in second quantized form, creation operators always occur on the left of annihilation operators. When this operator acts on any state, annihilation operators act first, followed by creation operators. We therefore assign to creation operators a time that is infinitesimally later than the time assigned to annihilation operators. We thus interpret $\langle T\,\hat{c}^{\dagger}_{\mathbf{k}'\sigma'}(\tau_1)\hat{c}_{\mathbf{k}'\sigma'}(\tau_1)\rangle_0$ as follows:

$$\langle T\,\hat{c}^{\dagger}_{\mathbf{k}'\sigma'}(\tau_1)\hat{c}_{\mathbf{k}'\sigma'}(\tau_1)\rangle_0 = \langle T\,\hat{c}^{\dagger}_{\mathbf{k}'\sigma'}(\tau_1^+)\hat{c}_{\mathbf{k}'\sigma'}(\tau_1)\rangle_0 = \langle \hat{c}^{\dagger}_{\mathbf{k}'\sigma'}(\tau_1^+)\hat{c}_{\mathbf{k}'\sigma'}(\tau_1)\rangle_0$$
$$= -\langle T\,\hat{c}_{\mathbf{k}'\sigma'}(\tau_1)\hat{c}^{\dagger}_{\mathbf{k}'\sigma'}(\tau_1^+)\rangle_0 = g^0(\mathbf{k}'\sigma', 0^-).$$

With this in mind, we can write

$$A = -g^0(\mathbf{k}\sigma, \tau)g^0(\mathbf{k}'\sigma', 0^-) + g^0(\mathbf{k}\sigma, \tau - \tau_1)g^0(\mathbf{k}\sigma, \tau_1)\delta_{\mathbf{k}\mathbf{k}'}\delta_{\sigma\sigma'}.$$

We have used the fact that an expression such as $\langle \hat{c}_{\mathbf{k}\sigma}(\tau)\hat{c}^{\dagger}_{\mathbf{k}'\sigma'}(\tau_1)\rangle_0$ vanishes unless $\mathbf{k} = \mathbf{k}'$ and $\sigma = \sigma'$; this is easily verified by writing the trace as a sum over diagonal elements and introducing one resolution of identity.

8.8.4 Some useful results

(a) First we prove that the anticommutator of two interaction picture single-particle fermion operators is a number. Since $\hat{c}_{\mathbf{k}\sigma}(\tau) = e^{\bar{H}_0\tau/\hbar}c_{\mathbf{k}\sigma}e^{-\bar{H}_0\tau/\hbar}$,

$$\frac{d}{d\tau}\hat{c}_{\mathbf{k}\sigma}(\tau) = \frac{1}{\hbar}[\bar{H}_0, \hat{c}_{\mathbf{k}\sigma}(\tau)] = \frac{1}{\hbar}\left[\sum \bar{\epsilon}_{\mathbf{k}\sigma}\hat{c}^{\dagger}_{\mathbf{k}\sigma}(\tau)\hat{c}_{\mathbf{k}\sigma}(\tau), \hat{c}_{\mathbf{k}\sigma}(\tau)\right].$$

The commutator is given by Eq. (8.42). Hence,

$$\hat{c}_{\mathbf{k}\sigma}(\tau) = \hat{c}_{\mathbf{k}\sigma}(0)e^{-\bar{\epsilon}_{\mathbf{k}\sigma}\tau/\hbar}, \quad \hat{c}^{\dagger}_{\mathbf{k}\sigma}(\tau) = \hat{c}^{\dagger}_{\mathbf{k}\sigma}(0)e^{\bar{\epsilon}_{\mathbf{k}\sigma}\tau/\hbar}.$$

Thus,

$$\{\hat{c}_{\mathbf{k}\sigma}(\tau_1), \hat{c}_{\mathbf{k}'\sigma'}(\tau_2)\} = \{\hat{c}^{\dagger}_{\mathbf{k}\sigma}(\tau_1), \hat{c}^{\dagger}_{\mathbf{k}'\sigma'}(\tau_2)\} = 0 \quad (8.63)$$

$$\{\hat{c}_{\mathbf{k}\sigma}(\tau_1), \hat{c}^{\dagger}_{\mathbf{k}'\sigma'}(\tau_2)\} = e^{\bar{\epsilon}_{\mathbf{k}\sigma}(\tau_2-\tau_1)/\hbar}\delta_{\mathbf{k}\mathbf{k}'}\delta_{\sigma\sigma'}. \quad (8.64)$$

(b) Let $\hat{B}$ be any operator; e.g., $\hat{B}$ could be a product of fermion creation and annihilation operators. Consider the ensemble average of the anticommutator

$$\begin{aligned}\langle\{\hat{c}_{\mathbf{k}\sigma}, \hat{B}\}\rangle_o &= Z_{G,0}^{-1}\,\mathrm{Tr}\,[e^{-\beta\bar{H}_0}\hat{c}_{\mathbf{k}\sigma}\hat{B} + e^{-\beta\bar{H}_0}\hat{B}\hat{c}_{\mathbf{k}\sigma}] \\ &= Z_{G,0}^{-1}\left(\mathrm{Tr}[e^{-\beta\bar{H}_0}\hat{c}_{\mathbf{k}\sigma}\hat{B}] + \mathrm{Tr}[\hat{c}_{\mathbf{k}\sigma}e^{-\beta\bar{H}_0}\hat{B}]\right). \qquad (8.65)\end{aligned}$$

The second equality results from the invariance of the trace under cyclic permutations. We now evaluate $\hat{c}_{\mathbf{k}\sigma}e^{-\beta\bar{H}_0}$. Since $[\hat{c}_{\mathbf{k}\sigma}, \bar{H}_0] = \bar{\epsilon}_{\mathbf{k}\sigma}\hat{c}_{\mathbf{k}\sigma}$,

$$\begin{aligned}&\hat{c}_{\mathbf{k}\sigma}\bar{H}_0 = (\bar{H}_0 + \bar{\epsilon}_{\mathbf{k}\sigma})\hat{c}_{\mathbf{k}\sigma} \\ &\hat{c}_{\mathbf{k}\sigma}\bar{H}_0^2 = \hat{c}_{\mathbf{k}\sigma}\bar{H}_o\bar{H}_0 = (\bar{H}_0 + \bar{\epsilon}_{\mathbf{k}\sigma})\hat{c}_{\mathbf{k}\sigma}\bar{H}_0 = (\bar{H}_0 + \bar{\epsilon}_{\mathbf{k}\sigma})^2\hat{c}_{\mathbf{k}\sigma} \\ &\vdots \\ &\hat{c}_{\mathbf{k}\sigma}\bar{H}_0^n = (\bar{H}_0 + \bar{\epsilon}_{\mathbf{k}\sigma})^n\hat{c}_{\mathbf{k}\sigma}.\end{aligned}$$

Consequently,

$$\begin{aligned}\hat{c}_{\mathbf{k}\sigma}e^{-\beta\bar{H}_o} &= \sum_{n=0}^{\infty}\frac{(-\beta)^n}{n!}\hat{c}_{\mathbf{k}\sigma}\bar{H}_0^n = \sum_{n=0}^{\infty}\frac{(-\beta)^n}{n!}(\bar{H}_o + \bar{\epsilon}_{\mathbf{k}\sigma})^n\hat{c}_{\mathbf{k}\sigma} \\ &= e^{-\beta(\bar{H}_o+\bar{\epsilon}_{\mathbf{k}\sigma})}\hat{c}_{\mathbf{k}\sigma} = e^{-\beta\bar{\epsilon}_{\mathbf{k}\sigma}}e^{-\beta\bar{H}_0}\hat{c}_{\mathbf{k}\sigma}.\end{aligned}$$

The last equality follows from the relation $e^{A+B} = e^Ae^B$, which is true if $[A, B] = 0$. In our case, $\beta\bar{H}_o$ commutes with $\beta\bar{\epsilon}_{\mathbf{k}\sigma}$ since the latter is simply a number. Equation (8.65) now becomes

$$\begin{aligned}\langle\{\hat{c}_{\mathbf{k}\sigma}, \hat{B}\}\rangle_0 &= Z_{G,0}^{-1}\mathrm{Tr}[e^{-\beta\bar{H}_0}\hat{c}_{\mathbf{k}\sigma}\hat{B} + e^{-\beta\bar{\epsilon}_{\mathbf{k}\sigma}}e^{-\beta\bar{H}_0}\hat{c}_{\mathbf{k}\sigma}\hat{B}] \\ &= (1 + e^{-\beta\bar{\epsilon}_{\mathbf{k}\sigma}})\langle\hat{c}_{\mathbf{k}\sigma}\hat{B}\rangle_0. \qquad (8.66a)\end{aligned}$$

Following exactly the same steps as above, we can show that

$$\langle\{\hat{c}_{\mathbf{k}\sigma}^\dagger, \hat{B}\}\rangle_0 = (1 + e^{\beta\bar{\epsilon}_{\mathbf{k}\sigma}})\langle\hat{c}_{\mathbf{k}\sigma}^\dagger\hat{B}\rangle_0. \qquad (8.66b)$$

(c) Let $b_1, b_2, \ldots, b_{2n}$ be interaction picture fermion operators. We want to find the anticommutator $\{b_1, b_2b_3\ldots b_{2n}\}$. First consider the case $n = 2$,

$$\{b_1, b_2b_3b_4\} = b_1b_2b_3b_4 + b_2b_3b_4b_1 = \{b_1, b_2\}b_3b_4 - b_2b_1b_3b_4 + b_2b_3b_4b_1.$$

We have replaced $b_1 b_2$ with $\{b_1, b_2\} - b_2 b_1$. Next we replace $b_1 b_3$ with $\{b_1, b_3\} - b_3 b_1$,

$$\begin{aligned}\{b_1, b_2 b_3 b_4\} &= \{b_1, b_2\} b_3 b_4 - b_2 \{b_1, b_3\} b_4 + b_2 b_3 b_1 b_4 + b_2 b_3 b_4 b_1 \\ &= \{b_1, b_2\} b_3 b_4 - b_2 \{b_1, b_3\} b_4 + b_2 b_3 \{b_1, b_4\} \\ &= \{b_1, b_2\} b_3 b_4 - \{b_1, b_3\} b_2 b_4 + \{b_1, b_4\} b_2 b_3.\end{aligned}$$

In the last step we used the result from (a) above, namely that the anticommutator is a number, so that we could move it to the left.

We can generalize the above result to any positive integer n,

$$\left\{ b_1, \prod_{k=2}^{2n} b_k \right\} = \sum_{m=2}^{2n} (-1)^m \{b_1, b_m\} \prod_{k=2}^{2n}{}' \, b_k \tag{8.67}$$

where the prime on $\prod$ means that $k = m$ is excluded. The general result given above is proven by mathematical induction. The result is true for $n = 2$ as shown above (and trivially so for $n = 1$). We assume that it is true for $n = r$, and show that it is true for $n = r + 1$. We thus consider

$$I = \{b_1, b_2 \dots b_{2r} b_{2r+1} b_{2r+2}\} = \{b_1, P b_{2r+1} b_{2r+2}\}$$

where $P = b_2 \dots b_{2r}$. Using $\{A, BC\} = \{A, B\}C - B[A, C]$, and $[A, BC] = \{A, B\}C - B\{A, C\}$, we can write,

$$\begin{aligned}I &= \{b_1, P\} b_{2r+1} b_{2r+2} - P[b_1, b_{2r+1} b_{2r+2}] \\ &= \{b_1, P\} b_{2r+1} b_{2r+2} - P\{b_1, b_{2r+1}\} b_{2r+2} + P b_{2r+1} \{b_1, b_{2r+2}\} \\ &= \{b_1, P\} b_{2r+1} b_{2r+2} - \{b_1, b_{2r+1}\} P b_{2r+2} + \{b_1, b_{2r+2}\} P b_{2r+1}.\end{aligned}$$

Since the result is assumed to be true for $n = r$, it applies to $\{b_1, P\}$. Therefore,

$$\begin{aligned}\{b_1, b_2 \dots b_{2r+2}\} &= \sum_{m=2}^{2r} (-1)^m \{b_1, b_m\} b_2 \dots b_{m-1} b_{m+1} \dots b_{2r} b_{2r+1} b_{2r+2} \\ &\quad - \{b_1, b_{2r+1}\} b_2 \dots b_{2r} b_{2r+2} + \{b_1, b_{2r+2}\} b_1 \dots b_{2r} b_{2r+1} \\ &= \sum_{m=2}^{2r+2} (-1)^m \{b_1, b_m\} \prod_{k=2}^{2r+2}{}' \, b_k\end{aligned}$$

which shows that Eq. (8.67), if assumed to be true for $n = r$, will be true for $n = r + 1$.

8.8.5 Proof of Wick's theorem

We prove Wick's theorem by mathematical induction. For $n = 1$,

$$\langle T \prod_{i=1}^{2} a_i \rangle_0 = \langle T a_1 a_2 \rangle_0.$$

Wick's theorem is true in this case since there is only one pair. We now assume that the theorem is true for $n - 1$, i.e.;

$$\langle T \prod_{i=1}^{2n-2} a_i \rangle_0 = \sum (-1)^P \prod \langle T a_i a_j \rangle_0$$

and we prove that the theorem is true for n.

Let $b_1, b_2, \ldots, b_{2n}$ be a permutation P_1 of $a_1, a_2, \ldots, a_{2n}$ such that $b_1, b_2, \ldots, b_{2n}$ are arranged in descending time order from left to right. Then

$$\langle T \prod_{i=1}^{2n} a_i \rangle_0 = (-1)^{P_1} \langle \prod_{k=1}^{2n} b_k \rangle_0 = (-1)^{P_1} \langle b_1 \prod_{k=2}^{2n} b_k \rangle_0.$$

Using Eq. (8.66) with $B = \prod_{k=2}^{2n} b_k$, we can write

$$\begin{aligned}
\langle T \prod_{i=1}^{2n} a_i \rangle_0 &= (-1)^{P_1} \left(1 + e^{\pm\beta\bar{\epsilon}_k}\right)^{-1} \langle \{b_1, \prod_{k=2}^{2n} b_k\} \rangle_0 \\
&= (-1)^{P_1} \left(1 + e^{\pm\beta\bar{\epsilon}_k}\right)^{-1} \left\langle \sum_{m=2}^{2n} (-1)^m \{b_1, b_m\} \prod_{k=2}^{2n}{}' b_k \right\rangle_0 .
\end{aligned}$$

In $e^{\pm\beta\bar{\epsilon}_k}$, the $-(+)$ sign is for the case when b_1 is an annihilation (a creation) operator. In the last step, we have used Eq. (8.67). Since $\{b_1, b_m\}$ is a number, it can be moved outside the ensemble average,

$$\langle T \prod_{i=1}^{2n} a_i \rangle_0 = (-1)^{P_1} \left(1 + e^{\pm\beta\bar{\epsilon}_k}\right)^{-1} \sum_{m=2}^{2n} (-1)^m \{b_1, b_m\} \langle \prod_{k=2}^{2n}{}' b_k \rangle_0.$$

Being a number, $\{b_1, b_m\}$ may be replaced by its ensemble average,

$$\{b_1, b_m\} = \langle \{b_1, b_m\} \rangle_0 = \left(1 + e^{\pm\beta\bar{\epsilon}_k}\right) \langle b_1 b_m \rangle_0 = \left(1 + e^{\pm\beta\bar{\epsilon}_k}\right) \langle T\, b_1 b_m \rangle_0$$

where use is made of Eq. (8.66), and the fact that $b_1, b_2, \ldots$ are arranged in descending time order. Thus,

$$\langle T \prod_{i=1}^{2n} a_i \rangle_0 = (-1)^{P_1} \sum_{m=2}^{2n} (-1)^m \langle T\, b_1 b_m \rangle_0 \langle T \prod_{k=2}^{2n}{}' b_k \rangle_0. \tag{8.68}$$

The introduction of T into the last factor is justified, since $b_2, \ldots, b_{2n}$ are arranged in descending time order.

Note that $\langle T \prod_{k=2}^{2n}{}' b_k \rangle_o$ is the ensemble average of the time-ordered product of $2n-2$ operators (recall that $k = m$ is excluded by the prime on $\prod$). By assumption, Wick's theorem is true for such a product. In other words, $\langle T \prod_{k=2}^{2n}{}' b_k \rangle_o$ is the sum over all contracted pairs that can be formed from the $2n-2$ operators. By summing over m in Eq. (8.68), we exhaust all pairs that can be formed from the $2n$ operators. Therefore, Wick's theorem is true for n, assuming that it is true for $n-1$, as long as the sign of each term in the sum over m is the right sign. To show that $(-1)^{P_1}(-1)^m$ is the right sign in each term, consider the original arrangement $a_1, a_2, \ldots, a_{2n}$. First we rearranged the operators to $b_1, b_2, \ldots, b_{2n}$, which brought about $(-1)^{P_1}$. Then, in order to form the contraction $\overline{b_1\, b_m}$, b_1 must be moved $m-2$ steps to the right (or b_m moved $m-2$ steps to the left), which brings about a factor $(-1)^{m-2} = (-1)^m$. Thus, $(-1)^{P_1}(-1)^m$ is indeed the correct sign in each term in the summation.

8.8.6 Some remarks on Wick's theorem

We state without proof the following remarks regarding Wick's theorem:

(a) We have proved the theorem for the case when each of $a_1, a_2, \ldots$ is either a fermion creation or annihilation operator. It is not difficult to extend the theorem to the case when each of $a_1, a_2, \ldots$ is a linear combination of a creation and an annihilation operator.
(b) Although the proof is given for fermion operators, the same steps may be followed to show that Wick's theorem is also valid for boson operators; in that case, one has to replace anticommutators with commutators.
(c) The theorem is also valid if some of the operators $a_1, a_2, \ldots$ are fermion operators while the rest are boson operators; in this case, the permutation P in the factor $(-1)^P$ is the permutation of the fermion operators.

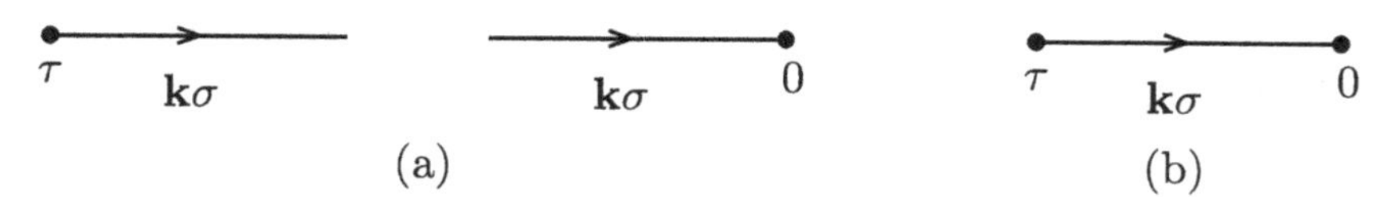

Figure 8.1 (a) A particle is created at time 0 in state $|\mathbf{k}\sigma\rangle$, and a particle in state $|\mathbf{k}\sigma\rangle$ is annihilated at time τ. (b) $g^0(\mathbf{k}\sigma, \tau)$ is represented by a solid line directed from point τ to point 0.

8.9 Case study: first-order interaction

As an example, let us use Wick's theorem to calculate the imaginary-time Green's function, to first order in the interaction, in a translationally invariant system of fermions interacting via pairwise interactions. The interaction term is given by

$$\hat{V}(\tau) = \frac{1}{2}\sum_{\mathbf{q}}\sum_{\mathbf{k}_1\sigma_1}\sum_{\mathbf{k}_2\sigma_2} v_{\mathbf{q}}\hat{c}^{\dagger}_{\mathbf{k}_1+\mathbf{q}\sigma_1}(\tau)\hat{c}^{\dagger}_{\mathbf{k}_2-\mathbf{q}\sigma_2}(\tau)\hat{c}_{\mathbf{k}_2\sigma_2}(\tau)\hat{c}_{\mathbf{k}_1\sigma_1}(\tau).$$

In the perturbation expansion for Green's function, Eq. (8.60), let us first consider the numerator. The zeroth order term, $n = 0$, gives

$$-\langle T\ \hat{c}_{\mathbf{k}\sigma}(\tau)\hat{c}^{\dagger}_{\mathbf{k}\sigma}(0)\rangle_0 = g^0(\mathbf{k}\sigma, \tau).$$

We can give a graphical representation to $g^0(\mathbf{k}\sigma, \tau)$. In the time-ordered product $-\langle T\ \hat{c}_{\mathbf{k}\sigma}(\tau)\hat{c}^{\dagger}_{\mathbf{k}\sigma}(0)\rangle_0$, the time arguments 0 and τ of the $\hat{c}$-operators are represented by two points arranged horizontally, where the point with time τ is on the left. At time 0, a particle is created in state $|\mathbf{k}\sigma\rangle$; this process is represented by an arrow entering the point with time argument 0. At time τ, a particle in state $|\mathbf{k}\sigma\rangle$ is annihilated; this is represented by an arrow leaving the point with time argument τ. These processes are depicted in Figure 8.1a. The noninteracting Green's function, $g^0(\mathbf{k}\sigma, \tau)$, can then be represented by a solid line directed from point τ to point 0, as shown in Figure 8.1b. We note that another convention is sometimes used where $g^0(\mathbf{k}\sigma, \tau)$ is represented by a line directed from 0 to τ; here we follow the convention used by Abrikosov, Gorkov, and Dzyaloshinski (1963).

The first-order term, $n = 1$, gives

$$\delta g^{(1)}_{num} = \frac{1}{\hbar V}\int_0^{\beta\hbar} d\tau_1 \sum_{\mathbf{q}}\sum_{\mathbf{k}_1\sigma_1}\sum_{\mathbf{k}_2\sigma_2}\frac{1}{2}v_{\mathbf{q}}$$
$$\times \left\langle T\,\hat{c}_{\mathbf{k}\sigma}(\tau)\hat{c}^{\dagger}_{\mathbf{k}\sigma}(0)\hat{c}^{\dagger}_{\mathbf{k}_1+\mathbf{q}\sigma_1}(\tau_1)\hat{c}^{\dagger}_{\mathbf{k}_2-\mathbf{q}\sigma_2}(\tau_1)\hat{c}_{\mathbf{k}_2\sigma_2}(\tau_1)\hat{c}_{\mathbf{k}_1\sigma_1}(\tau_1)\right\rangle_0$$

where the superscript and subscript on δg indicate that this is the first-order correction in the numerator. Applying Wick's theorem, we find

$$\begin{aligned}
&\langle T\ \hat{c}_{\mathbf{k}\sigma}(\tau)\hat{c}^{\dagger}_{\mathbf{k}\sigma}(0)\hat{c}^{\dagger}_{\mathbf{k}_1+\mathbf{q}\sigma_1}(\tau_1)\hat{c}^{\dagger}_{\mathbf{k}_2-\mathbf{q}\sigma_2}(\tau_1)\hat{c}_{\mathbf{k}_2\sigma_2}(\tau_1)\hat{c}_{\mathbf{k}_1\sigma_1}(\tau_1)\rangle_0 = \\
&\quad -\langle T\,\hat{c}_{\mathbf{k}\sigma}(\tau)\hat{c}^{\dagger}_{\mathbf{k}\sigma}(0)\rangle_0\langle T\,\hat{c}^{\dagger}_{\mathbf{k}_1+\mathbf{q}\sigma_1}(\tau_1)\hat{c}_{\mathbf{k}_2\sigma_2}(\tau_1)\rangle_0\langle T\,\hat{c}^{\dagger}_{\mathbf{k}_2-\mathbf{q}\sigma_2}(\tau_1)\hat{c}_{\mathbf{k}_1\sigma_1}(\tau_1)\rangle_0 \quad (a) \\
&\quad +\langle T\,\hat{c}_{\mathbf{k}\sigma}(\tau)\hat{c}^{\dagger}_{\mathbf{k}\sigma}(0)\rangle_0\langle T\,\hat{c}^{\dagger}_{\mathbf{k}_1+\mathbf{q}\sigma_1}(\tau_1)\hat{c}_{\mathbf{k}_1\sigma_1}(\tau_1)\rangle_0\langle T\,\hat{c}^{\dagger}_{\mathbf{k}_2-\mathbf{q}\sigma_2}(\tau_1)\hat{c}_{\mathbf{k}_2\sigma_2}(\tau_1)\rangle_0 \quad (b) \\
&\quad -\langle T\,\hat{c}_{\mathbf{k}\sigma}(\tau)\hat{c}^{\dagger}_{\mathbf{k}_2-\mathbf{q}\sigma_2}(\tau_1)\rangle_0\langle T\,\hat{c}^{\dagger}_{\mathbf{k}\sigma}(0)\hat{c}_{\mathbf{k}_2\sigma_2}(\tau_1)\rangle_0\langle T\,\hat{c}^{\dagger}_{\mathbf{k}_1+\mathbf{q}\sigma_1}(\tau_1)\hat{c}_{\mathbf{k}_1\sigma_1}(\tau_1)\rangle_0 \quad (c) \\
&\quad +\langle T\,\hat{c}_{\mathbf{k}\sigma}(\tau)\hat{c}^{\dagger}_{\mathbf{k}_1+\mathbf{q}\sigma_1}(\tau_1)\rangle_0\langle T\,\hat{c}^{\dagger}_{\mathbf{k}\sigma}(0)\hat{c}_{\mathbf{k}_2\sigma_2}(\tau_1)\rangle_0\langle T\,\hat{c}^{\dagger}_{\mathbf{k}_2-\mathbf{q}\sigma_2}(\tau_1)\hat{c}_{\mathbf{k}_1\sigma_1}(\tau_1)\rangle_0 \quad (d) \\
&\quad -\langle T\,\hat{c}_{\mathbf{k}\sigma}(\tau)\hat{c}^{\dagger}_{\mathbf{k}_1+\mathbf{q}\sigma_1}(\tau_1)\rangle_0\langle T\,\hat{c}^{\dagger}_{\mathbf{k}\sigma}(0)\hat{c}_{\mathbf{k}_1\sigma_1}(\tau_1)\rangle_0\langle T\,\hat{c}^{\dagger}_{\mathbf{k}_2-\mathbf{q}\sigma_2}(\tau_1)\hat{c}_{\mathbf{k}_2\sigma_2}(\tau_1)\rangle_0 \quad (e) \\
&\quad +\langle T\,\hat{c}_{\mathbf{k}\sigma}(\tau)\hat{c}^{\dagger}_{\mathbf{k}_2-\mathbf{q}\sigma_2}(\tau_1)\rangle_0\langle T\,\hat{c}^{\dagger}_{\mathbf{k}\sigma}(0)\hat{c}_{\mathbf{k}_1\sigma_1}(\tau_1)\rangle_0\langle T\,\hat{c}^{\dagger}_{\mathbf{k}_1+\mathbf{q}\sigma_1}(\tau_1)\hat{c}_{\mathbf{k}_2\sigma_2}(\tau_1)\rangle_0 \quad (f)
\end{aligned}$$

$$\begin{aligned}
&= g^{o}(\mathbf{k}\sigma,\tau)g^{0}(\mathbf{k}_2\sigma_2,0)g^{0}(\mathbf{k}_1\sigma_1,0)\delta_{\sigma_1\sigma_2}\delta_{\mathbf{k}_2,\mathbf{k}_1+\mathbf{q}} && (8.69a)\\
&\quad - g^{0}(\mathbf{k}\sigma,\tau)g^{0}(\mathbf{k}_1\sigma_1,0)g^{0}(\mathbf{k}_2\sigma_2,0)\delta_{\mathbf{q},\mathbf{0}} && (8.69b)\\
&\quad + g^{0}(\mathbf{k}\sigma,\tau-\tau_1)g^{0}(\mathbf{k}\sigma,\tau_1)g^{0}(\mathbf{k}_1\sigma_1,0)\delta_{\sigma\sigma_2}\delta_{\mathbf{k}\mathbf{k}_2}\delta_{\mathbf{q},\mathbf{0}} && (8.69c)\\
&\quad - g^{0}(\mathbf{k}\sigma,\tau-\tau_1)g^{0}(\mathbf{k}\sigma,\tau_1)g^{0}(\mathbf{k}_1\sigma_1,0)\delta_{\sigma\sigma_1}\delta_{\sigma\sigma_2}\delta_{\mathbf{k},\mathbf{k}_1+\mathbf{q}}\delta_{\mathbf{k}\mathbf{k}_2} && (8.69d)\\
&\quad + g^{0}(\mathbf{k}\sigma,\tau-\tau_1)g^{0}(\mathbf{k}\sigma,\tau_1)g^{0}(\mathbf{k}_2\sigma_2,0)\delta_{\sigma\sigma_1}\delta_{\mathbf{k}\mathbf{k}_1}\delta_{\mathbf{q},\mathbf{0}} && (8.69e)\\
&\quad - g^{0}(\mathbf{k}\sigma,\tau-\tau_1)g^{0}(\mathbf{k}\sigma,\tau_1)g^{0}(\mathbf{k}_2\sigma_2,0)\delta_{\sigma\sigma_1}\delta_{\sigma\sigma_2}\delta_{\mathbf{k}\mathbf{k}_1}\delta_{\mathbf{k},\mathbf{k}_2-\mathbf{q}}. && (8.69f)
\end{aligned}$$

We can represent the above terms graphically. The time arguments τ, τ_1, and 0 are arranged from left to right. At time 0, a particle is created in state $|\mathbf{k}\sigma\rangle$. At time τ, a particle in state $|\mathbf{k}\sigma\rangle$ is annihilated. At time τ_1, two particles in states $|\mathbf{k}_1\sigma_1\rangle$ and $|\mathbf{k}_2\sigma_2\rangle$ are annihilated, while two particles are created in states $|\mathbf{k}_1+\mathbf{q}\sigma_1\rangle$ and $|\mathbf{k}_2-\mathbf{q}\sigma_2\rangle$. These processes are depicted in Figure 8.2. The interaction process at time τ_1 is represented by a dashed line carrying wave-vector $\mathbf{q}$, two solid lines going out, and two solid lines coming in. A contraction of an annihilation operator and a creation operator corresponds to connecting an arrow that is leaving a point to an arrow that is entering the same or a different point, forming a directed solid line. The line represents a noninteracting Green's function. Wick's theorem tells us to sum over all possible ways of picking pairs for contraction, from amongst all the operators. Each possible way corresponds to connecting arrows in pairs. A pair consists of one arrow leaving a point and one arrow entering a point. Each possible way thus results in a particular diagram, known as a Feynman diagram or a Feynman graph. Representation of perturbation theory in terms of diagrams was originally developed by Feynman in his work on quantum electrodynamics (Feynman, 1949a, 1949b). Hence, we can restate Wick's theorem pictorially as

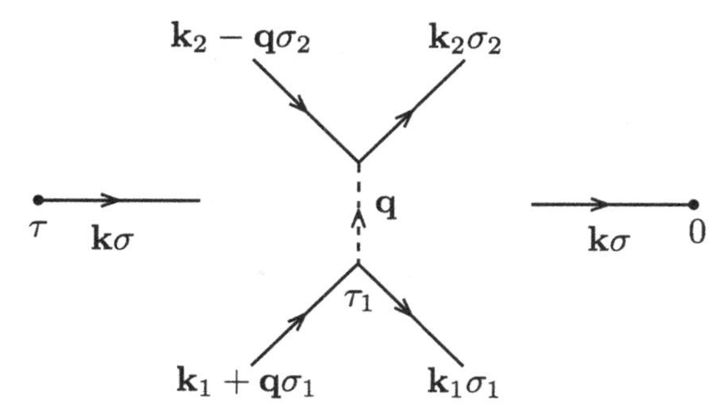

Figure 8.2 A particle is created in state $|\mathbf{k}\sigma\rangle$ at time 0, while a particle in state $|\mathbf{k}\sigma\rangle$ is annihilated at time τ. At time τ_1, two particles in states $|\mathbf{k}_1\sigma_1\rangle$ and $|\mathbf{k}_2\sigma_2\rangle$ are annihilated, while two particles are created in states $|\mathbf{k}_1 + \mathbf{q}\sigma_1\rangle$ and $|\mathbf{k}_2 - \mathbf{q}\sigma_2\rangle$.

follows: connect all arrows in pairs (an arrow leaving a point to an arrow entering a point), and do so in all possible ways; each possible way produces a diagram, then add all the diagrams. Contemplation of Figure 8.2 reveals that we can form six Feynman diagrams (see Figure 8.3), corresponding to the six different ways of contracting the pairs.

In Chapter 9, we will develop rules for translating the diagrams into algebraic expressions. For now, we only note the following:

- Each diagram in Figure 8.3 contains three solid lines; hence the analytical expression for each diagram contains a product of three noninteracting Green's functions. This is in conformity with Eq. (8.69).
- The Kronecker deltas which appear in Eq. (8.69) can be directly read off the diagrams, since each solid line may contain only one wave vector and one spin projection. For example, looking at Figure 8.3c, the following product of Kronecker deltas can be read directly off the diagram: $\delta_{\mathbf{k},\mathbf{k}_2-\mathbf{q}}\delta_{\mathbf{k},\mathbf{k}_2}\delta_{\mathbf{k}_1,\mathbf{k}_1+\mathbf{q}}\delta_{\sigma,\sigma_2} = \delta_{\mathbf{k},\mathbf{k}_2}\delta_{\mathbf{q},\mathbf{0}}\delta_{\sigma,\sigma_2}$.
- With the products of the three Green's functions and the Kronecker deltas determined, the six terms in Eq. (8.69) can be written directly from the diagrams. The only remaining question is the sign: diagrams a, c, and e produce a positive sign, while diagrams b, d, and f produce a negative sign. For now, we state without proof that the sign is given by $(-1)^{n+F}$, where n is the perturbation order ($n = 1$ in this case) and F is the number of closed fermion loops: $(-1)^{1+1} = +1$ for diagrams a, c, and e. Diagram b has two fermion loops, and diagrams d and f have zero fermion loops; hence their negative signs.
- Diagrams a and b are disconnected, whereas diagrams c, d, e, and f are connected.

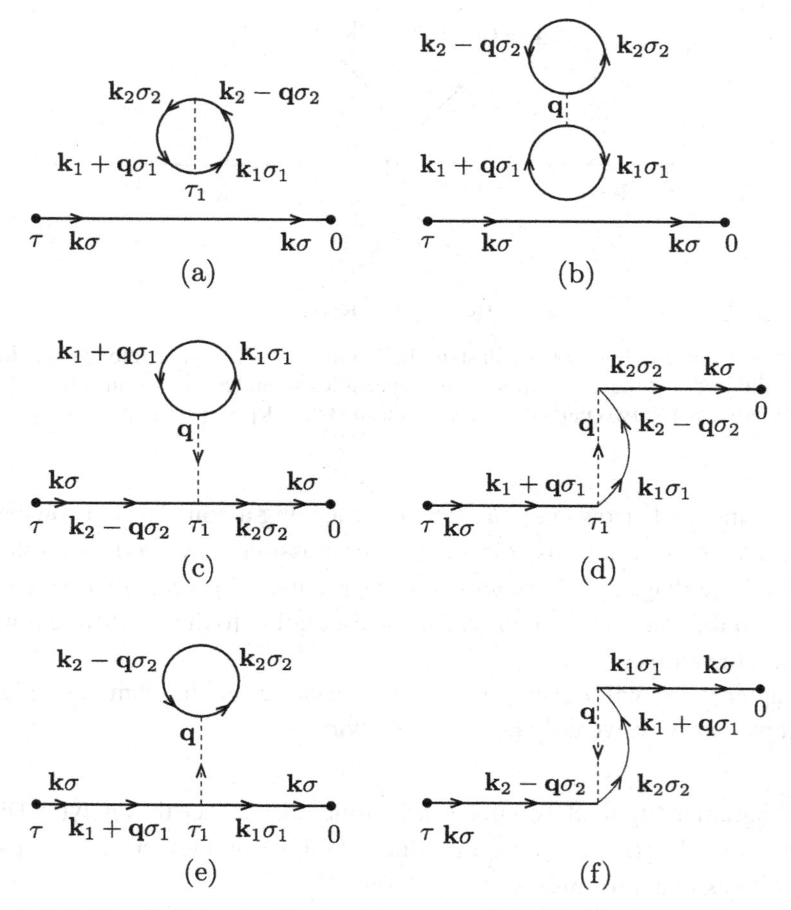

Figure 8.3 The six Feynman diagrams corresponding to the six terms in Eq. (8.69).

In the remainder of this section, we will show that, to first order in the interaction, the contribution of the disconnected diagrams to $\delta g^{(1)}_{\text{num}}$ is cancelled by the denominator in Eq. (8.60).

To first order (n = 1), the numerator $\mathcal{N}$ in Eq. (8.60) is given by

$$\mathcal{N} = g^0(\mathbf{k}\sigma,\tau) + \frac{1}{2\hbar V}\int_0^{\beta\hbar} d\tau_1 \sum_{\mathbf{q}}\sum_{\mathbf{k}_1\sigma_1}\sum_{\mathbf{k}_2\sigma_2} v_{\mathbf{q}} \times \langle T\,\hat{c}_{\mathbf{k}\sigma}(0)\hat{c}^{\dagger}_{\mathbf{k}\sigma}(\tau)\hat{c}^{\dagger}_{\mathbf{k}_1+\mathbf{q}\sigma_1}(\tau_1)\hat{c}^{\dagger}_{\mathbf{k}_2-\mathbf{q}\sigma_2}(\tau_1)\hat{c}_{\mathbf{k}_2\sigma_2}(\tau_1)\hat{c}_{\mathbf{k}_1\sigma_1}(\tau_1)\rangle_0.$$

We have already seen that the time-ordered product decomposes into six terms, the first two of which correspond to disconnected diagrams:

$$\begin{aligned}\mathcal{N} &= g^0(\mathbf{k}\sigma,\tau) + \frac{1}{2\hbar V}\int_0^{\beta\hbar} d\tau_1 \sum_{\mathbf{q}}\sum_{\mathbf{k}_1\sigma_1}\sum_{\mathbf{k}_2\sigma_2} v_{\mathbf{q}}\left[-\langle T\,\hat{c}_{\mathbf{k}\sigma}(\tau)\hat{c}^\dagger_{\mathbf{k}\sigma}(0)\rangle_0\right]\\
&\times\Big\{\langle T\,\hat{c}^\dagger_{\mathbf{k}_1+\mathbf{q}\sigma_1}(\tau_1)\hat{c}_{\mathbf{k}_2\sigma_2}(\tau_1)\rangle_0\langle T\,\hat{c}^\dagger_{\mathbf{k}_2-\mathbf{q}\sigma_2}(\tau_1)\hat{c}_{\mathbf{k}_1\sigma_1}(\tau_1)\rangle_0\\
&-\langle T\,\hat{c}^\dagger_{\mathbf{k}_1+\mathbf{q}\sigma_1}(\tau_1)\hat{c}_{\mathbf{k}_1\sigma_1}(\tau_1)\rangle_0\langle T\,\hat{c}^\dagger_{\mathbf{k}_2-\mathbf{q}\sigma_2}(\tau_1)\hat{c}_{\mathbf{k}_2\sigma_2}(\tau_1)\rangle_0\Big\} + \text{connected diagrams}\\
&= g^0(\mathbf{k}\sigma,\tau)\left[1+\frac{1}{2\hbar V}\int_0^{\beta\hbar} d\tau_1 \sum_{\mathbf{q}}\sum_{\mathbf{k}_1\sigma_1}\sum_{\mathbf{k}_2\sigma_2} v_{\mathbf{q}} I(\mathbf{k}_1\mathbf{k}_2\mathbf{q}\sigma_1\sigma_2,\tau_1)\right] + \text{conn. dgs.}\end{aligned}$$

We have explicitly written the first two terms that correspond to the disconnected diagrams, lumped together the remaining four terms as "connected diagrams," and denoted the term in braces by $I(\mathbf{k}_1\mathbf{k}_2\mathbf{q}\sigma_1\sigma_2,\tau_1)$. It is clear, using Wick's theorem, that

$$I(\mathbf{k}_1\mathbf{k}_2\mathbf{q}\sigma_1\sigma_2,\tau_1) = -\langle T\,\hat{c}^\dagger_{\mathbf{k}_1+\mathbf{q}\sigma_1}(\tau_1)\hat{c}^\dagger_{\mathbf{k}_2-\mathbf{q}\sigma_2}(\tau_1)\hat{c}_{\mathbf{k}_2\sigma_2}(\tau_1)\hat{c}_{\mathbf{k}_1\sigma_1}(\tau_1)\rangle_0.$$

We thus deduce that

$$\frac{1}{2V}\sum_{\mathbf{q}}\sum_{\mathbf{k}_1\sigma_1}\sum_{\mathbf{k}_2\sigma_2} v_{\mathbf{q}} I(\mathbf{k}_1\mathbf{k}_2\mathbf{q}\sigma_1\sigma_2,\tau_1) = -\langle T\hat{V}(\tau_1)\rangle_0.$$

Noting that, to first order in the interaction, the denominator is given by

$$\mathcal{D} = 1 - \frac{1}{\hbar}\int_0^{\beta\hbar} d\tau_1\langle T\hat{V}(\tau_1)\rangle_0,$$

we conclude that, to first order in $\hat{V}$

$$g(\mathbf{k}\sigma,\tau) = \frac{\mathcal{N}}{\mathcal{D}} = g^0(\mathbf{k}\sigma,\tau) + \text{connected diagrams}. \tag{8.70}$$

Dividing the connected diagrams (which are already first order in $\hat{V}$) by the denominator gives the same connected diagrams plus diagrams of higher order in $\hat{V}$; hence, to first order in $\hat{V}$, Eq. (8.70) is exact.

Thus, to first order in the interaction, the denominator cancels out the disconnected diagrams in $g(\mathbf{k}\sigma,\tau)$. As we show in the next section, this cancellation persists to all orders in the perturbation expansion.

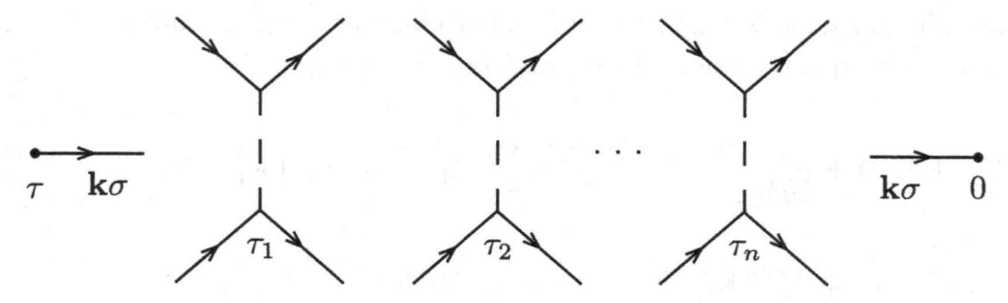

Figure 8.4 External points at times τ and 0, and internal points at times $\tau_1, \tau_2, \ldots \tau_n$. It is assumed that $\hat{V}$ is a two-body operator.

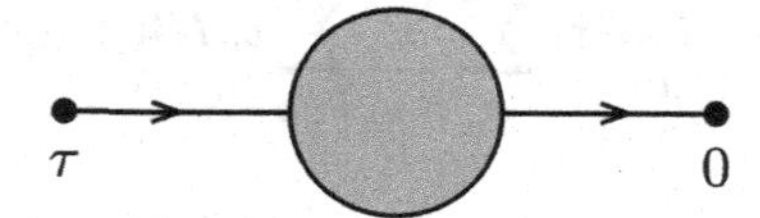

Figure 8.5 A connected Feynman diagram.

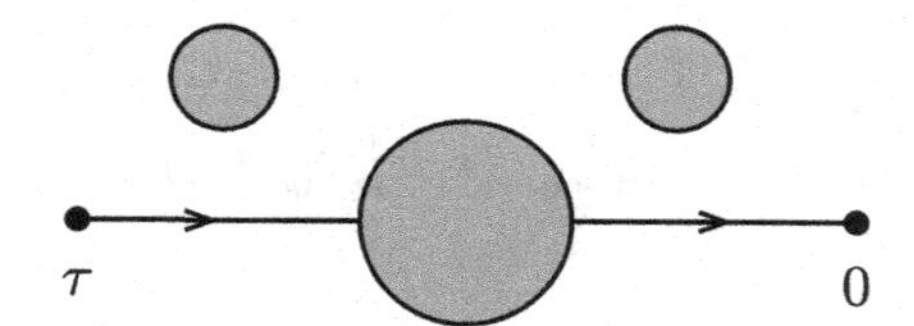

Figure 8.6 A disconnected Feynman diagram.

8.10 Cancellation of disconnected diagrams

All Feynman diagrams (graphs) are either connected or disconnected. For the nth order term in the expansion of g, we draw points representing time coordinates. There are two external points, one at time τ on the far left, and one at time 0 on the far right. In between are internal points at times $\tau_1, \tau_2, \ldots, \tau_n$. A line starts at the external point τ and a line ends at the external point 0. At each internal point, one line goes in and one line goes out if $\hat{V}$ is a one-body operator. If $\hat{V}$ is a two-body operator, two lines go in and two lines go out at each internal time. This is depicted in Figure 8.4 for the case when $\hat{V}$ is a two-body operator.

By a "connected" diagram we mean a diagram in which every internal point is connected, via a series of connected lines, to the external points. A connected diagram looks like Figure 8.5, while a disconnected diagram looks like Figure 8.6. In a connected diagram, $\hat{c}(\tau)$ is paired with a $\hat{c}^\dagger(\tau_i)$, a $\hat{c}(\tau_i)$ is paired with a $\hat{c}^\dagger(\tau_j)$, a $\hat{c}(\tau_j)$ is paired with a $\hat{c}^\dagger(\tau_k)$, and so on, until we reach $\hat{c}^\dagger(0)$ without missing any internal τ-point. The diagrams in which one or more internal points are not connected to the external points are "disconnected."

Regarding the order-n correction to Green's function, diagrams in which m internal points are connected to the external points, while $n-m$ are not, arise from the following expression in the numerator of Eq. (8.60)

$$-\frac{1}{n!}\left(-\frac{1}{\hbar}\right)^n \int d\tau_1 \ldots \int d\tau_m \langle T\,\hat{c}_{\mathbf{k}\sigma}(\tau)\hat{c}^\dagger_{\mathbf{k}\sigma}(0)\hat{V}(\tau_1)\ldots\hat{V}(\tau_m)\rangle_{0,c}$$
$$\times \int d\tau_{m+1}\ldots\int d\tau_n \langle T\,\hat{V}(\tau_{m+1})\ldots\hat{V}(\tau_n)\rangle_0.$$

The subscript c means "connected." We have chosen the internal points $\tau_1, \tau_2, \ldots, \tau_m$ to be the ones connected to the external points. However, had we chosen any other set of m internal points, the expression above would not change, since it would simply amount to a relabeling of the integration variables. The number of such identical expressions is equal to the number of ways of picking m points out of n points; this is $n!/m!(n-m)!$. Hence, the nth order correction in the numerator of Eq. (8.60) is

$$\delta g^{(n)}_{num} = -\sum_{m=0}^{n}\frac{1}{m!(n-m)!}\left(-\frac{1}{\hbar}\right)^n \int_0^{\beta\hbar} d\tau_1 \ldots \int_0^{\beta\hbar} d\tau_m$$
$$\times \langle T\,\hat{c}_{\mathbf{k}\sigma}(\tau)\hat{c}^\dagger_{\mathbf{k}\sigma}(0)\hat{V}(\tau_1)\ldots\hat{V}(\tau_m)\rangle_{0,c}\int_0^{\beta\hbar} d\tau_{m+1}\ldots\int_0^{\beta\hbar} d\tau_n \langle T\,\hat{V}(\tau_{m+1})\ldots\hat{V}(\tau_n)\rangle_0.$$

Diagrams in which all internal points are connected to the two external points correspond to $m=n$ in the above sum, while diagrams in which only one internal point is not connected to the two external points correspond to $m=n-1$. Diagrams in which none of the internal points are connected to the external points correspond to $m=0$. Summing over m from 0 to n generates all the diagrams, as required by Wick's theorem.

The summation over m in the above expansion may be rewritten as

$$\sum_{m=0}^{n}\frac{1}{m!(n-m)!}\cdots = \sum_{m=0}^{\infty}\sum_{j=0}^{\infty}\delta_{n,m+j}\frac{1}{m!j!}\cdots.$$

The Kronecker delta ensures that the summand is nonvanishing only when $j = n-m$ and $m \le n$. Summing over all orders, the numerator in Eq. (8.60) becomes

$$\mathcal{N} = -\sum_{n=0}^{\infty}\sum_{m=0}^{\infty}\sum_{j=0}^{\infty}\delta_{n,m+j}\frac{1}{m!j!}\left(-\frac{1}{\hbar}\right)^n \int_0^{\beta\hbar} d\tau_1 \ldots \int_0^{\beta\hbar} d\tau_m$$
$$\times \langle T\,\hat{c}_{\mathbf{k}\sigma}(0)\hat{c}^\dagger_{\mathbf{k}\sigma}(\tau)\hat{V}(\tau_1)\ldots\hat{V}(\tau_m)\rangle_{0,c}\int_0^{\beta\hbar} d\tau_1\ldots\int_0^{\beta\hbar} d\tau_j \langle T\,\hat{V}(\tau_1)\ldots\hat{V}(\tau_j)\rangle_0.$$

We relabeled $\tau_{m+1}, \ldots, \tau_n$ as $\tau_1, \ldots, \tau_j$, a step made possible by the fact that in summing over j, only terms with $j = n - m$ make a nonzero contribution. Summing first over n amounts to removing the Kronecker delta and replacing n by $m + j$, so that $(-1/\hbar)^n = (-1/\hbar)^m(-1/\hbar)^j$; thus

$$\mathcal{N} = -\sum_{m=0}^{\infty}\frac{1}{m!}\left(-\frac{1}{\hbar}\right)^m \int_0^{\beta\hbar} d\tau_1 \ldots \int_0^{\beta\hbar} d\tau_m \langle T\,\hat{c}_{\mathbf{k}\sigma}(\tau)\hat{c}^{\dagger}_{\mathbf{k}\sigma}(0)\hat{V}(\tau_1)\ldots\hat{V}(\tau_m)\rangle_{0,c}$$

$$\times\sum_{j=0}^{\infty}\frac{1}{j!}\left(-\frac{1}{\hbar}\right)^j \int_0^{\beta\hbar} d\tau_1 \ldots \int_0^{\beta\hbar} d\tau_j \langle T\,\hat{V}(\tau_1)\ldots\hat{V}(\tau_j)\rangle_0.$$

But the second factor, $\sum_{j=0}^{\infty}\cdots$ is simply the denominator in Eq. (8.60); hence

$$g(\mathbf{k}\sigma,\tau) = -\sum_{n=0}^{\infty}\frac{1}{n!}\left(-\frac{1}{\hbar}\right)^n \int_0^{\beta\hbar} d\tau_1 \ldots \int_0^{\beta\hbar} d\tau_n \langle T\,\hat{c}_{\mathbf{k}\sigma}(\tau)\hat{c}^{\dagger}_{\mathbf{k}\sigma}(0)\hat{V}(\tau_1)\ldots\hat{V}(\tau_n)\rangle_{0,c}. \tag{8.71}$$

We have proven that the denominator in Eq. (8.60) cancels out the disconnected diagrams of the numerator. Importantly, we have shown that Green's function is obtained by summing over all connected diagrams. Of course, there remains the question of how to translate Feynman diagrams into analytical expressions. We will take this up in the next chapter.

Further reading

Abrikosov, A.A., Gorkov, L.P., and Dzyaloshinski, I.E. (1963). *Methods of Quantum Field Theory in Statistical Physics*. New York: Dover Publications.

Fetter, A.L. and Walecka, J.D. (1971). *Quantum Theory of Many-Particle Systems*. New York: McGraw-Hill.

Mills, R. (1969). *Propagators for Many-Particle Systems*. New York: Gordon and Breach Science Publishers, Inc.

Problems

8.1 $\langle V\rangle$ *and* $\langle E\rangle$. For a translationally invariant system of interacting particles, assume that $v(\mathbf{r}_1 - \mathbf{r}_2)$ is spin-independent. The Hamiltonian is

$$\bar{H} = \sum_{\sigma}\int \Psi^{\dagger}_{\sigma}(\mathbf{r})\left(-\frac{\hbar^2}{2m}\nabla^2 - \mu\right)\Psi_{\sigma}(\mathbf{r})d^3r$$

$$+\frac{1}{2}\sum_{\sigma_1\sigma_2}\int d^3r_1\int d^3r_2\Psi^{\dagger}_{\sigma_1}(\mathbf{r}_1)\Psi^{\dagger}_{\sigma_2}(\mathbf{r}_2)v(\mathbf{r}_1-\mathbf{r}_2)\Psi_{\sigma_2}(\mathbf{r}_2)\Psi_{\sigma_1}(\mathbf{r}_1).$$

(a) Using the Heisenberg equation of motion, show that

$$\hbar\frac{\partial}{\partial\tau}\Psi_\sigma(\mathbf{r}\tau) = A(\mathbf{r},\tau)\Psi_\sigma(\mathbf{r}\tau)$$

where

$$A(\mathbf{r},\tau) = \left[\frac{\hbar^2}{2m}\nabla^2 + \mu - \sum_{\sigma_1}\int \Psi^\dagger_{\sigma_1}(\mathbf{r}_1\tau)v(\mathbf{r}-\mathbf{r}_1)\Psi_{\sigma_1}(\mathbf{r}_1\tau)d^3r_1\right].$$

(b) Using the result from part (a), show that

$$\left\langle \sum_{\sigma_1}\int d^3r_1 \Psi^\dagger_\sigma(\mathbf{r}'\tau)\Psi^\dagger_{\sigma_1}(\mathbf{r}_1\tau)v(\mathbf{r}-\mathbf{r}_1)\Psi_{\sigma_1}(\mathbf{r}_1\tau)\Psi_\sigma(\mathbf{r}\tau)\right\rangle$$

$$= \mp \lim_{\tau'\to\tau^+}\left(-\hbar\frac{\partial}{\partial\tau} + \frac{\hbar^2}{2m}\nabla^2 + \mu\right) g(\mathbf{r}\sigma\tau, \mathbf{r}'\sigma\tau').$$

(c) Take the limit $\mathbf{r}' \to \mathbf{r}$, sum over σ, multiply by $1/2$, and integrate over d^3r. Show that the result is Eq. (8.24):

$$\langle V\rangle = \mp\frac{1}{2}\int d^3r \lim_{\mathbf{r}'\to\mathbf{r}}\lim_{\tau'\to\tau^+}\left(-\hbar\frac{\partial}{\partial\tau} + \frac{\hbar^2}{2m}\nabla^2 + \mu\right)\sum_\sigma g(\mathbf{r}\sigma\tau, \mathbf{r}'\sigma\tau').$$

(d) Using the result from part (c), derive Eq. (8.25).

8.2 *Thermodynamic potential.* Define $\bar{H}(\lambda) = H_0 - \mu N + \lambda V = \bar{H}_0 + \lambda V$, $Z_G(\lambda) = \mathrm{Tr}\left[e^{-\beta\bar{H}(\lambda)}\right]$, $\Omega(\lambda) = -kT\, ln\, Z_G(\lambda)$.

(a) Show that

$$\frac{\partial}{\partial\lambda}\mathrm{Tr}\left[(\bar{H}_0 + \lambda V)^n\right] = n\,\mathrm{Tr}\left[(\bar{H}_0 + \lambda V)^{n-1}V\right].$$

(b) Using the above result, show that $\frac{\partial}{\partial\lambda}Z_G(\lambda) = -\beta Z_G(\lambda)\langle V\rangle_\lambda$ where $\langle V\rangle_\lambda = \mathrm{Tr}\left[e^{-\beta\bar{H}(\lambda)}V\right]/Z_G(\lambda)$.

(c) Show that $\frac{\partial}{\partial\lambda}\Omega(\lambda) = \langle V\rangle_\lambda = \frac{1}{\lambda}\langle\lambda V\rangle_\lambda$.

(d) Integrate both sides over λ from 0 to 1 and use the result of the previous problem to derive Eq. (8.26).

8.3 *Discontinuity in g^0.* For a system of noninteracting electrons, plot $g^0(\mathbf{k}\sigma,\tau)$ as a function of τ. Show that g^0 is discontinuous at $\tau = 0$ and that the discontinuity is equal to -1.

8.4 *Equation 8.66b.* Verify Eq. (8.66b).

8.5 *Wick's theorem: bosons.* Prove Wick's theorem for bosons.

8.6 *Wick's theorem.* Using Wick's theorem, evaluate $\langle \hat{N}(\tau)\hat{N}(\tau')\rangle_0$, where $\hat{N}$ is the total number of particles operator in the interaction picture.

8.7 *An equation for g^0.*

(a) Show that, for a translationally invariant system

$$\left(-\frac{\partial}{\partial\tau} + \frac{\hbar^2}{2m}\nabla^2 + \mu\right) g^0(\mathbf{r}-\mathbf{r}'\sigma, \tau-\tau') = \delta(\mathbf{r}-\mathbf{r}')\delta(\tau-\tau').$$

(b) By Fourier transforming, deduce the expression for $g^0(\mathbf{k}\sigma, \omega_n)$.

8.8 *Analytic continuation.* Given two operators A and B, let $\tilde{A} = A - \langle A\rangle$ and $\tilde{B} = B - \langle B\rangle$. Let

$$C^R_{AB}(t) = -i\theta(t)\langle[A(t), B(0)]\rangle, \quad c^T_{AB}(\tau) = -\langle A(\tau)B(0)\rangle.$$

It is clear that $C^R_{AB}(t) = C^R_{\tilde{A}\tilde{B}}(t)$ and $C^R_{AB}(\omega) = C^R_{\tilde{A}\tilde{B}}(\omega)$.

(a) Starting from the spectral representations of $C^R_{AB}(\omega)$ and $C^R_{\tilde{A}\tilde{B}}(\omega)$, show explicitly that

$$C^R_{AB}(\omega) = C^R_{\tilde{A}\tilde{B}}(\omega).$$

(b) Show that, if $\langle A\rangle$ and $\langle B\rangle$ do not vanish, then $c^T_{AB}(\tau) \neq c^T_{\tilde{A}\tilde{B}}(\tau)$.

(c) Find the relation between $c^T_{AB}(\omega_n)$ and $c^T_{\tilde{A}\tilde{B}}(\omega_n)$. Show that

$$C^R_{\tilde{A}\tilde{B}}(\omega) = c^T_{\tilde{A}\tilde{B}}(\omega_n)\Big|_{i\omega_n=\omega+i0^+}.$$

9

Diagrammatic techniques

So without these we all in vain shall try
To find the thing that gives them unity–
The thing to which each whispers, "Thou art thou"–
The soul which answers each, "And I am I."
–Titus Lucretius Carus, No Single Thing Abides
Translated by W. H. Mallock

We now consider in detail the rules for the construction and evaluation of Feynman diagrams. At the end of the previous chapter, we expressed the imaginary-time single-particle Green's function as a perturbation series in connected diagrams,

$$g(\mathbf{k}\sigma, \tau) = -\sum_{n=0}^{\infty} \frac{1}{n!} \left(-\frac{1}{\hbar}\right)^n \int_0^{\beta\hbar} \mathrm{d}\tau_1 \ldots \int_0^{\beta\hbar} \mathrm{d}\tau_n \left\langle T\, c_{\mathbf{k}\sigma}(\tau) c^{\dagger}_{\mathbf{k}\sigma}(0)\, V(\tau_1) \ldots V(\tau_n) \right\rangle_{0,c} . \tag{9.1}$$

To simplify notation, the hat "$\wedge$" above the operators is dropped. Additionally, throughout this chapter, all operators are interaction picture operators, unless stated otherwise. We show how to write $g(\mathbf{k}\sigma, \tau)$ as a sum of Feynman diagrams, and develop rules for translating diagrams into algebraic expressions.

To develop diagram rules, we begin by investigating, in sufficient detail, the second-order correction to $g(\mathbf{k}\sigma, \tau)$ in a system of fermions, where V is a two-particle interaction. We determine diagram rules, and later derive them for any order in the interaction.

9.1 Case study: second-order perturbation in a system of fermions

The second-order contribution to $g(\mathbf{k}\sigma, \tau)$ is given by

$$\delta g^{(2)}(\mathbf{k}\sigma, \tau) = -\frac{1}{2!} \left(-\frac{1}{\hbar}\right)^2 \int_0^{\beta\hbar} \mathrm{d}\tau_1 \int_0^{\beta\hbar} \mathrm{d}\tau_2 \left\langle T\, c_{\mathbf{k}\sigma}(\tau) c^{\dagger}_{\mathbf{k}\sigma}(0)\, V(\tau_1) V(\tau_2) \right\rangle_{0,c} \tag{9.2}$$

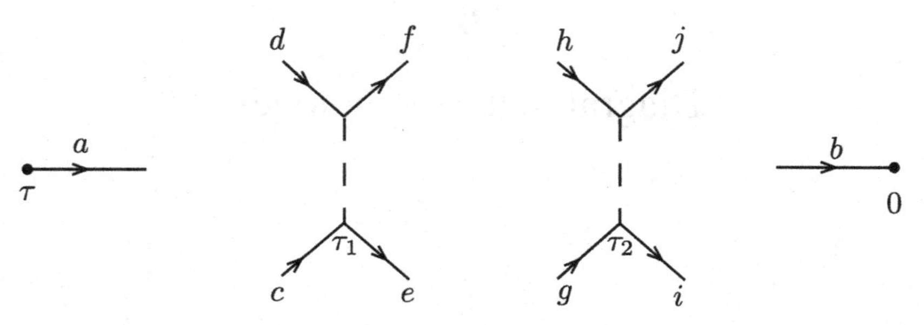

Figure 9.1 External fermion lines at τ and 0, and internal ones at τ_1 and τ_2. Dashed lines are interaction lines. a and b represent the coordinates $(\mathbf{k}\sigma)$, $c = (\mathbf{k_1} + \mathbf{q}\sigma_1)$, $d = (\mathbf{k_2} - \mathbf{q}\sigma_2)$, $e = (\mathbf{k_1}\sigma_1)$, $f = (\mathbf{k_2}\sigma_2)$, $g = (\mathbf{k_3} + \mathbf{q}'\sigma_3)$, $h = (\mathbf{k_4} - \mathbf{q}'\sigma_4)$, $i = (\mathbf{k_3}\sigma_3)$, and $j = (\mathbf{k_4}\sigma_4)$.

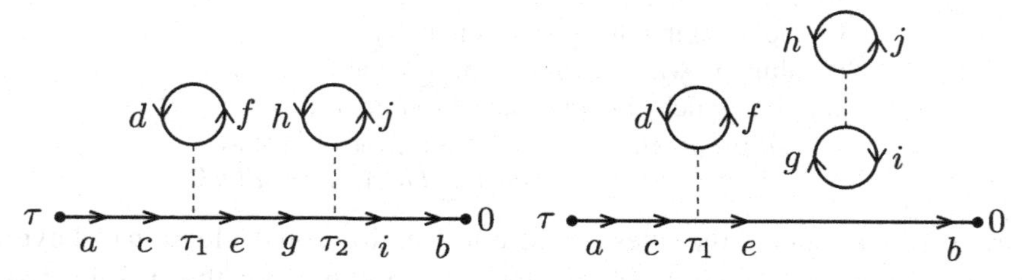

Figure 9.2 Connected and disconnected diagrams. The diagram on the left, denoted by $(ac)(eg)(fd)(ib)(jh)$, is connected. The diagram on the right, denoted by $(ac)(eb)(fd)(ig)(jh)$, is disconnected.

where, for a system of volume V,

$$V(\tau) = \frac{1}{V} \sum_{\mathbf{k}_1\sigma_1} \sum_{\mathbf{k}_2\sigma_2} \sum_{\mathbf{q}} \frac{1}{2} v_{\mathbf{q}} c^{\dagger}_{\mathbf{k}_1+\mathbf{q}\sigma_1}(\tau) c^{\dagger}_{\mathbf{k}_2-\mathbf{q}\sigma_2}(\tau) c_{\mathbf{k}_2\sigma_2}(\tau) c_{\mathbf{k}_1\sigma_1}(\tau). \tag{9.3}$$

We need to evaluate

$$\Big\langle T\, c_{\mathbf{k}\sigma}(\tau) c^{\dagger}_{\mathbf{k}\sigma}(0) c^{\dagger}_{\mathbf{k}_1+\mathbf{q}\sigma_1}(\tau_1) c^{\dagger}_{\mathbf{k}_2-\mathbf{q}\sigma_2}(\tau_1) c_{\mathbf{k}_2\sigma_2}(\tau_1) c_{\mathbf{k}_1\sigma_1}(\tau_1)$$
$$c^{\dagger}_{\mathbf{k}_3+\mathbf{q}'\sigma_3}(\tau_2) c^{\dagger}_{\mathbf{k}_4-\mathbf{q}'\sigma_4}(\tau_2) c_{\mathbf{k}_4\sigma_4}(\tau_2) c_{\mathbf{k}_3\sigma_3}(\tau_2) \Big\rangle_{0,c} .$$

The evaluation proceeds by using Wick's theorem: we sum over all possible contractions that result in connected diagrams. Referring to Figure 9.1 and its description, this corresponds to summing all connected diagrams obtained by connecting the directed fermion lines: one line leaving a vertex to another line entering the same, or a different, vertex. For example, one diagram (see Figure 9.2) results from connecting a to c, e to g, f to d, i to b, and j to h; this diagram is denoted by $(ac)(eg)(fd)(ib)(jh)$. On the other hand, a diagram such as $(ac)(eb)(fd)(ig)(jh)$ is disconnected and should not be counted.

Table 9.1 *Connected diagrams that can be drawn in Figure 9.1 by connecting a to either c or d.*

1. $(ac)(eb)(jd)(fh)(ig)$	21. $(ac)(ib)(eh)(fd)(jg)$
2. $(ac)(eb)(jd)(fg)(ih)$	22. $(ac)(ib)(eh)(fg)(jd)$
3. $(ac)(eb)(id)(fh)(jg)$	23. $(ad)(jb)(fc)(eh)(ig)$
4. $(ac)(eb)(id)(fg)(jh)$	24. $(ad)(jb)(fc)(eg)(ih)$
5. $(ac)(fb)(jd)(eh)(ig)$	25. $(ad)(jb)(fh)(ec)(ig)$
6. $(ac)(fb)(jd)(eg)(ih)$	26. $(ad)(jb)(fh)(eg)(ic)$
7. $(ac)(fb)(id)(eh)(jg)$	27. $(ad)(jb)(fg)(eh)(ic)$
8. $(ac)(fb)(id)(eg)(jh)$	28. $(ad)(jb)(fg)(ec)(ih)$
9. $(ad)(eb)(fh)(jc)(ig)$	29. $(ac)(jb)(fd)(eh)(ig)$
10. $(ad)(eb)(fh)(jg)(ic)$	30. $(ac)(jb)(fd)(eg)(ih)$
11. $(ad)(eb)(fg)(jc)(ih)$	31. $(ac)(jb)(fh)(ed)(ig)$
12. $(ad)(eb)(fg)(jh)(ic)$	32. $(ac)(jb)(fh)(eg)(id)$
13. $(ad)(fb)(eg)(jh)(ic)$	33. $(ac)(jb)(fg)(ed)(ih)$
14. $(ad)(fb)(eg)(jc)(ih)$	34. $(ac)(jb)(fg)(eh)(id)$
15. $(ad)(fb)(eh)(ic)(jg)$	35. $(ad)(ib)(fc)(eh)(jg)$
16. $(ad)(fb)(eh)(ig)(jc)$	36. $(ad)(ib)(fc)(eg)(jh)$
17. $(ac)(ib)(ed)(fh)(jg)$	37. $(ad)(ib)(fh)(ec)(jg)$
18. $(ac)(ib)(ed)(fg)(jh)$	38. $(ad)(ib)(fh)(eg)(jc)$
19. $(ac)(ib)(eg)(fd)(jh)$	39. $(ad)(ib)(fg)(ec)(jh)$
20. $(ac)(ib)(eg)(fh)(jd)$	40. $(ad)(ib)(fg)(eh)(jc)$

What is the total number of connected diagrams that can be drawn in Figure 9.1? The external line a can be connected to c, d, g, or h, and line b can be connected to e, f, i, or j. We start by counting all connected diagrams that can be obtained by connecting a to either c or d; there are 40 such diagrams, listed in Table 9.1.

What about diagrams that can be obtained by connecting a to g or h? It is clear that there are also 40 such diagrams, which can be obtained from the 40 diagrams listed in Table 9.1 by the following interchanges: $\tau_1 \leftrightarrow \tau_2$, $(\mathbf{k_1}\sigma_1) \leftrightarrow (\mathbf{k_3}\sigma_3)$, $(\mathbf{k_2}\sigma_2) \leftrightarrow (\mathbf{k_4}\sigma_4)$, and $\mathbf{q} \leftrightarrow \mathbf{q}'$. Since τ_1 and τ_2 are integrated over, and $\mathbf{k_1}\sigma_1$, $\mathbf{k_2}\sigma_2$, $\mathbf{k_3}\sigma_3$, $\mathbf{k_4}\sigma_4$, $\mathbf{q}$, and $\mathbf{q}'$ are summed over, the 40 diagrams obtained by connecting a to g or h make exactly the same contributions as the 40 diagrams enumerated in Table 9.1. Hence, we may consider only the 40 diagrams in Table 9.1 and cancel the factor 2! in the denominator of Eq. (9.1).

Next, we observe that when we draw the 40 Feynman diagrams listed in Table 9.1, we find that there are only 10 generically different (topologically distinct) diagrams. Each of these 10 diagrams occurs four times. The ten connected, topologically distinct diagrams are shown in Figure 9.3. The diagrams in Figure 9.3 are sometimes drawn in a different but equivalent way (see Figure 9.4). Table 9.2 lists the diagrams from Table 9.1 that are topologically equivalent to the diagrams in Figure 9.3.

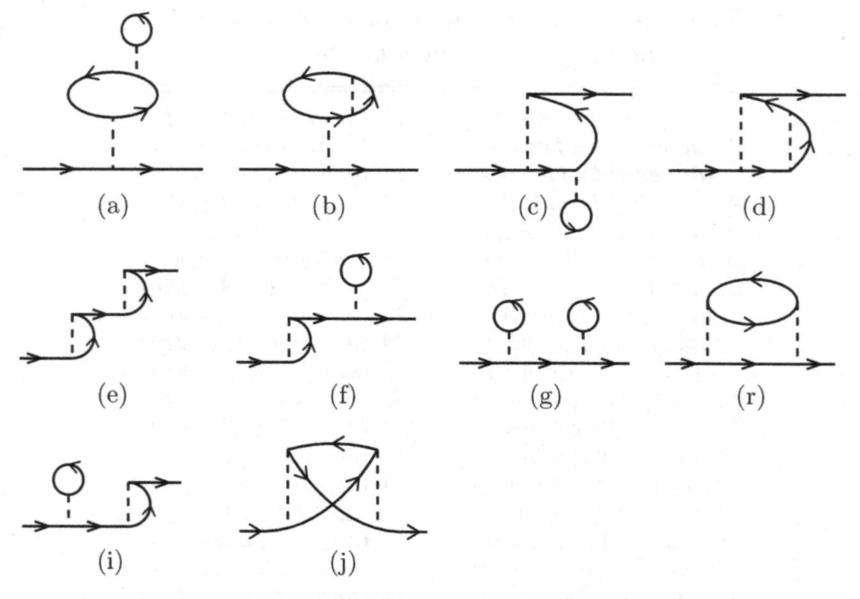

Figure 9.3 The ten connected, topologically distinct diagrams that arise in second order perturbation theory. Solid lines are fermion lines, while dashed lines are interaction lines.

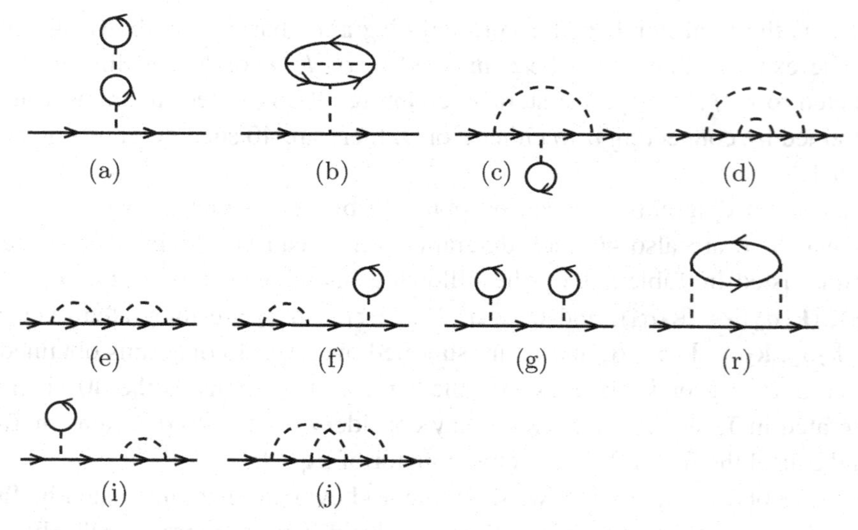

Figure 9.4 An alternative way of drawing the diagrams from Figure 9.3.

Our next observation is that topologically equivalent diagrams make the same contribution to $\delta g^{(2)}$, since they differ from each other only by virtue of a relabeling of their internal time, wave vector, and spin coordinates, which are integrated over. For example, diagram #20 in Table 9.1 makes the following

Table 9.2 *Topologically equivalent diagrams listed in Table 9.1.*

Diagrams in Table 9.1	Diagram in Fig. 9.3
1, 4, 13, 16	A
2, 3, 14, 15	B
5, 8, 9, 12	C
6, 7, 10, 11	D
17, 24, 33, 35	E
18, 23, 31, 36	F
19, 25, 29, 39	G
20, 26, 34, 40	R
21, 28, 30, 37	I
22, 27, 32, 38	J

contribution to $\delta g^{(2)}$:

$$\delta g^{(2)}_{\#20} = -\left(-\frac{1}{\hbar}\right)^2 \frac{1}{V^2} \int_0^{\beta\hbar} \mathrm{d}\tau_1 \int_0^{\beta\hbar} \mathrm{d}\tau_2 \sum_{\mathbf{q}\mathbf{q}'} \left(\frac{1}{2} v_{\mathbf{q}}\right) \left(\frac{1}{2} v_{\mathbf{q}'}\right) \sum_{\mathbf{k}_1\sigma_1} \sum_{\mathbf{k}_2\sigma_2} \sum_{\mathbf{k}_3\sigma_3} \sum_{\mathbf{k}_4\sigma_4}$$

$$\langle T\, c_{\mathbf{k}\sigma} c^\dagger_{\mathbf{k}\sigma} c^\dagger_{\mathbf{k}_1+\mathbf{q}\sigma_1} c^\dagger_{\mathbf{k}_2-\mathbf{q}\sigma_2} c_{\mathbf{k}_2\sigma_2} c_{\mathbf{k}_1\sigma_1} c^\dagger_{\mathbf{k}_3+\mathbf{q}'\sigma_3} c^\dagger_{\mathbf{k}_4-\mathbf{q}'\sigma_4} c_{\mathbf{k}_4\sigma_4} c_{\mathbf{k}_3\sigma_3} \rangle_0. \tag{9.4}$$

The time arguments of the operators are not shown explicitly: the first operator has argument τ, the second has 0, the next four have τ_1, and the last four have τ_2. We have ignored the factor $1/2!$, since, as noted earlier, this factor cancels out if we restrict ourselves to the 40 diagrams listed in Table 9.1. The ensemble-averaged term gives

$$\begin{aligned}\langle \cdots \rangle_0 = &-\langle T\, c_{\mathbf{k}\sigma}(\tau) c^\dagger_{\mathbf{k}_1+\mathbf{q}\sigma_1}(\tau_1)\rangle_0 \langle T\, c_{\mathbf{k}_3\sigma_3}(\tau_2) c^\dagger_{\mathbf{k}\sigma}(0)\rangle_0 \langle T\, c_{\mathbf{k}_4\sigma_4}(\tau_2) c^\dagger_{\mathbf{k}_2-\mathbf{q}\sigma_2}(\tau_1)\rangle_0 \\ &\langle T\, c_{\mathbf{k}_2\sigma_2}(\tau_1) c^\dagger_{\mathbf{k}_4-\mathbf{q}'\sigma_4}(\tau_2)\rangle_0 \langle T\, c_{\mathbf{k}_1\sigma_1}(\tau_1) c^\dagger_{\mathbf{k}_3+\mathbf{q}'\sigma_3}(\tau_2)\rangle_0.\end{aligned}$$

Using the relation

$$-\langle T\, c_{\mathbf{k}\sigma}(\tau) c^\dagger_{\mathbf{k}'\sigma'}(\tau')\rangle_0 = \delta_{\mathbf{k}\mathbf{k}'} \delta_{\sigma\sigma'}\, g^0(\mathbf{k}\sigma, \tau - \tau'),$$

we find

$$\begin{aligned}\delta g^{(2)}_{\#20} = &-\left(-\frac{1}{\hbar V}\right)^2 \int_0^{\beta\hbar} \mathrm{d}\tau_1 \int_0^{\beta\hbar} \mathrm{d}\tau_2 \sum_{\mathbf{q}} \frac{1}{4} v_{\mathbf{q}} v_{-\mathbf{q}} \sum_{\mathbf{k}'\sigma'} g^0(\mathbf{k}\sigma, \tau - \tau_1) g^0(\mathbf{k}\sigma, \tau_2) \\ &g^0(\mathbf{k}'\sigma', \tau_2 - \tau_1) g^0(\mathbf{k}' + \mathbf{q}\sigma', \tau_1 - \tau_2) g^0(\mathbf{k} - \mathbf{q}\sigma, \tau_1 - \tau_2).\end{aligned} \tag{9.5}$$

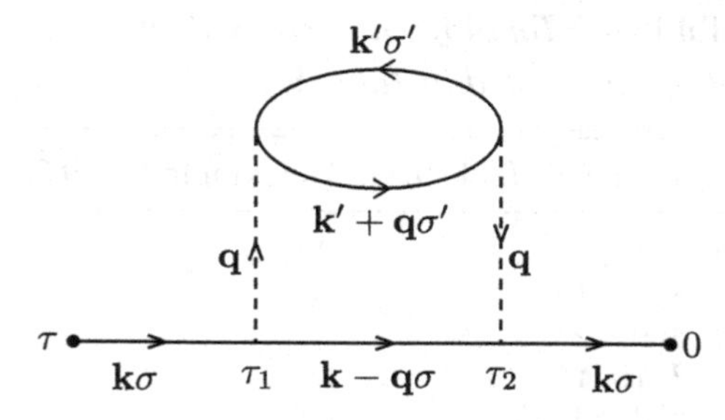

Figure 9.5 Diagram (R), a ring diagram.

Upon evaluating $\delta g^{(2)}_{\#34}$, as an illustrative example, we obtain the same expression as above, except that $v_{\mathbf{q}}^2$ replaces $v_{\mathbf{q}} v_{-\mathbf{q}}$. Since $v(\mathbf{r_1} - \mathbf{r_2}) = v(\mathbf{r_2} - \mathbf{r_1})$, it follows that $v_{\mathbf{q}} = v_{-\mathbf{q}}$; hence, diagram #34 makes exactly the same contribution to $\delta g^{(2)}$ as diagram #20. The fact that topologically equivalent diagrams yield the same algebraic expression means that we may sum only the topologically distinct diagrams and drop the factor $1/4$ in front of $v_{\mathbf{q}}^2$. In other words, we sum only connected, topologically distinct diagrams, but we replace a dashed line carrying a wave vector $\mathbf{q}$ with $(v_{\mathbf{q}}/V)$ instead of $(1/2)v_{\mathbf{q}}/V$. Thus, diagram (R), reproduced in Figure 9.5, is given by

$$(\mathrm{R}) = -\left(-\frac{1}{\hbar V}\right)^2 \int_0^{\beta\hbar} \mathrm{d}\tau_1 \int_0^{\beta\hbar} \mathrm{d}\tau_2 \sum_{\mathbf{q}} \sum_{\mathbf{k}'\sigma'} v_{\mathbf{q}}^2 g^0(\mathbf{k}\sigma, \tau - \tau_1) g^0(\mathbf{k} - \mathbf{q}\sigma, \tau_1 - \tau_2)$$
$$g^0(\mathbf{k}'\sigma', \tau_2 - \tau_1) g^0(\mathbf{k}' + \mathbf{q}\sigma', \tau_1 - \tau_2) g^0(\mathbf{k}\sigma, \tau_2). \tag{9.6}$$

This expression may be written from Figure 9.5 if we adopt the following rules:

(1) To each fermion line with coordinates $(\mathbf{k}\sigma)$, running from τ_i to τ_j, assign the noninteracting single-particle Green's function $g^0(\mathbf{k}\sigma, \tau_i - \tau_j)$.
(2) To each dashed line with wave vector $\mathbf{q}$, assign the factor $v_{\mathbf{q}}/V$.
(3) Conserve wave vector and spin at each vertex.
(4) Sum over all internal coordinates (in Figure 9.5, these are $\mathbf{k}'$, σ', and $\mathbf{q}$).
(5) Integrate over internal times from 0 to $\beta\hbar$ (τ and 0 are external times).
(6) Multiply by $(-1/\hbar)^n$, where n is the order of the interaction ($n = 2$ in Figure 9.5).
A factor of -1 is needed to reproduce Eq. (9.6). We note that diagram (R) has one closed fermion loop. The last rule is:
(7) Multiply by $(-1)^F$, where F is the number of closed fermion loops.

We are usually interested in calculating $g(\mathbf{k}\sigma, \omega_n)$. To find the contribution of diagram (R) in Figure 9.5 to $\delta g^{(2)}(\mathbf{k}\sigma, \omega_n)$, we write

$$\delta g_{\mathrm{R}}^{(2)}(\mathbf{k}\sigma, \tau) = \frac{1}{\beta\hbar} \sum_n \delta g_{\mathrm{R}}^{(2)}(\mathbf{k}\sigma, \omega_n) e^{-i\omega_n \tau}, \quad \omega_n = (2n+1)\pi/\beta\hbar.$$

On the RHS of Eq. (9.6), we also Fourier-expand the Green's functions,

$$\begin{aligned} \mathrm{RHS} = &- \left(-\frac{1}{\hbar}\right)^2 \left(\frac{1}{\beta\hbar}\right)^5 \sum_{\mathbf{q}\mathbf{k}'\sigma'} \left(\frac{v_{\mathbf{q}}}{V}\right)^2 \int_0^{\beta\hbar} \mathrm{d}\tau_1 \int_0^{\beta\hbar} \mathrm{d}\tau_2 \sum_n g^0(\mathbf{k}\sigma, \omega_n) e^{-i\omega_n(\tau-\tau_1)} \\ &\times \sum_{n_1} g^0(\mathbf{k}-\mathbf{q}\sigma, \omega_{n_1}) e^{-i\omega_{n_1}(\tau_1-\tau_2)} \sum_{n_2} g^0(\mathbf{k}'\sigma', \omega_{n_2}) e^{-i\omega_{n_2}(\tau_2-\tau_1)} \\ &\times \sum_{n_3} g^0(\mathbf{k}'+\mathbf{q}\sigma', \omega_{n_3}) e^{-i\omega_{n_3}(\tau_1-\tau_2)} \sum_{n_4} g^0(\mathbf{k}\sigma, \omega_{n_4}) e^{-i\omega_{n_4}\tau_2}. \end{aligned}$$

Collecting the exponentials, we find

$$I = e^{-i\omega_n\tau} \int_0^{\beta\hbar} \mathrm{d}\tau_1\, e^{i(\omega_n - \omega_{n_1} + \omega_{n_2} - \omega_{n_3})\tau_1} \int_0^{\beta\hbar} \mathrm{d}\tau_2\, e^{i(\omega_{n_1} - \omega_{n_2} + \omega_{n_3} - \omega_{n_4})\tau_2}.$$

Since the frequencies are odd, $\omega_n - \omega_{n_1} + \omega_{n_2} - \omega_{n_3}$ and $\omega_{n_1} - \omega_{n_2} + \omega_{n_3} - \omega_{n_4}$ are both even; hence

$$I = (\beta\hbar)^2 e^{-i\omega_n\tau} \delta_{\omega_n+\omega_{n_2},\, \omega_{n_1}+\omega_{n_3}} \delta_{\omega_{n_4}+\omega_{n_2},\, \omega_{n_1}+\omega_{n_3}}.$$

Thus, $I = 0$ unless $\omega_{n_4} = \omega_n$. Setting $\omega_n - \omega_{n_1} = \omega_m$ (ω_m is even), we find $\omega_{n_1} = \omega_n - \omega_m$ and $\omega_{n_3} = \omega_{n_2} + \omega_m$. Relabeling ω_{n_2} as $\omega_{n'}$, we obtain

$$\begin{aligned} \delta g_{\mathrm{R}}^{(2)}(\mathbf{k}\sigma, \omega_n) = &- \left(\frac{-1}{\beta\hbar^2}\right)^2 \sum_{\mathbf{q}\mathbf{k}'\sigma'} \left(\frac{v_{\mathbf{q}}}{V}\right)^2 \sum_{m,n'} g^0(\mathbf{k}\sigma, \omega_n) g^0(\mathbf{k}-\mathbf{q}\sigma, \omega_n - \omega_m) \\ &\times g^0(\mathbf{k}'\sigma', \omega_{n'}) g^0(\mathbf{k}'+\mathbf{q}\sigma', \omega_{n'} + \omega_m) g^0(\mathbf{k}\sigma, \omega_n). \qquad (9.7) \end{aligned}$$

In Figure 9.6, we redraw diagram (R), this time in momentum-frequency space. The above expression for $\delta g_{\mathrm{R}}^{(2)}(\mathbf{k}\sigma, \omega_n)$ can be read off Figure 9.6 if we adopt the following rules:

(1) Assign coordinates $(\mathbf{k}\sigma, \omega_n)$ to the two external fermion lines. To each interaction line, assign a wave vector and an even frequency.
(2) To each internal fermion line, assign wave vector, spin, and frequency coordinates. At each vertex, conserve wave vector, spin, and frequency.
(3) To each fermion line with coordinates $(\mathbf{k}\sigma, \omega_n)$, assign $g^0(\mathbf{k}\sigma, \omega_n)$. To each interaction (dashed) line with coordinates $(\mathbf{q}, \omega_m)$, assign $v_{\mathbf{q}}/V$. Form the product of all the g^0's and $(v_{\mathbf{q}}/V)$'s.

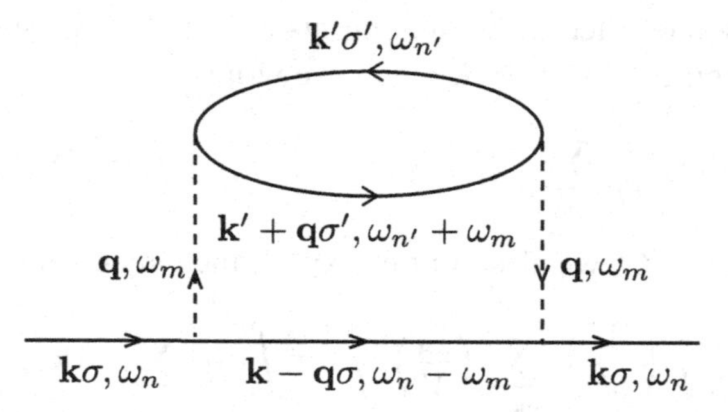

Figure 9.6 Diagram (R) in momentum-frequency space.

(4) Sum over all internal coordinates (wave vector, spin, and frequency).
(5) Multiply the resulting expression by $\left(-1/\beta\hbar^2\right)^n (-1)^F$, where n is the order of the interaction and F is the number of closed fermion loops.

9.2 Feynman rules in momentum-frequency space

For a system of fermions with spin-independent two-particle interaction, the order n correction to $g(\mathbf{k}\sigma, \tau)$ is

$$\delta g^{(n)}(\mathbf{k}\sigma, \tau) = -\frac{1}{n!}\left(-\frac{1}{\hbar}\right)^n \int_0^{\beta\hbar} \mathrm{d}\tau_1 \ldots \int_0^{\beta\hbar} \mathrm{d}\tau_n$$
$$\left\langle T\, c_{\mathbf{k}\sigma}(\tau)\, c^\dagger_{\mathbf{k}\sigma}(0)\, V(\tau_1)V(\tau_2)\ldots V(\tau_n)\right\rangle_{0,c} . \tag{9.8}$$

We make the following observations:

(a) Each connected diagram results from a particular set of contractions, e.g., $\overline{c(\tau)c^\dagger}(\tau_1)$, $\overline{c(\tau_2)c^\dagger}(0)$,..., $\overline{c(\tau_m)c^\dagger}(\tau_n)$. Let $i_1, i_2, \ldots, i_n$ be a permutation of $1, 2, \ldots, n$. The diagram which results from the set of contractions $\overline{c(\tau)c^\dagger}(\tau_{i_1})$, $\overline{c(\tau_{i_2})c^\dagger}(0)$, ..., $\overline{c(\tau_{i_m})c^\dagger}(\tau_{i_n})$ is topologically equivalent to the first diagram mentioned above, since the two diagrams differ only by a relabeling of their time indices. The two diagrams have the same numerical value since the contribution of the second diagram differs from that of the first only by a replacement of the product $V(\tau_1)V(\tau_2)\ldots V(\tau_n)$ by $V(\tau_{i_1})V(\tau_{i_2})\ldots V(\tau_{i_n})$ in the integral, and the dummy time variables are integrated over. Furthermore, no minus sign is incurred by this rearrangement since V contains an even number of fermion operators. Because there are $n!$ permutations of $1, 2, \ldots, n$, we conclude that there are $n!$ topologically equivalent diagrams which differ only by the permutation of their time indices, and that all of them have the same algebraic value.

Therefore, we consider only one diagram from this set and cancel the $n!$ factor in Eq. (9.8).

(b) Going back to the example used in Section 9.1, we note that the 40 diagrams listed in Table 9.2 are divided into ten groups, each of which contains four diagrams. The four diagrams in each group are topologically equivalent and have the same algebraic value. Consider, for example, these four diagrams: #20, #26, #34, and #40, which are all of type (R). Diagram #34 is obtained from diagram #20 by interchanging the two vertices of the interaction line at τ_1. Diagram #40, in turn, is obtained from diagram #20 by interchanging the two vertices of the interaction line at τ_2. Diagram #26 is obtained from #20 by interchanging the two vertices at τ_1 as well as the two vertices at τ_2. The four diagrams make the same contribution to $g(\mathbf{k}\sigma, \tau)$ because they differ only by a relabeling of their internal momentum and spin coordinates, which are summed over.

Similarly, at order n, any interchange of the two vertices of a given interaction (dashed) line yields a diagram that is topologically equivalent to the original diagram, with the same algebraic value. Since there are n interaction lines, each of which has two vertices, there are 2^n topologically equivalent diagrams with the same algebraic value that differ only by a relabeling of their internal momentum and spin indices. If we were to construct a table of these order n diagrams, similar to Table 9.2, we would find that each of the groups consists of 2^n diagrams (in Table 9.2, $n = 2$). Since a factor $(v_{\mathbf{q}}/2V)$ appears for each interaction line of wave vector $\mathbf{q}$, we may consider only one diagram from each group and assign $v_{\mathbf{q}}/V$ to each interaction line, rather than $(1/2)v_{\mathbf{q}}/V$.

(c) Let

$$P_g = -\langle T\, c_{\mathbf{k}\sigma}(\tau) c^\dagger_{\mathbf{k}\sigma}(0) c^\dagger_1 c^\dagger_{1'} c_{1'} c_1 \cdots c^\dagger_n c^\dagger_{n'} c_{n'} c_n \rangle_{0,c}.$$

Here, we have dropped the internal momentum, spin, and time coordinates, and the notation adopted is as follows. The interaction line at τ_i has two vertices, denoted by i and i'. The creation (annihilation) operator associated with vertex i is denoted by $c^\dagger_i$ (c_i). The operators can be rearranged such that

$$P_g = -\langle T\, c_{\mathbf{k}\sigma}(\tau) c^\dagger_1 c_1 c^\dagger_{1'} c_{1'} \cdots c^\dagger_n c_n c^\dagger_{n'} c_{n'} c^\dagger_{\mathbf{k}\sigma}(0) \rangle_{0,c}.$$

In applying Wick's theorem to evaluate P_g, we sum over all possible ways of contracting pairs of operators. Let us consider the following particular way: every annihilation operator is contracted with the creation operator immediately on its right side ($c_{\mathbf{k}\sigma}$ is contracted with $c^\dagger_1$, c_1 is contracted with $c^\dagger_{1'}$, and so on). This way of contracting operators produces a term in P_g which is a product of $2n + 1$ g^0's with an overall positive sign; the corresponding Feynman diagram has no loops. Other diagrams without any closed loops can be obtained by interchanging two internal vertices and following the same contraction procedure. For example, if we interchange $c_{1'}^{\dagger}c_{1'}$ and $c_2^{\dagger}c_2$ and contract every annihilation operator with the creation operator immediately to its right, we obtain another diagram without any loops. Since no minus sign is introduced by interchanging of one pair of operators with another, every diagram without loops that is generated by applying Wick's theorem to P_g is a product of $2n + 1$ single-particle Green's functions with an overall positive sign.

(d) A closed fermion loop is formed if a fermion line leaves a vertex and then reenters the same vertex, possibly after entering and leaving a number of other vertices. For example, let us consider again the time-ordered product P_g, and let us interchange c_1 and $c_{1'}$. This interchange introduces a minus sign; hence,

$$P_g = \langle T c_{\mathbf{k}\sigma}(\tau) c_1^{\dagger} c_{1'} c_{1'}^{\dagger} c_1 \cdots c_n^{\dagger} c_n c_{n'}^{\dagger} c_{n'} c_{\mathbf{k}\sigma}^{\dagger}(0)\rangle_{0,c}.$$

We now follow the same contraction procedure outlined earlier: every annihilation operator is contracted with the creation operator immediately on its right side. This way of contracting pairs of operators produces a term in P_g which is a product of $2n + 1$ g^0's with an overall negative sign. The corresponding Feynman diagram contains one fermion loop which results from contracting $c_{1'}$ with $c_{1'}^{\dagger}$. As another example, if we interchange c_1 and c_2 and follow the same contraction procedure as above, we end up with a Feynman diagram with one loop, where a line starts at vertex $1'$, runs to vertex 2, and back to $1'$. Once again, the corresponding term in P_g contains a product of $2n + 1$ g^0's with an overall negative sign. Thus, a factor of -1 is assigned to each closed fermion loop.

If a fermion line closes on itself or is joined by the same interaction line, it corresponds to the contraction $\overbrace{c_{\mathbf{k}'\sigma'}^{\dagger}(\tau')c_{\mathbf{k}''\sigma''}}(\tau')$. Since a contraction at equal times is ill-defined, and since in $V(\tau')$ the creation operators always occur to the left of the annihilation operators, the contraction at equal times is interpreted as

$$\begin{aligned}\overbrace{c_{\mathbf{k}'\sigma'}^{\dagger}(\tau')c_{\mathbf{k}''\sigma''}}(\tau') &= \overbrace{c_{\mathbf{k}'\sigma'}^{\dagger}(\tau'+0^+)c_{\mathbf{k}''\sigma''}}(\tau') = \langle T c_{\mathbf{k}'\sigma'}^{\dagger}(\tau'+0^+)c_{\mathbf{k}''\sigma''}(\tau')\rangle_0 \\ &= -\langle T c_{\mathbf{k}''\sigma''}(\tau')c_{\mathbf{k}'\sigma'}^{\dagger}(\tau'+0^+)\rangle_0 = \delta_{\sigma'\sigma''}\delta_{\mathbf{k}'\mathbf{k}''}g^0(\mathbf{k}'\sigma', 0^-).\end{aligned}$$

When Fourier-transformed, it yields

$$(\beta\hbar)^{-1}\sum_{n'} g^0(\mathbf{k}'\sigma', \omega_{n'})e^{-i\omega_{n'}0^-} = (\beta\hbar)^{-1}\sum_{n'} g^0(\mathbf{k}'\sigma', \omega_{n'})e^{i\omega_{n'}0^+}.$$

(e) Each interaction (dashed) line of wave vector $\mathbf{q}$, occurring at time τ', has two vertices. At each vertex, one fermion line enters and another one leaves, such

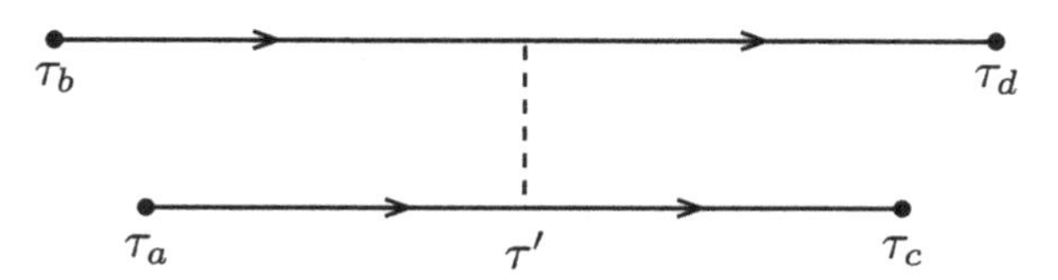

Figure 9.7 Two fermion lines, coming from vertices at τ_a and τ_b, enter the interaction line at τ' and then leave to vertices at τ_c and τ_d.

that momentum and spin are conserved. Let us represent the two fermion lines entering vertices 1 and 2 of the interaction by $g^0(\tau_a - \tau')$ and $g^0(\tau_b - \tau')$, respectively, for some τ_a and τ_b. The two fermion lines leaving vertices 1 and 2 of the interaction are represented by $g^0(\tau' - \tau_c)$ and $g^0(\tau' - \tau_d)$, respectively, for some τ_c and τ_d (see Figure 9.7). In terms of frequency, we write

$$g^0(\tau_a - \tau') = (\beta\hbar)^{-1} \sum_{n_1} g^0(\omega_{n_1}) e^{-i\omega_{n_1}(\tau_a - \tau')}$$

$$g^0(\tau_b - \tau') = (\beta\hbar)^{-1} \sum_{n_2} g^0(\omega_{n_2}) e^{-i\omega_{n_2}(\tau_b - \tau')}$$

$$g^0(\tau' - \tau_c) = (\beta\hbar)^{-1} \sum_{n_3} g^0(\omega_{n_3}) e^{-i\omega_{n_3}(\tau' - \tau_c)}$$

$$g^0(\tau' - \tau_d) = (\beta\hbar)^{-1} \sum_{n_4} g^0(\omega_{n_4}) e^{-i\omega_{n_4}(\tau' - \tau_d)}.$$

Since τ' is integrated over, we obtain a factor of

$$\int_0^{\beta\hbar} d\tau' e^{i(\omega_{n_1}+\omega_{n_2}-\omega_{n_3}-\omega_{n_4})\tau'} = \beta\hbar\delta_{\omega_{n_1}+\omega_{n_2},\ \omega_{n_3}+\omega_{n_4}}.$$

Setting $\omega_{n_1} - \omega_{n_3} = \omega_m$ (even), we find that $\omega_{n_3} = \omega_{n_1} - \omega_m$ and $\omega_{n_4} = \omega_{n_2} + \omega_m$. Thus, we can associate a wave vector $\mathbf{q}$ and an even frequency ω_m with an interaction line, and demand momentum, spin, and frequency conservation at each vertex, as shown in Figure 9.8.

(f) Each diagram occurring in the expansion of $\delta g^{(n)}(\mathbf{k}\sigma, \tau)$ has $2n + 1$ fermion lines and n interaction lines at $\tau_1, \tau_2, \ldots, \tau_n$. Each fermion line, when Fourier-transformed, produces a factor of $(\beta\hbar)^{-1}$, while the integration over $\tau_1, \tau_2, \ldots, \tau_n$ produces a factor of $(\beta\hbar)^n$. Hence on the RHS of the expression for $\delta g^{(n)}(\mathbf{k}\sigma, \tau)$, there is a factor of $(\beta\hbar)^{-n-1}$. Since $\delta g^{(n)}(\mathbf{k}\sigma, \tau) = (\beta\hbar)^{-1} \sum_n \delta g^{(n)}(\mathbf{k}\sigma, \omega_n) e^{-i\omega_n\tau}$, the expression for $\delta g^{(n)}(\mathbf{k}\sigma, \omega_n)$ contains the factor $(\beta\hbar)^{-n}$. Combining this factor with the prefactor $(-1/\hbar)^n$ which occurs in the expansion of $\delta g^{(n)}(\mathbf{k}\sigma, \tau)$, we conclude that the expression for $\delta g^{(n)}(\mathbf{k}\sigma, \omega_n)$ contains the factor $(-1/\beta\hbar^2)^n$.

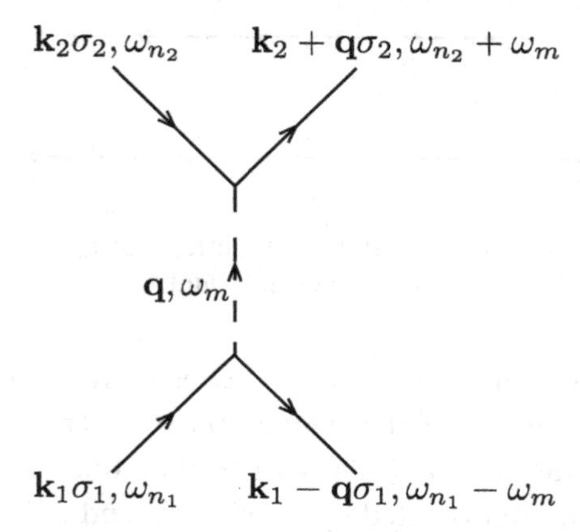

Figure 9.8 Momentum, spin, and frequency conservation at each vertex of an interaction (dashed) line.

We are now in a position to state Feynman rules, in momentum-frequency space, for the construction and evaluation of diagrams that contribute to the correction $\delta g^{(n)}(\mathbf{k}\sigma, \omega_n)$, in order n, to Green's function $g(\mathbf{k}\sigma, \omega_n)$. The rules for two-particle interaction are as follows:

1. Draw all connected, topologically distinct diagrams with n interaction lines and $2n + 1$ directed fermion lines. Two of the fermion lines are external lines, and the rest $(2n - 1)$ are internal lines.
2. The two external fermion lines have coordinates $(\mathbf{k}\sigma, \omega_n)$. With each internal fermion line, associate momentum, spin, and frequency coordinates. To each interaction line assign a direction, a wave vector, and an even frequency. Conserve wave vector, spin, and frequency at each vertex.
3. Assign $g^0(\mathbf{k}\sigma, \omega_n)$ to each of the two external fermion lines. To each internal fermion line of coordinates $(\mathbf{k}'\sigma', \omega_{n'})$, assign $g^0(\mathbf{k}'\sigma', \omega_{n'})$. For each fermion line with frequency $\omega_{n'}$ that closes on itself or runs from one vertex of an interaction line to the other vertex, insert the factor $e^{i\omega_{n'}0^+}$.
4. Assign to each interaction line with wave vector $\mathbf{q}$ and frequency ω_m the factor $v_\mathbf{q}/V$, where V is the system's volume.
5. Form the product of all the g^0's and all the $(v_\mathbf{q}/V)$'s, and then sum over all internal wave vectors, spins, and frequencies.
6. Multiply by the factor $(-1/\beta\hbar^2)^n(-1)^F$, where F is the number of closed fermion loops.

Finally, we consider the following question: when we draw the connected, topologically distinct diagrams, how do we know whether we have exhausted them all? To restate the question, what is the number $C_{TD}(n)$ of connected, topologically distinct diagrams that contribute to $\delta g^{(n)}(\mathbf{k}\sigma, \tau)$? According to the discussion above, two diagrams are topologically equivalent if one is obtained from the other either by a relabeling of internal time indices, or by interchanging the two vertices of one or more interaction lines; therefore,

$$C_{TD}(n) = \frac{C(n)}{n!2^n} \tag{9.9}$$

where $C(n)$ is the number of connected diagrams of order n. In calculating $\delta g^{(n)}(\mathbf{k}\sigma, \tau)$, the ensemble average is taken over a product involving $2n + 1$ creation operators and $2n + 1$ annihilation operators. In forming contractions, an annihilation operator is contracted with a creation operator. Each possible way of contracting the $2n + 1$ annihilation operators with the $2n + 1$ creation operators produces one diagram. The first annihilation operator can be contracted with any of the $2n + 1$ creation operators; the second annihilation operator can be contracted with any of the remaining $2n$ creation operators, and so on. The total number of diagrams, connected or not, is therefore equal to $(2n + 1)!$. A disconnected diagram is obtained if the operators in one or more V-operators are contracted among themselves.

Consider the case where the creation and annihilation operators in m V-operators are contracted among themselves. We note the following:

(a) In these m V-operators, there are $2m$ annihilation operators and $2m$ creation operators. The number of diagrams formed by contractions of these operators is $(2m)!$.
(b) The number of connected diagrams formed by the remaining $2(n - m) + 1$ annihilation operators and $2(n - m) + 1$ creation operators is $C(n - m)$; it is simply the number of connected diagrams of order $n - m$.
(c) There are $n!/m!(n - m)!$ ways of choosing m V's from among n V's.

The number of connected diagrams is thus given by the recursion formula

$$C(n) = (2n + 1)! - \sum_{m=1}^{n} \frac{n!}{m!(n - m)!}(2m)!C(n - m) \tag{9.10}$$

along with

$$C(0) = 1. \tag{9.11}$$

Equations (9.9–9.11) determine the number of connected, topologically distinct diagrams. For $n = 1$,

$$C(1) = 3! - 2!C(0) = 4, \quad C_{TD}(1) = \frac{C(1)}{2} = 2.$$

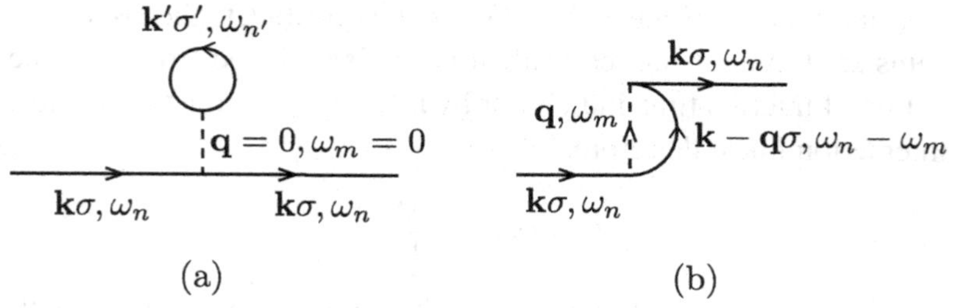

Figure 9.9 The two first-order connected, topologically distinct diagrams arising from the correction to $g(\mathbf{k}\sigma, \omega_n)$ in first order of the interaction ($n = 1$). (a) is a direct interaction diagram and (b) is an exchange interaction diagram.

For $n = 2$,

$$C(2) = 5! - \frac{2!}{1!1!}2!C(1) - \frac{2!}{2!0!}4!C(0) = 80, \quad C_{TD}(2) = \frac{80}{2^2 2!} = 10.$$

For $n = 3$, we find $C_{TD}(3) = 74$. Clearly, the number of connected, topologically distinct diagrams grows rapidly with increasing perturbation order.

9.3 An example of how to apply Feynman rules

Consider the correction to $g(\mathbf{k}\sigma, \omega_n)$ in first order of the interaction ($n = 1$). There is one interaction line and three fermion lines, two of which are external lines with coordinates $(\mathbf{k}\sigma, \omega_n)$. There are two connected, topologically distinct diagrams, shown in Figure 9.9. Using the Feynman rules, we can readily write the contributions of these two diagrams:

$$\delta g_a^{(1)}(\mathbf{k}\sigma, \omega_n) = -\left(\frac{-1}{\beta\hbar^2}\right)\frac{v_0}{V}[g^0(\mathbf{k}\sigma, \omega_n)]^2 \sum_{\mathbf{k}'\sigma'}\sum_{n'} g^0(\mathbf{k}'\sigma', \omega_{n'})e^{i\omega_{n'}0^+} \quad (9.12)$$

$$\delta g_b^{(1)}(\mathbf{k}\sigma, \omega_n) = \left(\frac{-1}{\beta\hbar^2}\right)[g^0(\mathbf{k}\sigma, \omega_n)]^2 \sum_{\mathbf{q}m}\frac{v_{\mathbf{q}}}{V} g^0(\mathbf{k}-\mathbf{q}\sigma, \omega_n - \omega_m)e^{i(\omega_n-\omega_m)0^+}$$

$$\delta g^{(1)}(\mathbf{k}\sigma, \omega_n) = \delta g_a^{(1)}(\mathbf{k}\sigma, \omega_n) + \delta g_b^{(1)}(\mathbf{k}\sigma, \omega_n). \quad (9.13)$$

The first minus sign in $\delta g_a^{(1)}$ results from the existence of a closed fermion loop in diagram (a). The convergence factor $e^{i\omega_{n'}0^+}$ is inserted because the fermion line with coordinates $(\mathbf{k}'\sigma', \omega_{n'})$ closes in on itself. On the other hand, the convergence factor $e^{i(\omega_n-\omega_m)0^+}$ in $\delta g_b^{(1)}$ arises because the fermion line with coordinates $(\mathbf{k}-\mathbf{q}\sigma, \omega_n - \omega_m)$ is joined by an interaction line. These convergence factors are important; without them, the summation over frequencies would diverge (see Problem 9.3).

Frequency sums, as in the above expression for $\delta g_a^{(1)}(\mathbf{k}\sigma, \omega_n)$ and $\delta g_b^{(1)}(\mathbf{k}\sigma, \omega_n)$, often arise in applications of the finite temperature Green's function. Here we record the following formula:

$$\sum_{n=-\infty}^{\infty} \frac{e^{i\omega_n 0^+}}{i\omega_n - \bar{\epsilon}/\hbar} = \begin{cases} -\beta\hbar n_{\bar{\epsilon}} & \text{bosons} \\ \beta\hbar f_{\bar{\epsilon}} & \text{fermions} \end{cases} \tag{9.14}$$

(see Problem 9.3). In the above equation, $n_{\bar{\epsilon}}$ and $f_{\bar{\epsilon}}$ are the Bose–Einstein and Fermi–Dirac distribution functions, respectively.

9.4 Feynman rules in coordinate space

The perturbation expansion for Green's function, derived in the previous chapter, applies to both momentum and coordinate space; hence

$$g(\mathbf{r}\sigma\tau, \mathbf{r}'\sigma'\tau') = -\sum_{n=0}^{\infty} \frac{1}{n!}(-1/\hbar)^n \int_0^{\beta\hbar} d\tau_1 \ldots \int_0^{\beta\hbar} d\tau_n$$

$$\langle T\Psi_\sigma(\mathbf{r}\tau)\Psi_{\sigma'}^\dagger(\mathbf{r}'\tau')V(\tau_1)\ldots V(\tau_n)\rangle_{0,c}. \tag{9.15}$$

All operators are interaction picture operators. For two-particle interactions (see Eq. [3.53]),

$$V(\tau) = \frac{1}{2}\sum_{\lambda\lambda'}\sum_{\mu\mu'} \int d^3r \int d^3r' \Psi_\lambda^\dagger(\mathbf{r}\tau)\Psi_\mu^\dagger(\mathbf{r}'\tau)v_{\lambda\mu,\lambda'\mu'}(\mathbf{r}, \mathbf{r}')\Psi_{\mu'}(\mathbf{r}'\tau)\Psi_{\lambda'}(\mathbf{r}\tau)$$

where λ, λ', μ, and μ' are spin projection indices and $v_{\lambda\mu,\lambda'\mu'}(\mathbf{r}, \mathbf{r}') = \langle\lambda\mu|v(\mathbf{r}\sigma, \mathbf{r}'\sigma')|\lambda'\mu'\rangle$. If v is spin-independent, $v_{\lambda\mu,\lambda'\mu'}(\mathbf{r}, \mathbf{r}') = v(\mathbf{r}, \mathbf{r}')\delta_{\lambda\lambda'}\delta_{\mu\mu'}$. Let $U(\mathbf{r}\sigma\tau, \mathbf{r}'\sigma'\tau') = v(\mathbf{r}\sigma, \mathbf{r}'\sigma')\delta(\tau - \tau')$, where $0 < \tau, \tau' < \beta\hbar$. $V(\tau)$ may be written as

$$V(\tau) = \frac{1}{2}\sum_{\lambda\lambda'\mu\mu'} \int d^3r \int d^3r' \int_0^{\beta\hbar} d\tau' \Psi_\lambda^\dagger(\mathbf{r}\tau)\Psi_\mu^\dagger(\mathbf{r}'\tau')U_{\lambda\mu,\lambda'\mu'}(\mathbf{r}\tau, \mathbf{r}'\tau')$$

$$\times \Psi_{\mu'}(\mathbf{r}'\tau')\Psi_{\lambda'}(\mathbf{r}\tau). \tag{9.16}$$

The interaction is depicted in Figure 9.10. The two vertices of the interaction line are assigned coordinates $(\mathbf{r}\tau)$ and $(\mathbf{r}'\tau')$. The first-order correction is given by

$$\delta g^{(1)}(\mathbf{r}\sigma\tau, \mathbf{r}'\sigma'\tau') = \frac{1}{2\hbar}\int d^3r_1 \int_0^{\beta\hbar} d\tau_1 \int d^3r_1' \int_0^{\beta\hbar} d\tau_1' \sum_{\lambda\lambda'}\sum_{\mu\mu'} \langle T\Psi_\sigma(\mathbf{r}\tau)\Psi_{\sigma'}^\dagger(\mathbf{r}'\tau')$$

$$\psi_\lambda^\dagger(\mathbf{r}_1\tau_1)\psi_\mu^\dagger(\mathbf{r}_1'\tau_1')U_{\lambda\mu,\lambda'\mu'}(\mathbf{r}_1\tau_1, \mathbf{r}_1'\tau_1')\psi_{\mu'}(\mathbf{r}_1'\tau_1')\psi_{\lambda'}(\mathbf{r}_1\tau_1)\rangle_{0,c}. \tag{9.17}$$

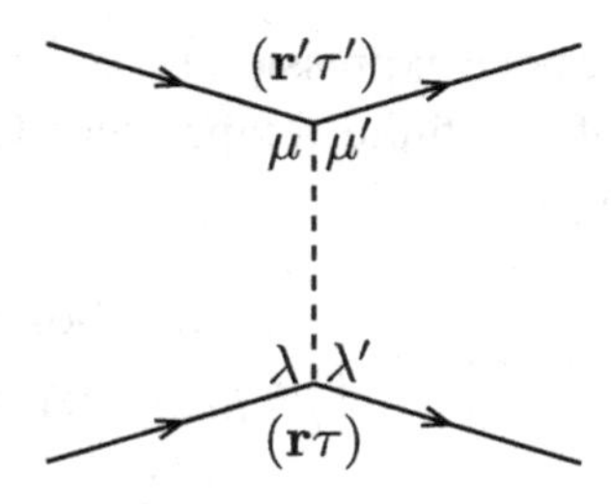

Figure 9.10 Graphical representation of the two-particle interaction in coordinate space.

The notation can be simplified by introducing the four-dimensional coordinate $x = (\mathbf{r}, \tau)$. Setting $\int d^3r \int_0^{\beta\hbar} d\tau = \int dx$, Eq. (9.17) reduces to

$$\delta g^{(1)}_{\sigma\sigma'}(x, x') = (1/2\hbar) \int dx_1 \int dx'_1 \sum_{\lambda\lambda'} \sum_{\mu\mu'} \langle T \psi_\sigma(x) \Psi^\dagger_{\sigma'}(x') \Psi^\dagger_\lambda(x_1) \Psi^\dagger_\mu(x'_1)$$

$$U_{\lambda\mu,\lambda'\mu'}(x_1, x'_1) \Psi_{\mu'}(x'_1) \Psi_{\lambda'}(x_1) \rangle_{0,c}.$$

Similar to the momentum space, the following observations apply:

(a) Diagrams that differ from each other by a permutation of $\tau_1, \tau_2, \ldots, \tau_n$ are topologically equivalent; they make identical contributions to $\delta g^{(n)}_{\sigma\sigma'}(x, x')$. There are $n!$ such diagrams. We select one diagram from this set and cancel the factor $1/n!$ in Eq. (9.15).

(b) In order n, there are 2^n topologically equivalent diagrams that differ only by some interchange of the interaction vertices $(\mathbf{r}_i \tau_i) \leftrightarrow (\mathbf{r}'_i \tau'_i)$, along with the corresponding spin indices. These diagrams make identical contributions to $\delta g^{(n)}$, since the interaction is symmetric under such an interchange:

$$U_{\lambda\mu,\lambda'\mu'}(x_i, x'_i) = U_{\lambda'\mu',\lambda\mu}(x'_i, x_i).$$

The symmetry occurs because the particles comprising the system are indistinguishable: since the particles are identical, interchanging the position and spin coordinates of any two particles does not change the interaction between them. Hence, it is sufficient to count each topologically distinct diagram only once, assigning to each interaction line with vertices x and x' the factor $U_{\lambda\mu,\lambda'\mu'}(x, x')$ rather than $(1/2)U_{\lambda\mu,\lambda'\mu'}(x, x')$.

(c) For each fermion loop, a factor of -1 is assigned.

Following these observations, we write below the Feynman rules for calculating the correction, of order n, to the Green's function $g(\mathbf{r}\sigma\tau, \mathbf{r}'\sigma'\tau')$:

(1) Draw all connected, topologically distinct diagrams with $2n$ vertices (i.e., n interaction, or dashed lines) and two external fermion lines. At each vertex,

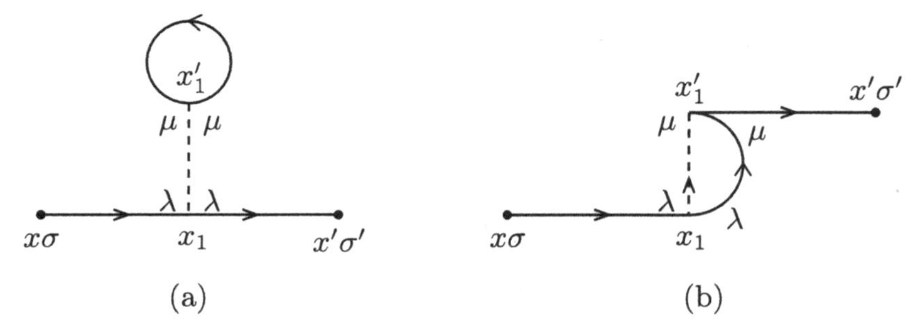

Figure 9.11 Connected, topologically distinct diagrams that contribute to the first-order correction to Green's function.

one fermion line enters and another one leaves. If the interaction is spin-independent, conserve spin at each vertex.

(2) To each fermion line directed from $(\mathbf{r}\sigma\tau)$ to $(\mathbf{r}'\sigma'\tau')$, assign $g^0(\mathbf{r}\sigma\tau, \mathbf{r}'\sigma'\tau')$.
(3) To each interaction line with vertices $(\mathbf{r}_i\,\tau_i)$ and $(\mathbf{r}_i'\,\tau_j')$, assign the matrix element $U_{\lambda\mu,\lambda'\mu'}(\mathbf{r}_i\,\tau_i, \mathbf{r}_i'\,\tau_i')$.
(4) Integrate over all vertex coordinates ($2n$ space and $2n$ time integrations).
(5) Sum over all internal spin indices.
(6) Multiply the resulting factor by $(-1/\hbar)^n(-1)^F$, where F is the number of fermion loops.
(7) Any Green's function at equal times is interpreted as $g^0(\mathbf{r}\sigma\tau, \mathbf{r}'\sigma'\tau^+)$.

As an example, consider a system of fermions with spin-independent two-particle interaction: $U_{\lambda\mu,\lambda'\mu'}(x, x') = U_{\lambda\mu,\lambda\mu}(x, x')\delta_{\lambda\lambda'}\delta_{\mu\mu'}$. The first order correction $\delta g^{(1)}(\mathbf{r}\sigma\tau, \mathbf{r}'\sigma'\tau')$ is obtained below. There are two connected, topologically distinct, diagrams (see Figure 9.11). Using the diagram rules,

$$\delta g^{(1)}(\mathbf{r}\sigma\tau, \mathbf{r}'\sigma'\tau') \equiv \delta g^{(1)}_{\sigma\sigma'}(x, x') = \delta g^{(1)}_{\sigma\sigma',a}(x, x') + \delta g^{(1)}_{\sigma\sigma',b}(x, x') \equiv A + B$$

$$\begin{aligned} A &= \frac{1}{\hbar}\int dx_1 \int dx_1' \sum_{\lambda\mu} g^0_{\sigma\lambda}(x, x_1)U_{\lambda\mu,\lambda\mu}(x_1, x_1')g^0_{\mu\mu}(x_1', x_1')g^0_{\lambda\sigma'}(x_1, x') \\ &= \delta_{\sigma\sigma'}\frac{1}{\hbar}\int dx_1 \int dx_1' \sum_{\mu} g^0_{\sigma\sigma}(x, x_1)U_{\sigma\mu,\sigma\mu}(x_1, x_1')g^0_{\mu\mu}(x_1', x_1')g^0_{\sigma\sigma}(x_1, x') \end{aligned} \tag{9.18}$$

where $U_{\sigma\mu,\sigma\mu}(x_1, x_1') = v(\mathbf{r}_1, \mathbf{r}_1')\delta(\tau_1 - \tau_1')$ and $g^0_{\mu\mu}(x_1', x_1') = g^0_{\mu\mu}(\mathbf{r}_1'\tau_1', \mathbf{r}_1'\tau_1'^+)$.

$$\begin{aligned} B &= \frac{-1}{\hbar}\int dx_1 \int dx_1' \sum_{\lambda\mu} g^0_{\sigma\lambda}(x, x_1)U_{\lambda\mu,\lambda\mu}(x_1, x_1')g^0_{\lambda\mu}(x_1, x_1')g^0_{\mu\sigma'}(x_1', x') \\ &= \delta_{\sigma\sigma'}\frac{-1}{\hbar}\int dx_1 \int dx_1' g^0_{\sigma\sigma}(x, x_1)U_{\sigma\sigma,\sigma\sigma}(x_1, x_1')g^0_{\sigma\sigma}(x_1, x_1')g^0_{\sigma\sigma}(x_1', x'). \end{aligned} \tag{9.19}$$

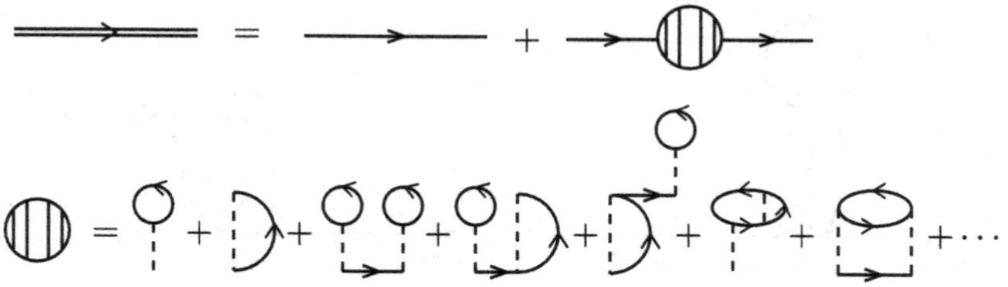

Figure 9.12 Graphical representation of Green's function in terms of the self energy. The directed solid double-line represents $g(\mathbf{k}\sigma, \omega_n)$, the directed solid single line represents $g^0(\mathbf{k}\sigma, \omega_n)$, and the dashed line is the two-particle interaction. The hatched circle represents the self energy $\Sigma(\mathbf{k}\sigma, \omega_n)$.

9.5 Self energy and Dyson's equation

It is clear from our previous analysis that every connected diagram in the perturbation expansion for Green's function contains two external particle lines at its ends. Thus, every connected diagram has an algebraic value given by $g^0(\mathbf{k}\sigma, \omega_n)B(\mathbf{k}\sigma, \omega_n)g^0(\mathbf{k}\sigma, \omega_n)$, for some $B(\mathbf{k}\sigma, \omega_n)$ that is determined by the structure of the diagram. Hence, we may write

$$g(\mathbf{k}\sigma, \omega_n) = g^0(\mathbf{k}\sigma, \omega_n) + g^0(\mathbf{k}\sigma, \omega_n)\Sigma(\mathbf{k}\sigma, \omega_n)g^0(\mathbf{k}\sigma, \omega_n)$$

where $\Sigma(\mathbf{k}\sigma, \omega_n)$, known as the particle's self energy, is obtained by summing $B(\mathbf{k}\sigma, \omega_n)$ over all connected, topologically distinct diagrams of all orders in the perturbation. Graphically, the equation for $g(\mathbf{k}\sigma, \omega_n)$ is represented in Figure 9.12, where the self energy $\Sigma(\mathbf{k}\sigma, \omega_n)$ is written as an infinite sum of terms, called self-energy terms.

An examination of the diagrams appearing in the expansion of $\Sigma(\mathbf{k}\sigma, \omega_n)$ reveals that there are two types of diagrams: those that can be separated into two pieces by cutting a single-particle line (third, fourth, and fifth diagrams), and those that cannot. If we add up all the self energy diagrams that cannot be separated into two pieces by cutting a single-particle line, we obtain what is known as the proper, or irreducible, self energy $\Sigma^*(\mathbf{k}\sigma, \omega_n)$. It is clear that

$$\Sigma(\mathbf{k}\sigma, \omega_n) = \Sigma^*(\mathbf{k}\sigma, \omega_n) + \Sigma^*(\mathbf{k}\sigma, \omega_n)g^0(\mathbf{k}\sigma, \omega_n)\Sigma^*(\mathbf{k}\sigma, \omega_n) + \cdots .$$

This is depicted graphically in Figure 9.13. It follows that Green's function may be written as

$$\begin{aligned} g(\mathbf{k}\sigma, \omega_n) &= g^0(\mathbf{k}\sigma, \omega_n) + g^0(\mathbf{k}\sigma, \omega_n)\Sigma^*(\mathbf{k}\sigma, \omega_n)g^0(\mathbf{k}\sigma, \omega_n) \\ &\quad + g^0(\mathbf{k}\sigma, \omega_n)\Sigma^*(\mathbf{k}\sigma, \omega_n)g^0(\mathbf{k}\sigma, \omega_n)\Sigma^*(\mathbf{k}\sigma, \omega_n)g^0(\mathbf{k}\sigma, \omega_n) + \cdots \\ \Longrightarrow g(\mathbf{k}\sigma, \omega_n) &= g^0(\mathbf{k}\sigma, \omega_n) + g^0(\mathbf{k}\sigma, \omega_n)\Sigma^*(\mathbf{k}\sigma, \omega_n)g(\mathbf{k}\sigma, \omega_n). \end{aligned} \tag{9.20}$$

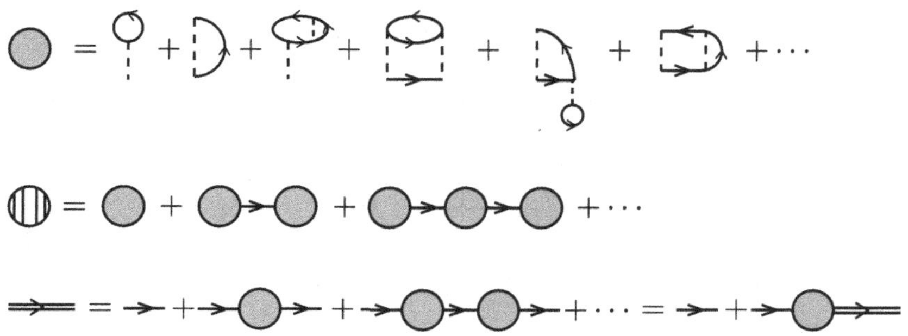

Figure 9.13 The shaded circle is the proper self energy $\Sigma^*(\mathbf{k}\sigma, \omega_n)$, while the hatched circle is the self energy $\Sigma(\mathbf{k}\sigma, \omega_n)$.

This is Dyson's equation (Dyson, 1949a, 1949b). An exact expression for Green's function, in terms of the proper self energy, follows:

$$g(\mathbf{k}\sigma, \omega_n) = \frac{g^0(\mathbf{k}\sigma, \omega_n)}{1 - g^0(\mathbf{k}\sigma, \omega_n)\Sigma^*(\mathbf{k}\sigma, \omega_n)} = \frac{1}{g^{0^{-1}}(\mathbf{k}\sigma, \omega_n) - \Sigma^*(\mathbf{k}\sigma, \omega_n)}$$
$$= \frac{1}{i\omega_n - \bar{\epsilon}_{\mathbf{k}\sigma}/\hbar - \Sigma^*(\mathbf{k}\sigma, \omega_n)}. \tag{9.21}$$

where $\bar{\epsilon}_{\mathbf{k}\sigma} = \epsilon_{\mathbf{k}\sigma} - \mu$. We should note that even though the above expression is exact, the calculation of $\Sigma^*(\mathbf{k}\sigma, \omega_n)$ is generally a formidable task; in practice, $\Sigma^*(\mathbf{k}\sigma, \omega_n)$ is approximated by a few diagrams.

9.6 Energy shift and the lifetime of excitations

We recall, from Section 8.4, that the retarded Green's function is obtained from its imaginary-time counterpart by using the replacement $i\omega_n \to \omega + i0^+$. Therefore,

$$G^R(\mathbf{k}\sigma, \omega) = \frac{1}{\omega - \bar{\epsilon}_{\mathbf{k}\sigma}/\hbar - Re\Sigma^{*R}(\mathbf{k}\sigma, \omega) - i\,Im\Sigma^{*R}(\mathbf{k}\sigma, \omega)}$$

where $\Sigma^{*R}(\mathbf{k}\sigma, \omega) = \Sigma^*(\mathbf{k}\sigma, i\omega_n \to \omega + i0^+)$ is the retarded proper self energy. We assume that $Im\Sigma^{*R}(\mathbf{k}\sigma, \omega)$ is nonzero so that the additional $i0^+$ in the denominator can be neglected. The spectral density function, equal to $-2\,Im\,G^R$ (see Eq. [6.42]), is given by

$$A(\mathbf{k}\sigma, \omega) = \frac{-2\,Im\Sigma^{*R}(\mathbf{k}\sigma, \omega)}{[\omega - \bar{\epsilon}_{\mathbf{k}\sigma}/\hbar - Re\Sigma^{*R}(\mathbf{k}\sigma, \omega)]^2 + [Im\Sigma^{*R}(\mathbf{k}\sigma, \omega)]^2}.$$

For fermions, $A(\mathbf{k}\sigma, \omega) \geq 0$ (see Eq. [6.35]). Hence, $Im\Sigma^{*R}(\mathbf{k}\sigma, \omega) \leq 0$ for all values of ω. While $A(\mathbf{k}\sigma, \omega)$ is a Dirac-delta function for a noninteracting system, the above expression shows that, in the presence of interactions,

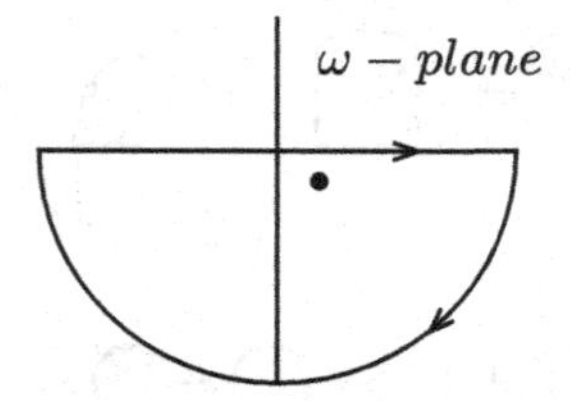

Figure 9.14 The contour C in Eq. (9.22) consists of the real axis, from $-\infty$ to $+\infty$, and a semicircle at infinity in the lower half-plane.

$A(\mathbf{k}\sigma, \omega)$ is a Lorentzian with a shifted center and half-width at full maximum given by $(1/2)Im\Sigma_R^*(\mathbf{k}\sigma, \omega_0)$, where ω_0 is the solution of the equation $\omega - \bar{\epsilon}_{\mathbf{k}\sigma}/\hbar - Re\Sigma^{*R}(\mathbf{k}\sigma, \omega) = 0$.

Going next to the time domain,

$$G^R(\mathbf{k}\sigma, t) = \frac{1}{2\pi}\int_{-\infty}^{\infty} G^R(\mathbf{k}\sigma, \omega)e^{-i\omega t}d\omega = \frac{1}{2\pi}\int_{-\infty}^{\infty} \frac{e^{-i\omega t}d\omega}{\omega - \epsilon'_{\mathbf{k}\sigma}/\hbar + i\gamma}$$

where $\epsilon'_{\mathbf{k}\sigma} \simeq \bar{\epsilon}_{\mathbf{k}\sigma} + \hbar Re\Sigma_R^*(\mathbf{k}\sigma, \bar{\epsilon}_{\mathbf{k}\sigma}/\hbar)$ and $\gamma \simeq -Im\Sigma_R^*(\mathbf{k}\sigma, \epsilon'_{\mathbf{k}\sigma}/\hbar) > 0$. Since $t > 0$ ($G^R(\mathbf{k}\sigma, t)$ vanishes for $t < 0$), the above integral vanishes if taken over the semicircle at infinity in the lower half ω-plane. Hence,

$$G^R(\mathbf{k}\sigma, t) = \frac{1}{2\pi}\int_C \frac{e^{-i\omega t}d\omega}{\omega - \epsilon'_{\mathbf{k}\sigma}/\hbar + i\gamma} \tag{9.22}$$

where C is the closed contour shown in Figure 9.14. The position of the pole is given by $\omega_{pole} = \epsilon'_{\mathbf{k}\sigma}/\hbar - i\gamma$. By the residue theorem, we obtain

$$G^R(\mathbf{k}\sigma, t) = -i\theta(t)e^{-i\epsilon'_{\mathbf{k}\sigma}t/\hbar}e^{-\gamma t}.$$

The minus sign arises because we go around the contour in a clockwise direction, and the step function ensures that the retarded function vanishes for $t < 0$. In fact, if $t < 0$, we consider a contour consisting of the real axis and a semicircle at infinity in the upper half-plane. The contour integral then vanishes, since the pole is outside the contour. Furthermore, for $t < 0$, the integral over the semicircle at infinity in the upper half-plane also vanishes; hence, the integral along the real axis vanishes, and $G^R(\mathbf{k}\sigma, t < 0) = 0$.

The retarded function of the noninteracting system is given by

$$G^{R,0}(\mathbf{k}\sigma, t) = -i\theta(t)e^{-i\bar{\epsilon}_{\mathbf{k}\sigma}t/\hbar}$$

(see Eq.[6.61]). A comparison of the expressions for $G^R(\mathbf{k}\sigma, t)$ and $G^{R,0}(\mathbf{k}\sigma, t)$ shows that the effects of the interaction are shifting the energy of the single-particle excitation by $\hbar Re\Sigma_R^*(\mathbf{k}\sigma, \bar{\epsilon}_{\mathbf{k}\sigma}/\hbar)$ and causing a damping of this excitation.

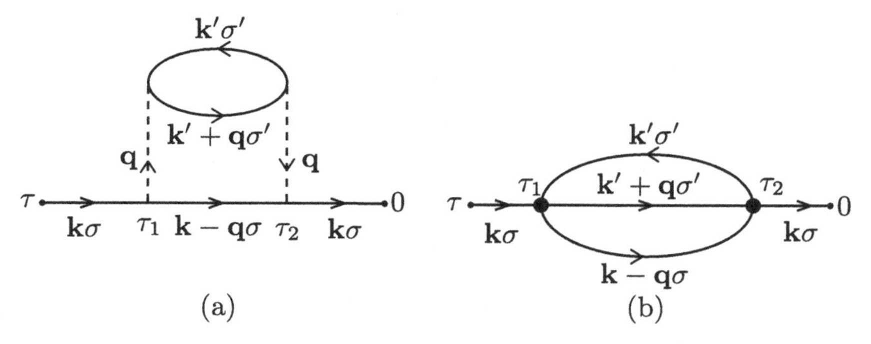

Figure 9.15 Two equivalent drawings of the ring diagram.

The lifetime of the excitation, τ, is defined by $\gamma\tau = 1/2$; hence,

$$\tau = \frac{-1}{2\, Im\, \Sigma_R^*(\mathbf{k}\sigma, \epsilon'_{\mathbf{k}\sigma}/\hbar)}.$$

9.7 Time-ordered diagrams: a case study

For a system of interacting fermions, let us consider the ring diagram again, shown in Figure 9.15. Diagram (b) is an equivalent way of representing the ring diagram if we assign $v_{\mathbf{q}}/\mathrm{V}$ to each filled circle. Its contribution is given by

$$\delta g^{(2)} = c(2)\int_0^{\beta\hbar} d\tau_1 \int_0^{\beta\hbar} d\tau_2 \sum_{\mathbf{k}'\sigma'\mathbf{q}} \frac{v_{\mathbf{q}}^2}{V^2} g^0(\mathbf{k}\sigma, \tau - \tau_1) g^0(\mathbf{k}\sigma, \tau_2) F(\mathbf{k}\mathbf{k}'\mathbf{q}\sigma\sigma', \tau_1 - \tau_2)$$

where $c(2) = -(-1/\hbar)^2$, and

$$F(\mathbf{k}\mathbf{k}'\mathbf{q}\sigma\sigma', \tau_1 - \tau_2) = g^0(\mathbf{k} - \mathbf{q}\sigma, \tau_1 - \tau_2) g^0(\mathbf{k}'\sigma', \tau_2 - \tau_1) g^0(\mathbf{k}' + \mathbf{q}\sigma', \tau_1 - \tau_2).$$

Fourier-expanding the external Green's functions,

$$g^0(\mathbf{k}\sigma, \tau - \tau_1) = (\beta\hbar)^{-1} \sum_{n_1} g^0(\mathbf{k}\sigma, \omega_{n_1}) e^{-i\omega_{n_1}(\tau - \tau_1)},$$

$$g^0(\mathbf{k}\sigma, \tau_2) = (\beta\hbar)^{-1} \sum_{n_2} g^0(\mathbf{k}\sigma, \omega_{n_2}) e^{-i\omega_{n_2}\tau_2},$$

we obtain

$$\delta g^{(2)} = c(2)(\beta\hbar)^{-2} \sum_{\mathbf{k}'\sigma'\mathbf{q}} \frac{v_{\mathbf{q}}^2}{V^2} \sum_{n_1 n_2} g^0(\mathbf{k}\sigma, \omega_{n_1}) g^0(\mathbf{k}\sigma, \omega_{n_2}) e^{-i\omega_{n_1}\tau} I$$

$$I = \int_0^{\beta\hbar} d\tau_1 \int_0^{\beta\hbar} d\tau_2\, e^{i\omega_{n_1}\tau_1} e^{-i\omega_{n_2}\tau_2} F(\mathbf{k}\mathbf{k}'\mathbf{q}\sigma\sigma', \tau_1 - \tau_2).$$

Noting that

$$\int_0^{\beta\hbar} d\tau_1 \int_0^{\beta\hbar} d\tau_2 \cdots = \int_0^{\beta\hbar} d\tau_1 \int_0^{\tau_1} d\tau_2 \cdots + \int_0^{\beta\hbar} d\tau_2 \int_0^{\tau_2} d\tau_1 \ldots,$$

the expression for I is written as

$$I = I_1 + I_2$$
$$I_1 = \int_0^{\beta\hbar} d\tau_1 \int_0^{\tau_1} d\tau_2\, e^{i\omega_{n_1}\tau_1} e^{-i\omega_{n_2}\tau_2} F^{><>}$$
$$I_2 = \int_0^{\beta\hbar} d\tau_2 \int_0^{\tau_2} d\tau_1\, e^{i\omega_{n_1}\tau_1} e^{-i\omega_{n_2}\tau_2} F^{<><}$$

where

$$F^{><>} = g^{0>}(\mathbf{k}-\mathbf{q}\sigma, \tau_1 - \tau_2) g^{0<}(\mathbf{k}'\sigma', \tau_2 - \tau_1) g^{0>}(\mathbf{k}'+\mathbf{q}\sigma', \tau_1 - \tau_2)$$
$$F^{<><} = g^{0<}(\mathbf{k}-\mathbf{q}\sigma, \tau_1 - \tau_2) g^{0>}(\mathbf{k}'\sigma', \tau_2 - \tau_1) g^{0<}(\mathbf{k}'+\mathbf{q}\sigma', \tau_1 - \tau_2).$$

Here, $g^{0>}(\mathbf{k}\sigma, \tau) = g^0(\mathbf{k}\sigma, \tau > 0)$ and $g^{0<}(\mathbf{k}\sigma, \tau) = g^0(\mathbf{k}\sigma, \tau < 0)$.

We recall some results from Chapter 8,

$$g^{0>}(\mathbf{k}\sigma, \tau) = \int_{-\infty}^{\infty} P^{0>}(\mathbf{k}\sigma, \epsilon) e^{-\epsilon\tau} \frac{d\epsilon}{2\pi}, \quad P^{0>}(\mathbf{k}\sigma, \epsilon) = -(1 - f_\epsilon) A^0(\mathbf{k}\sigma, \epsilon)$$

$$g^{0<}(\mathbf{k}\sigma, \tau) = \int_{-\infty}^{\infty} P^{0<}(\mathbf{k}\sigma, \epsilon) e^{-\epsilon\tau} \frac{d\epsilon}{2\pi}, \quad P^{0<}(\mathbf{k}\sigma, \epsilon) = f_\epsilon A^0(\mathbf{k}\sigma, \epsilon)$$

(see Eqs (8.28) and (8.32–8.34)). $A^0(\mathbf{k}\sigma, \epsilon)$ is the spectral density function for the noninteracting system. Using the above expressions for $g^{0>}$ and $g^{0<}$, we can write

$$I_1 = \int_{-\infty}^{\infty} \frac{d\epsilon_1}{2\pi} \int_{-\infty}^{\infty} \frac{d\epsilon_2}{2\pi} \int_{-\infty}^{\infty} \frac{d\epsilon_3}{2\pi} P^{0>}(\mathbf{k}-\mathbf{q}\sigma, \epsilon_1) P^{0<}(\mathbf{k}'\sigma', \epsilon_2) P^{0>}(\mathbf{k}'+\mathbf{q}\sigma', \epsilon_3)$$
$$\times \int_0^{\beta\hbar} d\tau_1 e^{(i\omega_{n_1} - \epsilon_1 + \epsilon_2 - \epsilon_3)\tau_1} \int_0^{\tau_1} d\tau_2 e^{(-i\omega_{n_2} + \epsilon_1 - \epsilon_2 + \epsilon_3)\tau_2}. \tag{9.23}$$

The time integrals are easily evaluated,

$$\int_0^{\beta\hbar} d\tau_1 e^{(i\omega_{n_1} - \epsilon_1 + \epsilon_2 - \epsilon_3)\tau_1} \int_0^{\tau_1} d\tau_2 e^{(-i\omega_{n_2} + \epsilon_1 - \epsilon_2 + \epsilon_3)\tau_2}$$
$$= \frac{\beta\hbar}{-i\omega_{n_1} + \epsilon_1 - \epsilon_2 + \epsilon_3} \delta_{\omega_{n_1}\omega_{n_2}} - \int_0^{\beta\hbar} d\tau_1 \frac{e^{(i\omega_{n_1} - \epsilon_1 + \epsilon_2 - \epsilon_3)\tau_1}}{-i\omega_{n_2} + \epsilon_1 - \epsilon_2 + \epsilon_3}. \tag{9.24}$$

Therefore,

$$I_1 = \beta\hbar\delta_{\omega_{n_1}\omega_{n_2}} \int_{-\infty}^{\infty}\frac{d\epsilon_1}{2\pi}\int_{-\infty}^{\infty}\frac{d\epsilon_2}{2\pi}\int_{-\infty}^{\infty}\frac{d\epsilon_3}{2\pi}\frac{B}{-i\omega_{n_1}+\epsilon_1-\epsilon_2+\epsilon_3}$$
$$-\int_{-\infty}^{\infty}\frac{d\epsilon_1}{2\pi}\int_{-\infty}^{\infty}\frac{d\epsilon_2}{2\pi}\int_{-\infty}^{\infty}\frac{d\epsilon_3}{2\pi}B\int_0^{\beta\hbar}d\tau_1\frac{e^{(i\omega_{n_1}-\epsilon_1+\epsilon_2-\epsilon_3)\tau_1}}{-i\omega_{n_2}+\epsilon_1-\epsilon_2+\epsilon_3} \tag{9.25}$$

where

$$B = P^{0>}(\mathbf{k}-\mathbf{q}\sigma,\epsilon_1)P^{0<}(\mathbf{k}'\sigma',\epsilon_2)P^{0>}(\mathbf{k}'+\mathbf{q}\sigma',\epsilon_3). \tag{9.26}$$

Similarly, defining D by

$$D = P^{0<}(\mathbf{k}-\mathbf{q}\sigma,\epsilon_1)P^{0>}(\mathbf{k}'\sigma',\epsilon_2)P^{0<}(\mathbf{k}'+\mathbf{q}\sigma',\epsilon_3) \tag{9.27}$$

we can write an expression for I_2

$$I_2 = \beta\hbar\delta_{\omega_{n_1}\omega_{n_2}} \int_{-\infty}^{\infty}\frac{d\epsilon_1}{2\pi}\int_{-\infty}^{\infty}\frac{d\epsilon_2}{2\pi}\int_{-\infty}^{\infty}\frac{d\epsilon_3}{2\pi}\frac{D}{i\omega_{n_1}-\epsilon_1+\epsilon_2-\epsilon_3}$$
$$-\int_{-\infty}^{\infty}\frac{d\epsilon_1}{2\pi}\int_{-\infty}^{\infty}\frac{d\epsilon_2}{2\pi}\int_{-\infty}^{\infty}\frac{d\epsilon_3}{2\pi}D\int_0^{\beta\hbar}d\tau_2\frac{e^{(-i\omega_{n_2}+\epsilon_1-\epsilon_2+\epsilon_3)\tau_2}}{i\omega_{n_1}-\epsilon_1+\epsilon_2-\epsilon_3}. \tag{9.28}$$

The expression for $\delta g^{(2)}$ now becomes

$$\delta g^{(2)}(\mathbf{k}\sigma,\tau) = \delta g_a^{(2)}(\mathbf{k}\sigma,\tau)+\delta g_b^{(2)}(\mathbf{k}\sigma,\tau)$$

$$\delta g_a^{(2)}(\mathbf{k}\sigma,\tau) = c(2)(\beta\hbar)^{-1}\sum_{\mathbf{k}'\sigma'\mathbf{q}}(v_{\mathbf{q}}/V)^2\sum_n[g^0(\mathbf{k}\sigma,\omega_n)]^2e^{-i\omega_n\tau}\int_{-\infty}^{\infty}\frac{d\epsilon_1}{2\pi}$$
$$\times\int_{-\infty}^{\infty}\frac{d\epsilon_2}{2\pi}\int_{-\infty}^{\infty}\frac{d\epsilon_3}{2\pi}\left[\frac{B}{-i\omega_n+\epsilon_1-\epsilon_2+\epsilon_3}+\frac{D}{i\omega_n-\epsilon_1+\epsilon_2-\epsilon_3}\right] \tag{9.29}$$

$$\delta g_b^{(2)}(\mathbf{k}\sigma,\tau) = -c(2)(\beta\hbar)^{-2}\sum_{\mathbf{k}'\sigma'\mathbf{q}}(v_{\mathbf{q}}/V)^2\sum_{n_1n_2}g^0(\mathbf{k}\sigma,\omega_{n_1})g^0(\mathbf{k}\sigma,\omega_{n_2})e^{-i\omega_{n_1}\tau}$$
$$\times\int_{-\infty}^{\infty}\frac{d\epsilon_1}{2\pi}\int_{-\infty}^{\infty}\frac{d\epsilon_2}{2\pi}\int_{-\infty}^{\infty}\frac{d\epsilon_3}{2\pi}\left[B\int_0^{\beta\hbar}d\tau_1\frac{e^{(i\omega_{n_1}-\epsilon_1+\epsilon_2-\epsilon_3)\tau_1}}{-i\omega_{n_2}+\epsilon_1-\epsilon_2+\epsilon_3}\right.$$
$$\left.+D\int_0^{\beta\hbar}d\tau_2\frac{e^{(-i\omega_{n_2}+\epsilon_1-\epsilon_2+\epsilon_3)\tau_2}}{i\omega_{n_1}-\epsilon_1+\epsilon_2-\epsilon_3}\right]. \tag{9.30}$$

Denoting the term in brackets in Eq. (9.30) by J, we find

$$J = -\frac{B[1+e^{\beta\hbar(-\epsilon_1+\epsilon_2-\epsilon_3)}]+D[1+e^{\beta\hbar(\epsilon_1-\epsilon_2+\epsilon_3)}]}{(i\omega_{n_1}-\epsilon_1+\epsilon_2-\epsilon_3)(-i\omega_{n_2}+\epsilon_1-\epsilon_2+\epsilon_3)} \equiv -\frac{X}{Y}.$$

To arrive at this expression, we used $exp(i\omega_{n_1}\beta\hbar) = exp(i\omega_{n_2}\beta\hbar) = -1$. Replacing B and D by their values given in Eqs (9.26) and (9.27), we find

$$\begin{aligned}
X &= P^{0>}(\mathbf{k}-\mathbf{q}\sigma,\epsilon_1)P^{0<}(\mathbf{k}'\sigma',\epsilon_2)P^{0>}(\mathbf{k}'+\mathbf{q}\sigma',\epsilon_3)[1+e^{\beta\hbar(-\epsilon_1+\epsilon_2-\epsilon_3)}] \\
&\quad + P^{0<}(\mathbf{k}-\mathbf{q}\sigma,\epsilon_1)P^{0>}(\mathbf{k}'\sigma',\epsilon_2)P^{0<}(\mathbf{k}'+\mathbf{q}\sigma',\epsilon_3)[1+e^{\beta\hbar(\epsilon_1-\epsilon_2+\epsilon_3)}] \\
&= A^0(\mathbf{k}-\mathbf{q}\sigma,\epsilon_1)A^0(\mathbf{k}'\sigma',\epsilon_2)A^0(\mathbf{k}'+\mathbf{q}\sigma',\epsilon_3) \\
&\quad \times \left\{(1-f_{\epsilon_1})f_{\epsilon_2}(1-f_{\epsilon_3})[1+e^{\beta\hbar(-\epsilon_1+\epsilon_2-\epsilon_3)}] - f_{\epsilon_1}(1-f_{\epsilon_2})f_{\epsilon_3}[1+e^{\beta\hbar(\epsilon_1-\epsilon_2+\epsilon_3)}]\right\}.
\end{aligned} \tag{9.31}$$

Miraculously, the term in braces vanishes, which can be easily verified. In fact, the vanishing of this term is not coincidental; it follows from general considerations: only the term in $\delta g^{(2)}$ which is proportional to $\delta_{\omega_{n_1}\omega_{n_2}}$ should survive. We conclude that $\delta g_b^{(2)} = 0$ and $\delta g^{(2)} = \delta g_a^{(2)}$. Fourier-expanding $\delta g^{(2)}$,

$$\delta g^{(2)}(\mathbf{k}\sigma,\tau) = (\beta\hbar)^{-1}\sum_n \delta g^{(2)}(\mathbf{k}\sigma\omega_n)e^{-i\omega_n\tau},$$

we finally obtain

$$\begin{aligned}
\delta g^{(2)}(\mathbf{k}\sigma,\omega_n) &= c(2)\sum_{\mathbf{k}'\sigma'\mathbf{q}}(v_\mathbf{q}/V)^2[g^0(\mathbf{k}\sigma,\omega_n)]^2\int_{-\infty}^{\infty}\frac{d\epsilon_1}{2\pi}\int_{-\infty}^{\infty}\frac{d\epsilon_2}{2\pi}\int_{-\infty}^{\infty}\frac{d\epsilon_3}{2\pi} \\
&\quad \times\left[\frac{P^{0>}(\mathbf{k}-\mathbf{q}\sigma,\epsilon_1)P^{0<}(\mathbf{k}'\sigma',\epsilon_2)P^{0>}(\mathbf{k}'+\mathbf{q}\sigma',\epsilon_3)}{-i\omega_n+\epsilon_1-\epsilon_2+\epsilon_3}\right. \\
&\quad \left.+\frac{P^{0<}(\mathbf{k}-\mathbf{q}\sigma,\epsilon_1)P^{0>}(\mathbf{k}'\sigma',\epsilon_2)P^{0<}(\mathbf{k}'+\mathbf{q}\sigma',\epsilon_3)}{i\omega_n-\epsilon_1+\epsilon_2-\epsilon_3}\right].
\end{aligned} \tag{9.32}$$

We recall the spectral representation of $g(\mathbf{k}\sigma,\omega_n)$ and $G^R(\mathbf{k}\sigma,\omega)$:

$$g(\mathbf{k}\sigma,\omega_n) = \int_{-\infty}^{\infty}\frac{A(\mathbf{k}\sigma,\epsilon)}{i\omega_n-\epsilon}\frac{d\epsilon}{2\pi}, \quad G^R(\mathbf{k}\sigma,\omega) = \int_{-\infty}^{\infty}\frac{A(\mathbf{k}\sigma,\epsilon)}{\omega-\epsilon+i0^+}\frac{d\epsilon}{2\pi}.$$

$G^R(\mathbf{k}\sigma,\omega)$ is analytic and well-defined everywhere in the upper half ω–plane; it is also well-defined on the positive imaginary axis. Thus,

$$G^R(\mathbf{k}\sigma,i\omega_n) = g(\mathbf{k}\sigma,\omega_n), \quad \omega_n > 0.$$

Now consider the function $F(\mathbf{k}\sigma,\omega)$ defined by

$$F(\mathbf{k}\sigma,\omega) = \delta g^{(2)}(\mathbf{k}\sigma,i\omega_n\to\omega+i0^+).$$

This function is analytic everywhere above the real axis, as can be seen from Eq. (9.32). Furthermore, it coincides with $\delta G^{R,(2)}(\mathbf{k}\sigma,\omega)$ on an infinite sequence of points, along the positive imaginary axis, whose limit lies in the region of analyticity (the limit point is at infinity on the positive imaginary axis). A theorem

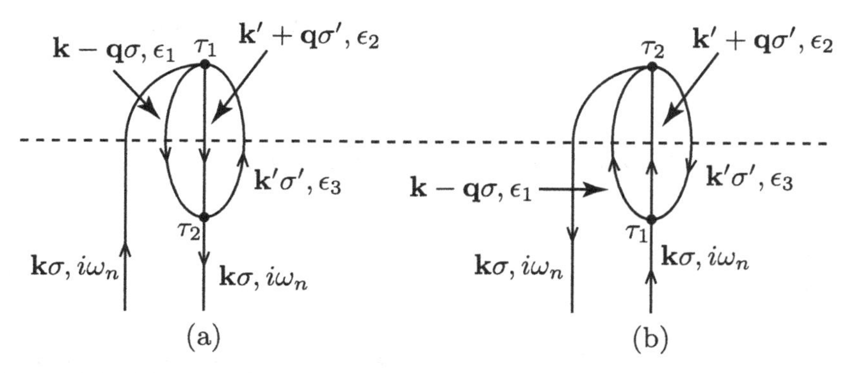

Figure 9.16 The two time-ordered diagrams corresponding to the diagram in Figure 9.15: (a) $\tau_1 > \tau_2$. (b) $\tau_2 > \tau_1$.

to that effect, in the theory of complex functions, assures us that the functions $F(\mathbf{k}\sigma, \omega)$ and $\delta G^{R,(2)}(\mathbf{k}\sigma, \omega)$ coincide everywhere in the upper half-plane. Therefore, the expression for $\delta G^{R,(2)}(\mathbf{k}\sigma, \omega)$ is obtained from $\delta g^{(2)}(\mathbf{k}\sigma, \omega_n)$ by simply replacing $i\omega_n$ by $\omega + i0^+$. We conclude that by using the approach outlined in this section, resolving the problem of analytic continuation, which was mentioned in the previous chapter, is straightforward: obtain $g(\mathbf{k}\sigma, \omega_n)$ as a sum of time-ordered diagrams and replace $i\omega_n$ by $\omega + i0^+$ to obtain $G^R(\mathbf{k}\sigma, \omega)$.

There remains the problem of how to develop a set of rules for writing the algebraic expression that corresponds to any particular diagram. It took a great effort to arrive at the expression for $\delta g^{(2)}(\mathbf{k}\sigma, \omega_n)$, and this approach becomes worthless if a similar effort is required for the evaluation of each diagram.

We observe that the expression for $\delta g^{(2)}(\mathbf{k}\sigma, \omega_n)$ can be written directly if we redraw Figure 9.15 as in Figure 9.16, and if we adopt the following rules. Arrange the time coordinates of the vertices so that they decrease from top to bottom. Consider all possible time orderings; in this case, there are two time orderings: $\tau_1 > \tau_2$ (Figure 9.16a) and $\tau_2 > \tau_1$ (Figure 9.16b). The two external lines are drawn vertically. To each external line, assign $g^0(\mathbf{k}\sigma, \omega_n)$ and frequency $i\omega_n$. To each vertex which corresponds to an interaction with wave vector $\mathbf{q}$, assign $v_\mathbf{q}/V$. To each internal line, assign the coordinates $(\mathbf{k}_i\sigma_i, \epsilon_i)$ such that momentum and spin are conserved at each vertex. Draw a horizontal dashed line, called a section, which separates one vertex from the one below it. To each internal line with coordinates $(\mathbf{k}_i\sigma_i, \epsilon_i)$ which crosses a section, assign $P^{0>}(\mathbf{k}_i\sigma_i, \epsilon_i)$ if it is directed downward and $P^{0<}(\mathbf{k}_i\sigma_i, \epsilon_i)$ if it is directed upward. To each section, assign a denominator equal to the sum of the frequencies of the lines that intersect the section, with each frequency carrying a plus sign if the line is directed downward and a minus sign if it is directed upward. Sum over the internal momentum and spin coordinates and integrate over ϵ_i's, the internal frequency coordinates. Finally, multiply the resulting expression by $c(n)$, which is determined by the original Feynman diagram (here, $n = 2$).

9.8 Time-ordered diagrams: Dzyaloshinski's rules

We indicated in the example given in the previous section that the vanishing of $\delta g_b^{(2)}$ is not coincidental. In fact, it follows from the general property that Green's function, along with all Feynman diagrams, depends only on a difference of time coordinates. Consider any Feynman diagram, denoted by Γ_N, of order N. One external line runs from τ to τ_N, and another runs from τ_1 to τ'. The algebraic expression corresponding to this diagram is

$$\delta g^{\Gamma_N}(\tau, \tau') = \int_0^{\beta\hbar} d\tau_1 \ldots \int_0^{\beta\hbar} d\tau_N g^0(\tau - \tau_N) g^0(\tau_1 - \tau') A(\tau_1, \ldots, \tau_N)$$

where $A(\tau_1, \ldots, \tau_N)$ is some function of the internal time coordinates, and the wave vector and spin arguments are suppressed. Fourier-expanding the external lines,

$$\delta g^{\Gamma_N}(\tau, \tau') = \frac{1}{(\beta\hbar)^2} \sum_{n,n'} g^0(\omega_n) g^0(\omega_{n'}) \int_0^{\beta\hbar} d\tau_1 \ldots \int_0^{\beta\hbar} d\tau_N e^{-i\omega_n(\tau - \tau_N)} e^{-i\omega_{n'}(\tau_1 - \tau')}$$
$$\times\ A(\tau_1, \ldots, \tau_N) = \frac{1}{(\beta\hbar)^2} \sum_{n,n'} g^0(\omega_n) g^0(\omega_{n'}) e^{-i\omega_n(\tau - \tau')} e^{-i(\omega_n - \omega_{n'})\tau'}$$
$$\times \int_0^{\beta\hbar} d\tau_1 \ldots \int_0^{\beta\hbar} d\tau_N e^{i\omega_n \tau_N} e^{-i\omega_{n'}\tau_1} A(\tau_1, \ldots, \tau_N). \tag{9.33}$$

Since δg^{Γ_N} depends on $\tau - \tau'$ and not on τ and τ' separately, it follows that

$$\int_0^{\beta\hbar} d\tau_1 \ldots \int_0^{\beta\hbar} d\tau_N e^{i\omega_n \tau_N} e^{-i\omega_{n'}\tau_1} A(\tau_1, \ldots, \tau_N) \propto \delta_{n,n'}.$$

That is, since τ and τ' do not appear in the above integral, the integral must vanish unless $\omega_n = \omega_{n'}$; if it did not, then δg^{Γ_N} would depend on τ' because of the term $exp[-i(\omega_n - \omega_{n'})\tau']$. In other words, the frequencies of the two external lines, along with their momentum and spin coordinates, have to be the same; we have assumed this all along.

The derivation of the rules for the time-ordered diagrams is given below. These rules were enunciated, but not derived, in a paper by Dzyaloshinski (Dzyaloshinski, 1962). A similar set of rules were derived by Baym and Sessler (Baym and Sessler, 1963). An alternative derivation is provided in the remainder of this section. The

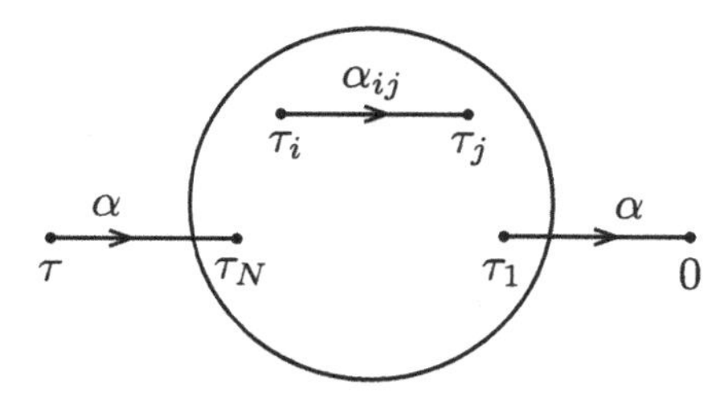

Figure 9.17 A general Feynman diagram, of order N, denoted by Γ_N. The external lines, directed from τ to τ_N and from τ_1 to 0, have momentum and spin coordinates denoted by α. An internal line, directed from τ_i to τ_j, has momentum and spin coordinates denoted by α_{ij}.

reader who is interested only in applying the rules may go to the end of the section, where the rules are listed.

Consider any Feynman diagram Γ_N of order N. We can represent it as in Figure 9.17. An external line is directed from τ to some vertex, whose time coordinate we call τ_N. A second external line is directed from some vertex, whose time coordinate we call τ_1, to $\tau = 0$. Inside the circle, there are N interaction time vertices, and internal lines run between these vertices. The total number of internal lines depends on the system under consideration and on the type of interaction involved. For a system of fermions with two-particle interaction, the total number of internal lines is $2N - 1$. For other types of interactions, such as scattering due to impurities, or for other kinds of systems, such as interacting electrons and phonons, the number differs from $2N - 1$.

The algebraic expression corresponding to the diagram in Figure 9.17 is

$$\delta g^{\Gamma_N}(\tau) = c_{\Gamma_N} \sum_{\{\alpha_{ij}\}} M^{\Gamma_N}_{\{\alpha_{ij}\}} \int_0^{\beta\hbar} d\tau_N \ldots \int_0^{\beta\hbar} d\tau_1 g^0_\alpha(\tau - \tau_N) g^0_\alpha(\tau_1) \prod_{i,j} F_{[\alpha_{ij}]}(\tau_i - \tau_j).$$

c_{Γ_N} is a counting factor that depends on the structure of the diagram, and it is determined by the Feynman rules. $M^{\Gamma_N}_{\{\alpha_{ij}\}}$ is the product of the interaction matrix elements (determined by the types of interactions at the vertices), and it depends on internal momentum and spin coordinates, denoted collectively by $\{\alpha_{ij}\}$, which are summed over. The quantity $M^{\Gamma_N}_{\{\alpha_{ij}\}}$ contains Kronecker deltas that express momentum and spin conservation at the vertices. In the above expression, $F_{[\alpha_{ij}]}(\tau_i - \tau_j)$ is the product of Green's functions corresponding to the lines directed from τ_i to τ_j. If only one line with coordinates α_{ij} is directed from τ_i to τ_j, then $F_{[\alpha_{ij}]}(\tau_i - \tau_j) = g^0_{\alpha_{ij}}(\tau_i - \tau_j)$. If two lines, having coordinates α_{ij} and α'_{ij} are directed from τ_i to τ_j, then $F_{[\alpha_{ij}]}(\tau_i - \tau_j) = g^0_{\alpha_{ij}}(\tau_i - \tau_j) g^0_{\alpha'_{ij}}(\tau_i - \tau_j)$. If no lines are directed from τ_i to τ_j, then $F_{[\alpha_{ij}]}(\tau_i - \tau_j) = 1$. Fourier-expanding the external lines, we can write

$$\delta g^{\Gamma_N}(\tau) = c_{\Gamma_N}(\beta\hbar)^{-2} \sum_{\{\alpha_{ij}\}} M^{\Gamma_N}_{\{\alpha_{ij}\}} \sum_{n_1 n_2} g^0_\alpha(\omega_{n_1}) g^0_\alpha(\omega_{n_2}) e^{-i\omega_{n_1}\tau} I_1 \tag{9.34}$$

$$I_1 = \int_0^{\beta\hbar} d\tau_N \ldots \int_0^{\beta\hbar} d\tau_1 \, e^{i\omega_{n_1}\tau_N} e^{-i\omega_{n_2}\tau_1} \prod_{i,j} F_{[\alpha_{ij}]}(\tau_i - \tau_j). \tag{9.35}$$

Next, I_1 is written as a sum of $N!$ integrals corresponding to the $N!$ permutations of $\tau_1, \tau_2, \ldots, \tau_N$:

$$I_1 = \sum_P \int_0^{\beta\hbar} d\tau_{P_N} \int_0^{\tau_{P_N}} d\tau_{P_{N-1}} \ldots \int_0^{\tau_{P_2}} d\tau_{P_1} e^{i\omega_{n_1}\tau_N} e^{-i\omega_{n_2}\tau_1} \prod_{P_i, P_j} F_{[\alpha_{P_i P_j}]}(\tau_{P_i} - \tau_{P_j})$$

where $P_1, P_2, \ldots, P_N$ is a permutation of $1, 2, \ldots, N$, and the sum over P is a sum over all such permutations. Now we make use of the following:

$$g^{0>}(\tau) = \int_{-\infty}^{\infty} P^{0>}(\epsilon) e^{-\epsilon\tau} \frac{d\epsilon}{2\pi}, \quad g^{0<}(\tau) = \int_{-\infty}^{\infty} P^{0<}(\epsilon) e^{-\epsilon\tau} \frac{d\epsilon}{2\pi}$$

where, for bosons (B) and fermions (F),

$$P^{0>}(\epsilon) = \begin{cases} -(1 + n_\epsilon) A^0(\epsilon) & \text{B} \\ -(1 - f_\epsilon) A^0(\epsilon) & \text{F} \end{cases} \qquad P^{0<}(\epsilon) = \begin{cases} -n_\epsilon A^0(\epsilon) & \text{B} \\ f_\epsilon A^0(\epsilon) & \text{F.} \end{cases} \tag{9.36}$$

n_ϵ and f_ϵ are the Bose and Fermi distribution functions, respectively. To each internal line directed from τ_i to τ_j we assign a frequency ϵ_{ij} and replace $g^0_{\alpha_{ij}}(\tau_i - \tau_j)$ by an integral over ϵ_{ij}, as in the above expressions for $g^{0>}$ and $g^{0<}$. We obtain

$$\begin{aligned} I_1 = \sum_P \int_{-\infty}^{\infty} \cdots \int_{-\infty}^{\infty} \frac{d\{\epsilon_{P_i P_j}\}}{(2\pi)^{N_L}} F^P_{\{\alpha_{P_i P_j}\}}(\{\epsilon_{P_i P_j}\}) \int_0^{\beta\hbar} d\tau_{P_N} \int_0^{\tau_{P_N}} d\tau_{P_{N-1}} \cdots \\ \times \int_0^{\tau_{P_2}} d\tau_{P_1} e^{i\omega_{n_1}\tau_N} e^{-i\omega_{n_2}\tau_1} \prod_{P_i, P_j} e^{-\epsilon_{P_i P_j}(\tau_{P_i} - \tau_{P_j})} \end{aligned} \tag{9.37}$$

where N_L is the number of internal lines, and

$$F^P_{\{\alpha_{P_i P_j}\}}(\{\epsilon_{P_i P_j}\}) = \prod_{P_i} \left[\prod_{P_j < P_i} P^{0>}_{\alpha_{P_i P_j}}(\epsilon_{P_i P_j}) \prod_{P_j > P_i} P^{0<}_{\alpha_{P_i P_j}}(\epsilon_{P_i P_j}) \right].$$

I_1 may now be written as

$$I_1 = \sum_P \int_{-\infty}^{\infty} \cdots \int_{-\infty}^{\infty} \frac{d\{\epsilon_{P_i P_j}\}}{(2\pi)^{N_L}} F^P_{\{\alpha_{P_i P_j}\}}(\{\epsilon_{P_i P_j}\}) I_{2P}$$

$$\begin{aligned} I_{2P} &= \int_0^{\beta\hbar} d\tau_{P_N} \int_0^{\tau_{P_N}} d\tau_{P_{N-1}} \cdots \int_0^{\tau_{P_2}} d\tau_{P_1} e^{i\omega_{n_1}\tau_N} e^{-i\omega_{n_2}\tau_1} \prod_{P_i, P_j} e^{-\epsilon_{P_i P_j}(\tau_{P_i} - \tau_{P_j})} \\ &= \int_0^{\beta\hbar} d\tau_{P_N} e^{\epsilon_{P_N}\tau_{P_N}} \int_0^{\tau_{P_N}} d\tau_{P_{N-1}} e^{\epsilon_{P_{N-1}}\tau_{P_{N-1}}} \cdots \int_0^{\tau_{P_2}} d\tau_{P_1} e^{\epsilon_{P_1}\tau_{P_1}} \end{aligned} \quad (9.38)$$

where

$$\epsilon_{P_i} = -\sum_{P_j} \epsilon_{P_i P_j} + \sum_{P_j} \epsilon_{P_j P_i} + i\omega_{n_1}\delta_{P_i,N} - i\omega_{n_2}\delta_{P_i,1}. \quad (9.39)$$

Notice the form ϵ_{P_i}: it involves a sum over all the lines that enter or leave vertex τ_{P_i}. If a line leaves τ_{P_i}, it carries a negative frequency; if it enters τ_{P_i}, it carries a positive frequency. Also note that

$$\sum_{P_i} \epsilon_{P_i} = i\omega_{n_1} - i\omega_{n_2}. \quad (9.40)$$

Since the N integrands are all exponentials, the evaluation of I_{2P} is straightforward. The evaluation must begin with the integral on the far right, then move left, one step at a time. The integral on the far right gives

$$\int_0^{\tau_{P_2}} d\tau_{P_1} e^{\epsilon_{P_1}\tau_{P_1}} = \frac{e^{\epsilon_{P_1}\tau_{P_2}} - 1}{\epsilon_{P_1}}.$$

The next integral is

$$\int_0^{\tau_{P_3}} d\tau_{P_2} e^{\epsilon_{P_2}\tau_{P_2}} \int_0^{\tau_{P_2}} d\tau_{P_1} e^{\epsilon_{P_1}\tau_{P_1}} = \frac{e^{(\epsilon_{P_1}+\epsilon_{P_2})\tau_{P_3}}}{\epsilon_{P_1}(\epsilon_{P_1}+\epsilon_{P_2})} + \text{other}.$$

Continuing in this fashion, and using Eq. (9.40), we obtain

$$\begin{aligned} I_{2P} &= \frac{e^{\beta\hbar(\epsilon_{P_1}+\epsilon_{P_2}+\cdots+\epsilon_{P_N})} - 1}{\epsilon_{P_1}(\epsilon_{P_1}+\epsilon_{P_2})\ldots(\epsilon_{P_1}+\epsilon_{P_2}+\cdots+\epsilon_{P_N})} + \text{others} \\ &= \frac{e^{\beta\hbar(i\omega_{n_1}-i\omega_{n_2})} - 1}{(i\omega_{n_1}-i\omega_{n_2})\epsilon_{P_1}(\epsilon_{P_1}+\epsilon_{P_2})\ldots(\epsilon_{P_1}+\epsilon_{P_2}+\cdots+\epsilon_{P_{N-1}})} + \text{others}. \end{aligned}$$

The first term in I_{2P} is obtained by keeping only the upper limits when integrating over $\tau_{P_1}, \tau_{P_2}, \ldots, \tau_{P_{N-1}}$. The second term, called "others," consists of the rest of the terms. If $\omega_{n_1} \neq \omega_{n_2}$, then $e^{\beta\hbar(i\omega_{n_1}-i\omega_{n_2})} - 1 = 0$, and the first term in I_{2P} vanishes.

For $\omega_{n_1} = \omega_{n_2}$, L'Hopital's rule gives

$$\frac{e^{\beta\hbar(i\omega_{n_1}-i\omega_{n_2})} - 1}{(i\omega_{n_1} - i\omega_{n_2})} = \beta\hbar \Longrightarrow$$

$$I_{2P} = \frac{\beta\hbar\delta_{n_1,n_2}}{\epsilon_{P_1}(\epsilon_{P_1} + \epsilon_{P_2})\ldots(\epsilon_{P_1} + \epsilon_{P_2} + \cdots + \epsilon_{P_{N-1}})} + \text{others}.$$

We showed in the previous section, by explicit calculations, that the terms represented by "others" combine to give a vanishing result. We can show that this is true in general, for any order of the interaction (see Appendix B). For now, we assume that this is indeed the case. I_{2P} is then given by

$$I_{2P} = \frac{\beta\hbar\delta_{n_1,n_2}}{\prod_{i<N}\tilde{\epsilon}_{Pi}}, \qquad \tilde{\epsilon}_{Pi} = \sum_{j=1}^{i}\epsilon_{Pj}. \tag{9.41}$$

The final expression for the Feynman diagram of Figure 9.17 can now be written. Fourier expanding: $\delta g^{\Gamma_N}(\tau) = (1/\beta\hbar)\sum_n \delta g^{\Gamma_N}(\omega_n)e^{-i\omega_n\tau}$, we obtain

$$\delta g^{\Gamma_N}(\omega_n) = [g^0_\alpha(\omega_n)]^2 c_{\Gamma_N} \sum_{\{\alpha_{ij}\}} M^{\Gamma_N}_{\{\alpha_{ij}\}} \sum_P \int \frac{F^P_{\{\alpha_{P_iP_j}\}}(\{\epsilon_{P_iP_j}\})}{\prod_{i<N}\tilde{\epsilon}_{Pi}} \frac{d\{\epsilon_{P_iP_j}\}}{(2\pi)^{N_L}}. \tag{9.42}$$

The rules for time-ordered diagrams follow directly from Eq. (9.42). Before writing the rules, however, let us clarify the meaning of $\tilde{\epsilon}_{P_i}$. Note that

$$\tilde{\epsilon}_{P_i} = \sum_{j=1}^{i}\epsilon_{Pj} = \sum_{j=1}^{i}\left[-\sum_{P_k}\epsilon_{P_jP_k} + \sum_{P_k}\epsilon_{P_kP_j} + i\omega_n\delta_{P_j.N} - i\omega_n\delta_{P_j.1}\right].$$

Since $\sum_{P_k}\cdots = \sum_k\cdots = \sum_{k=1}^{i} + \sum_{k>i}$, the expression for $\tilde{\epsilon}_{P_i}$ reduces to

$$\tilde{\epsilon}_{P_i} = -\sum_{j=1}^{i}\sum_{k>i}\epsilon_{P_jP_k} + \sum_{j=1}^{i}\sum_{k>i}\epsilon_{P_kP_j} + i\omega_n\sum_{j=1}^{i}\delta_{P_j.N} - i\omega_n\sum_{j=1}^{i}\delta_{P_j.1}. \tag{9.43}$$

Let us now arrange the time vertices vertically such that time decreases as we move down. Since $\tau_{P_N} > \tau_{P_{N-1}} > \cdots > \tau_{P_1}$, the resulting arrangement is that shown in Figure 9.18. A horizontal dashed line, called a section, is drawn between $\tau_{P_{i+1}}$ and τ_{P_i}.

Regarding the above expression for $\tilde{\epsilon}_{P_i}$,

(1) The first term is the sum of the frequencies of all internal lines that start at $\tau_{P_i}, \tau_{P_{i-1}}, \ldots, \tau_{P_1}$ and end at the vertices above τ_{P_i}. That is, it is the sum of the frequencies of the internal lines that are directed upward and that intersect the section. These frequencies carry a minus sign.

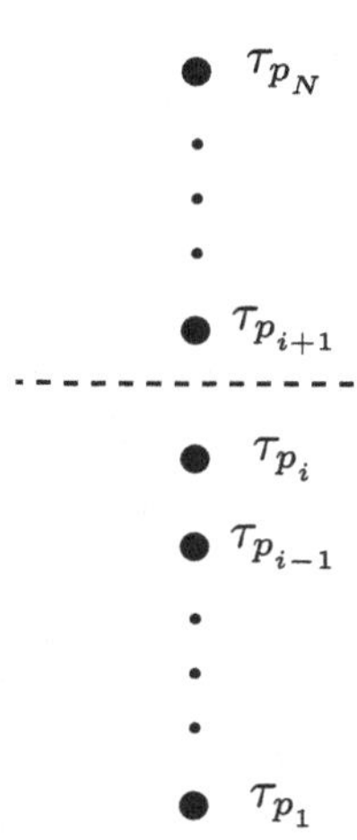

Figure 9.18 Arrangement of time vertices such that time decreases from top to bottom. A horizontal dashed line between $\tau_{P_{i+1}}$ and τ_{P_i} is a section.

(2) The second term is the sum of the frequencies of the internal lines that start at $\tau_{P_{i+1}}, \tau_{P_{i+2}}, \ldots, \tau_{P_N}$ and end at $\tau_{P_i}, \tau_{P_{i-1}}, \ldots, \tau_{P_1}$. That is, it is the sum of the frequencies of the internal lines that are directed downward and that intersect the section. These frequencies carry a positive sign.

(3) The third and fourth terms relate to the external lines. One external line enters at τ_N and one leaves at τ_1. We draw these two lines vertically as in Figure 9.16. Note that if both lines enter and leave below the section, the combined contribution of the third and fourth terms is $i\omega_n - i\omega_n = 0$. Similarly, if both lines enter and leave above the section, the contribution of each of the third and fourth terms is zero. If one external line enters below the section ($\tau_N \in \{\tau_{P_1}, \tau_{P_2}, \ldots, \tau_{P_i}\}$) and the other line leaves from a vertex above the section ($\tau_1 \in \{\tau_{P_{i+1}}, \ldots, \tau_{P_N}\}$), the combined contribution of the third and fourth terms is $i\omega_n$. On the other hand, if one external line enters a vertex above the section and the other line leaves from a vertex below the section, the combined contribution of the third and fourth terms is $-i\omega_n$.

These observations can be summarized as follows. If we assign a frequency $i\omega_n$ to each of the two external lines, and a frequency $\epsilon_{P_i P_j}$ to each internal line directed from vertex τ_{P_i} to vertex τ_{P_j}, then $\tilde{\epsilon}_{P_i}$ is the sum of all the frequencies of the lines (internal and external) that intersect the section between $\tau_{P_{i+1}}$ and τ_{P_i}. The frequency carries a plus sign if the line is directed downward and a minus sign if the line is directed upward.

We are now in a position to state Dzyaloshinski's rules for time-ordered diagrams:

(1) Construct $N!$ time-ordered diagrams corresponding to the $N!$ permutations of $\tau_1, \tau_2, \ldots, \tau_N$. In each diagram, time decreases as we go downward. External lines must always be drawn vertically. Assign coordinates $(\mathbf{k}\sigma, i\omega_n)$ to each of the two external lines, and coordinates $(\mathbf{k}_i\sigma_i, \epsilon_i)$ to the ith internal line. Draw a horizontal dashed line (a section) that separates each vertex from the one below it (there are $N-1$ sections for each time-ordered diagram).
(2) Assign to each vertex a matrix element M that depends on the momentum and spin coordinates of the lines that meet at the vertex. Conserve momentum and spin at each vertex.
(3) Assign $g^0(\mathbf{k}\sigma, \omega_n)$ to each of the two external lines. To each internal line with coordinates $(\mathbf{k}_i\sigma_i, \epsilon_i)$, assign $(1/2\pi)P^{0>}(\mathbf{k}_i\sigma_i, \epsilon_i)$ if it is directed down and $(1/2\pi)P^{0<}(\mathbf{k}_i\sigma_i, \epsilon_i)$ if it is directed up.
(4) To each section assign a denominator equal to the sum of the frequencies of the lines intersected by the section; line frequencies carry a plus sign if a line is directed down and a minus sign if a line is directed up.
(5) Multiply all the factors in rules 2, 3, and 4, sum over all internal momentum and spin coordinates, and integrate over the frequencies (ϵ's) of the internal lines.
(6) Sum all the contributions of the $N!$ time-ordered diagrams.
(7) Multiply the resulting expression by c_{Γ_N}, the counting factor that corresponds to the original Feynman diagram.

Further reading

Abrikosov, A.A., Gorkov, L.P., and Dzyaloshinski, I.E. (1963). *Methods of Quantum Field Theory in Statistical Physics*. New York: Dover Publications.

Bruus, H. and Flensberg, K. (2004). *Many-Body Quantum Theory in Condensed Matter Physics*. Oxford: Oxford University Press.

Fetter, A.L. and Walecka, J.D. (1971). *Quantum Theory of Many-Particle Systems*. New York: McGraw-Hill.

Mahan, G.D. (2000). *Many-Particle Physics*, 3rd edn. New York: Kluwer Academic/Plenum Publishers.

Problems

9.1 *A vanishing sum*. Show that

$$\sum_{n=-\infty}^{\infty} e^{i\omega_n 0^+} = 0.$$

9.2 *Thermodynamic potential*. Using Eq. (8.26) for the thermodynamic potential, show that, for a system of interacting electrons,

$$\Omega(T, V, \mu) = \Omega_0(T, V, \mu) + \frac{1}{2\beta}\int_0^1 \frac{d\lambda}{\lambda} \sum_{\mathbf{k}\sigma} \sum_{n=-\infty}^{\infty} e^{i\omega_n 0^+} \Sigma^{*\lambda}(\mathbf{k}\sigma, \omega_n) g^\lambda(\mathbf{k}\sigma, \omega_n).$$

9.3 *Frequency sums.* This problem shows how to evaluate $\sum_{n=-\infty}^{\infty} \frac{e^{i\omega_n 0^+}}{i\omega_n - \bar{\epsilon}/\hbar}$.
Consider the contour integral

$$I = \lim_{\eta \to 0^+} \int_C \frac{e^{\eta z}}{z - \bar{\epsilon}/\hbar} \frac{dz}{e^{\beta\hbar z} \pm 1}$$

where C is a circle of infinite radius centered at $z = 0$. As $|z| \to \infty$, if $Re\, z > 0$, the absolute value of the integrand is of order $(1/|z|)e^{-\beta\hbar\, Re\, z}$. If $Re\, z < 0$, the absolute value of the integrand is of order $(1/|z|)e^{\eta\, Re\, z}$. The integrand is thus exponentially small as $|z| \to \infty$; therefore, $I = 0$.

(a) For the case of bosons, consider

$$I = \lim_{\eta \to 0^+} \int_C \frac{e^{\eta z}}{z - \bar{\epsilon}/\hbar} \frac{dz}{e^{\beta\hbar z} - 1} = 0.$$

The poles of the integrand occur at $z = 2n\pi i/\beta\hbar,\ n \in \mathbb{Z}$, and at $z = \bar{\epsilon}/\hbar$. Use the residue theorem to show that

$$\sum_{n=-\infty}^{\infty} \frac{e^{i\omega_n 0^+}}{i\omega_n - \bar{\epsilon}/\hbar} = -\beta\hbar n_{\bar{\epsilon}}$$

where $\omega_n = 2n\pi/\beta\hbar$ and $n_{\bar{\epsilon}} = \left(e^{\beta\hbar\bar{\epsilon}} - 1\right)^{-1}$.

(b) For the case of fermions, consider

$$I = \lim_{\eta \to 0^+} \int_C \frac{e^{\eta z}}{z - \bar{\epsilon}/\hbar} \frac{dz}{e^{\beta\hbar z} + 1} = 0.$$

The poles of the integrand are at $z = (2n+1)\pi i/\beta\hbar,\ n \in \mathbb{Z}$, and at $z = \bar{\epsilon}/\hbar$. Use the residue theorem to show that

$$\sum_{n=-\infty}^{\infty} \frac{e^{i\omega_n 0^+}}{i\omega_n - \bar{\epsilon}/\hbar} = \beta\hbar f_{\bar{\epsilon}}$$

where $\omega_n = (2n+1)\pi/\beta\hbar$, and $f_{\bar{\epsilon}} = \left(e^{\beta\hbar\bar{\epsilon}} + 1\right)^{-1}$.

9.4 *An alternative method.* Noting that $\langle c^\dagger_{\mathbf{k}\sigma} c_{\mathbf{k}\sigma}\rangle_0 = n_{\mathbf{k}\sigma}\ (f_{\mathbf{k}\sigma})$ for bosons (fermions), derive Eq. (9.14) for the frequency sum.

9.5 *External potential.* For a system of noninteracting particles in the presence of a spin-independent static external potential, the Hamiltonian is

$$\bar{H} = \sum_{\mathbf{k}\sigma} \bar{\epsilon}_{\mathbf{k}\sigma} c^\dagger_{\mathbf{k}\sigma} c_{\mathbf{k}\sigma} + \sum_{\mathbf{k}\mathbf{q}\sigma} v_{\mathbf{q}} c^\dagger_{\mathbf{k}+\mathbf{q}\sigma} c_{\mathbf{k}\sigma}.$$

(a) Using Wick's theorem, evaluate $g(\mathbf{k}\sigma, \tau)$ to second order in the perturbation.

(b) Calculate $g(\mathbf{k}\sigma, \omega_n)$ to second order in the perturbation.
(c) Deduce the Feynman rules in momentum-frequency space.

9.6 *Impurity in a metal.* Consider an impurity in a metal host. As a model Hamiltonian we take $H = H_0 + H'$, where

$$H_0 = \epsilon \sum_{\sigma} d_\sigma^\dagger d_\sigma + \sum_{\mathbf{k}\sigma} \epsilon_{\mathbf{k}} c_{\mathbf{k}\sigma}^\dagger c_{\mathbf{k}\sigma}, \quad H' = \sum_{\mathbf{k}\sigma} \left(V_{\mathbf{k}} c_{\mathbf{k}\sigma}^\dagger d_\sigma + V_{\mathbf{k}}^* d_\sigma^\dagger c_{\mathbf{k}\sigma} \right).$$

We ignore the onsite Coulomb repulsion that results when two electrons occupy the impurity orbital.
(a) Write, graphically, Dyson's equation for the impurity Green's function $g(d\sigma, \tau) = -\langle T d_\sigma(\tau) d_\sigma^\dagger(0) \rangle$.
(b) By Fourier transforming, determine $g(d\sigma, \omega_n)$.

9.7 *An exchange diagram.* Using the Feynman rules in momentum-frequency space, write the algebraic value of diagram J in Figure 9.3.

9.8 *Time-ordered diagrams.* Using Dzyaloshinski's rules for time-ordered diagrams, write the algebraic value of diagram J in Figure 9.3.

9.9 *A frequency sum.* Evaluate the frequency sum over n' in the expression for the ring diagram given in Eq. (9.7).

9.10 *Diagrams without loops.* For a system of interacting fermions (V is a two-particle interaction), show that, at order n in the perturbation, the number of connected, topologically distinct diagrams without any closed loops is $(2n)!/(n!2^n)$.

10

Electron gas: a diagrammatic approach

A subtle chain of countless rings
The next unto the farthest brings
–Ralph Waldo Emerson
Nature: Addresses and Lectures

In this chapter we apply diagram rules to the study of an interacting electron gas in the high density limit. We saw in Chapter 4 that, in this limit, the Coulomb repulsion between electrons is small compared to their kinetic energy, and that it can be treated as a perturbation added to otherwise free electrons. We now show that perturbation theory must be carried out to infinite order to yield meaningful results (we previously caught a glimpse of this notion in Chapter 4). This is due to the long-range nature of the Coulomb interaction: even though the Coulomb energy between two electrons, e^2/r, decreases with increasing distance between the electrons, the number of electrons in a spherical shell of radius r and thickness dr is proportional to r^2dr, so the interaction of one electron with electrons far away from it is still important. We then use perturbation theory to calculate the linear response of an interacting electron gas to an external field, and apply this technique to graphene.

10.1 Model Hamiltonian

Our model system consists of an interacting electron gas in the presence of a uniform positive background, the so-called jellium model, which we first encountered in Chapter 4. The Hamiltonian is

$$\bar{H} = \sum_{\mathbf{k}\sigma} \bar{\varepsilon}_{\mathbf{k}\sigma} c^\dagger_{\mathbf{k}\sigma} c_{\mathbf{k}\sigma} + \frac{1}{2V} \sum_{\mathbf{k}_1\sigma_1} \sum_{\mathbf{k}_2\sigma_2} \sum_{\mathbf{q}}{}' \frac{4\pi e^2}{q^2} c^\dagger_{\mathbf{k}_1+\mathbf{q}\sigma_1} c^\dagger_{\mathbf{k}_2-\mathbf{q}\sigma_2} c_{\mathbf{k}_2\sigma_2} c_{\mathbf{k}_1\sigma_1} \tag{10.1}$$

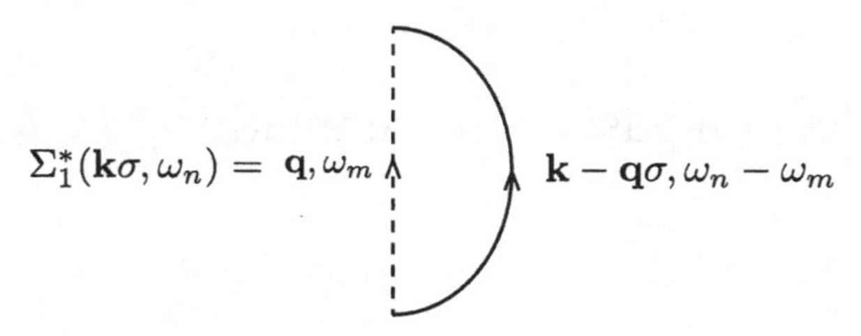

Figure 10.1 Proper self energy of an interacting electron gas in first-order perturbation theory.

where $\bar{\epsilon}_{\mathbf{k}\sigma} = \epsilon_{\mathbf{k}\sigma} - \mu = \hbar^2k^2/2m - \mu$ is the energy of an electron in the single-particle state $|\mathbf{k}\sigma\rangle$, measured relative to the chemical potential μ, and V is the system's volume. The prime on the summation over $\mathbf{q}$ indicates that the $\mathbf{q} = \mathbf{0}$ term is excluded. We exclude this term because of the background–background and electron–background interactions.

10.2 The need to go beyond first-order perturbation theory

The thermodynamic potential of the electron gas at temperature T is given by

$$\Omega(T, V, \mu) = \Omega_0(T, V, \mu) + \frac{1}{2\beta}\int_0^1 \frac{d\lambda}{\lambda} \sum_{\mathbf{k}\sigma n} e^{i\omega_n 0^+} \Sigma^{*\lambda}(\mathbf{k}\sigma, \omega_n) g^{\lambda}(\mathbf{k}\sigma, \omega_n) \tag{10.2}$$

(see Problem 9.2). Here, Ω_0 is the thermodynamic potential of the noninteracting electron gas, $\beta = 1/k_BT$, $\Sigma^{*\lambda}(\mathbf{k}\sigma, \omega_n)$ is the proper self energy when the interaction is $\lambda v_{\mathbf{q}}$, and $g^{\lambda}(\mathbf{k}\sigma, \omega_n)$ is the corresponding imaginary-time Green's function. We have found in the previous chapter that, to first order in the interaction, $\Sigma^*(\mathbf{k}\sigma, \omega_n)$ is a sum of two diagrams (see Figure 9.13). However, one diagram has $v_{\mathbf{q}=\mathbf{0}}$, and since the $\mathbf{q} = \mathbf{0}$ term is excluded in Eq. (10.1), we are left with only one diagram, as shown in Figure 10.1.

The expression for Σ_1^* is readily written using the diagram rules,

$$\Sigma_1^*(\mathbf{k}\sigma, \omega_n) = \left(-\frac{1}{\beta\hbar^2 V}\right) \sum_{\mathbf{q}m} v_{\mathbf{q}} e^{i(\omega_n - \omega_m)0^+} g^o(\mathbf{k} - \mathbf{q}\sigma, \omega_n - \omega_m). \tag{10.3}$$

There is one internal wave vector $\mathbf{q}$ and one internal frequency ω_m, and they are summed over. The interaction line is replaced by $v_{\mathbf{q}}/V$, and the fermion line by $g^o(\mathbf{k} - \mathbf{q}\sigma, \omega_n - \omega_m)$. The factor $e^{i(\omega_n - \omega_m)0^+}$ arises because the fermion line connects two vertices of the same interaction line, and the whole expression is multiplied by $(-1/\beta\hbar^2)^n$, where $n = 1$ is the order of the interaction. Defining $\omega_{n'}$

by $\omega_{n'} = \omega_n - \omega_m$, we obtain

$$\Sigma_1^*(\mathbf{k}\sigma, \omega_n) = \left(-\frac{1}{\beta\hbar^2 V}\right) \sum_{\mathbf{q}} v_{\mathbf{q}} \sum_{n'} e^{i\omega_{n'}0^+} g^o(\mathbf{k} - \mathbf{q}\sigma, \omega_{n'}).$$

The summation over n' is given in Eq. (9.14); we obtain

$$\Sigma_1^*(\mathbf{k}\sigma, \omega_n) = -\frac{1}{\hbar V} \sum_{\mathbf{q}} v_{\mathbf{q}}\, f_{\mathbf{k}-\mathbf{q}} \tag{10.4}$$

where $f_{\mathbf{k}-\mathbf{q}} = (e^{\beta\bar{\varepsilon}_{\mathbf{k}-\mathbf{q}}} + 1)^{-1}$ is the Fermi–Dirac distribution function. It follows that

$$\Sigma_1^{*\lambda}(\mathbf{k}\sigma, \omega_n) = -\frac{\lambda}{\hbar V} \sum_{\mathbf{q}} v_{\mathbf{q}}\, f_{\mathbf{k}-\mathbf{q}}.$$

Thus, to first order in the interaction, the thermodynamic potential of the electron gas is

$$\begin{aligned}\Omega(T, V, \mu) &= \Omega_0(T, V, \mu) - \frac{1}{2\beta\hbar V} \int_0^1 d\lambda \sum_{\mathbf{k}\mathbf{q}\sigma} v_{\mathbf{q}} f_{\mathbf{k}-\mathbf{q}} \sum_n e^{i\omega_n 0^+} g^0(\mathbf{k}\sigma, \omega_n) \\ &= \Omega_0(T, V, \mu) - \frac{1}{V} \sum_{\mathbf{k},\mathbf{q}} v_{\mathbf{q}}\, f_{\mathbf{k}}\, f_{\mathbf{k}+\mathbf{q}} \end{aligned} \tag{10.5}$$

where Eq. (9.14) is used again, and $\mathbf{q}$ is changed to $-\mathbf{q}$, taking advantage of the fact that $v_{-\mathbf{q}} = v_{\mathbf{q}}$. Note that in order to calculate $\Omega(T, V, \mu)$ to first order in the interaction, $g^\lambda(\mathbf{k}\sigma, \omega_n)$ in Eq. (10.2) is replaced by the non-interacting Green's function $g^0(\mathbf{k}\sigma, \omega_n)$, since $\Sigma_1^{*\lambda}(\mathbf{k}\sigma, \omega_n)$ is already of first order in the interaction.

What is the problem with stopping at first order in the interaction? It turns out that doing so leads to some anomalous predictions about the behavior of the electron gas at low temperatures:

(a) As $T \to 0$, the proper self energy Σ_1^* becomes

$$\Sigma_1^*(\mathbf{k}\sigma, \omega_n) = -\frac{e^2 k_F}{\pi\hbar}\left[1 + \frac{1 - x^2}{x} \ln\left|\frac{1+x}{1-x}\right|\right], \tag{10.6}$$

where k_F is the Fermi wave vector and $x = k/k_F$ (see Problem 10.1). Thus

$$\frac{d\,\Sigma_1^*(\mathbf{k}\sigma; \omega_n)}{dx} = -\frac{e^2 k_F}{\pi\hbar}\left[\frac{1}{x} - \frac{1 + x^2}{2x^2} \ln\left|\frac{1+x}{1-x}\right|\right] \tag{10.7}$$

which diverges logarithmically at $x = 1$, i.e., at the Fermi surface. Since the energy is shifted by $\hbar Re\Sigma_{1,\text{ret}}^*$, and Σ_1^* is real and independent of ω_n, $Re\Sigma_{1,\text{ret}}^* = \Sigma_1^*$. Therefore, the derivative of the energy, $dE_{\mathbf{k}\sigma}/dk$, diverges logarithmically

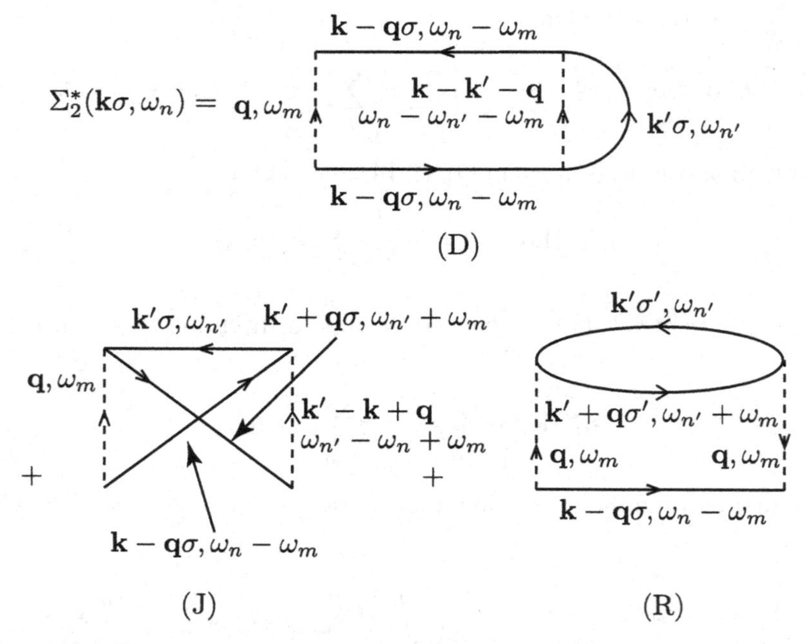

Figure 10.2 The three second-order diagrams that contribute to the proper self energy of an interacting electron gas.

at the Fermi surface. It follows that the density of states at the Fermi surface vanishes (see Problem 2.7). However, no such behavior is observed in metals: the density of states at the Fermi surface of metals is actually nonzero.

(b) From the expression for the thermodynamic potential, Eq. (10.5), the specific heat of the electron gas at constant volume, C_V, can be evaluated. It is found that, as $T \to 0$, $C_V \to T \ln T$ (Bardeen, 1936; Horovitz and Thieberger, 1974; Glasser, 1981; Glasser and Boersma, 1983). The logarithmic dependence on temperature of the specific heat, a measurable quantity, is not observed in metals at low temperatures; in fact, the electronic specific heat varies linearly with T.

The above discussion shows that, to obtain meaningful results, it is insufficient to expand $g(\mathbf{k}\sigma, \omega_n)$ to first order in the interaction; we must go to higher orders.

10.3 Second-order perturbation theory: still inadequate

The proper self energy $\Sigma^*(\mathbf{k}\sigma, \omega_n)$ was given in Figure 9.13. Ignoring diagrams that contain $v_{\mathbf{q}=\mathbf{0}}$ (the $\mathbf{q} = \mathbf{0}$ term is excluded from the Hamiltonian), we are left with three second-order diagrams (see Figure 10.2). Using the Feynman diagram rules, we can write the algebraic expressions corresponding to these diagrams:

$$\Sigma^*_{2,(D)}(\mathbf{k}\sigma,\omega_n) = \left(-\frac{1}{\beta\hbar^2 V}\right)^2 \sum_{\mathbf{q}m}\sum_{\mathbf{k}'n'} v_{\mathbf{q}} v_{\mathbf{k}-\mathbf{k}'-\mathbf{q}} e^{i\omega_{n'}0^+} g^0(\mathbf{k}-\mathbf{q}\sigma,\omega_n-\omega_m)$$
$$\times\, g^0(\mathbf{k}'\sigma,\omega_{n'}) g^0(\mathbf{k}-\mathbf{q}\sigma,\omega_n-\omega_m)$$
$$\Sigma^*_{2,(J)}(\mathbf{k}\sigma,\omega_n) = \left(-\frac{1}{\beta\hbar^2 V}\right)^2 \sum_{\mathbf{q}m}\sum_{\mathbf{k}'n'} v_{\mathbf{q}} v_{\mathbf{k}'-\mathbf{k}+\mathbf{q}}\, g^0(\mathbf{k}-\mathbf{q}\sigma,\omega_n-\omega_m)$$
$$\times\, g^0(\mathbf{k}'\sigma,\omega_{n'})\, g^0(\mathbf{k}'+\mathbf{q}\sigma,\omega_{n'}+\omega_m)$$
$$\Sigma^*_{2,(R)}(\mathbf{k}\sigma,\omega_n) = -\left(-\frac{1}{\beta\hbar^2 V}\right)^2 \sum_{\mathbf{q}m}\sum_{\mathbf{k}'\sigma'n'} v_q^2\, g^0(\mathbf{k}-\mathbf{q}\sigma,\omega_n-\omega_m)$$
$$\times\, g^0(\mathbf{k}'\sigma',\omega_{n'})\, g^0(\mathbf{k}'+\mathbf{q}\sigma',\omega_{n'}+\omega_m). \tag{10.8}$$

In the expression for $\Sigma^*_{2,(D)}$, a factor $e^{i\omega_{n'}0^+}$ is inserted because the line with coordinates $(\mathbf{k}'\sigma,\omega_{n'})$ connects two vertices of the same interaction line. In $\Sigma^*_{2,(R)}$, a factor of -1 results from the presence of one fermion loop.

In the expressions written above, if summation over the frequencies were to be carried out, it would result in Fermi and Bose distribution functions. Summations over wave vectors are replaced by integrals; e.g.,

$$\sum_{\mathbf{q}} \to \frac{V}{(2\pi)^3}\int d^3q.$$

A close investigation of the integrations over the wave vectors, reminiscent of the one carried out in Chapter 4, shows that $\Sigma^*_{2,(R)}$ is divergent, while $\Sigma^*_{2,(D)}$ and $\Sigma^*_{2,(J)}$ are not. In $\Sigma^*_{2,(R)}$, there is a term $v_q^2 = (4\pi e^2)^2/q^4$ and one integration over $\mathbf{q}$: $\int d^3q = \int q^2 dq \int d\cos\theta \int d\phi$. We are left with an integral $\int dq/q^2 \ldots$, and as $q \to 0$, the integral can be shown to diverge. This situation does not occur in diagrams (D) and (J), where the two interaction lines have different wave vectors.

The fact that $\Sigma^*_2(\mathbf{k}\sigma,\omega_n)$ is divergent (due to the divergence of the ring diagram) means that it is insufficient to carry out a perturbation expansion to second order. Were we to stop at second order, the energy of an electron in the electron gas would be infinite, and this is certainly not true. In fact, many diagrams in higher order also yield divergent contributions. Among the diagrams at a given order of perturbation, the most divergent diagram is the most important one. In what follows, our approach will be to classify the diagrams at each order in the interaction according to their degree of divergence, select the most divergent diagram at each order, and sum only those most divergent diagrams.

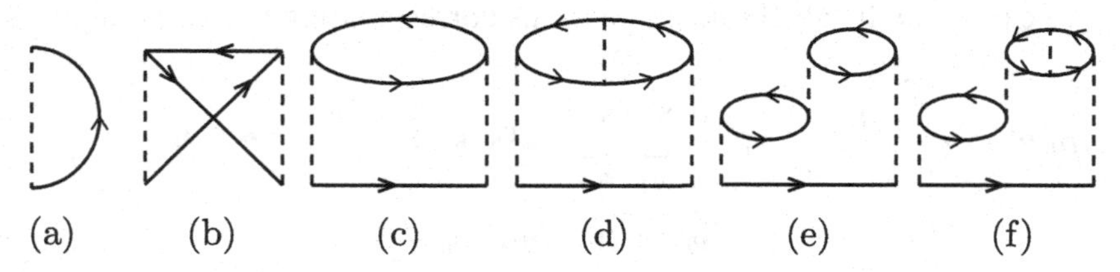

Figure 10.3 A collection of self energy diagrams.

10.4 Classification of diagrams according to the degree of divergence

Consider the self energy diagrams shown in Figure 10.3. How do we decide which of these diagrams should be included in the proper self energy? We have seen in the previous section that one criterion is the power of q in the denominator. Diagram (a) is an integral $\int d^3q/q^2 \ldots$, and diagram (b) is a similar integral with the same power of q in the denominator. If we include diagram (a) in Σ^*, should we also include diagram (b)? Diagram (c) is an integral $\int d^3q/q^4 \ldots$, which has q^4 in the denominator, but so does diagram (d). If we include diagram (c) in Σ^*, should we also include diagram (d)? We define the degree of divergence (DoD) of a given diagram as the largest number of interaction lines, in the diagram, that have the same wave vector $\mathbf{q}$. Thus $DoD(a) = DoD(b) = 1$, $DoD(c) = DoD(d) = 2$, and $DoD(e) = DoD(f) = 3$. The following analysis answers the questions raised above.

We assume that the electron gas is in the high density limit, $r_s \to 0$. The dimensionless quantity r_s is defined by the relation: $4\pi(r_s a_0)^3/3 = V/N$, where V is the system's volume, N is the number of electrons, and a_0 is the Bohr radius. It is easy to verify that $r_s = (9\pi/4)^{(1/3)}/a_0 k_F$, where k_F is the Fermi wave vector (see Section 4.3). We now look at the contribution of each diagram and determine its dependence on r_s.

Every self-energy diagram of order n has n interaction lines and $2n - 1$ fermion lines (the total number of fermion lines in $\delta g^{(n)}$ is $2n + 1$; the number of external lines is 2). Each diagram of order n also has n internal wave vectors and n internal frequencies. Denoting the contribution of the ith self energy diagram of order n by $\Sigma^*_{n,(i)}$, we can write

$$\Sigma^*_{n,(i)} \propto \beta^{-n} \int d^3p_1 \ldots \int d^3p_n \, v_{\mathbf{q}_1} v_{\mathbf{q}_2} \cdots v_{\mathbf{q}_n} \sum_{\omega_{n_1} \ldots \omega_{n_n}} \prod_{j=1}^{2n-1} g^0(\mathbf{k}_j\sigma_j, \bar{\omega}_j).$$

Here, the internal wave vectors are denoted by $\mathbf{p}_1, \ldots, \mathbf{p}_n$. The wave vectors $\mathbf{q}_1, \ldots, \mathbf{q}_n$, and $\mathbf{k}_1, \ldots, \mathbf{k}_{2n-1}$ depend on the external wave vector $\mathbf{k}$ and the internal wave vectors. Similarly, the frequencies $\bar{\omega}_1, \ldots, \bar{\omega}_{2n-1}$ depend on the external frequency ω_n and the n internal frequencies $\omega_{n_1}, \ldots, \omega_{n_n}$. To find the dependence of $\Sigma^*_{n,(i)}$ on r_s, we rewrite $\Sigma^*_{n,(i)}$ as a factor that depends on r_s times a dimensionless integral. This is accomplished by the following changes of variables:

$$p_j \to k_F p'_j,\ q_j \to k_F q'_j,\ k_j \to k_F k'_j,\ 1/\beta \to \epsilon_F/\beta',\ \bar{\epsilon}_{\mathbf{k}\sigma} \to \epsilon_F \bar{\epsilon}'_{\mathbf{k}\sigma},\ \hbar\omega_n \to \epsilon_F \omega'_n$$

where $\epsilon_F = \hbar^2 k_F^2/2m$ is the Fermi energy. The primed quantities are all dimensionless. Thus,

$$\beta^{-n} \propto k_F^{2n} \beta'^{-n}, \quad \int d^3 p_j = k_F^3 \int d^3 p'_j, \quad v_{q_j} = k_F^{-2} v_{q'_j}$$

$$g^0(\mathbf{k}_j\sigma_j, \bar{\omega}_j) = (i\bar{\omega}_j - \bar{\epsilon}_{\mathbf{k}_j\sigma_j})^{-1} \propto k_F^{-2} g^0(\mathbf{k}'_j\sigma_j, \bar{\omega}'_j)$$

$$\Rightarrow \Sigma^*_{n,(i)} \propto k_F^{2n} k_F^{3n} k_F^{-2n} k_F^{-2(2n-1)} (\text{D. I.}) = k_F^{2-n} (\text{D. I.}).$$

The dimensionless integral (D. I.) has no dependence on k_F. Since $r_s \propto k_F^{-1}$, the dimensionless integral is independent of r_s. We conclude that

$$\Sigma^*_{n,(i)} \propto r_s^{n-2}.$$

Thus, as $r_s \to 0$ (high density limit), given two diagrams with the same degree of divergence (DoD), the diagram of lower order in the interaction makes a much larger contribution to the self energy. For example, diagrams (a) and (b) in Figure 10.3 have the same DoD, but $\Sigma^*_{1,(a)} \propto r_s^{-1}$, while $\Sigma^*_{2,(b)} \propto r_s^0$; hence, as $r_s \to 0$, $|\Sigma^*_{1,(a)}| \gg |\Sigma^*_{2,(b)}|$. Similarly, $|\Sigma^*_{2,(c)}| \gg |\Sigma^*_{3,(d)}|$ and $|\Sigma^*_{3,(e)}| \gg |\Sigma^*_{4,(f)}|$. Thus, for any set of diagrams with the same DoD, we retain only the one with the lowest order in the interaction. From the set of diagrams in Figure 10.3, we retain only diagrams (a), (c), and (e).

The above conclusion may be stated differently: at any order of the interaction, only the diagram with the highest degree of divergence is retained. All diagrams with the same order of interaction have the same r_s dependence; hence, from among these diagrams, the one with the highest degree of divergence makes the largest contribution to the self energy.

10.5 Self energy in the random phase approximation (RPA)

On the basis of the above discussion, we can represent the proper self energy of an interacting electron gas as an infinite sum of diagrams (see Figure 10.4). At each order of the interaction, only the diagram with the highest degree of divergence is retained. Besides the first-order exchange diagram, the proper self energy is an infinite series of ring diagrams. Although each one of these ring diagrams is divergent, the infinite sum turns out to be convergent. The expression for Σ^* as

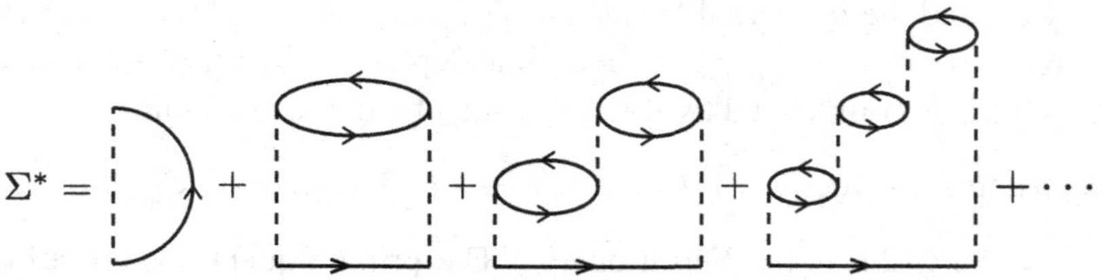

$$\Sigma^* = \;\cdots\; + \;\cdots\; + \;\cdots\; + \;\cdots\; + \cdots$$

Figure 10.4 The proper self energy of an interacting electron gas in the random phase approximation.

a sum of the first-order exchange diagram and the ring diagrams is clearly an approximation, since other diagrams are not included; it is known as the random phase approximation (RPA).

10.6 Summation of the ring diagrams

Using the Feynman rules, the contribution of the ring diagrams to the proper self energy can be readily written,

$$\begin{aligned}\Sigma^*_{\text{ring}}(\mathbf{k}\sigma, \omega_n) = &-\left(-\frac{1}{\beta\hbar^2 V}\right)^2 \sum_{\mathbf{q}m} v_{\mathbf{q}}^2 g^0(\mathbf{k}-\mathbf{q}\sigma, \omega_n - \omega_m) B(\mathbf{q}, \omega_m) \\ &+ \left(-\frac{1}{\beta\hbar^2 V}\right)^3 \sum_{\mathbf{q}m} v_{\mathbf{q}}^3 g^0(\mathbf{k}-\mathbf{q}\sigma, \omega_n - \omega_m) B^2(\mathbf{q}, \omega_m) + \cdots\end{aligned} \tag{10.9}$$

where

$$B(\mathbf{q}, \omega_m) = \sum_{\mathbf{k}'\sigma'n'} g^0(\mathbf{k}'\sigma', \omega_{n'}) g^0(\mathbf{k}'+\mathbf{q}\sigma', \omega_{n'}+\omega_m).$$

Defining the bare pair bubble $\Pi^0(\mathbf{q}, \omega_m)$ by

$$\Pi^0(\mathbf{q}, \omega_m) = \frac{1}{\beta\hbar^2 V} \sum_{\mathbf{k}'\sigma'n'} g^0(\mathbf{k}'\sigma', \omega_{n'}) g^0(\mathbf{k}'+\mathbf{q}\sigma', \omega_{n'}+\omega_m), \tag{10.10}$$

we can write

$$\Sigma^*_{\text{ring}}(\mathbf{k}\sigma, \omega_n) = -\frac{1}{\beta\hbar^2 V} \sum_{\mathbf{q}m} v_{\mathbf{q}}^2 g^0(\mathbf{k}-\mathbf{q}\sigma, \omega_n - \omega_m) \Pi_{\text{RPA}}(\mathbf{q}, \omega_m), \tag{10.11}$$

where

$$\Pi_{\text{RPA}}(\mathbf{q}, \omega_m) = \Pi^0(\mathbf{q}, \omega_m) + v_{\mathbf{q}} \left[\Pi^0(\mathbf{q}, \omega_m)\right]^2 + v_{\mathbf{q}}^2 \left[\Pi^0(\mathbf{q}, \omega_m)\right]^3 + \cdots \tag{10.12}$$

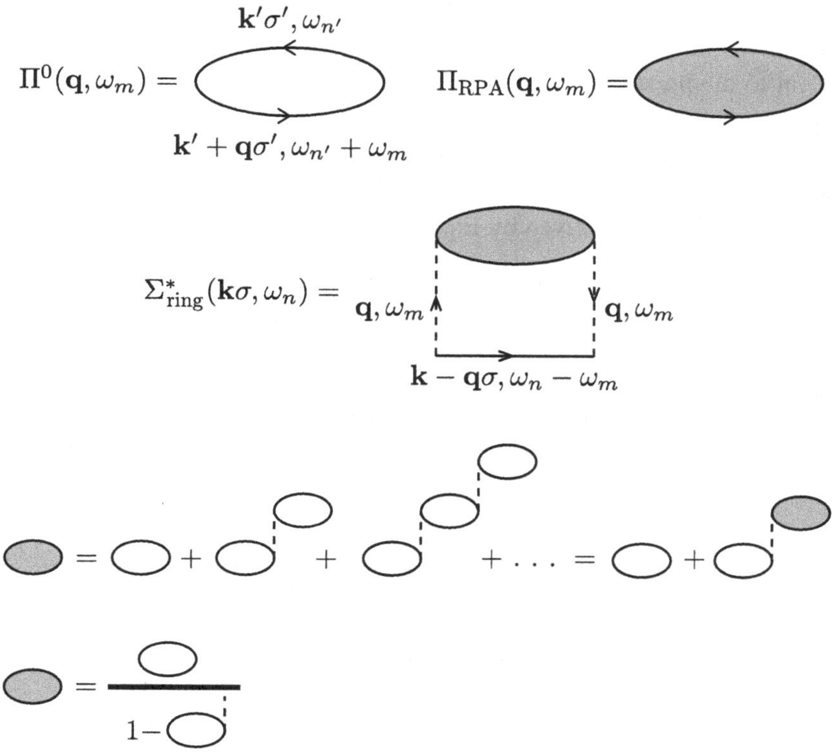

$\mathbf{k}'\sigma', \omega_{n'}$

$\Pi^0(\mathbf{q}, \omega_m) =$ $\qquad$ $\Pi_{\text{RPA}}(\mathbf{q}, \omega_m) =$

$\mathbf{k}' + \mathbf{q}\sigma', \omega_{n'} + \omega_m$

$\Sigma^*_{\text{ring}}(\mathbf{k}\sigma, \omega_n) =$ $\quad \mathbf{q}, \omega_m \quad \mathbf{q}, \omega_m$

$\mathbf{k} - \mathbf{q}\sigma, \omega_n - \omega_m$

$= \; + \; + \; + \ldots = \; +$

$= \; 1-$

Figure 10.5 The bare pair bubble, the dressed pair bubble in RPA, and the proper self energy that results from summing over ring diagrams.

is the dressed pair bubble. Equations (10.11) and (10.12) are represented graphically in Figure 10.5. The dressed pair bubble, in random phase approximation, is given by

$$\begin{aligned}
\Pi_{\text{RPA}}(\mathbf{q}, \omega_m) &= \Pi^0(\mathbf{q}, \omega_m) + \Pi^0(\mathbf{q}, \omega_m) v_{\mathbf{q}} \\
&\quad \times \left[\Pi^0(\mathbf{q}, \omega_m) + v_{\mathbf{q}} [\Pi^0(\mathbf{q}, \omega_m)]^2 + v_{\mathbf{q}}^2 [\Pi^0(\mathbf{q}, \omega_m)]^3 + \cdots \right] \\
&= \Pi^0(\mathbf{q}, \omega_m) + \Pi^0(\mathbf{q}, \omega_m) v_{\mathbf{q}} \Pi_{\text{RPA}}(\mathbf{q}, \omega_m) \\
\Longrightarrow \Pi_{\text{RPA}}(\mathbf{q}, \omega_m) &= \frac{\Pi^0(\mathbf{q}, \omega_m)}{1 - v_{\mathbf{q}} \Pi^0(\mathbf{q}, \omega_m)}.
\end{aligned} \tag{10.13}$$

The dressed bubble is also known as the polarizability of the interacting electron gas, while the bare bubble is the polarizability of the noninteracting electron gas. The nomenclature results from observing that a pair bubble represents a virtual process (energy is not conserved) in which an electron–hole pair is created and then annihilated. An electron in state $|k'\sigma'\rangle$ below the Fermi surface absorbs momentum $\hbar\mathbf{q}$ and moves above the Fermi surface; its absence from the Fermi sphere is equivalent to the presence of a hole. The electron then surrenders the momentum $\hbar\mathbf{q}$, and recombines with the hole. The electron and the hole, being of opposite charges, their creation is tantamount to the creation of a dipole moment, which causes the medium to become polarized.

The bare pair bubble is given by Eq. (10.10); thus,

$$\Pi^0(\mathbf{q}, \omega_m) = \frac{2}{\beta\hbar^2 V}\sum_{\mathbf{k}}\sum_{n=-\infty}^{\infty}\frac{1}{(i\omega_n - \bar{\epsilon}_{\mathbf{k}}/\hbar)(i\omega_n + i\omega_m - \bar{\epsilon}_{\mathbf{k+q}}/\hbar)}.$$

The factor 2 results from summing over the spin index. As $n \to \pm\infty$, the summand $\to -1/\omega_n^2$; hence, the series is convergent, and we are justified in introducing a convergence factor $e^{i\omega_n 0^+}$ (redundant in this case). This allows us to evaluate $\Pi^0(\mathbf{q}, \omega_m)$ by the method of partial fractions:

$$\begin{aligned}\Pi^0(\mathbf{q}, \omega_m) &= \frac{2}{\beta\hbar^2 V}\sum_{\mathbf{k}}\sum_{n=-\infty}^{\infty}\frac{e^{i\omega_n 0^+}}{(i\omega_n - \bar{\epsilon}_{\mathbf{k}}/\hbar)(i\omega_n + i\omega_m - \bar{\epsilon}_{\mathbf{k+q}}/\hbar)}\\ &= \frac{2}{\beta\hbar^2 V}\sum_{\mathbf{k}}\frac{1}{i\omega_m - (\bar{\epsilon}_{\mathbf{k+q}} - \bar{\epsilon}_{\mathbf{k}})/\hbar}\\ &\quad\times\sum_{n=-\infty}^{\infty}\left(\frac{e^{i\omega_n 0^+}}{i\omega_n - \bar{\epsilon}_{\mathbf{k}}/\hbar} - \frac{e^{i\omega_n 0^+}}{i\omega_n + i\omega_m - \bar{\epsilon}_{\mathbf{k+q}}/\hbar}\right).\end{aligned}$$

The sum over n (see Eq. [9.14]) is now evaluated,

$$\sum_{n=-\infty}^{\infty}\frac{e^{i\omega_n 0^+}}{i\omega_n - \bar{\epsilon}_{\mathbf{k}}/\hbar} = \beta\hbar f_{\mathbf{k}}, \quad \sum_{n=-\infty}^{\infty}\frac{e^{i\omega_n 0^+}}{i\omega_n + i\omega_m - \bar{\epsilon}_{\mathbf{k+q}}/\hbar} = \frac{\beta\hbar}{e^{\beta\bar{\epsilon}_{\mathbf{k+q}}}e^{-i\beta\hbar\omega_m} + 1},$$

$$e^{-i\beta\hbar\omega_m} = 1 \quad (\omega_m = 2m\pi/\beta\hbar).$$

The polarizability of the noninteracting electron gas reduces to

$$\Pi^0(\mathbf{q}, \omega_m) = \frac{2}{V}\sum_{\mathbf{k}}\frac{f_{\mathbf{k}} - f_{\mathbf{k+q}}}{i\hbar\omega_m + \epsilon_{\mathbf{k}} - \epsilon_{\mathbf{k+q}}}. \tag{10.14}$$

This is the Lindhard function which we encountered earlier in Chapter 6.

10.7 Screened Coulomb interaction

The contribution of ring diagrams to the proper self energy may be written in a way that differs from, but is equivalent to, the way presented in Figure 10.5. This is

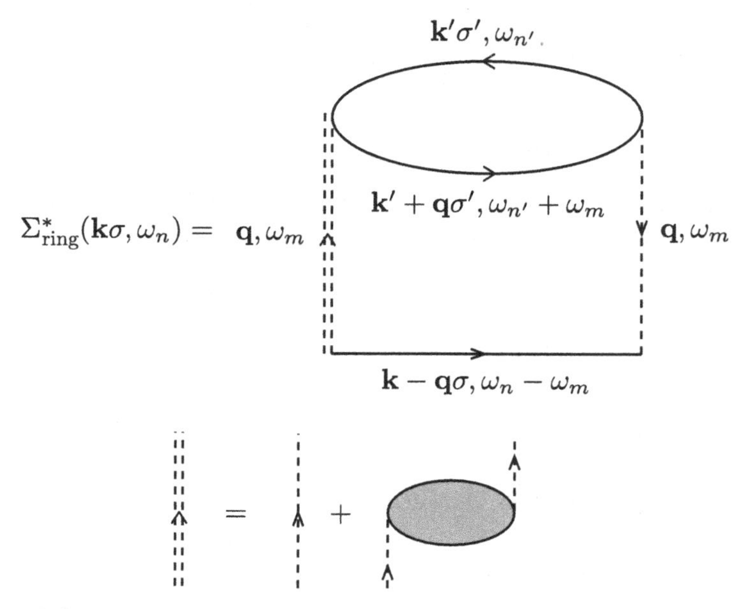

Figure 10.6 Ring contribution to the proper self energy of an interacting electron gas in terms of the screened Coulomb interaction.

shown in Figure 10.6. The double-dashed line is known as the screened Coulomb interaction $V(\mathbf{q}, \omega_m)$, while the single-dashed line is now called the bare Coulomb interaction $v_{\mathbf{q}}$. The expression for $V(\mathbf{q}, \omega_m)$ can be read directly from Figure 10.6:

$$\begin{aligned} V(\mathbf{q}, \omega_m) &= v_{\mathbf{q}} + v_{\mathbf{q}} \Pi_{\text{RPA}}(\mathbf{q}, \omega_m) v_{\mathbf{q}} = v_{\mathbf{q}} \left[1 + v_{\mathbf{q}} \Pi_{\text{RPA}}(\mathbf{q}, \omega_m)\right] \\ &= v_{\mathbf{q}} \left[1 + \frac{v_{\mathbf{q}} \Pi^0(\mathbf{q}, \omega_m)}{1 - v_{\mathbf{q}} \Pi^0(\mathbf{q}, \omega_m)}\right] = \frac{v_{\mathbf{q}}}{1 - v_{\mathbf{q}} \Pi^0(\mathbf{q}, \omega_m)} = \frac{v_{\mathbf{q}}}{\varepsilon(\mathbf{q}, \omega_m)}. \end{aligned} \tag{10.15}$$

The expression for $\Pi_{\text{RPA}}(\mathbf{q}, \omega_m)$ in Eq. (10.13) was used. $\varepsilon(\mathbf{q}, \omega_m)$ is the dielectric function; as we will show later in the chapter, it measures the response of the interacting electron gas to an external electric potential.

10.8 Collective electronic density fluctuations

In Section 6.7 we introduced the retarded density-density correlation function

$$D^R(\mathbf{q}, t) = -i\theta(t)\frac{1}{V}\langle[n_{\bar{H}}(\mathbf{q}, t), n_{\bar{H}}(-\mathbf{q}, 0)]\rangle = -i\theta(t)\frac{1}{V}\langle[\tilde{n}_{\bar{H}}(\mathbf{q}, t), \tilde{n}_{\bar{H}}(-\mathbf{q}, 0)]\rangle$$

where $\tilde{n}_{\bar{H}}(\mathbf{q}, t) = n_{\bar{H}}(\mathbf{q}, t) - \langle n_H(\mathbf{q}, t)\rangle$ is the deviation of the electronic density from its ensemble average. In the above equation, the last equality follows since

$\langle n_{\bar{H}}(\mathbf{q}, t)\rangle$ is simply a number, and numbers commute with operators. We have already calculated $D^{R,0}(\mathbf{q}, \omega)$ for the noninteracting electron gas (see Eq. [6.90]). Note that

$$D^{R,0}(\mathbf{q}, \omega) = \hbar\Pi^{o}(\mathbf{q}, i\omega_m \to \omega + i0^+). \tag{10.16}$$

In the presence of interactions, the retarded correlation function may be obtained from the corresponding imaginary-time correlation function

$$D^{R}(\mathbf{q}, \omega) = D(\mathbf{q}, i\omega_m \to \omega + i0^+).$$

The imaginary-time correlation function $D(\mathbf{q}, \tau)$ is given by

$$\begin{aligned} D(\mathbf{q}, \tau) &= -\frac{1}{V}\langle T\tilde{n}_{\bar{H}}(\mathbf{q}, \tau)\tilde{n}_{\bar{H}}(-\mathbf{q}, 0)\rangle \\ &= -\frac{1}{V}\langle T\{[n_{\bar{H}}(\mathbf{q}, \tau) - \langle n_{\bar{H}}(\mathbf{q}, \tau)\rangle]\,[n_{\bar{H}}(-\mathbf{q}, 0) - \langle n_{\bar{H}}(-\mathbf{q}, 0)\rangle]\}\rangle \\ &= -\frac{1}{V}\langle Tn_{\bar{H}}(\mathbf{q}, \tau)n_{\bar{H}}(-\mathbf{q}, 0)\rangle + \frac{1}{V}\langle n_{\bar{H}}(\mathbf{q}, \tau)\rangle\langle n_{\bar{H}}(-\mathbf{q}, 0)\rangle \end{aligned} \tag{10.17}$$

where use is made of the fact that

$$\langle Tn_{\bar{H}}(\mathbf{q}, \tau)\rangle = \sum_{\mathbf{k}\sigma}\langle Tc^{\dagger}_{\mathbf{k}\sigma}(\tau)c_{\mathbf{k}+\mathbf{q}\sigma}(\tau)\rangle = \sum_{\mathbf{k}\sigma}\langle c^{\dagger}_{\mathbf{k}\sigma}(\tau)c_{\mathbf{k}+\mathbf{q}\sigma}(\tau)\rangle = \langle n_{\bar{H}}(\mathbf{q}, \tau)\rangle. \tag{10.18}$$

In Eq. (10.18), the second equality holds because whenever a creation and an annihilation operator share the same time argument, it is assumed that the time argument of the creation operator is infinitesimally greater than that of the annihilation operator. This is because creation operators occur on the left in the Hamiltonian, while annihilation operators occur on the right (see Eq. [10.1]). The correlation function $D(\mathbf{q}, \tau)$ is known as the collective electronic density fluctuations. Note that, in order to obtain D^R from D by analytic continuation, D must be given in terms of $\tilde{n}_{\bar{H}}$, as in Eq. (10.17), and not in terms of $n_{\bar{H}}$ (see Problem 8.8).

Employing the perturbation expansion for the time-ordered product of modified Heisenberg picture operators, we can write

$$D(\mathbf{q}, \tau) - \frac{1}{V}\langle n_{\bar{H}}(\mathbf{q}, \tau)\rangle\langle n_{\bar{H}}(-\mathbf{q}, 0)\rangle = -\frac{1}{V}\sum_{\mathbf{k}\sigma}\sum_{\mathbf{k}'\sigma'}\sum_{n=0}^{\infty}\frac{1}{n!}\left(-\frac{1}{\hbar}\right)^{n}\int_0^{\beta\hbar}d\tau_1\ldots$$
$$\int_0^{\beta\hbar}d\tau_n\,\langle\hat{c}^{\dagger}_{\mathbf{k}\sigma}(\tau)\hat{c}_{\mathbf{k}+\mathbf{q}\sigma}(\tau)\hat{c}^{\dagger}_{\mathbf{k}'\sigma'}(0)\hat{c}_{\mathbf{k}'-\mathbf{q}\sigma'}(0)\,\hat{V}(\tau_1)\ldots\hat{V}(\tau_n)\rangle_{0,c}. \tag{10.19}$$

The operators in the expansion are now interaction-picture operators. We have used Eq. (3.25) to express the number-density operators in terms of creation and annihilation operators. When we consider the connected, topologically distinct

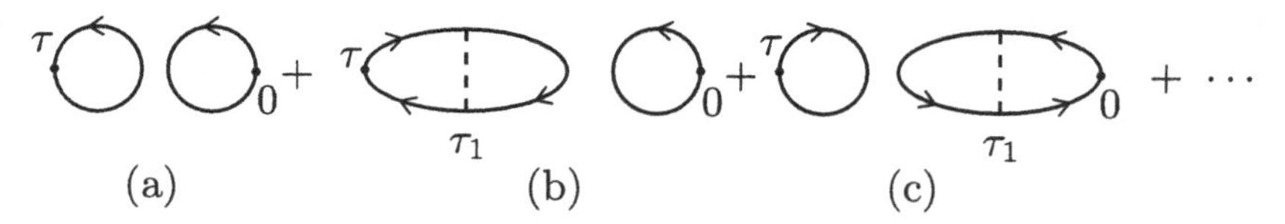

Figure 10.7 Connected but disjoint diagrams that arise from the perturbation expansion in Eq. (10.19).

diagrams, we encounter two different sets. One set consists of connected but disjoint diagrams (see Figure 10.7). We emphasize that these diagrams *are* connected, since every solid line is connected to one of the exterior points at τ and 0.

Upon inspecting any connected, disjoint diagram of order n, we observe that it results from two sets of contractions. One set involves creation and annihilation operators in $\hat{n}(\mathbf{q},\tau)$ and r $\hat{V}$-operators $\hat{V}(\tau_{i_1}),\ldots,\hat{V}(\tau_{i_r})$ being contracted among themselves. The other set involves contractions among $\hat{n}(-\mathbf{q},0)$ and the remaining $s=n-r$ $\hat{V}$-operators. For example, in diagram (b) of Figure 10.7, the operators in $\hat{n}(\mathbf{q},\tau)$ and $\hat{V}(\tau_1)$ are contracted among themselves, while the creation and annihilation operators in $\hat{n}(-\mathbf{q},0)$ are contracted together. Thus, for diagram (b), $n=1$, $r=1$, and $s=0$. For diagram (c), $n=1$, $r=0$, and $s=1$, while for diagram (a), $n=r=s=0$. Since all such diagrams must be summed, and since there are $n!/r!s!$ ways of choosing r $\hat{V}$-operators from among n $\hat{V}$-operators, the contribution of all the connected, disjoint diagrams (*cdd*) on the RHS of Eq. (10.19) is given by

$$cdd = -\frac{1}{V}\sum_{n=0}^{\infty}\sum_{r=0}^{n}\frac{1}{n!}\left(-\frac{1}{\hbar}\right)^{n}\frac{n!}{r!s!}\int_0^{\beta\hbar}d\tau_1\ldots\int_0^{\beta\hbar}d\tau_r\langle T\hat{n}(\mathbf{q},\tau)\hat{V}(\tau_1)\ldots\hat{V}(\tau_r)\rangle_{0,c}$$
$$\times\int_0^{\beta\hbar}d\tau_{r+1}\ldots\int_0^{\beta\hbar}d\tau_n\langle T\hat{n}(-\mathbf{q},0)\hat{V}(\tau_{r+1})\ldots\hat{V}(\tau_n)\rangle_{0,c}.$$

The above expression results from the fact that the arrangement of the density and interaction operators in the time-ordered product is immaterial; since each of these operators consists of an even number of fermion operators, no minus sign is incurred upon any reordering of the operators. Since $s=n-r$, the above expression may be recast in the following form:

$$cdd = -\frac{1}{V}\sum_{n=0}^{\infty}\sum_{r=0}^{\infty}\sum_{s=0}^{\infty}\left(-\frac{1}{\hbar}\right)^{n}\frac{1}{r!s!}\delta_{n,r+s}\int_0^{\beta\hbar}d\tau_1\ldots\int_0^{\beta\hbar}d\tau_r$$
$$\langle T\hat{n}(\mathbf{q},\tau)\hat{V}(\tau_1)\ldots\hat{V}(\tau_r)\rangle_{0,c}\int_0^{\beta\hbar}d\tau_1\ldots\int_0^{\beta\hbar}d\tau_s\langle T\hat{n}(-\mathbf{q},0)\hat{V}(\tau_1)\ldots\hat{V}(\tau_s)\rangle_{0,c}.$$

The Kronecker delta ensures that $r + s = n$, allowing the summation over r and s to extend to infinity. Summing over n first has the effect of removing the Kronecker delta and replacing $(-1/\hbar)^n$ by $(-1/\hbar)^r(-1/\hbar)^s$. Hence,

$$
\begin{aligned}
cdd &= -\frac{1}{V}\sum_r \frac{1}{r!}\left(-\frac{1}{\hbar}\right)^r \int_0^{\beta\hbar} d\tau_1 \dots \int_0^{\beta\hbar} d\tau_r \, \langle T\hat{n}(\mathbf{q},\tau)\hat{V}(\tau_1)\dots\hat{V}(\tau_r)\rangle_{0,c} \\
&\quad \times \sum_s \frac{1}{s!}\left(-\frac{1}{\hbar}\right)^s \int_0^{\beta\hbar} d\tau_1 \dots \int_0^{\beta\hbar} d\tau_s \, \langle T\hat{n}(-\mathbf{q},0)\hat{V}(\tau_1)\dots\hat{V}(\tau_s)\rangle_{0,c} \\
&= -\frac{1}{V}\langle T\, n_{\bar{H}}(\mathbf{q},\tau)\rangle\langle T\, n_{\bar{H}}(-\mathbf{q},0)\rangle = -\frac{1}{V}\langle n_{\bar{H}}(\mathbf{q},\tau)\rangle\langle n_{\bar{H}}(-\mathbf{q},0)\rangle.
\end{aligned}
\tag{10.20}
$$

Thus, the sum over all connected, disjoint diagrams, which appears on the RHS of Eq. (10.19), exactly cancels the second term on the LHS of that equation. We conclude that $D(\mathbf{q},\tau)$ is the sum of all connected (c), nondisjoint (nd) diagrams that result from the expansion on the RHS of Eq. (10.19). The zeroth order contribution to $D(\mathbf{q},\tau)$ is thus

$$
\begin{aligned}
D^0(\mathbf{q},\tau) &= -\frac{1}{V}\sum_{\mathbf{k}\sigma}\sum_{\mathbf{k}'\sigma'}\langle T\,\hat{c}^\dagger_{\mathbf{k}\sigma}(\tau)\hat{c}_{\mathbf{k}+\mathbf{q}\sigma}(\tau)\hat{c}^\dagger_{\mathbf{k}'\sigma'}(0)\hat{c}_{\mathbf{k}'-\mathbf{q}\sigma'}(0)\rangle_{0,c,nd} \\
&= \frac{1}{V}\sum_{\mathbf{k}\sigma}\sum_{\mathbf{k}'\sigma'}\langle T\,\hat{c}_{\mathbf{k}+\mathbf{q}\sigma}(\tau)\hat{c}^\dagger_{\mathbf{k}'\sigma'}(0)\rangle_0\,\langle T\,\hat{c}_{\mathbf{k}'-\mathbf{q}\sigma'}(0)\hat{c}^\dagger_{\mathbf{k}\sigma}(\tau)\rangle_0 \\
&= \frac{1}{V}\sum_{\mathbf{k}\sigma} g^0(\mathbf{k}+\mathbf{q}\sigma,\tau)g^0(\mathbf{k}\sigma,-\tau).
\end{aligned}
$$

Going to the frequency domain,

$$
\frac{1}{\beta\hbar}\sum_m D^0(\mathbf{q},\omega_m)e^{-i\omega_m\tau} = \frac{1}{V(\beta\hbar)^2}\sum_{\mathbf{k}\sigma nn'} g^0(\mathbf{k}+\mathbf{q}\sigma,\omega_{n'})g^0(\mathbf{k}\sigma,\omega_n)e^{-i(\omega_{n'}-\omega_n)\tau}.
$$

Therefore, $\omega_m = \omega_{n'} - \omega_n$, and

$$
D^0(\mathbf{q},\omega_m) = \frac{1}{\beta\hbar V}\sum_{\mathbf{k}\sigma}\sum_n g^0(\mathbf{k}+\mathbf{q}\sigma,\omega_n+\omega_m)g^0(\mathbf{k}\sigma,\omega_n) = \hbar\Pi^0(\mathbf{q},\omega_m).
\tag{10.21}
$$

As expected, $(1/\hbar)D^0(\mathbf{q},\omega_m)$ is the bare pair bubble, which is the zeroth order connected, nondisjoint diagram. In order to calculate $D(\mathbf{q},\omega_m)$, we sum over all connected, nondisjoint diagrams. This is carried out in Figure 10.8. In the random phase approximation,

$$
D(\mathbf{q},\omega_m) = \hbar\Pi_{\text{RPA}}(\mathbf{q},\omega_m).
\tag{10.22}
$$

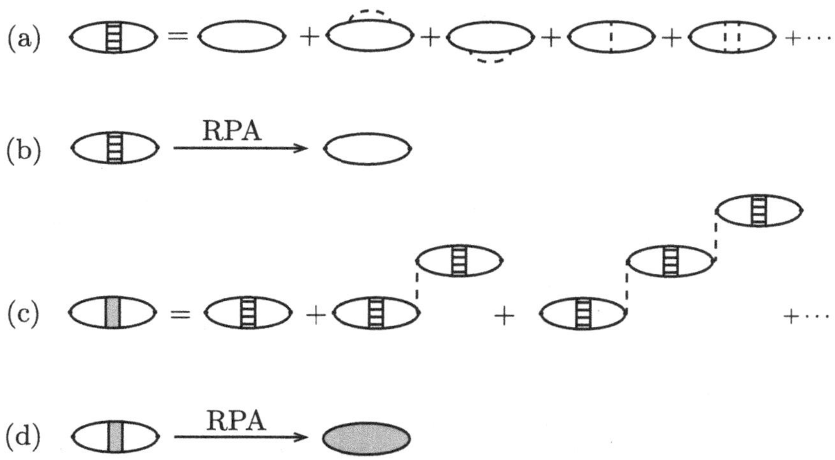

(a) = + + + + + ⋯

(b) RPA

(c) = + + +⋯

(d) RPA

Figure 10.8 (a) The irreducible bubble is obtained by summing diagrams containing one bubble with all interaction lines connected to its legs. (b) In random phase approximation, the irreducible bubble is replaced by the bare bubble. (c) Collective electronic density fluctuations. (d) In random phase approximation, the collective electronic density fluctuations are given by the dressed bubble.

The retarded correlation function, in RPA, is thus given by

$$D^R(\mathbf{q},\omega) = \hbar\Pi_{\text{RPA}}(\mathbf{q}, i\omega_m \to \omega + i0^+). \tag{10.23}$$

10.9 How do electrons interact?

As we saw earlier, in first-order perturbation theory, the proper self energy arises from the exchange term, while in higher orders ($n \geq 2$), the dominant contribution to the proper self energy arises from direct processes involving $n-1$ bubbles. In higher orders, many diagrams involving only exchange interactions, or both Coulomb direct and exchange interactions, make contributions to the self energy; these contributions are dominated by those which arise from purely direct processes. In order to come up with a reasonable classification scheme in which diagrams are classified as either Coulomb direct or exchange diagrams, we take a closer look at proper self energy diagrams of up to third order that involve one or more exchange interactions (Figure 10.9). In third order, only diagrams containing a pair bubble are retained, since these diagrams have a higher degree of divergence than diagrams which contain three purely exchange interactions.

An examination of diagrams (a), (b), and (d), shows that they are parts of a single diagram, similar to diagram (a), but with $g^0(\mathbf{k}-\mathbf{q}\sigma, \omega_n - \omega_m)$ replaced by $g(\mathbf{k}-\mathbf{q}\sigma, \omega_n - \omega_m)$, as shown in Figure 10.10a. Even though this diagram contains both Coulomb direct and exchange interactions, we classify it as an

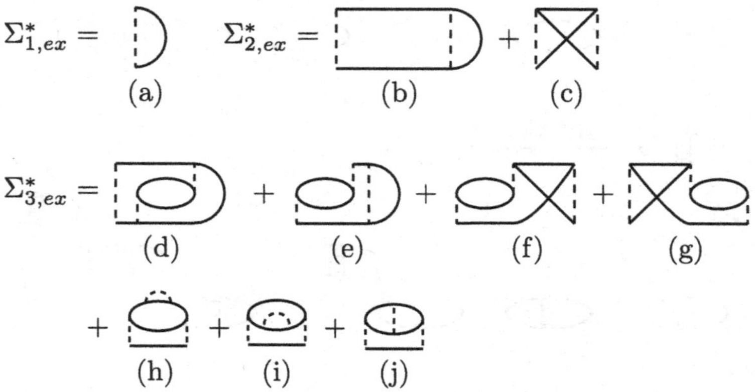

Figure 10.9 Proper selfenergy diagrams involving at least one exchange interaction. In third order, we show only diagrams containing one bubble.

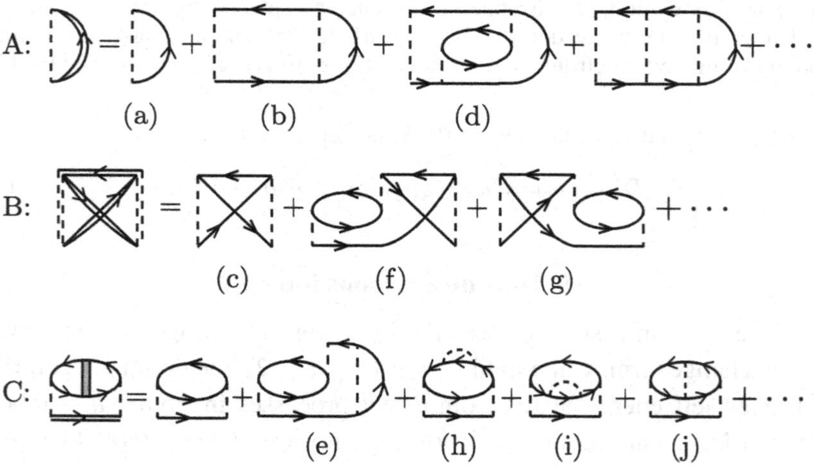

Figure 10.10 Classification of diagrams: (A) and (B) are classified as exchange diagrams, while diagrams in (C) are classified as Coulomb direct diagrams.

exchange diagram. Diagrams (c), (f), and (g) are also parts of diagram 10.10B, so we classify them as exchange diagrams as well. However, diagrams (e), (h), (i), and (j) are part of diagram 10.10C, so they are classified as Coulomb direct diagrams.

According to the above discussion, we may classify self energy diagrams of an interacting electron gas as Coulomb direct or as exchange diagrams. This is summarized in Figure 10.11.

Let us now take up the question of how electrons interact. The conventional picture is that, since electron–electron scattering is a pairwise interaction, electrons

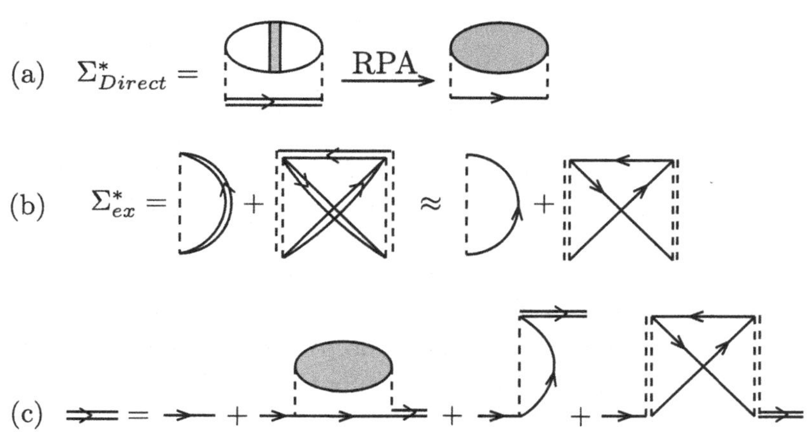

Figure 10.11 (a) Proper self energy resulting from Coulomb direct processes. (b) Proper self energy arising from exchange processes. Replacing the dressed Green's functions with bare ones amounts to retaining only the dominant diagrams at each order of interaction. (c) The dressed Green's function.

scatter off each other directly and in a pairwise manner. Allowance is made, however, for the collective motion of a dense electron gas by assuming that the pairwise scattering potential is screened by the electronic dielectric function. Figure 10.11, however, suggests an alternative, more subtle picture, namely that, in the dominant Coulomb direct interaction processes, an electron can scatter off the fluctuating potential generated by the collective electronic density fluctuations. This scattering is caused by the bare fluctuating potential, as seen in Figure 10.11. The conventional picture also holds, but only for the exchange scattering processes of order $n \geq 2$, as Figure 10.11 shows; these exchange processes constitute only a small correction to the dominant form (Das and Jishi, 1990).

10.10 Dielectric function

The dielectric function was introduced in Section 6.8 as a measure of the response of a system to an external electric potential. For an electron gas, we found that

$$\varepsilon(\mathbf{q}, \omega) = \left[1 + \frac{1}{\hbar} v_q D^R(\mathbf{q}, \omega)\right]^{-1} \tag{10.24}$$

where $D^R(\mathbf{q}, \omega)$ is the retarded density-density correlation function. In random phase approximation, $D^R(\mathbf{q}, \omega) = \hbar \Pi_{\text{RPA}}(\mathbf{q}, \omega)$. Using Eq. (10.13),

$$\varepsilon^{RPA}(\mathbf{q}, \omega) = \left[1 + \frac{v_\mathbf{q} \Pi^o(\mathbf{q}, \omega)}{1 - v_\mathbf{q} \Pi^o(\mathbf{q}, \omega)}\right]^{-1} = 1 - v_\mathbf{q} \Pi^o(\mathbf{q}, \omega). \tag{10.25}$$

The screened Coulomb interaction, given in Eq. (10.15), is

$$V(\mathbf{q}, \omega_m) = \frac{v_\mathbf{q}}{1 - v_\mathbf{q}\Pi^o(\mathbf{q}, \omega_m)} = \frac{v_\mathbf{q}}{\varepsilon^{RPA}(\mathbf{q}, \omega_m)}. \tag{10.26}$$

The dielectric function is thus a measure of both the screening of the Coulomb potential in an interacting electron gas, and the response of an interacting electron gas to an external electric potential

To obtain an expression in closed form for the dielectric function, we need to carry out the sum over $\mathbf{k}$ in the expression for $\Pi^o(\mathbf{q}, \omega)$, given in Eq. (10.14). At high temperatures, this is very hard to do. At low temperatures ($k_B T \ll \epsilon_F$, where ϵ_F is the Fermi energy), an expression for $\Pi^0(\mathbf{q}, \omega)$ is not difficult to obtain. $\Pi^0(\mathbf{q}, \omega)$ is given by $\Pi^0(\mathbf{q}, i\omega_m \to \omega + i0^+)$:

$$\begin{aligned}\Pi^0(\mathbf{q}, \omega) &= \frac{2}{V}\sum_\mathbf{k} \frac{f_\mathbf{k} - f_{\mathbf{k}+\mathbf{q}}}{\hbar\omega + \epsilon_\mathbf{k} - \epsilon_{\mathbf{k}+\mathbf{q}} + i0^+} \\ &= \frac{2}{V}\sum_\mathbf{k} \frac{f_\mathbf{k}}{\hbar\omega + \epsilon_\mathbf{k} - \epsilon_{\mathbf{k}+\mathbf{q}} + i0^+} + \frac{2}{V}\sum_\mathbf{k} \frac{f_{\mathbf{k}+\mathbf{q}}}{-\hbar\omega + \epsilon_{\mathbf{k}+\mathbf{q}} - \epsilon_\mathbf{k} - i0^+}.\end{aligned}$$

Replacing $\mathbf{k}$ by $-\mathbf{k} - \mathbf{q}$ in the second term, and noting that $f_{-\mathbf{k}} = f_\mathbf{k}$ and $\epsilon_{-\mathbf{k}} = \epsilon_\mathbf{k} = \hbar^2 k^2/2m$,

$$\Pi^0(\mathbf{q}, \omega) = \frac{2}{V}\sum_\mathbf{k} \frac{f_\mathbf{k}}{\hbar\omega + \epsilon_\mathbf{k} - \epsilon_{\mathbf{k}+\mathbf{q}} + i0^+} + \frac{2}{V}\sum_\mathbf{k} \frac{f_\mathbf{k}}{-\hbar\omega + \epsilon_\mathbf{k} - \epsilon_{\mathbf{k}+\mathbf{q}} - i0^+}. \tag{10.27}$$

First we evaluate the real part of $\Pi^0(\mathbf{q}, \omega)$,

$$\begin{aligned}Re\,\Pi^0(\mathbf{q}, \omega) &= \frac{2}{V}\sum_\mathbf{k} \frac{f_\mathbf{k}}{\hbar\omega + \epsilon_\mathbf{k} - \epsilon_{\mathbf{k}+\mathbf{q}}} + \frac{2}{V}\sum_\mathbf{k} \frac{f_\mathbf{k}}{-\hbar\omega + \epsilon_\mathbf{k} - \epsilon_{\mathbf{k}+\mathbf{q}}} \\ &\equiv A(\mathbf{q}, \omega) + A(\mathbf{q}, -\omega).\end{aligned} \tag{10.28}$$

For $k_B T \ll \epsilon_F$, we may replace $f_\mathbf{k}$ by the step function $\theta(\epsilon_F - \epsilon_\mathbf{k})$,

$$A(\mathbf{q}, \omega) = \frac{2}{V}\frac{V}{(2\pi)^3}\int d^3k \frac{1}{\hbar\omega - (\hbar^2/m)\mathbf{k}\cdot\mathbf{q} - (\hbar^2/2m)q^2}$$

where the integration is over the Fermi sphere. Replacing $\mathbf{k}\cdot\mathbf{q}$ by $kq\cos\theta$, $\int d^3k$ by $2\pi \int_0^{k_F} k^2 dk \int_{-1}^{1} d\cos\theta$, and defining $x = k/k_F$, the integration over $\cos\theta$ is first carried out; it yields

$$A(\mathbf{q}, \omega) = \frac{mk_F^2}{2\pi^2\hbar^2 q}\int_0^1 x \ln\left|\frac{x + u_-}{x - u_-}\right| dx$$

where $u_\pm = \omega/qv_F \pm q/2k_F$, and $v_F = \hbar k_F/m$ is the Fermi velocity. Using

$$\int x \ln|x+b|dx = \frac{x^2-b^2}{2}\ln|x+b| - \frac{1}{4}(x-b)^2, \tag{10.29}$$

we obtain

$$A(\mathbf{q},\omega) = d(\epsilon_F)\frac{k_F}{2q}\left[u_- + \frac{1-u_-^2}{2}\ln\left|\frac{1+u_-}{1-u_-}\right|\right], \tag{10.30}$$

where $d(\epsilon_F) = mk_F/\pi^2\hbar^2$ is the density of states, per unit volume, at the Fermi surface (see Problem 2.2). Equations (10.28) and (10.30) give

$$Re\,\Pi^o(\mathbf{q},\omega) = -d(\epsilon_F)\left[\frac{1}{2} - \frac{1-u_-^2}{4q/k_F}\ln\left|\frac{1+u_-}{1-u_-}\right| + \frac{1-u_+^2}{4q/k_F}\ln\left|\frac{1+u_+}{1-u_+}\right|\right]. \tag{10.31}$$

In the static limit, $\omega = 0$, $u_\pm = \pm q/2k_F$, and Eq. (10.31) reduces to

$$Re\,\Pi^0(\mathbf{q},0) = -d(\epsilon_F)\left[\frac{1}{2} + \frac{4k_F^2-q^2}{8k_Fq}\ln\left|\frac{2k_F+q}{2k_F-q}\right|\right] = -d(\epsilon_F)g(q') \tag{10.32}$$

where $q' = q/2k_F$, and

$$g(q') = \frac{1}{2} + \frac{1-q'^2}{4q'}\ln\left|\frac{1+q'}{1-q'}\right|. \tag{10.33}$$

Next, we evaluate the imaginary part of $\Pi^0(\mathbf{q},\omega)$. From Eq. (10.27), we find

$$Im\ \Pi^0(\mathbf{q},\omega) = B_1(\mathbf{q},\omega) + B_2(\mathbf{q},\omega) \tag{10.34}$$

where

$$B_1(\mathbf{q},\omega) = -\frac{2\pi}{V}\sum_{\mathbf{k}} f_{\mathbf{k}}\,\delta(\hbar\omega + \epsilon_{\mathbf{k}} - \epsilon_{\mathbf{k}+\mathbf{q}}), \quad B_2(\mathbf{q},\omega) = -B_1(\mathbf{q},-\omega). \tag{10.35}$$

In writing $B_1(\mathbf{q},\omega)$ and $B_2(\mathbf{q},\omega)$, we have used $Im(\frac{1}{x\pm i0^+}) = \mp\pi\delta(x)$. The Dirac-delta function is given by

$$\begin{aligned}\delta(\hbar\omega + \epsilon_{\mathbf{k}} - \epsilon_{\mathbf{k}+\mathbf{q}}) &= \delta(\hbar\omega - \frac{\hbar^2}{m}kq\cos\theta - \frac{\hbar^2q^2}{2m})\\ &= \frac{m}{\hbar^2kq}\delta\left(\frac{\omega}{\hbar kq/m} - \frac{q}{2k} - \cos\theta\right).\end{aligned} \tag{10.36}$$

Replacing f_k by $\theta(\epsilon_F - \epsilon_{\mathbf{k}})$, and $\sum_{\mathbf{k}}$ by $\frac{V}{(2\pi)^3}2\pi\int k^2dk\int_{-1}^{1}d\cos\theta$, we obtain

$$B_1(\mathbf{q},\omega) = -\frac{m}{2\pi\hbar^2q}\int_0^{k_F}kdk\int_{-1}^{1}d\cos\theta\,\delta\left(\frac{\omega}{\hbar kq/m} - \frac{q}{2k} - \cos\theta\right). \tag{10.37}$$

The integral over $\cos\theta$ vanishes if $\left(\frac{\omega}{\hbar kq/m} - \frac{q}{2k}\right)^2 > 1$, and is equal to unity if $\left(\frac{\omega}{\hbar kq/m} - \frac{q}{2k}\right)^2 < 1$; hence, it is the step function $\theta\left[1 - \left(\frac{\omega}{\hbar kq/m} - \frac{q}{2k}\right)^2\right]$. Thus,

$$
\begin{aligned}
B_1(\mathbf{q}, \omega) &= -\frac{m}{2\pi\hbar^2 q}\int_0^{k_F} k\,\theta\left[1 - \left(\frac{\omega}{\hbar kq/m} - \frac{q}{2k}\right)^2\right] dk \\
&= -\frac{m}{2\pi\hbar^2 q}\int_0^{k_F} k\,\theta\left[1 - \frac{k_F^2}{k^2}\left(\frac{\omega}{qv_F} - \frac{q}{2k_F}\right)^2\right] dk \\
&= -\frac{mk_F^2}{2\pi\hbar^2 q}\int_0^{1} x\,\theta(1 - u_-^2/x^2)dx \qquad (10.38)
\end{aligned}
$$

where a change of variable from k to $x = k/k_F$ is made. If $u_-^2 > 1$, then certainly u_-^2/x^2 will be greater than 1 (since x varies from 0 to 1) and the integral will vanish. On the other hand, if $u_-^2 < 1$, the integral is nonvanishing,

$$
\int_0^1 x\,\theta(1 - u_-^2/x^2)dx = \begin{cases} 0 & u_-^2 > 1 \\ \int_{|u_-|}^1 x dx = (1 - u_-^2)/2 & u_-^2 < 1. \end{cases} \qquad (10.39)
$$

Therefore,

$$
B_1(\mathbf{q}, \omega) = -\frac{mk_F^2}{4\pi\hbar^2 q}(1 - u_-^2)\theta(1 - u_-^2) = -d(\epsilon_F)\frac{\pi k_F}{4q}(1 - u_-^2)\theta(1 - u_-^2) \qquad (10.40)
$$

$$
B_2(\mathbf{q}, \omega) = -B_1(\mathbf{q}, -\omega) = d(\epsilon_F)\frac{\pi k_F}{4q}(1 - u_+^2)\theta(1 - u_+^2) \qquad (10.41)
$$

$$
Im\,\Pi^0(\mathbf{q}, \omega) = -d(\epsilon_F)\frac{\pi k_F}{4q}\left[(1 - u_-^2)\theta(1 - u_-^2) - (1 - u_+^2)\theta(1 - u_+^2)\right]. \qquad (10.42)
$$

In the static limit, $\omega = 0$ and $u_-^2 = u_+^2$; hence, $Im\,\Pi^o(\mathbf{q}, 0) = 0$. To summarize, we collect below the results for $\Pi^o(\mathbf{q}, \omega)$ at very low temperatures:

$$
\begin{aligned}
Re\,\Pi^0(\mathbf{q}, \omega) &= -d(\epsilon_F)\left[\frac{1}{2} - \frac{1 - u_-^2}{4q/k_F}\ln\left|\frac{1 + u_-}{1 - u_-}\right| + \frac{1 - u_+^2}{4q/k_F}\ln\left|\frac{1 + u_+}{1 - u_+}\right|\right] \\
Re\,\Pi^0(\mathbf{q}, 0) &= -d(\epsilon_F)\left[\frac{1}{2} + \frac{4k_F^2 - q^2}{8k_F q}\ln\left|\frac{2k_F + q}{2k_F - q}\right|\right] \\
Im\,\Pi^0(\mathbf{q}, \omega) &= -d(\epsilon_F)\frac{\pi k_F}{4q}[(1 - u_-^2)\theta(1 - u_-^2) - (1 - u_+^2)\theta(1 - u_+^2)] \\
Im\,\Pi^0(\mathbf{q}, 0) &= 0 \\
u_\pm &= \omega/qv_F \pm q/2k_F, \quad d(\epsilon_F) = mk_F/\pi^2\hbar^2.
\end{aligned}
$$

The static dielectric function is given by

$$\varepsilon(\mathbf{q},0) = 1 - v_{\mathbf{q}}\Pi^0(\mathbf{q},0) = 1 + \frac{4\pi e^2 d(\epsilon_F)}{q^2}\left[\frac{1}{2} + \frac{4k_F^2 - q^2}{8k_F q}\ln\left|\frac{2k_F + q}{2k_F - q}\right|\right]. \tag{10.43}$$

This is known as the Lindhard dielectric function (Lindhard, 1954).

10.10.1 Thomas–Fermi screening model

In the Thomas–Fermi model (Thomas, 1927; Fermi, 1927), the dielectric function $\varepsilon(q,\omega)$ is replaced by its value in the static, long wavelength limit,

$$\varepsilon_{\text{TF}}(\mathbf{q},\omega) = \lim_{\mathbf{q}\to 0}\varepsilon^{RPA}(\mathbf{q},0) = \lim_{\mathbf{q}\to 0}[1 - v_{\mathbf{q}}\Pi^0(\mathbf{q},0)]. \tag{10.44}$$

Since $Im\,\Pi^0(\mathbf{q},0) = 0$, $\Pi^0(\mathbf{q},0)$ is real. Using $\lim_{x\to 0}\ln|1+x| = x$, we find that $\lim_{\mathbf{q}\to 0}\Pi^0(\mathbf{q},0) = -d(\epsilon_F)$. Hence,

$$\varepsilon_{\text{TF}}(\mathbf{q},\omega) = 1 + d(\epsilon_F)v_{\mathbf{q}}. \tag{10.45}$$

The screened Coulomb interaction in the Thomas–Fermi model is given by

$$V_{\text{TF}}(\mathbf{q},\omega) = \frac{4\pi e^2}{q^2[1 + 4\pi e^2 d(\epsilon_F)/q^2]} = \frac{4\pi e^2}{q^2 + q_{\text{TF}}^2} \tag{10.46}$$

where

$$q_{\text{TF}}^2 = 4\pi e^2 d(\epsilon_F) \tag{10.47}$$

is the square of the Thomas–Fermi wave number. At low temperatures, where electrons occupy the states below the Fermi surface, $d(\varepsilon_F) = mk_F/\pi^2\hbar^2 = k_F/\pi^2 e^2 a_0$, where a_0 is the Bohr radius. Thus, $q_{\text{TF}}^2 = 4k_F/\pi a_0$. Since $k_F \sim 1$ Å^{-1} in metals, we find that $q_{\text{TF}} \sim 1$ Å^{-1}.

In the Thomas–Fermi model, the screened Coulomb interaction in real space is the inverse Fourier transform of V_{TF},

$$v(\mathbf{r}_1 - \mathbf{r}_2) = \frac{e^2}{|\mathbf{r}_1 - \mathbf{r}_2|}e^{-q_{\text{TF}}|\mathbf{r}_1 - \mathbf{r}_2|}. \tag{10.48}$$

Suppose that an impurity of charge Ze is placed in a metal at the origin. The bare Coulomb potential produced by the charged impurity is

$$V_{\text{bare}}(\mathbf{r}) = Ze/r = \frac{4\pi Ze}{(2\pi)^3}\int \frac{1}{q^2}e^{i\mathbf{q}\cdot\mathbf{r}}d^3q. \tag{10.49}$$

In writing the above equation, we have used $1/r = (1/V)\sum_{\mathbf{q}}(4\pi/q^2)e^{i\mathbf{q}.\mathbf{r}}$. Since the charge is static ($\omega = 0$), the screened Coulomb potential produced by the impurity is

$$V_{sc}(\mathbf{r}) = \frac{4\pi Ze}{(2\pi)^3} \int \frac{e^{i\mathbf{q}\cdot\mathbf{r}}}{q^2\varepsilon(\mathbf{q},0)} d^3q. \tag{10.50}$$

The difference between the two potentials is caused by the induced charge density ρ_{ind} in the medium; hence, Poisson's equation gives

$$4\pi\rho_{\text{ind}}(\mathbf{r}) = -\nabla^2[V_{sc}(\mathbf{r}) - V_{\text{bare}}(\mathbf{r})]. \tag{10.51}$$

Using Eqs. (10.49) and (10.50), along with $\nabla^2 e^{i\mathbf{q}\cdot\mathbf{r}} = -q^2 e^{i\mathbf{q}\cdot\mathbf{r}}$, we obtain

$$\rho_{\text{ind}}(\mathbf{r}) = \frac{Ze}{(2\pi)^3} \int \left[\frac{1}{\varepsilon(\mathbf{q},0)} - 1\right] e^{i\mathbf{q}\cdot\mathbf{r}} d^3q. \tag{10.52}$$

The total induced charge is

$$\begin{aligned} Q_{\text{ind}} &= \int \rho_{\text{ind}}(\mathbf{r}) d^3r = Ze \int d^3q \left[\frac{1}{\varepsilon(\mathbf{q},0)} - 1\right] \frac{1}{(2\pi)^3} \int e^{i\mathbf{q}\cdot\mathbf{r}} d^3r \\ &= Ze \int d^3q \left[\frac{1}{\varepsilon(\mathbf{q},0)} - 1\right] \delta(\mathbf{q}) = Ze \left[\frac{1}{\varepsilon(0,0)} - 1\right]. \end{aligned} \tag{10.53}$$

Since $\varepsilon(0,0) = 1 + d(\epsilon_F)v_{\mathbf{q}=0} = \infty$, it follows that $Q_{\text{ind}} = -Ze$; the screening of the charge impurity is complete. This result is reasonable. However, there is a defect in the Thomas–Fermi model, namely that $\rho_{\text{ind}}(\mathbf{r})$ diverges at $\mathbf{r} = 0$. Using $V_{sc}(\mathbf{r}) = (Ze/r)e^{-q_{\text{TF}}r}$, $V_{\text{bare}}(\mathbf{r}) = Ze/r$, and $\nabla^2 = (1/r)\partial^2/\partial r^2\, r$, Eq. (10.51) gives

$$\rho_{\text{ind}}(\mathbf{r}) = -ed(\epsilon_F)\frac{Ze^2}{r} e^{-q_{\text{TF}}r}, \tag{10.54}$$

which is infinite at $r = 0$. Significantly, no such singularities are observed in experiments that probe the electronic density near charged impurities. This defect is remedied by using the Lindhard dielectric function (see Eq. [10.43]) instead of the Thomas–Fermi dielectric function.

10.11 Plasmons and Landau damping

A dense electron gas is capable of supporting high-frequency longitudinal oscillatory modes known as plasmons. They can be observed when energetic electrons scatter from a metallic crystal. When an energetic electron strikes a metal, it may excite a plasmon, whose energy is ~ 10 eV; the scattered electron would then be downshifted in energy by an equal amount relative to the incident electron.

A classical treatment illustrates how plasmons can be formed. In the jellium model, consider a small time-dependent density fluctuation: each electron at $\mathbf{r}$ is given a small displacement $\mathbf{u}(\mathbf{r}, t)$. In an infinitesimal volume d^3r centered on $\mathbf{r}$,

nd^3r electrons are each displaced by $u(\mathbf{r}, t)$, where $n = N/V$ is the electron number density at equilibrium. The induced dipole moment in d^3r is $-ne\mathbf{u}(\mathbf{r}, t)d^3r$; hence, the induced polarization in the medium (the dipole moment per unit volume) is $\mathbf{P}(\mathbf{r}, t) = -en\mathbf{u}(\mathbf{r}, t)$. The induced charge density is

$$\rho_{\text{ind}}(\mathbf{r}, t) = -\nabla.\mathbf{P} = ne\nabla.\mathbf{u}(\mathbf{r}, t). \tag{10.55}$$

If we Fourier-expand $\mathbf{u}(\mathbf{r}, t)$:

$$\mathbf{u}(\mathbf{r}, t) = \frac{1}{V}\sum_{\mathbf{q}} \mathbf{u}_{\mathbf{q}}(t)e^{i\mathbf{q}\cdot\mathbf{r}} \tag{10.56}$$

then

$$\nabla.\mathbf{u} = \frac{i}{V}\sum_{\mathbf{q}} \mathbf{q}.\mathbf{u}_{\mathbf{q}}(t)e^{i\mathbf{q}\cdot\mathbf{r}} \tag{10.57}$$

which is nonzero for longitudinal modes ($\mathbf{q}\|\mathbf{u}_{\mathbf{q}}$). The induced electric field is, by Gauss's law,

$$\nabla.\mathbf{E} = 4\pi\rho_{\text{ind}} = 4\pi ne\nabla.\mathbf{u}(\mathbf{r}, t). \tag{10.58}$$

The above equation is to be solved subject to the boundary condition that $\mathbf{E} = 0$ if $\mathbf{u} = 0$; hence $\mathbf{E} = 4\pi ne\mathbf{u}(\mathbf{r}, t)$. Newton's second law now gives

$$m\ddot{\mathbf{u}} = -e\mathbf{E} \Rightarrow \ddot{\mathbf{u}} = (-4\pi ne^2/m)\mathbf{u}. \tag{10.59}$$

Thus, the motion of the electrons is oscillatory, with a frequency of

$$\omega_p = (4\pi ne^2/m)^{1/2}, \tag{10.60}$$

which is the plasmon frequency. For metals, $\hbar\omega_p = 10$–20 eV.

10.11.1 Plasmons

From a quantum mechanical point of view, the retarded correlation function is

$$C^R_{AB}(\omega) = \hbar\, Z_G^{-1} \sum_{n,m} \frac{\langle n|A|m\rangle\langle m|B|n\rangle(e^{-\beta\bar{E}_n} \mp e^{-\beta\bar{E}_m})}{\hbar\omega - (\bar{E}_m - \bar{E}_n) + i0^+} \tag{10.61}$$

(see Eq. [6.47]). The $+(-)$ sign corresponds to the occurrence of fermionic (bosonic) operators A and B. Setting

$$A = \tilde{n}_{\mathbf{q}} = n_{\mathbf{q}} - \langle n_{\mathbf{q}}\rangle, \quad B = \tilde{n}_{-\mathbf{q}} = n_{-\mathbf{q}} - \langle n_{-\mathbf{q}}\rangle$$

and noting that $n_{-\mathbf{q}} = \sum_{\mathbf{k}\sigma} c^\dagger_{\mathbf{k}\sigma} c_{\mathbf{k}-\mathbf{q}} = \sum_{\mathbf{k}\sigma} c^\dagger_{\mathbf{k}+\mathbf{q}} c_{\mathbf{k}} = n^\dagger_{\mathbf{q}}$, and that the density operator is bosonic, we obtain the retarded density–density correlation function

$$D^R(\mathbf{q}, \omega) = \hbar Z_G^{-1} \sum_{n,m} \frac{|\langle m|\tilde{n}_{\mathbf{q}}|n\rangle|^2 (e^{-\beta \bar{E}_n} - e^{-\beta \bar{E}_m})}{\hbar\omega - (\bar{E}_m - \bar{E}_n) + i0^+}. \tag{10.62}$$

As this expression shows, the poles of $D^R(\mathbf{q}, \omega)$ are the excitation energies of the system. Note that these energies are not the excitation energies of an added particle; $\tilde{n}_{\mathbf{q}}$ represents electron density fluctuations, and it conserves the number of particles. The poles of $D^R(\mathbf{q}, \omega)$ are thus the excitation energies of the density fluctuations of the electron gas. In the random phase approximation,

$$D^R(\mathbf{q}, \omega) = \hbar \Pi_{\mathrm{RPA}}(\mathbf{q}, \omega) = \frac{\hbar \Pi^0(\mathbf{q}, \omega)}{1 - v_{\mathbf{q}} \Pi^0(\mathbf{q}, \omega)}$$
$$= \frac{\hbar \Pi^0(\mathbf{q}, \omega)}{1 - v_{\mathbf{q}} Re\, \Pi^0(\mathbf{q}, \omega) - i v_{\mathbf{q}} Im\, \Pi^0(\mathbf{q}, \omega)}. \tag{10.63}$$

The imaginary part of $\Pi^0(\mathbf{q}, \omega)$ gives rise to damping of the excitation modes. To search for well-defined, long-lived excitations, we consider the region where $Im\, \Pi^0(\mathbf{q}, \omega) = 0$. This occurs when $\omega/qv_F > 1 + q/2k_F$ (see Eq. [10.42]). The poles are obtained by setting $1 - v_{\mathbf{q}} Re\, \Pi^o(\mathbf{q}, \omega) = 0$. We evaluate $Re\, \Pi^0(\mathbf{q}, \omega)$ in the low temperature limit ($k_B T \ll \epsilon_F$), long wavelength limit ($q \ll k_F$), and high frequency limit ($\omega \gg qv_F$). The expression for $Re\, \Pi^0(\mathbf{q}, \omega)$, given in Eq. (10.31), can be written as

$$Re\, \Pi^0(\mathbf{q}, \omega) = -d(\epsilon_F) \left[\frac{1}{2} - \frac{1 - (\omega/qv_F - q/2k_F)^2}{4q/k_F} \ln \left| \frac{1 + x_-}{1 - x_+} \right| \right.$$
$$\left. + \frac{1 - (\omega/qv_F + q/2k_F)^2}{4q/k_F} \ln \left| \frac{1 + x_+}{1 - x_-} \right| \right] \tag{10.64}$$

where

$$x_\pm = \frac{qv_F}{\omega}(1 \pm q/2k_F).$$

In the high frequency, long wavelength limit, $x_\pm \ll 1$. By expanding

$$\ln|1 + x| = x - \frac{x^2}{2} + \frac{x^3}{3} - \frac{x^4}{4} + \frac{x^5}{5} - \frac{x^6}{6} + \cdots$$

and carrying out tedious calculations, we find

$$Re\, \Pi^o(\mathbf{q}, \omega) = \frac{n}{m} \left(\frac{q}{\omega} \right)^2 \left[1 + \frac{3}{5} \left(\frac{qv_F}{\omega} \right)^2 + \cdots \right]. \tag{10.65}$$

Another method for obtaining the above result is outlined in Problem 10.4.

The poles of $D^R(\mathbf{q}, \omega)$ are obtained by solving

$$1 - \frac{4\pi ne^2}{m\omega^2}\left[1 + \frac{3}{5}\left(\frac{qv_F}{\omega}\right)^2\right] = 0$$

$$\Rightarrow \omega^2 = \omega_p^2\left[1 + \frac{3}{5}\left(\frac{qv_F}{\omega}\right)^2 + \cdots\right] = \omega_p^2\left[1 + \frac{3}{5}\left(\frac{qv_F}{\omega_p}\right)^2 + \cdots\right]$$

$$\Rightarrow \omega(\mathbf{q}) = \omega_p\left[1 + \frac{3}{10}\left(\frac{v_F}{\omega_p}\right)^2 q^2 + \cdots\right]. \tag{10.66}$$

At $q = 0$, $\omega = \omega_p$; the quantum mechanical treatment reproduces the classical result, and, in addition, yields the dispersion of the plasmon mode.

10.11.2 Landau damping

The plasmon mode is damped if $Im\,\Pi^0(\mathbf{q}, \omega) \neq 0$. From Eq. (10.42), this occurs if $u_-^2 < 1$ or $u_+^2 < 1$. Since $u_+^2 > u_-^2$, it is necessary and sufficient that $u_-^2 < 1$ for $Im\,\Pi^0(\mathbf{q}, \omega)$ to be nonzero:

$$u_-^2 < 1 \Rightarrow -1 < \frac{\omega}{qv_F} - \frac{q}{2k_F} < 1 \Rightarrow -qv_F + \frac{q^2v_F}{2k_F} < \omega < qv_F + \frac{q^2v_F}{2k_F}$$

$$\Rightarrow (q/k_F)^2 - 2q/k_F < \hbar\omega/\epsilon_F < (q/k_F)^2 + 2q/k_F.$$

In the shaded region of the q-ω plane (see Figure 10.12), $Im\,\Pi^0(\mathbf{q}, \omega) \neq 0$. The plasmon mode dispersion is also shown. For $q > q_c$, the plasmon mode is damped, and it becomes difficult to observe due to its short lifetime. This damping is known as Landau damping. The shaded region is the region of single-particle excitations, whereby an electron below the Fermi surface is excited to above the Fermi surface. Outside this region, it is not possible to conserve energy and wave vector in a single-particle excitation process. We can understand the situation as follows. Suppose an external field with wave vector $\mathbf{q}$ and frequency ω impinges on a metal at low temperature. Under what circumstances would it be possible for an electron to absorb momentum $\hbar\mathbf{q}$ and energy $\hbar\omega$ (supplied by the field) that would allow it to move from beneath to above the Fermi surface? For any given $\mathbf{q}$, the maximum energy that can be absorbed corresponds to a transition in which an electron at the Fermi surface with wave vector $\mathbf{k}\|\mathbf{q}$, $|\mathbf{k}| = k_F$, transitions to a state with wave vector $\mathbf{k} + \mathbf{q}$, where $|\mathbf{k} + \mathbf{q}| = k_F + q$ (depicted in Figure 10.13a). The absorbed energy is $\hbar\omega = \hbar^2q^2/2m + \hbar^2k_Fq/m$. If $\hbar\omega > \hbar^2q^2/2m + \hbar^2k_Fq/m$ (corresponding to points to the left of the left-hand parabola in Figure 10.12), then conservation of energy is not possible for any single-particle excitation.

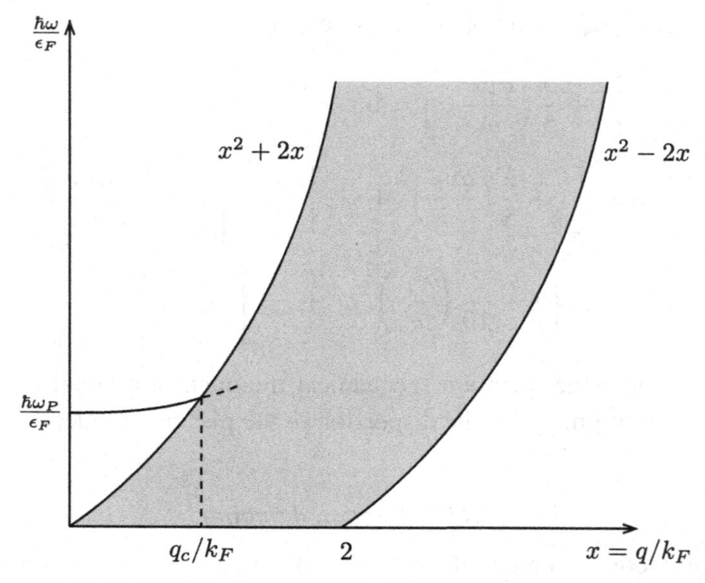

Figure 10.12 Plasmon damping: the shaded region is the region of $\omega - q$ plane where single-particle excitations are possible. For $x > q_c/k_F$, the plasmon decays by exciting electron–hole pairs.

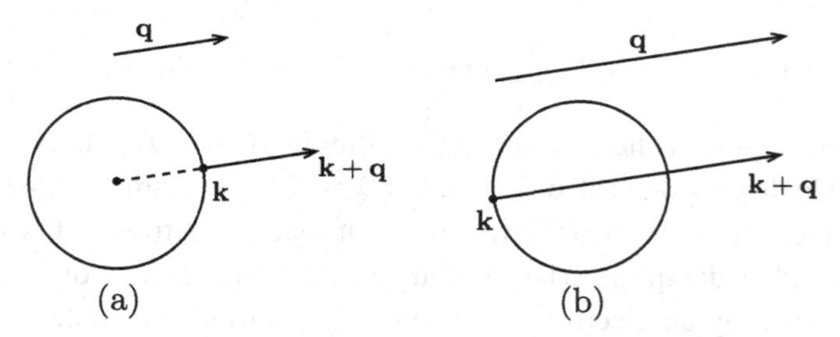

Figure 10.13 (a) A single-particle excitation in which maximum energy is absorbed, and (b) a single-particle excitation in which minimum energy is absorbed.

Similarly, for any given **q**, the minimum energy that can be absorbed in a single-particle excitation corresponds to a situation where an electron at the Fermi surface, having a wave vector **k** in a direction opposite to that of **q**, transitions to a state with wave vector $\mathbf{k} + \mathbf{q}$, $|\mathbf{k} + \mathbf{q}| = q - k_F$ (depicted in Figure 10.13b). The absorbed energy is $\hbar\omega = (\hbar^2/2m)(q^2 - 2k_F q)$. If $\hbar\omega < \hbar^2 q^2/2m - \hbar^2 k_F q/m$ (corresponding to points to the right of the right-hand parabola in Figure 10.12), then no single-particle excitation is possible. Clearly, if $q < 2k_F$, the minimum energy absorbed is zero.

We conclude that if an external field with wave vector $\mathbf{q}$ and frequency ω were to strike a metal, where the point $(q/k_F, \hbar\omega/\epsilon_F)$ lies outside the shaded region shown in Figure 10.12, then the energy and momentum carried by the field could not be absorbed through single-particle excitations. If the field's wave vector and frequency were to match those of the plasmon mode, then the plasmon mode would be excited.

10.12 Case study: dielectric function of graphene

In this section we use results obtained in Problems 2.4, 2.5, 2.6, and 3.6. The reader is advised to study the results of these problems before proceeding with this section. We shall calculate the dielectric function for pure, undoped graphene. A more general treatment that includes doped graphene is also possible (Hwang and Das Sarma, 2007).

There are two valleys in the electronic band structure of graphene, one near point $K = (2\pi/\sqrt{3}a, 2\pi/3a)$ and one near point $K' = (2\pi/\sqrt{3}a, -2\pi/3a)$ in the first Brillouin zone (FBZ). In the vicinity of these points, the energy dispersion is linear:

$$E_{\mathbf{k}} = \pm\hbar v_F k. \tag{10.67}$$

The minus (plus) sign refers to the valence (conduction) band, $\mathbf{k}$ is measured from K (or K'), and $k = |\mathbf{k}|$. In undoped, pure graphene at low temperatures, the valence band is full while the conduction band is empty. We assume that q is small, so that we can ignore intervalley scattering. The dielectric function is given by

$$\varepsilon(q, \omega) = 1 - v_{\mathbf{q}}\Pi^0(\mathbf{q}, \omega) \tag{10.68}$$

where $v_{\mathbf{q}} = 2\pi e^2/q$, since graphene is two-dimensional (see Problem 4.4), and $\Pi^0(\mathbf{q}, \omega) = (1/\hbar)D^{R,0}(\mathbf{q}, \omega)$ is the polarizability of the noninteracting system. First we evaluate $D^0(\mathbf{q}, \omega_m)$, from which the retarded density–density correlation function $D^{R,0}(\mathbf{q}, \omega)$ is obtained by $i\omega_m \to \omega + i0^+$. We have

$$D^0(\mathbf{q}, \tau) = -\frac{1}{A}\langle T\, n(\mathbf{q}, \tau)n(-\mathbf{q}, 0)\rangle_{0,\,\text{conn, nondisjoint}} \tag{10.69}$$

where A is the area of the system. Consider the valley near K (or K'). The number-density operator (see Problem 3.6) is given by

$$n(\mathbf{q}) = \sum_{\mathbf{k}\sigma}\sum_{ss'}\langle\psi^s_{\mathbf{k}}|e^{-i\mathbf{q}\cdot\mathbf{r}}|\psi^{s'}_{\mathbf{k}+\mathbf{q}}\rangle c^\dagger_{s\mathbf{k}\sigma}c_{s'\mathbf{k}+\mathbf{q}\sigma}. \tag{10.70}$$

Here, s and s' are band indices: $s, s' = v$ (valence) or c (conduction). Note that

$$n(\mathbf{r}) = \frac{1}{A}\sum_{\mathbf{q}} n(\mathbf{q})e^{i\mathbf{q}.\mathbf{r}} = \frac{1}{A}\sum_{\mathbf{q}} n(-\mathbf{q})e^{-i\mathbf{q}.\mathbf{r}}. \tag{10.71}$$

Since $n(\mathbf{r})$ is a Hermitian operator,

$$n(\mathbf{r}) = n^\dagger(\mathbf{r}) = \frac{1}{A}\sum_{\mathbf{q}} n^\dagger(\mathbf{q})e^{-i\mathbf{q}\cdot\mathbf{r}} \Rightarrow n^\dagger(\mathbf{q}) = n(-\mathbf{q}). \tag{10.72}$$

Thus,

$$\begin{aligned} D^0(\mathbf{q},\tau) &= -\frac{1}{A}\langle T\, n(\mathbf{q},\tau)n^\dagger(\mathbf{q},0)\rangle_{0,\mathrm{c,nd}} \\ &= -\frac{1}{A}\sum_{\mathbf{k}\sigma ss'}\sum_{\mathbf{k}'\sigma' rr'} \langle\psi^s_{\mathbf{k}}|e^{-i\mathbf{q}\cdot\mathbf{r}}|\psi^{s'}_{\mathbf{k}+\mathbf{q}}\rangle\langle\psi^r_{\mathbf{k}'}|e^{-i\mathbf{q}\cdot\mathbf{r}}|\psi^{r'}_{\mathbf{k}'+\mathbf{q}}\rangle^* \\ &\quad\times \langle T\, c^\dagger_{s\mathbf{k}\sigma}(\tau)c_{s'\mathbf{k}+\mathbf{q}\sigma}(\tau)c^\dagger_{r'\mathbf{k}'+\mathbf{q}\sigma'}(0)c_{r\mathbf{k}'\sigma'}(0)\rangle_{0,\mathrm{c,nd}}. \end{aligned} \tag{10.73}$$

The subscripts c and nd mean connected and nondisjoint, respectively. The τ-ordered product is evaluated by means of Wick's theorem; it is equal to $-\langle T\, c_{r\mathbf{k}'\sigma'}(0)c^\dagger_{s\mathbf{k}\sigma}(\tau)\rangle_0\langle T\, c_{s'\mathbf{k}+\mathbf{q}\sigma}(\tau)c^\dagger_{r'\mathbf{k}'+\mathbf{q}\sigma'}(0)\rangle_0$, which, in turn, is equal to $-g^0(s\mathbf{k}\sigma,-\tau)g^0(s'\mathbf{k}+\mathbf{q}\sigma,\tau)\delta_{sr}\delta_{s'r'}\delta_{\sigma\sigma'}\delta_{\mathbf{k}\mathbf{k}'}$. Hence,

$$D^0(\mathbf{q},\tau) = \frac{1}{A}\sum_{\mathbf{k}\sigma}\sum_{ss'} |\langle\psi^s_{\mathbf{k}}|e^{-i\mathbf{q}\cdot\mathbf{r}}|\psi^{s'}_{\mathbf{k}+\mathbf{q}}\rangle|^2 g^0(s\mathbf{k}\sigma,-\tau)g^0(s'\mathbf{k}+\mathbf{q}\sigma,\tau). \tag{10.74}$$

Fourier transforming, we obtain

$$\begin{aligned} \frac{1}{\beta\hbar}\sum_m D^0(\mathbf{q},\omega_m)e^{-i\omega_m\tau} &= \frac{1}{(\beta\hbar)^2 A}\sum_{\mathbf{k}\sigma}\sum_{ss'} F_{ss'}(\mathbf{k},\mathbf{q}) \\ &\quad\times\sum_{nn'} g^0(s\mathbf{k}\sigma,\omega_n)g^0(s'\mathbf{k}+\mathbf{q}\sigma,\omega_{n'})e^{-i(\omega_{n'}-\omega_n)\tau} \end{aligned} \tag{10.75}$$

where

$$F_{ss'}(\mathbf{k},\mathbf{q}) = \frac{1}{2}\left(1 + ss'\frac{k + q\cos\phi}{|\mathbf{k}+\mathbf{q}|}\right) \tag{10.76}$$

(see Problem 2.6). Here, ϕ is the angle between $\mathbf{k}$ and $\mathbf{q}$, and $s, s' = +1(-1)$ if $s, s' = c(v)$. It follows that $\omega_{n'} = \omega_n + \omega_m$. The summation over n was carried out in Section 10.6; we therefore have

$$D^0(\mathbf{q},\omega_m) = \frac{1}{A}\sum_{\mathbf{k}\sigma}\sum_{ss'} F_{ss'}(\mathbf{k},\mathbf{q})\frac{f_{s\mathbf{k}} - f_{s'\mathbf{k}+\mathbf{q}}}{i\omega_m + (\epsilon_{s\mathbf{k}} - \epsilon_{s'\mathbf{k}+\mathbf{q}})/\hbar}. \tag{10.77}$$

The bare polarizability $\Pi^0(\mathbf{q}, \omega) = (1/\hbar)D^{R,0}(q, \omega)$ is thus given by

$$\Pi^0(\mathbf{q}, \omega) = \frac{4}{A} \sum_{\mathbf{k}ss'} F_{ss'}(\mathbf{k}, \mathbf{q}) \frac{f_{s\mathbf{k}} - f_{s'\mathbf{k}+\mathbf{q}}}{\hbar\omega + \epsilon_{s\mathbf{k}} - \epsilon_{s'\mathbf{k}+\mathbf{q}} + i0^+}. \tag{10.78}$$

A factor of 2 arises from the existence of two valleys, and another factor of 2 arises from summing over the spin index; hence we have a factor of 4 in the above equation. Denoting $f_{c\mathbf{k}}$ by $f_{\mathbf{k}+}$, $f_{v\mathbf{k}}$ by $f_{\mathbf{k}-}$, $\epsilon_{c\mathbf{k}}$ by $\epsilon_{\mathbf{k}+}$, $\epsilon_{v\mathbf{k}}$ by $\epsilon_{\mathbf{k}-}$, and summing over band indices, we obtain

$$\begin{aligned}\Pi^0(\mathbf{q}, \omega) &= \frac{4}{A} \sum_{\mathbf{k}} \left[\frac{(f_{\mathbf{k}+} - f_{\mathbf{k}+\mathbf{q}+})F_{++}(\mathbf{k}, \mathbf{q})}{\hbar\omega + \epsilon_{\mathbf{k}+} - \epsilon_{\mathbf{k}+\mathbf{q}+} + i0^+} + \frac{f_{\mathbf{k}+}F_{+-}(\mathbf{k}, \mathbf{q})}{\hbar\omega + \epsilon_{\mathbf{k}+} - \epsilon_{\mathbf{k}+\mathbf{q}-} + i0^+} \right. \\ &\quad \left. - \frac{f_{\mathbf{k}+\mathbf{q}+}F_{-+}(\mathbf{k}, \mathbf{q})}{\hbar\omega + \epsilon_{\mathbf{k}-} - \epsilon_{\mathbf{k}+\mathbf{q}+} + i0^+} \right] + \frac{4}{A} \sum_{\mathbf{k}} \left[\frac{(f_{\mathbf{k}-} - f_{\mathbf{k}+\mathbf{q}-})F_{--}(\mathbf{k}, \mathbf{q})}{\hbar\omega + \epsilon_{\mathbf{k}-} - \epsilon_{\mathbf{k}+\mathbf{q}-} + i0^+} \right. \\ &\quad \left. + \frac{f_{\mathbf{k}-}F_{-+}(\mathbf{k}, \mathbf{q})}{\hbar\omega + \epsilon_{\mathbf{k}-} - \epsilon_{\mathbf{k}+\mathbf{q}+} + i0^+} - \frac{f_{\mathbf{k}+\mathbf{q}-}F_{+-}(\mathbf{k}, \mathbf{q})}{\hbar\omega + \epsilon_{\mathbf{k}+} - \epsilon_{\mathbf{k}+\mathbf{q}-} + i0^+} \right] \\ &\equiv \Pi^{0,+}(\mathbf{q}, \omega) + \Pi^{0,-}(\mathbf{q}, \omega). \end{aligned} \tag{10.79}$$

We restrict our calculations to the case of undoped, pure graphene at low temperatures. Under these conditions, the conduction band is empty and the valence band is full: $f_{\mathbf{k}+} = f_{\mathbf{k}+\mathbf{q}+} = 0$, and $f_{\mathbf{k}-} = f_{\mathbf{k}+\mathbf{q}-} = 1$. Hence $\Pi^{0,+}(\mathbf{q}, \omega) = 0$, and

$$\begin{aligned}\Pi^0(\mathbf{q}, \omega) &= \frac{2}{A} \sum_{\mathbf{k}} \left(1 - \frac{k + q\cos\phi}{|\mathbf{k} + \mathbf{q}|}\right) \\ &\quad \times \left[\frac{1}{\hbar\omega + \epsilon_{\mathbf{k}-} - \epsilon_{\mathbf{k}+\mathbf{q}+} + i0^+} - \frac{1}{\hbar\omega + \epsilon_{\mathbf{k}+} - \epsilon_{\mathbf{k}+\mathbf{q}-} + i0^+} \right]. \end{aligned} \tag{10.80}$$

First, we evaluate the imaginary part of $\Pi^0(\mathbf{q}, \omega)$,

$$\begin{aligned}Im\, \Pi^0(\mathbf{q}, \omega) &= -\frac{2\pi}{A} \sum_{\mathbf{k}} \left(1 - \frac{k + q\cos\phi}{|\mathbf{k} + \mathbf{q}|}\right) \\ &\quad \times \{\delta\left[\hbar\omega - \hbar v_F(k + |\mathbf{k} + \mathbf{q}|)\right] - \delta\left[\hbar\omega + \hbar v_F(k + |\mathbf{k} + \mathbf{q}|)\right]\} \end{aligned} \tag{10.81}$$

where we assume that q is small, so that the linear energy dispersion will be a good approximation. Since $\delta(x) = \delta(-x)$, the above expression implies that

$$Im\, \Pi^0(\mathbf{q}, -\omega) = -Im\, \Pi^0(\mathbf{q}, \omega). \tag{10.82}$$

Note further that, from Eq. (10.80), we have

$$Re\,\Pi^0(\mathbf{q}, -\omega) = Re\,\Pi^0(\mathbf{q}, \omega). \tag{10.83}$$

It is thus sufficient to evaluate $\Pi^0(\mathbf{q}, \omega)$ for $\omega > 0$. In this case, the second Dirac-delta function in Eq. (10.81) vanishes, and we end up with

$$Im\,\Pi^0(\mathbf{q}, \omega > 0) = -\frac{1}{2\pi\hbar}\iint k\left(1 - \frac{k + q\cos\phi}{|\mathbf{k}+\mathbf{q}|}\right)\delta(\omega - v_F k - v_F|\mathbf{k}+\mathbf{q}|)\,dk d\phi. \tag{10.84}$$

We have used $\delta(ax) = \delta(x)/|a|$ and made the replacement

$$\sum_{\mathbf{k}} \rightarrow \frac{A}{(2\pi)^2}\int k dk \int_0^{2\pi} d\phi.$$

Consider the argument $f(\cos\phi)$ of the Dirac-delta function

$$f(\cos\phi) = \omega - v_F k - v_F(k^2 + q^2 + 2kq\cos\phi)^{1/2}. \tag{10.85}$$

For the Dirac-delta function $\delta[f(\cos\phi)]$ to be nonvanishing, ω must be greater than or equal to $v_F k$: $\omega \geq v_F k$. The root of $f(\cos\phi)$ is

$$f(\cos\phi) = 0 \Rightarrow \cos\phi = (\omega^2 - 2v_F k\omega - v_F^2 q^2)/2v_F^2 kq \tag{10.86}$$

and

$$|\partial f/\partial \cos\phi|_{\text{root}} = v_F^2 kq/(\omega - v_F k), \quad |\mathbf{k}+\mathbf{q}|_{\text{root}} = (\omega - v_F k)/v_F. \tag{10.87}$$

Using

$$\delta[f(x)] = \sum_i \frac{\delta[x - x_i]}{|\partial f/\partial x|_{x_i}},$$

where x_i's are the roots of $f(x)$, we can write

$$\delta(\omega - v_F k - v_F|\mathbf{k}+\mathbf{q}|) = \frac{|\mathbf{k}+\mathbf{q}|}{v_F kq}\delta\left(\cos\phi - \frac{\omega^2 - 2v_F k\omega - v_F^2 q^2}{2v_F^2 kq}\right). \tag{10.88}$$

Since $-1 \leq \cos\phi \leq 1$, for the Dirac-delta function to be nonzero, we should have $-1 \leq (\omega^2 - 2v_F k\omega - v_F^2 q^2)/2v_F^2 kq \leq 1$. This is satisfied if the following two conditions are satisfied:

(a) $\omega \geq v_F q$

(b) $v_F(2k - q) \leq \omega \leq v_F(2k + q)$.

We also note that

$$\int_0^{2\pi} d\phi \cdots = \int_0^{\pi} d\phi \cdots + \int_{\pi}^{2\pi} d\phi \cdots = -\int_{\phi=0}^{\phi=\pi} \frac{d\cos\phi}{\sin\phi} \cdots - \int_{\phi=\pi}^{\phi=2\pi} \frac{d\cos\phi}{\sin\phi} \cdots, \tag{10.89}$$

and that $sin\phi > 0$ for $0 < \phi < \pi$ while $sin\phi < 0$ for $\pi < \phi < 2\pi$; hence

$$\int_0^{2\pi} d\phi \cdots = \int_{-1}^{1} \frac{d\cos\phi}{|\sin\phi|} \cdots + \int_{-1}^{1} \frac{d\cos\phi}{|\sin\phi|} \cdots = 2\int_{-1}^{1} \frac{d\cos\phi}{|\sin\phi|} \cdots. \tag{10.90}$$

Finally, we note that, at the root of $f(\cos\phi)$

$$|\sin\phi|_{\text{root}} = \left[(2v_F^2 kq)^2 - (\omega^2 - 2v_F k\omega - v_F^2 q^2)^2\right]^{1/2} / 2v_F^2 kq. \tag{10.91}$$

The integration over ϕ can now be carried out; it gives

$$\begin{aligned} Im\Pi^0(\mathbf{q}, \omega > 0) = &-\frac{\theta(\omega - v_F q)}{\pi\hbar v_F\sqrt{\omega^2 - v_F^2 q^2}} \int \left[v_F^2 q^2 - (\omega - 2v_F k)^2\right]^{1/2} \\ &\times \{\theta\left[\omega - v_F(2k - q)\right] - \theta[\omega - v_F(2k + q)]\}\, dk. \end{aligned} \tag{10.92}$$

The step functions ensure that conditions (a) and (b), which were given earlier, are satisfied. Condition (b), enforced by the step functions inside the integral, implies that $\omega/2v_F - q/2 \le k \le \omega/2v_F + q/2$. Thus,

$$Im\Pi^0(\mathbf{q}, \omega > 0) = -\frac{\theta(\omega - v_F q)}{\pi\hbar v_F\sqrt{\omega^2 - v_F^2 q^2}} \int_{\frac{\omega}{2v_F} - \frac{q}{2}}^{\frac{\omega}{2v_F} + \frac{q}{2}} \left[v_F^2 q^2 - (\omega - 2v_F k)^2\right]^{1/2} dk. \tag{10.93}$$

By a change of variable: $\omega - 2v_F k \to x$, the integration is easily done,

$$Im\,\Pi^0(\mathbf{q}, \omega > 0) = -\frac{q^2\theta(\omega - v_F q)}{4\hbar\sqrt{\omega^2 - v_F^2 q^2}}. \tag{10.94}$$

As noted earlier, $Re\,\Pi^0(\mathbf{q}, \omega)$ is an even function of ω, while $Im\,\Pi^0(\mathbf{q}, \omega)$ is an odd function of ω. The poles of $\Pi^0(\mathbf{q}, \omega)$ are below the real axis, and $\Pi^0(\mathbf{q}, \omega) \to 0$ as $|\omega| \to \infty$. The Kramers–Kronig relations (see Problem 6.11) are thus applicable to $\Pi^0(\mathbf{q}, \omega)$:

$$\begin{aligned} Re\,\Pi^0(\mathbf{q}, \omega) &= \frac{2}{\pi} P \int_0^{\infty} \frac{\omega' Im\,\Pi^0(\mathbf{q}, \omega')}{\omega'^2 - \omega^2} d\omega' \\ &= -\frac{q^2}{2\pi\hbar} P \int_0^{\infty} \frac{\omega'\theta(\omega' - v_F q)}{(\omega'^2 - \omega^2)\sqrt{\omega'^2 - v_F^2 q^2}} d\omega'. \end{aligned} \tag{10.95}$$

The integral is carried out by making a change of variable,

$$\omega'^2 - \omega^2 = x \Rightarrow \omega' d\omega' = dx/2,$$

$$Re\,\Pi^0(\mathbf{q},\omega) = -\frac{q^2}{4\pi\hbar}\int_{v_F^2q^2-\omega^2}^{\infty}\frac{dx}{x\sqrt{x+\omega^2-v_F^2q^2}} \equiv -\frac{q^2}{4\pi\hbar}J. \tag{10.96}$$

This is a tabulated integral,

$$J = \begin{cases} \dfrac{1}{\sqrt{\omega^2 - v_F^2q^2}}\ln\left|\dfrac{\sqrt{x+\omega^2-v_F^2q^2}-\sqrt{\omega^2-v_F^2q^2}}{\sqrt{x+\omega^2-v_F^2q^2}+\sqrt{\omega^2-v_F^2q^2}}\right|_{v_F^2q^2-\omega^2}^{\infty} & \omega > v_Fq \\ \dfrac{2}{\sqrt{v_F^2q^2-\omega^2}}\,sec^{-1}\sqrt{\dfrac{x}{v_F^2q^2-\omega^2}}\Bigg|_{v_F^2q^2-\omega^2}^{\infty} & \omega < v_Fq \end{cases}$$

$$= \begin{cases} 0 & \omega > v_Fq \\ \dfrac{\pi}{\sqrt{v_F^2q^2-\omega^2}} & \omega < v_Fq. \end{cases} \tag{10.97}$$

Thus,

$$Re\ \Pi^0(\mathbf{q},\omega) = -\frac{q^2}{4\hbar}\frac{\theta(v_Fq-\omega)}{\sqrt{v_F^2q^2-\omega^2}} \tag{10.98}$$

and

$$\Pi^0(\mathbf{q},\omega>0) = -\frac{q^2}{4\hbar}\left[\frac{\theta(v_Fq-\omega)}{\sqrt{v_F^2q^2-\omega^2}} + i\frac{\theta(\omega-v_Fq)}{\sqrt{\omega^2-v_F^2q^2}}\right]. \tag{10.99}$$

For $\omega < 0$

$$\Pi^0(\mathbf{q},\omega<0) = [\Pi^0(\mathbf{q},\omega>0)]^*. \tag{10.100}$$

The dielectric function is

$$\varepsilon(\mathbf{q},\omega) = 1 - \frac{2\pi e^2}{q}\Pi^0(\mathbf{q},\omega). \tag{10.101}$$

Further reading

Bruus, H. and Flansberg, K. (2004). *Many-Body Quantum Theory in Condensed Matter*. Oxford: Oxford University Press.

Fetter, A.L. and Walecka, J.D. (1971). *Quantum Theory of Many-Particle Systems*. New York: McGraw-Hill.

Mahan, G.D. (2000). *Many-Particle Physics*, 3rd edn. New York: Kluwer Academic/Plenum Publishers.
Mattuck, R.D. (1976). *A Guide To Feynman Diagrams in the Many-Body Problem*, 2nd edn. New York: McGraw-Hill.

Problems

10.1 *First-order self energy.* Show that the first-order contribution to the self energy of an electron in an electron gas as $T \to 0$ is given by

$$\Sigma_1^*(\mathbf{k}\sigma, \omega_n) = -\frac{e^2 k_F}{\pi\hbar}\left[1 + \frac{1-x^2}{2x}\ln\left|\frac{1+x}{1-x}\right|\right]$$

where $x = k/k_F$. To obtain the above result, start from

$$\Sigma_1^*(\mathbf{k}\sigma, \omega_n) = -\frac{1}{\hbar V}\sum_{\mathbf{k}'}\frac{4\pi e^2}{|\mathbf{k}-\mathbf{k}'|^2} f_{\mathbf{k}'}.$$

As $T \to 0$, $f_{\mathbf{k}'} \to \theta(k_F - k')$. Replace sum over $\mathbf{k}'$ by integration, and use the formula

$$\int x \ln|x+a|dx = \frac{x^2 - a^2}{2}\ln|x+a| - \frac{1}{4}(x-a)^2.$$

10.2 *Proper self energy in two dimensions.* Calculate $\Sigma_1^*(\mathbf{k}\sigma, \omega_n)$ for a two-dimensional electron gas in the limit $T \to 0$. Show that, at $k = k_F$, it is given by $-2e^2 k_F/(\pi\hbar)$.

10.3 *High frequency limit of* $\varepsilon(\mathbf{q}, \omega)$. Show that the high frequency limit of the dielectric function of an electron gas is given by

$$\lim_{\omega\to\infty} \varepsilon(\mathbf{q}, \omega) = 1 - \omega_p^2/\omega^2$$

where $\omega_p = (4\pi n e^2/m)^{1/2}$ is the plasmon frequency.

10.4 *An alternative derivation of the plasmon dispersion.*

(a) Show that

$$Re\,\Pi^0(\mathbf{q}, \omega) = \frac{4}{V}\sum_{\mathbf{k}}\frac{f_{\mathbf{k}}(\epsilon_{\mathbf{k}+\mathbf{q}} - \epsilon_{\mathbf{k}})}{(\hbar\omega)^2 - (\epsilon_{\mathbf{k}+\mathbf{q}} - \epsilon_{\mathbf{k}})^2}.$$

(b) In the long-wavelength limit ($q \ll k_F$), and high-frequency limit ($\omega \gg q v_F$), we have $\hbar\omega \gg (\epsilon_{\mathbf{k}+\mathbf{q}} - \epsilon_{\mathbf{k}})$. Show that, in these limits,

$$Re\,\Pi^0(\mathbf{q}, \omega) = \frac{4}{V(\hbar\omega)^2}\sum_{\mathbf{k}} f_{\mathbf{k}}(\epsilon_{\mathbf{k}+\mathbf{q}} - \epsilon_{\mathbf{k}})\left[1 + \frac{(\epsilon_{\mathbf{k}+\mathbf{q}} - \epsilon_{\mathbf{k}})^2}{(\hbar\omega)^2} + \cdots\right].$$

(c) As $T \to 0$, $f_{\mathbf{k}} \to \theta(k_F - k)$. Using $\epsilon_{\mathbf{k}} = \hbar^2 k^2/2m$, show that $Re\,\Pi^0(\mathbf{q}, \omega)$ is given by Eq. (10.65), and hence, the plasmon mode dispersion is given by Eq. (10.66).

10.5 *Thomas–Fermi wave number in two dimensions.* Show that, in two dimensions, $q_{\text{TF}} = 2/a_0$, where $a_0 = \hbar^2/me^2$ is the Bohr radius.

10.6 *Plasmons in two dimensions.* Show that, in a two-dimensional electron gas with n electrons per unit area, the plasmon dispersion is given by

$$\omega_{\mathbf{q}} = \sqrt{\frac{2\pi n e^2 q}{m}} \left[1 + \frac{3qa_0}{8} + \cdots \right].$$

11

Phonons, photons, and electrons

> When the sky is illumined with crystal
> Then gladden my road and broaden my path
> And clothe me in light.
> *–From "The Book of the Dead," Ancient Egypt*
> *Translated by Robert Hillyer*

In this chapter we turn to phonons, photons, and their interactions with electrons. These interactions play an important role in condensed matter physics. At room temperature, the resistivity of metals results mainly from electron–phonon interaction. At low temperature, this interaction is responsible for the superconducting properties of many metals. On the other hand, the electron–photon interaction plays a dominant role in light scattering by solids, from which we derive a great deal of information about excitation modes in solids. Much of our knowledge about energy bands in crystals has been obtained through optical absorption experiments, whose interpretation relies on an understanding of how electrons and photons interact.

We begin by discussing lattice vibrations in crystals and show that, upon quantization, the vibrational modes are described in terms of phonons, which are particle-like excitations that carry energy and momentum. We will see that the effect of lattice vibrations on electronic states is to cause scattering, whereby electrons change their states by emitting or absorbing phonons. Similarly, the interaction of electrons with an electromagnetic field will be represented as scattering processes in which electrons emit or absorb photons.

A discussion of lattice vibrations in the general case of a three-dimensional crystal with a basis of more than one atom is somewhat complicated. To keep the presentation simple, we consider in detail the simplest case, a one-dimensional crystal with only one atom per unit cell. Next, we consider a diatomic chain, and then indicate briefly how things look in three dimensions. The reader interested in a treatment of the general case of a three-dimensional crystal with more than one atom per primitive cell will find a detailed presentation in Appendix C.

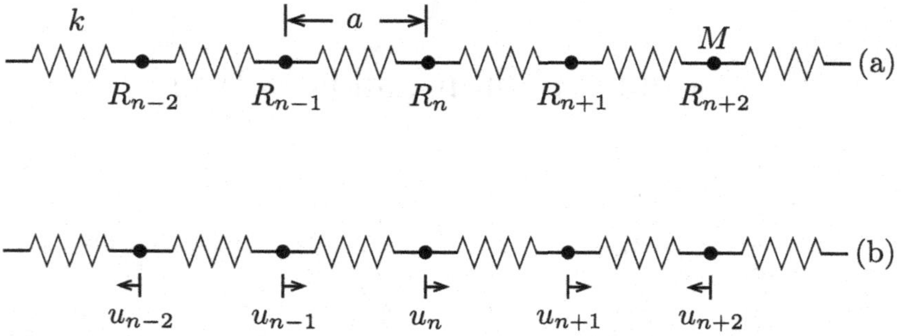

Figure 11.1 A line of atoms, each of mass M, connected by massless springs of force constant k. (a) The atoms sit at their equilibrium positions, with the equilibrium position of atom n being R_n. In equilibrium, the separation between neighboring atoms is a. (b) The atoms are displaced from equilibrium, with the displacement of atom n being u_n.

11.1 Lattice vibrations in one dimension

The simplest case we can deal with is a one-dimensional crystal with one atom per unit cell. Consider a line of N atoms ($N \gg 1$), each of mass M. In equilibrium, the position of atom n is $R_n = na$, and the separation between adjacent atoms is a. We model the interatomic interactions by massless springs, each of force constant k, which connect neighboring atoms (see Figure 11.1). When atoms vibrate, they are displaced from equilibrium. Let u_n be the displacement from equilibrium of atom n. We adopt periodic boundary conditions: $u_1 = u_{N+1}$. Newton's second law gives

$$M\ddot{u}_n = k(u_{n+1} - 2u_n + u_{n-1}). \tag{11.1}$$

This is a set of N coupled differential equations ($n = 1, 2, \ldots N$). The general approach to solving such a set of coupled equations is to first find the normal modes; the general solution is then obtained by writing the displacements as linear combinations of these modes. In a normal mode all atoms vibrate with the same wave vector and frequency. Denoting wave vector as q and frequency as ω_q, atom n in a normal mode has a displacement given by

$$u_n = A\exp[i(qR_n - \omega_q t)] = A\exp[i(qna - \omega_q t)] \tag{11.2}$$

where A is a constant. Inserting this into Eq. (11.1), we obtain

$$-M\omega_q^2 = k(e^{iqa} - 2 + e^{-iqa}) = 2k[\cos(qa) - 1].$$

Writing $\cos(qa) = 1 - 2\sin^2(qa/2)$, the frequency can be expressed as

$$\omega_q = \omega_m|\sin(qa/2)|, \quad \omega_m = (4k/M)^{1/2}. \tag{11.3}$$

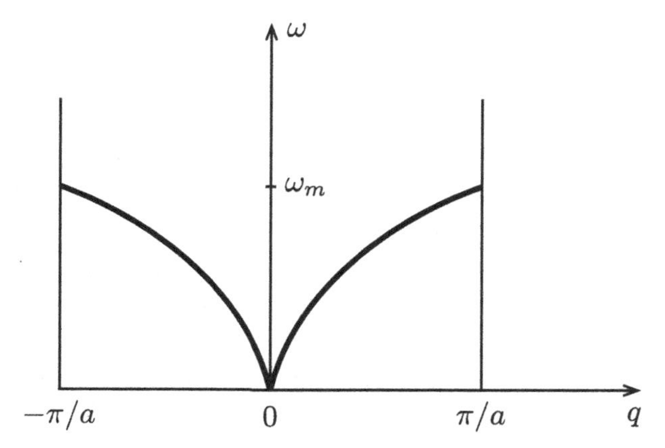

Figure 11.2 A plot of ω vs. q for values of q in the first Brillouin zone. The crystal is one-dimensional, with one atom per primitive cell.

The relation between ω_q and q is known as the dispersion relation. We note the following:

1. Periodic boundary conditions, applied to Eq. (11.2), give the allowed values for q, namely

$$q = 0, \pm 2\pi/L, \pm 4\pi/L, \ldots$$

   where $L = Na$ is the length of the line of atoms.
2. It follows from Eqs (11.2) and (11.3) that

$$\omega_q = \omega_{-q} = \omega_{q+2\pi/a}\,, \quad u_n(q) = u_n(q + 2\pi/a).$$

3. As $q \to 0$, $\omega_q = vq$, where $v = \omega_m a/2$.

The second remark implies that it is sufficient to restrict the values of q to the first Brillouin zone (FBZ): $-\pi/a < q \leqslant \pi/a$. The number of normal modes is equal to the number of q-points within the FBZ, which is exactly equal to N. Since $L \gg a$, the first remark means that the separation between neighboring values of q is too small compared to the width of the FBZ; hence, when plotting ω vs q, we may consider q to be continuous. Such a plot is shown in Figure 11.2. Regarding the third remark, the fact that $\omega \to 0$ as $q \to 0$ is obvious on physical grounds: as $q \to 0$, neighboring atoms undergo equal displacements during the vibration, and the restoring forces vanish. The fact that ω approaches zero linearly in q in the long wavelength limit ($q \to 0$) assigns the name "acoustic branch" to the branch in the dispersion in Figure 11.2; v is the speed of sound in this one-dimensional crystal. The general solution of the equation of motion, Eq. (11.1), is a linear combination of the normal modes,

$$u_n = \frac{1}{\sqrt{NM}} \sum_{q \in \text{FBZ}} Q_q e^{iqR_n}, \tag{11.4}$$

where the factor $e^{-i\omega_q t}$ is absorbed into the expansion coefficients Q_q and the factor $1/\sqrt{NM}$ is inserted for later convenience. In effect, Eq. (11.4) is a Fourier expansion of the displacement u_n. The expansion coefficients Q_q are called normal coordinates. They satisfy the relation $Q_q^* = Q_{-q}$, which is a consequence of the fact that the displacement u_n is real.

Our next task is to construct an expression for the energy of the line of atoms in terms of the normal coordinates. The kinetic energy is given by

$$T = (M/2) \sum_{n=1}^{N} \dot{u}_n^2 = \frac{1}{2N} \sum_n \sum_{qq'} \dot{Q}_q \dot{Q}_{q'} e^{i(q+q')R_n},$$

where $q, q' \in$ FBZ. Summing first over n ($\sum_n e^{i(q+q')R_n} = N\delta_{q',-q}$), we find

$$T = (1/2) \sum_q \dot{Q}_q \dot{Q}_{-q}. \tag{11.5}$$

The potential energy is the elastic energy of the springs,

$$V = (k/2) \sum_{n=1}^{N} (u_{n+1} - u_n)^2. \tag{11.6}$$

From Eq. (11.4), we can write

$$u_{n+1} - u_n = \frac{1}{\sqrt{NM}} \sum_q Q_q e^{iqR_n}(e^{iqa} - 1).$$

The potential energy is thus given by

$$V = \frac{k}{2NM} \sum_n \sum_{qq'} Q_q Q_{q'} (e^{iqa} - 1)(e^{iq'a} - 1) e^{i(q+q')R_n}.$$

Carrying out the summation over n first, we obtain

$$V = \frac{k}{2M} \sum_q Q_q Q_{-q} |e^{iqa} - 1|^2 = \frac{2k}{M} \sum_q Q_q Q_{-q} \sin^2(qa/2).$$

Using Eq. (11.3), the above expression becomes

$$V = (1/2) \sum_q \omega_q^2 Q_q Q_{-q}. \tag{11.7}$$

The Lagrangian $L = T - V$ is thus a function of the normal coordinates. The canonical momentum conjugate to Q_q is

$$P_q = \partial L/\partial \dot{Q}_q = \dot{Q}_{-q}. \qquad (11.8)$$

The Hamiltonian, in terms of the dynamical variables Q_q and P_q, is

$$H = \left(\sum_q P_q \dot{Q}_q - L\right)_{\dot{Q}_q = P_{-q}}. \qquad (11.9)$$

Substituting $T - V$ for L, we find

$$H = \frac{1}{2}\sum_q \left(P_q P_{-q} + \omega_q^2 Q_q Q_{-q}\right). \qquad (11.10)$$

The quantum theory of lattice vibrations of the one-dimensional monatomic crystal is obtained by treating the dynamical variables Q_q and P_q as operators that satisfy the commutation relations

$$[Q_q, Q_{q'}] = [P_q, P_{q'}] = 0, \quad [Q_q, P_{q'}] = i\hbar\delta_{qq'}. \qquad (11.11)$$

Analogous to the case of the harmonic oscillator (see Section 1.2), we introduce two new operators,

$$a_q = (2\hbar\omega_q)^{-1/2}(\omega_q Q_q + iP_{-q}), \quad a_q^\dagger = (2\hbar\omega_q)^{-1/2}(\omega_q Q_{-q} - iP_q). \quad (11.12)$$

These operators satisfy the commutation relations

$$[a_q, a_{q'}] = [a_q^\dagger, a_{q'}^\dagger] = 0, \quad [a_q, a_{q'}^\dagger] = \delta_{qq'}. \qquad (11.13)$$

It is straightforward to show that, in terms of these operators, the Hamiltonian is

$$H = \sum_q \hbar\omega_q(a_q^\dagger a_q + 1/2). \qquad (11.14)$$

The Hamiltonian is seen to be a collection of N independent harmonic oscillators. The eigenvalues are $\sum_q \hbar\omega_q(n_q + 1/2)$, where n_q is a non-negative integer. The ground state is obtained when $n_q = 0$ for all values of q. We interpret n_q as the number of particle-like excitations, called phonons, that occupy the normal mode specified by q; each phonon has energy $\hbar\omega_q$ and wave number q. The operator $a_q^\dagger(a_q)$ is interpreted as a creation (annihilation) operator that creates (annihilates) a phonon of wave number q and energy $\hbar\omega_q$. The commutation relations satisfied by a_q and $a_q^\dagger$ mean that phonons are bosonic particles. Since the quantum number $q \in$ FBZ completely specifies a vibrational mode, phonons are spinless particles.

A phonon of wave number q represents a traveling wave of wavelength $\lambda = 2\pi/|q|$. Therefore, a phonon of wave number $q = 0$ does not exist; the $q = 0$ normal mode represents a translation of the whole crystal, not a traveling wave.

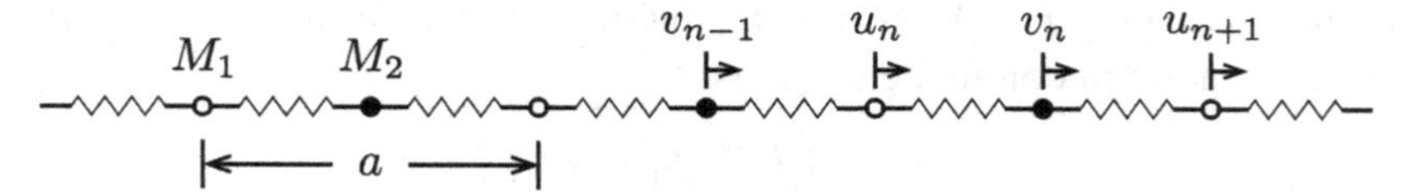

Figure 11.3 A linear diatomic chain with lattice constant a. The two different atoms have masses M_1 and M_2. Neighboring atoms are connected by springs of force constant k. In unit cell n, the displacements from equilibrium of the atoms of masses M_1 and M_2 are u_n and v_n, respectively.

11.2 One-dimensional diatomic lattice

We now consider a one-dimensional diatomic lattice (Figure 11.3). The two different atoms in a unit cell have masses M_1 and M_2, and their equilibrium separation is $a/2$. Neighboring atoms are assumed to be connected by massless springs, each of force constant k. We denote by u_n and v_n, respectively, the displacements from equilibrium of the atoms of masses M_1 and M_2, located in unit cell n. Newton's second law yields the following equations:

$$M_1 \ddot{u}_n = k(v_n - 2u_n + v_{n-1}) \tag{11.15}$$

$$M_2 \ddot{v}_n = k(u_{n+1} - 2v_n + u_n). \tag{11.16}$$

These constitute a set of $2N$ coupled differential equations, where N is the number of unit cells. To find the normal modes, we consider the trial solutions

$$u_n = u e^{i(qR_n - \omega_q t)}, \quad v_n = v e^{i(qR_n - \omega_q t)}, \tag{11.17}$$

where $R_n = na$. Inserting these solutions into Eqs (11.15) and (11.16), we obtain the following homogeneous algebraic equations involving u and v:

$$(-M_1\omega_q^2 + 2k)u - k(1 + e^{-iqa})v = 0 \tag{11.18}$$

$$-k(1 + e^{iqa})u + (-M_2\omega_q^2 + 2k)v = 0. \tag{11.19}$$

A nontrivial solution exists only if the determinant of the coefficients of u and v vanishes. The result is the following expression:

$$\omega_q^2 = \frac{k}{\mu} \pm k\sqrt{\frac{1}{\mu^2} - \frac{4\sin^2(qa/2)}{M_1 M_2}} \tag{11.20}$$

where $\mu = M_1 M_2/(M_1 + M_2)$ is the reduced mass of the two atoms in the unit cell. A plot of ω vs q reveals that the dispersion curves consist of two branches (see Figure 11.4). The lower branch is the acoustic branch, while the upper one is

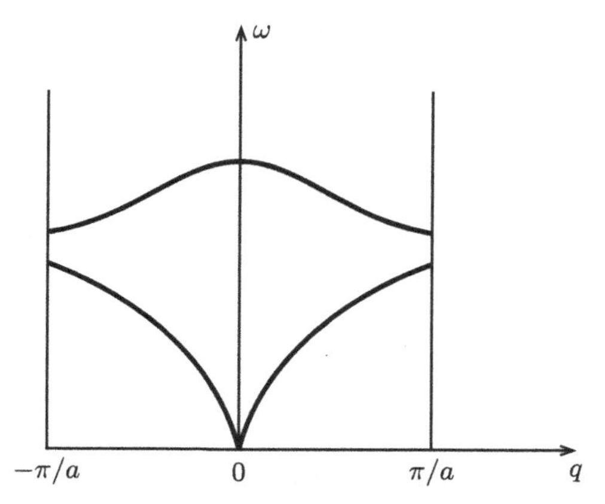

Figure 11.4 Dispersion curves for a linear diatomic chain with lattice constant a. The lower (upper) branch is the acoustic (optical) branch.

the optical branch. As $q \to 0$,

$$\omega = \sqrt{\frac{ka^2}{2(M_1 + M_2)}}\, q \quad \text{(acoustic)}, \qquad \omega = \sqrt{2k/\mu} \quad \text{(optical)}. \tag{11.21}$$

At the Brillouin zone edges ($q = \pm\pi/a$) we find, based on Eq. (11.20), that

$$\omega = \sqrt{\frac{2k}{\max(M_1, M_2)}} \quad \text{(acoustic)}, \qquad \omega = \sqrt{\frac{2k}{\min(M_1, M_2)}} \quad \text{(optical)} \tag{11.22}$$

where $\max(M_1, M_2)$ is the larger of M_1 and M_2, and $\min(M_1, M_2)$ is the smaller of the two masses. At $q = 0$, Eqs (11.18), (11.19), and (11.21) give

$$u/v = 1 \quad \text{(acoustic)}, \qquad u/v = -M_2/M_1 \quad \text{(optical)}. \tag{11.23}$$

At the Brillouin zone center ($q = 0$), all atoms vibrate in phase in the acoustic mode, undergoing equal displacements; the vanishing of the frequency results from the absence of any restoring forces. In the optical mode, on the other hand, adjacent atoms vibrate out of phase (see Figure 11.5). We note that the optical mode is excited by infrared light, hence the name "optical mode."

If we were to construct a quantum theory of lattice vibrations for the diatomic chain, we would find the following Hamiltonian:

$$H = \sum_{q \in \text{FBZ}} \sum_{\lambda=1}^{2} \hbar\omega_{q\lambda}(a_{q\lambda}^{\dagger} a_{q\lambda} + 1/2). \tag{11.24}$$

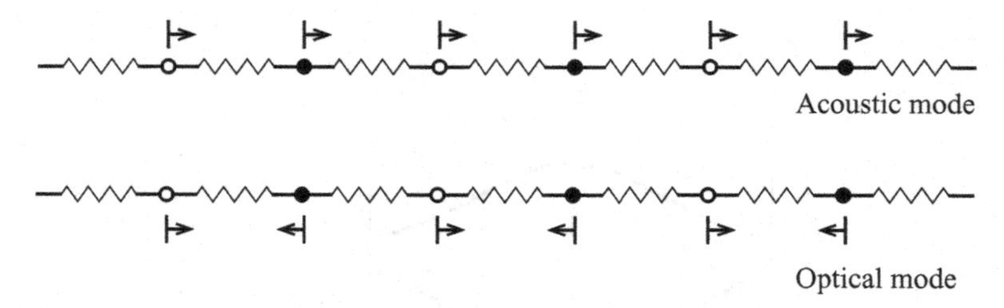

Figure 11.5 The acoustic (upper figure) and optical (lower figure) modes at the Brillouin zone center ($q = 0$) of a linear diatomic chain.

Here, the index λ refers to the phonon branch. There are two branches, an acoustic one and an optical one. The operator $a^{\dagger}_{q\lambda}$ ($a_{q\lambda}$) creates (annihilates) a phonon of wave number q, branch index λ, and energy $\hbar\omega_{q\lambda}$.

11.3 Phonons in three-dimensional crystals

We now briefly indicate how the one-dimensional case is generalized to three dimensions. A detailed account is given in Appendix C.

We consider a crystal consisting of N unit cells with a basis of r atoms. The displacement from equilibrium of atom l ($l = 1, 2, \ldots, r$) in unit cell n ($n = 1, 2, \ldots, N$) is denoted by $\mathbf{u}_{nl}$. Since there are Nr atoms in the crystal and each atom vibrates in three dimensions, there are $3Nr$ degrees of freedom; consequently, there is a total of $3Nr$ normal modes. In a normal mode, all atoms vibrate with the same wave vector and frequency. A normal mode is specified by a wave vector $\mathbf{q} \in$ FBZ (there are N such vectors) and a branch index $\lambda = 1, 2, \ldots, 3r$. In a normal mode with coordinates $(\mathbf{q}\lambda)$,

$$\mathbf{u}_{nl} \propto (M_l)^{-1/2}\boldsymbol{\epsilon}_{\lambda}^{(l)}(\mathbf{q})e^{i(\mathbf{q}.\mathbf{R}_n-\omega_{\mathbf{q}\lambda}t)},$$

where M_l is the mass of atom l and $\boldsymbol{\epsilon}_{\lambda}^{(l)}$ is a polarization vector that determines the direction of the displacement $\mathbf{u}_{nl}$ relative to the wave vector $\mathbf{q}$. In a purely longitudinal normal mode, $\boldsymbol{\epsilon}_{\lambda}^{(l)} \parallel \mathbf{q}$, while in a purely transverse mode, $\boldsymbol{\epsilon}_{\lambda}^{(l)} \perp \mathbf{q}$. The general solution of the equations of motion is a linear combination of the $3Nr$ normal modes,

$$\mathbf{u}_{nl} = (NM_l)^{-1/2}\sum_{\mathbf{q}\lambda} Q_{\mathbf{q}\lambda}\boldsymbol{\epsilon}_{\lambda}^{(l)}(\mathbf{q})e^{i\mathbf{q}.\mathbf{R}_n}. \tag{11.25}$$

The time-dependent coefficients $Q_{\mathbf{q}\lambda}$ are called normal coordinates. Since $\mathbf{u}^*_{nl} = \mathbf{u}_{nl}$ (displacements are real), it follows that $Q^*_{\mathbf{q}\lambda} = Q_{-\mathbf{q}\lambda}$ and $\boldsymbol{\epsilon}_{\lambda}^{(l)*}(\mathbf{q}) = \boldsymbol{\epsilon}_{\lambda}^{(l)}(-\mathbf{q})$. The Hamiltonian can be expressed in terms of the normal coordinates $Q_{\mathbf{q}\lambda}$ and their conjugate momenta $P_{\mathbf{q}\lambda}$:

$$H = (1/2)\sum_{\mathbf{q}\lambda}(P_{\mathbf{q}\lambda}P_{-\mathbf{q}\lambda} + \omega^2_{\mathbf{q}\lambda}Q_{\mathbf{q}\lambda}Q_{-\mathbf{q}\lambda}). \tag{11.26}$$

Passage to a quantum theory of lattice vibrations is accomplished by treating the dynamical variables $Q_{\mathbf{q}\lambda}$ and $P_{\mathbf{q}\lambda}$ as operators that satisfy the commutation relations

$$[Q_{\mathbf{q}\lambda}, Q_{\mathbf{q}'\lambda'}] = [P_{\mathbf{q}\lambda}, P_{\mathbf{q}'\lambda'}] = 0, \quad [Q_{\mathbf{q}\lambda}, P_{\mathbf{q}'\lambda'}] = i\hbar\,\delta_{\mathbf{q}\mathbf{q}'}\delta_{\lambda\lambda'}. \tag{11.27}$$

We introduce two new operators $a_{\mathbf{q}\lambda}$ and $a^\dagger_{\mathbf{q}\lambda}$ such that

$$Q_{\mathbf{q}\lambda} = \sqrt{\frac{\hbar}{2\omega_{\mathbf{q}\lambda}}}(a_{\mathbf{q}\lambda} + a^\dagger_{-\mathbf{q}\lambda}) \tag{11.28}$$

$$P_{\mathbf{q}\lambda} = i\sqrt{\frac{\hbar\omega_{\mathbf{q}\lambda}}{2}}(a^\dagger_{\mathbf{q}\lambda} - a_{-\mathbf{q}\lambda}). \tag{11.29}$$

Note that $Q_{-\mathbf{q}\lambda} = Q^\dagger_{\mathbf{q}\lambda}$. The new operators satisfy the commutation relations

$$[a_{\mathbf{q}\lambda}, a_{\mathbf{q}'\lambda'}] = [a^\dagger_{\mathbf{q}\lambda}, a^\dagger_{\mathbf{q}'\lambda'}] = 0, \quad [a_{\mathbf{q}\lambda}, a^\dagger_{\mathbf{q}'\lambda'}] = \delta_{\mathbf{q}\mathbf{q}'}\delta_{\lambda\lambda'}. \tag{11.30}$$

In terms of these operators, the Hamiltonian can be written as

$$H = \sum_{\mathbf{q}\lambda}\hbar\omega_{\mathbf{q}\lambda}(a^\dagger_{\mathbf{q}\lambda}a_{\mathbf{q}\lambda} + 1/2). \tag{11.31}$$

The operator $a^\dagger_{\mathbf{q}\lambda}$ ($a_{\mathbf{q}\lambda}$) is interpreted as a creation (annihilation) operator of a phonon of wave vector $\mathbf{q}$, branch index λ, and energy $\hbar\omega_{\mathbf{q}\lambda}$.

Finally, we note that there are three acoustic phonon branches with a zero frequency at the Brillouin zone center ($\mathbf{q} = 0$), and $3r - 3$ optical phonon branches with nonvanishing frequencies.

11.4 Phonon statistics

The Hamiltonian given in Eq. (11.31) describes a system of noninteracting phonons. Its eigenvalues are $\sum_{\mathbf{q}\lambda}(n_{\mathbf{q}\lambda} + 1/2)\hbar\omega_{\mathbf{q}\lambda}$, where $n_{\mathbf{q}\lambda} = 0, 1, 2\ldots$ is a non-negative integer, interpreted as the number of phonons of wave vector $\mathbf{q}$, branch index λ, and energy $\hbar\omega_{\mathbf{q}\lambda}$. In the ground state, $n_{\mathbf{q}\lambda} = 0$ for all values of $\mathbf{q}$ and λ. We now calculate $\langle n_{\mathbf{q}\lambda}\rangle_0$, the average number of phonons occupying the normal mode ($\mathbf{q}\lambda$), for a system of noninteracting phonons in equilibrium at temperature T. The subscript "0" refers to a noninteracting system.

When the system is in equilibrium at temperature T, any particular normal mode ($\mathbf{q}\lambda$) may be occupied by any number of phonons. The probability that n phonons occupy the mode $(\mathbf{q}\lambda)$ is $e^{-\beta n\hbar\omega_{\mathbf{q}\lambda}} / \sum_{n=0}^{\infty} e^{-\beta n\hbar\omega_{\mathbf{q}\lambda}}$. Hence,

$$\langle n_{\mathbf{q}\lambda}\rangle_0 = \sum_{n=0}^{\infty} n e^{-\beta n\hbar\omega_{\mathbf{q}\lambda}} \Big/ \sum_{n=0}^{\infty} e^{-\beta n\hbar\omega_{\mathbf{q}\lambda}}.$$

The evaluation of the above expression is straightforward: the denominator is a geometric series, while the numerator is proportional to the derivative, with respect to β, of the same series. We obtain

$$\langle n_{\mathbf{q}\lambda}\rangle_0 = \frac{1}{e^{\beta\hbar\omega_{\mathbf{q}\lambda}} - 1} \equiv n_{\omega_{\mathbf{q}\lambda}}. \tag{11.32}$$

As expected, phonons obey Bose–Einstein statistics. The important point here (and the reason for going through the derivation) is the absence of a chemical potential: $\mu = 0$, as Eq. (11.32) indicates. The vanishing of the chemical potential results from the fact that the number of phonons in the system is unrestricted: an arbitrary number of phonons can occupy a normal mode $(\mathbf{q}\lambda)$.

11.5 Electron–phonon interaction: rigid-ion approximation

The basic idea underlying the electron–phonon interaction is simple, as illustrated in Figure 11.6. When ions sit at their equilibrium positions, the state of an electron is described by a Bloch function of wave vector $\mathbf{k}$ (and spin projection σ and band index n). A phonon disturbs the lattice, and ions move out of their equilibrium positions. This causes a change in the potential seen by the electron (the potential no longer has the periodicity of the lattice). This change, in turn, scatters the electron into another state with wave vector $\mathbf{k}'$.

In this section, we calculate the electron–phonon interaction within the rigid-ion approximation: in it, the potential field of an ion is assumed to be rigidly attached to the ion as it moves. This is an approximation because, in reality, as a nucleus moves, it does not rigidly carry along the electronic charge that surrounds it. In the rigid-ion approximation, the interaction between an electron and an ion depends on the distance that separates them. This approximation is reasonable for simple metals, but it is not adequate for polar crystals, where ionic vibrations produce an electric field which acts on the electron.

For simplicity of notation, we assume that there is one atom per unit cell; the extension to a crystal with a basis is straightforward. The interaction of an electron at position $\mathbf{r}_j$ with the ions is given by

$$V_{e-i} = \sum_n V(\mathbf{r}_j - \mathbf{R}_n - \mathbf{u}_n) \simeq \sum_n V(\mathbf{r}_j - \mathbf{R}_n) - \sum_n \mathbf{u}_n . \nabla V(\mathbf{r}_j - \mathbf{R}_n).$$

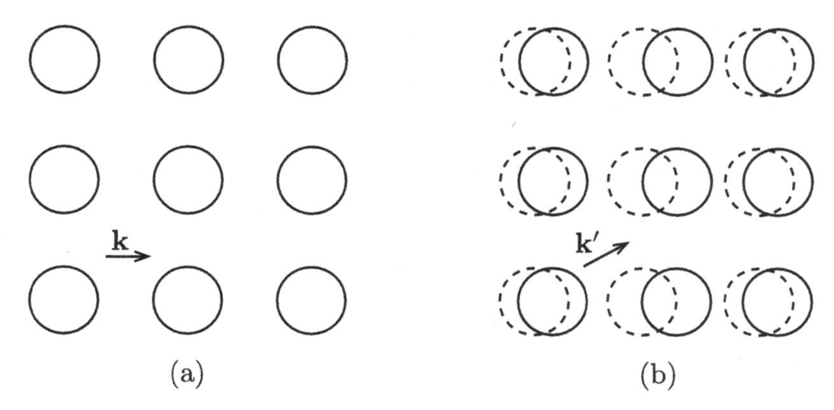

Figure 11.6 (a) An electron is in the Bloch state $|n\mathbf{k}\sigma\rangle$. As long as the lattice is static, the electron remains in this state. (b) In the presence of a phonon, the ions in the lattice vibrate, and the electron sees a different ionic potential than it does in (a). The change in the potential energy of the electron acts as a time-dependent perturbation that can scatter the electron into the stationary states $|n'\mathbf{k}'\sigma\rangle$.

$V(\mathbf{r}_j - \mathbf{R}_n - \mathbf{u}_n)$ is the interaction energy of the electron with the ion in unit cell n when the ion is displaced from equilibrium by $\mathbf{u}_n$. The gradient is with respect to the electron coordinates. The above equation is no more than a Taylor expansion of V_{e-i} to first order in the displacement; higher orders are ignored due to the smallness of the ionic displacement as compared to the spacing between neighboring ions. The first term in the expansion is the periodic potential energy which results from the interaction of the electron with the static ions at their equilibrium positions; when combined with the electron's kinetic energy, this term gives rise to the Bloch Hamiltonian whose eigenfunctions are the Bloch functions. The second term, when summed over all electrons, is the electron–phonon interaction:

$$H_{e-\text{phonon}} = -\sum_{jn} \mathbf{u}_n . \nabla V(\mathbf{r}_j - \mathbf{R}_n). \tag{11.33}$$

The displacement $\mathbf{u}_n$ is now written in terms of the normal coordinates (see Eq. [11.25]); we obtain

$$H_{e-\text{phonon}} = \frac{-1}{\sqrt{NM}} \sum_n \sum_{\mathbf{q}\lambda} Q_{\mathbf{q}\lambda} e^{i\mathbf{q}.\mathbf{R}_n} \sum_j \boldsymbol{\epsilon}_\lambda . \nabla V(\mathbf{r}_j - \mathbf{R}_n). \tag{11.34}$$

In this expression, the sum over the electrons is identified as a one-body operator of the form $\sum_j h(\mathbf{r}_j)$; its second quantized form is

$$\sum_j \boldsymbol{\epsilon}_\lambda . \nabla V(\mathbf{r}_j - \mathbf{R}_n) = \sum_{\mathbf{k}\mathbf{k}'\sigma} \left(\int \psi^*_{\mathbf{k}'\sigma}(\mathbf{r}) \boldsymbol{\epsilon}_\lambda . \nabla V(\mathbf{r} - \mathbf{R}_n) \psi_{\mathbf{k}\sigma}(\mathbf{r}) d^3r \right) c^\dagger_{\mathbf{k}'\sigma} c_{\mathbf{k}\sigma} \tag{11.35}$$

where $\psi_{\mathbf{k}\sigma}(\mathbf{r})$ and $\psi_{\mathbf{k}'\sigma}(\mathbf{r})$ are Bloch functions. In writing the above equation, we have assumed that there is only one partially filled band (as is the case in simple metals) and that electrons scatter within this band; we thus ignored the band index when writing the Bloch functions. Furthermore, no spin flip can occur when an electron is scattered by lattice vibrations.

We recall that, according to Bloch's theorem (Section 2.3),

$$\psi_{\mathbf{k}\sigma}(\mathbf{r}+\mathbf{R}_n) = e^{i\mathbf{k}.\mathbf{R}_n}\psi_{\mathbf{k}\sigma}(\mathbf{r}). \tag{11.36}$$

We can take advantage of this property: in the integral in Eq. (11.35), we replace the integration variable $\mathbf{r}$ with $\mathbf{r}+\mathbf{R}_n$; then

$$\sum_j \boldsymbol{\epsilon}_\lambda.\nabla V(\mathbf{r}_j - \mathbf{R}_n) = \sum_{\mathbf{k}\mathbf{k}'\sigma} e^{i(\mathbf{k}-\mathbf{k}').\mathbf{R}_n}\left(\int \psi^*_{\mathbf{k}'\sigma}(\mathbf{r})\boldsymbol{\epsilon}_\lambda.\nabla V(\mathbf{r})\psi_{\mathbf{k}\sigma}(\mathbf{r})d^3r\right) c^\dagger_{\mathbf{k}'\sigma}c_{\mathbf{k}\sigma}.$$

We insert this into Eq. (11.34) and carry out the summation over n:

$$\sum_n e^{i(\mathbf{k}-\mathbf{k}'+\mathbf{q}).\mathbf{R}_n} = N\sum_{\mathbf{G}} \delta_{\mathbf{k}',\mathbf{k}+\mathbf{q}+\mathbf{G}}, \tag{11.37}$$

where $\mathbf{G}$ is a reciprocal lattice vector. Equation (11.34) becomes

$$H_{e-\text{phonon}} = -\sqrt{N/M}\sum_{\mathbf{k}\sigma}\sum_{\mathbf{q}\lambda}\sum_{\mathbf{G}} \boldsymbol{\epsilon}_\lambda.\mathbf{T}(\mathbf{k},\mathbf{q},\mathbf{G}\sigma)c^\dagger_{\mathbf{k}+\mathbf{q}+\mathbf{G}\sigma}c_{\mathbf{k}\sigma}Q_{\mathbf{q}\lambda}. \tag{11.38}$$

Here,

$$\mathbf{T}(\mathbf{k},\mathbf{q},\mathbf{G}\sigma) = \int \psi^*_{\mathbf{k}+\mathbf{q}+\mathbf{G}\sigma}(\mathbf{r})\nabla V(\mathbf{r})\psi_{\mathbf{k}\sigma}(\mathbf{r})d^3r. \tag{11.39}$$

We can proceed a bit further if we adopt the effective mass approximation: $\psi_{\mathbf{k}}(\mathbf{r}) \simeq (1/\sqrt{V})e^{i\mathbf{k}.\mathbf{r}}$, $\epsilon_{\mathbf{k}} \simeq \hbar^2k^2/2m^*$, where V is the volume of the crystal and m^* is the effective electron mass. This approximation is adequate for simple metals. Expanding $V(\mathbf{r})$ in a Fourier series,

$$V(\mathbf{r}) = (1/V)\sum_{\mathbf{p}} V_{\mathbf{p}}e^{i\mathbf{p}.\mathbf{r}} \Rightarrow \nabla V = (i/V)\sum_{\mathbf{p}} \mathbf{p}V_{\mathbf{p}}e^{i\mathbf{p}.\mathbf{r}},$$

making use of the relation

$$\int e^{i(\mathbf{p}-\mathbf{q}-\mathbf{G}).\mathbf{r}}d^3r = V\delta_{\mathbf{p},\mathbf{q}+\mathbf{G}},$$

and writing $Q_{\mathbf{q}\lambda}$ in terms of phonon creation and annihilation operators (see Eq. [11.28]), we finally obtain

$$H_{e-\text{phonon}} = -\frac{i}{V}\sum_{\mathbf{k}\sigma}\sum_{\mathbf{q}\lambda}\sum_{\mathbf{G}}\sqrt{\frac{N\hbar}{2M\omega_{\mathbf{q}\lambda}}}V_{\mathbf{q}+\mathbf{G}}(\mathbf{q}+\mathbf{G}).\boldsymbol{\epsilon}_\lambda(\mathbf{q})c^\dagger_{\mathbf{k}+\mathbf{q}+\mathbf{G}\sigma}c_{\mathbf{k}\sigma}(a_{\mathbf{q}\lambda}+a^\dagger_{-\mathbf{q}\lambda}). \tag{11.40}$$

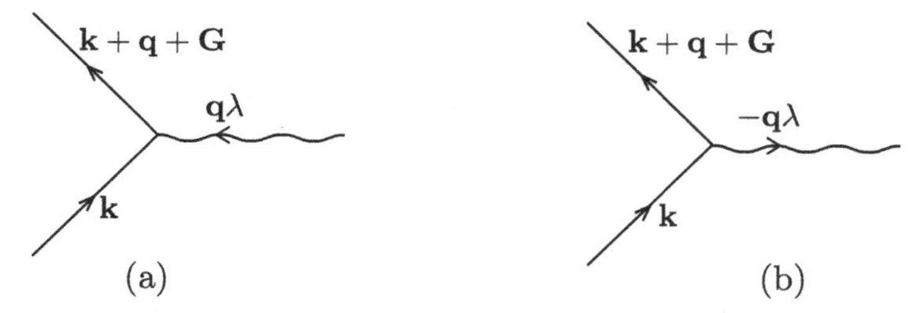

Figure 11.7 A pictorial representation of electron–phonon interaction. (a) An electron is scattered from state $|\mathbf{k}\sigma\rangle$ into state $|\mathbf{k}+\mathbf{q}+\mathbf{G}\sigma\rangle$ by absorbing a phonon of wave vector $\mathbf{q}$ and branch index λ. (b) Here, scattering occurs by the emission of a phonon with coordinates $(-\mathbf{q}\lambda)$.

For a crystal with a basis of r atoms: $l = 1, 2, \ldots, r$, the above expression is modified as follows: $M \to M_l$, $V_{\mathbf{q}+\mathbf{G}} \to V^{(l)}_{\mathbf{q}+\mathbf{G}}$, $\boldsymbol{\epsilon}_\lambda(\mathbf{q}) \to \boldsymbol{\epsilon}^{(l)}_\lambda(\mathbf{q})$, and an extra summation $\sum_{l=1}^{r}$ is carried out.

We make the following remarks regarding the electron–phonon interaction:

(1) The interaction is seen to be a sum of terms, with each term representing a scattering process in which an electron is scattered from state $|\mathbf{k}\sigma\rangle$ into state $|\mathbf{k}+\mathbf{q}+\mathbf{G}\sigma\rangle$ by either emitting or absorbing a phonon. The scattering process is depicted in Figure 11.7.
(2) The wave vectors $\mathbf{q}$, $\mathbf{k}$, and $\mathbf{k}' = \mathbf{k}+\mathbf{q}+\mathbf{G}$ must all lie in the first Brillouin zone (FBZ). Hence, in summing over $\mathbf{G}$, there is only one term in the summation for any fixed $\mathbf{k}$ and $\mathbf{q}$; $\mathbf{G}$ is the one reciprocal lattice vector which, when added to $\mathbf{k}+\mathbf{q}$, carries it back into the FBZ. If $\mathbf{k}+\mathbf{q} \in$ FBZ, then $\mathbf{G} = 0$; otherwise, $\mathbf{G} \neq 0$. Electron scattering processes (by the emission or absorption of a phonon) for which $\mathbf{G} = 0$ are called normal processes. A process for which $\mathbf{G} \neq 0$ is called an Umklapp process. Normal and Umklapp processes are depicted in Figure 11.8.
(3) We restrict further discussion to normal processes only: $\mathbf{G} = 0$. The factor $\mathbf{q}.\boldsymbol{\epsilon}_\lambda(\mathbf{q})$ in Eq. (11.40) implies that electrons interact only with longitudinal phonons. In isotropic media, phonon polarization vectors are actually either longitudinal or transverse.
(4) We write the electron–phonon interaction in the following form:

$$H_{e-\text{phonon}} = \sum_{\mathbf{k}\sigma}\sum_{\mathbf{q}\lambda} M_{\mathbf{q}\lambda} c^\dagger_{\mathbf{k}+\mathbf{q}\sigma} c_{\mathbf{k}\sigma}(a_{\mathbf{q}\lambda} + a^\dagger_{-\mathbf{q}\lambda}). \tag{11.41}$$

The matrix element $M_{\mathbf{q}\lambda}$ is a measure of the strength of the electron–phonon interaction. Its mathematical form depends on the kind of approximations one

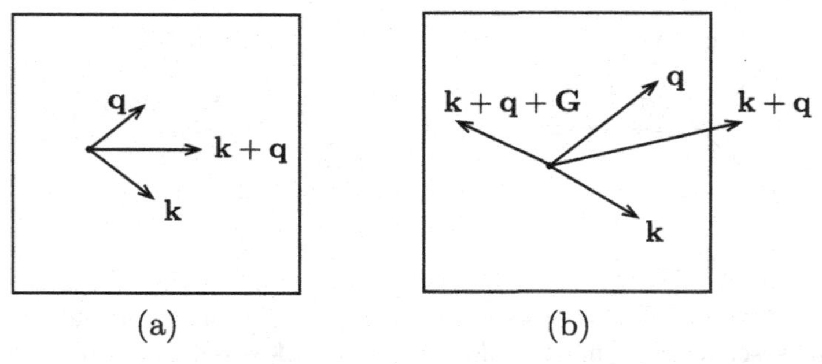

Figure 11.8 Normal and Umklapp processes in a two-dimensional square lattice of lattice constant a. The square shown in the figure is the FBZ; its side is $2\pi/a$. In the scattering process, an electron of wave vector $\mathbf{k}$ absorbs a phonon of wave vector $\mathbf{q}$ or emits a phonon of wave vector $-\mathbf{q}$. (a) The wave vector $\mathbf{k}+\mathbf{q} \in$ FBZ; the scattering process is normal. (b) $\mathbf{k}+\mathbf{q} \notin$ FBZ; a reciprocal lattice vector $\mathbf{G}$ must be added to carry $\mathbf{k}+\mathbf{q}$ back into the FBZ, so the scattering process is Umklapp.

makes (Ziman, 1960). Since H_{e-ph} is Hermitian, it follows that

$$M^*_{\mathbf{q}\lambda} = M_{-\mathbf{q}\lambda}. \tag{11.42}$$

Let us consider the case of an isotropic medium (a cubic crystal, for example) with only one atom per unit cell. In this case, there is one longitudinal acoustic branch and two transverse acoustic branches. Electrons interact only with the longitudinal phonons. Assuming that the crystal is a metal, the Coulomb potential of the ions is screened by the conduction electrons. Since ions move very slowly compared to electrons, we can assume that the screening is static, similar to the screening of a fixed charged impurity. The screened Coulomb potential of an ion is thus taken to be $4\pi Ze/q^2\varepsilon(\mathbf{q},0)$, where Ze is the ionic charge and $\varepsilon(\mathbf{q},0)$ is the static dielectric function. For small values of q, this is given by $4\pi Ze/(q^2 + q_{TF}^2) \simeq 4\pi Ze/q_{TF}^2$, where q_{TF} is the Thomas–Fermi wave number. The electron–ion interaction energy thus has the Fourier component $V_{\mathbf{q}} = -4\pi Ze^2/q_{TF}^2$. Under these assumptions, it follows from Eq. (11.40) that the electron–phonon interaction matrix element depends on $\mathbf{q}$ only and is given by

$$M_{\mathbf{q}} = \frac{i}{V}\sqrt{\frac{N\hbar}{2M\omega_{\mathbf{q}}}}\frac{4\pi Ze^2}{q_{TF}^2}\mathbf{q}.\boldsymbol{\epsilon}_\lambda(\mathbf{q}). \tag{11.43}$$

Note that because $\boldsymbol{\epsilon}_\lambda(-\mathbf{q}) = \boldsymbol{\epsilon}^*_\lambda(\mathbf{q})$, $M_{\mathbf{q}}$ satisfies the relation $M^*_{\mathbf{q}} = M_{-\mathbf{q}}$, as it should. Often in the literature $\mathbf{q}.\boldsymbol{\epsilon}_\lambda(\mathbf{q})$ is replaced by q, which is the magnitude of $\mathbf{q}$; this is not quite accurate, for then the equality $M^*_{\mathbf{q}} = M_{-\mathbf{q}}$ is not satisfied. Note further that $M_{\mathbf{q}=\mathbf{0}} = 0$: in a normal mode, with $\mathbf{q} = \mathbf{0}$, the periodicity of the lattice is preserved, Bloch states are still the stationary states of the system, and no scattering takes place.

11.6 Electron–LO phonon interaction in polar crystals

We indicated in the previous section that the rigid-ion approximation is not adequate for polar crystals, where the electric field associated with longitudinal optical (LO) vibrations acts on the electrons. We consider the simplest case of a cubic polar crystal with two ions per unit cell; the ions have equal but opposite charges. Moreover, we only consider vibrations in the long wavelength limit. Even in this case, the calculation of the electron–LO phonon coupling is not easy. We relegate the details to Appendix D. Here, we simply summarize the main results.

(1) There is no electric field associated with transverse optical (TO) modes.

(2) The electric field associated with a longitudinal optical (LO) mode exerts a restoring force on the ions, in addition to the short-range restoring forces that are present in the absence of an electric field. The result is that the LO–phonon frequency is higher than the TO–phonon frequency. In the long wavelength limit ($q \to 0$), we find

$$\omega_{\mathrm{LO}}^2 = \frac{\varepsilon(0)}{\varepsilon(\infty)}\omega_{\mathrm{TO}}^2\,, \tag{11.44}$$

where $\varepsilon(0) = \varepsilon(\mathbf{q} \to 0, \omega = 0)$ is the static dielectric constant (measured by applying a static electric field to the crystal) and $\varepsilon(\infty) = \varepsilon(\mathbf{q} = 0, \omega \gg \omega_{\mathrm{phonon}})$ is the high-frequency dielectric constant of the crystal (it is the square of the refractive index of the crystal). The above relation is known as the Lyddane–Sachs–Teller (LST) relation.

(3) The electron–LO phonon interaction takes the form:

$$H_{e\text{–LO}} = \sum_{ss'}\sum_{\mathbf{k}\sigma}\sum_{\mathbf{q}}{}' M_{\mathbf{q}}^{ss'} c_{s'\mathbf{k}+\mathbf{q}\sigma}^{\dagger} c_{s\mathbf{k}\sigma}(a_{\mathbf{q}} + a_{-\mathbf{q}}^{\dagger}). \tag{11.45}$$

The prime on the summation indicates that the $\mathbf{q} = 0$ term is excluded, $a_{\mathbf{q}}^{\dagger}(a_{\mathbf{q}})$ creates (annihilates) an LO–phonon of wave vector $\mathbf{q}$, and $c_{s\mathbf{k}\sigma}^{\dagger}(c_{s\mathbf{k}\sigma})$ creates (annihilates) an electron in state $|s\mathbf{k}\sigma\rangle$, where s is the band index. The matrix element $M_{\mathbf{q}}^{ss'}$ is given by

$$M_{\mathbf{q}}^{ss'} = i\omega_{\mathrm{LO}}\left[\frac{1}{\epsilon(\infty)} - \frac{1}{\epsilon(0)}\right]^{1/2}\left(\frac{2\pi\hbar e^2}{Vq^2\omega_{\mathrm{LO}}}\right)^{1/2} \boldsymbol{\epsilon}_L(\mathbf{q}).\hat{\mathbf{q}}\langle s'\mathbf{k} + \mathbf{q}\sigma|e^{i\mathbf{q}.\mathbf{r}}|s\mathbf{k}\sigma\rangle. \tag{11.46}$$

Here, V is the crystal volume, $\hat{\mathbf{q}}$ is the unit vector in the direction of $\mathbf{q}$, and $\epsilon_L(\mathbf{q})$ is the LO–phonon unit polarization vector.

If we assume that electron scattering by phonons takes place in only one band, the sum over s and s' in Eq. (11.45) is no longer there. If we also approximate the Bloch functions by plane waves, the matrix element $\langle s'\mathbf{k}+\mathbf{q}\sigma|e^{i\mathbf{q}.\mathbf{r}}|s\mathbf{k}\sigma\rangle$ becomes equal to unity.

11.7 Phonon Green's function

In previous chapters, we defined the retarded and advanced Green's functions for bosons in terms of the ensemble average of the commutator of an annihilation and a creation operator. The imaginary-time Green's function was also defined in terms of the ensemble average of the time-ordered product of an annihilation and a creation operator. Although a similar definition for the phonon Green's function may be adopted, this is not the most convenient one. This is because the linear combination $a_{\mathbf{q}\lambda} + a^{\dagger}_{-\mathbf{q}\lambda}$ appears in the electron–phonon interaction. It is this particular combination that is employed in the definition of the phonon Green's function.

11.7.1 Definitions

The phonon retarded Green's function is defined by

$$d^R(\mathbf{q}\lambda, t) = -i\theta(t)\langle[\phi_{\mathbf{q}\lambda}(t), \phi^{\dagger}_{\mathbf{q}\lambda}(0)]\rangle, \tag{11.47}$$

where $\theta(t)$ is the step function,

$$\phi_{\mathbf{q}\lambda} = a_{\mathbf{q}\lambda} + a^{\dagger}_{-\mathbf{q}\lambda} \tag{11.48}$$

is the phonon field operator, and $\phi_{\mathbf{q}\lambda}(t) = e^{iHt/\hbar}\phi_{\mathbf{q}\lambda}(0)e^{-iHt/\hbar}$. In Eq. (11.47), the average is over a canonical ensemble,

$$\langle\cdots\rangle = Tr(e^{-\beta H}\ldots)/Tr(e^{-\beta H}). \tag{11.49}$$

For phonons, canonical and grand canonical ensembles coincide because the chemical potential vanishes, as discussed in Section 11.4.

The phonon imaginary-time (Matsubara) Green's function is defined by

$$d(\mathbf{q}\lambda, \tau) = -\langle T\phi_{\mathbf{q}\lambda}(\tau)\phi^{\dagger}_{\mathbf{q}\lambda}(0)\rangle \tag{11.50}$$

Here, $\phi_{\mathbf{q}\lambda}(\tau) = e^{H\tau/\hbar}\phi_{\mathbf{q}\lambda}(0)e^{-H\tau/\hbar}$, and T is the time-ordering operator,

$$T\phi_{\mathbf{q}\lambda}(\tau)\phi^{\dagger}_{\mathbf{q}\lambda}(0) = \begin{cases} \phi_{\mathbf{q}\lambda}(\tau)\phi^{\dagger}_{\mathbf{q}\lambda}(0) & \tau > 0 \\ \phi^{\dagger}_{\mathbf{q}\lambda}(0)\phi_{\mathbf{q}\lambda}(\tau) & \tau < 0. \end{cases} \tag{11.51}$$

No minus sign is incurred upon interchanging the operators, since they are bosonic. We focus our attention on the imaginary-time Green's function; from it, the retarded function can be obtained by analytic continuation.

11.7.2 Periodicity

As discussed in Chapter 8, the time τ is restricted to the interval $[-\beta\hbar, \beta\hbar]$. Since $\phi_{\mathbf{q}\lambda}$ is a bosonic operator, it follows from the results of Chapter 8 that the phonon Green's function is periodic:

$$d(\mathbf{q}\lambda, \tau > 0) = d(\mathbf{q}\lambda, \tau - \beta\hbar). \tag{11.52}$$

The Fourier expansion of Green's function is given by

$$d(\mathbf{q}\lambda, \tau) = \frac{1}{\beta\hbar} \sum_{m=-\infty}^{\infty} d(\mathbf{q}\lambda, \omega_m) e^{-i\omega_m \tau}, \quad \omega_m = 2\pi m/\beta\hbar \tag{11.53}$$

and the Fourier transform is

$$d(\mathbf{q}\lambda, \omega_m) = \int_0^{\beta\hbar} d(\mathbf{q}\lambda, \tau) e^{i\omega_m \tau} d\tau. \tag{11.54}$$

11.8 Free-phonon Green's function

For a noninteracting system of phonons,

$$d^0(\mathbf{q}\lambda, \tau) = \theta(\tau) d^{0>}(\mathbf{q}\lambda, \tau) + \theta(-\tau) d^{0<}(\mathbf{q}\lambda, \tau). \tag{11.55}$$

The greater and lesser functions are given by

$$d^{0>}(\mathbf{q}\lambda, \tau) = -\langle \phi_{\mathbf{q}\lambda}(\tau) \phi^{\dagger}_{\mathbf{q}\lambda}(0) \rangle_0, \quad d^{0<}(\mathbf{q}\lambda, \tau) = -\langle \phi^{\dagger}_{\mathbf{q}\lambda}(0) \phi_{\mathbf{q}\lambda}(\tau) \rangle_0. \tag{11.56}$$

In terms of phonon creation and annihilation operators,

$$d^{0>}(\mathbf{q}\lambda, \tau) = -\left\langle \left(a_{\mathbf{q}\lambda}(\tau) + a^{\dagger}_{-\mathbf{q}\lambda}(\tau) \right) \left(a^{\dagger}_{\mathbf{q}\lambda}(0) + a_{-\mathbf{q}\lambda}(0) \right) \right\rangle_0 \tag{11.57}$$

where $a_{\mathbf{q}\lambda}(\tau) = e^{H\tau/\hbar} a_{\mathbf{q}\lambda}(0) e^{-H\tau/\hbar}$ and $a^{\dagger}_{-\mathbf{q}\lambda}(\tau) = e^{H\tau/\hbar} a^{\dagger}_{-\mathbf{q}\lambda}(0) e^{-H\tau/\hbar}$. Taking the derivative with respect to τ, we obtain

$$\dot{a}_{\mathbf{q}\lambda}(\tau) = (1/\hbar)[H(\tau), a_{\mathbf{q}\lambda}(\tau)], \quad \dot{a}^{\dagger}_{-\mathbf{q}\lambda}(\tau) = (1/\hbar)[H(\tau), a^{\dagger}_{-\mathbf{q}\lambda}(\tau)].$$

Note that $H(0) = H(\tau)$. Since $H = \sum_{\mathbf{q}\lambda} \hbar\omega_{\mathbf{q}\lambda}(a^{\dagger}_{\mathbf{q}\lambda} a_{\mathbf{q}\lambda} + 1/2)$, the commutators are evaluated easily; we find

$$a_{\mathbf{q}\lambda}(\tau) = e^{-\omega_{\mathbf{q}\lambda}\tau} a_{\mathbf{q}\lambda}, \quad a^{\dagger}_{-\mathbf{q}\lambda}(\tau) = e^{\omega_{\mathbf{q}\lambda}\tau} a^{\dagger}_{-\mathbf{q}\lambda}. \tag{11.58}$$

Inserting these into Eq. (11.57), and noting that the terms $\langle a_{\mathbf{q}\lambda}a_{-\mathbf{q}\lambda}\rangle_0$ and $\langle a^\dagger_{-\mathbf{q}\lambda}a^\dagger_{\mathbf{q}\lambda}\rangle_0$ vanish, we obtain

$$d^{0>}(\mathbf{q}\lambda,\tau) = -\left[e^{-\omega_{\mathbf{q}\lambda}\tau}\langle a_{\mathbf{q}\lambda}a^\dagger_{\mathbf{q}\lambda}\rangle_0 + e^{\omega_{\mathbf{q}\lambda}\tau}\langle a^\dagger_{-\mathbf{q}\lambda}a_{-\mathbf{q}\lambda}\rangle_0\right].$$

The commutation relation between the phonon annihilation and creation operators implies that $a_{\mathbf{q}\lambda}a^\dagger_{\mathbf{q}\lambda} = 1 + a^\dagger_{\mathbf{q}\lambda}a_{\mathbf{q}\lambda}$. Moreover, in thermal equilibrium, $\langle a^\dagger_{\mathbf{q}\lambda}a_{\mathbf{q}\lambda}\rangle_0$ is the occupation number $n_{\omega_{\mathbf{q}\lambda}}$ of the normal mode $(\mathbf{q}\lambda)$, given by Eq. (11.32). Since $\omega_{-\mathbf{q}\lambda} = \omega_{\mathbf{q}\lambda}$, we can write

$$d^{0>}(\mathbf{q}\lambda,\tau) = -\left[e^{-\omega_{\mathbf{q}\lambda}\tau}(1+n_{\omega_{\mathbf{q}\lambda}}) + e^{\omega_{\mathbf{q}\lambda}\tau}n_{\omega_{\mathbf{q}\lambda}}\right]. \tag{11.59}$$

The observation that $1 + n_{\omega_{\mathbf{q}\lambda}} = -n_{-\omega_{\mathbf{q}\lambda}}$ (easily verified) allows us to write the above expression in another way:

$$d^{0>}(\mathbf{q}\lambda,\tau) = -\left[n_{\omega_{\mathbf{q}\lambda}}e^{\omega_{\mathbf{q}\lambda}\tau} - n_{-\omega_{\mathbf{q}\lambda}}e^{-\omega_{\mathbf{q}\lambda}\tau}\right]. \tag{11.60}$$

Similarly, following the same steps, we can show that

$$d^{0<}(\mathbf{q}\lambda,\tau) = -\left[n_{\omega_{\mathbf{q}\lambda}}e^{-\omega_{\mathbf{q}\lambda}\tau} - n_{-\omega_{\mathbf{q}\lambda}}e^{\omega_{\mathbf{q}\lambda}\tau}\right]. \tag{11.61}$$

Before proceeding to calculate the free-phonon Green's function, let us rewrite the above expressions for $d^{0>}$ and $d^{0<}$ in the following way:

$$d^{0>}(\mathbf{q}\lambda,\tau) = \int_{-\infty}^{\infty} P_d^{0>}(\mathbf{q}\lambda,\epsilon)e^{-\epsilon\tau}\frac{d\epsilon}{2\pi} \tag{11.62a}$$

$$d^{0<}(\mathbf{q}\lambda,\tau) = \int_{-\infty}^{\infty} P_d^{0<}(\mathbf{q}\lambda,\epsilon)e^{-\epsilon\tau}\frac{d\epsilon}{2\pi} \tag{11.62b}$$

where

$$P_d^{0>}(\mathbf{q}\lambda,\epsilon) = 2\pi n_{-\epsilon}\left[\delta(\epsilon-\omega_{\mathbf{q}\lambda}) - \delta(\epsilon+\omega_{\mathbf{q}\lambda})\right] \tag{11.63a}$$

$$P_d^{0<}(\mathbf{q}\lambda,\epsilon) = -2\pi n_{\epsilon}\left[\delta(\epsilon-\omega_{\mathbf{q}\lambda}) - \delta(\epsilon+\omega_{\mathbf{q}\lambda})\right]. \tag{11.63b}$$

The Fourier transform of the free-phonon Green's function is

$$d^0(\mathbf{q}\lambda,\omega_m) = \int_0^{\beta\hbar} d^0(\mathbf{q}\lambda,\tau)e^{i\omega_m\tau}d\tau = \int_0^{\beta\hbar} d^{0>}(\mathbf{q}\lambda,\tau)e^{i\omega_m\tau}d\tau.$$

This is calculated by inserting the expression for $d^{0>}(\mathbf{q}\lambda,\tau)$ from either Eq. (11.60) or Eq. (11.62a); the result is

$$d^0(\mathbf{q}\lambda,\omega_m) = \frac{2\omega_{\mathbf{q}\lambda}}{(i\omega_m)^2 - \omega^2_{\mathbf{q}\lambda}}. \tag{11.64}$$

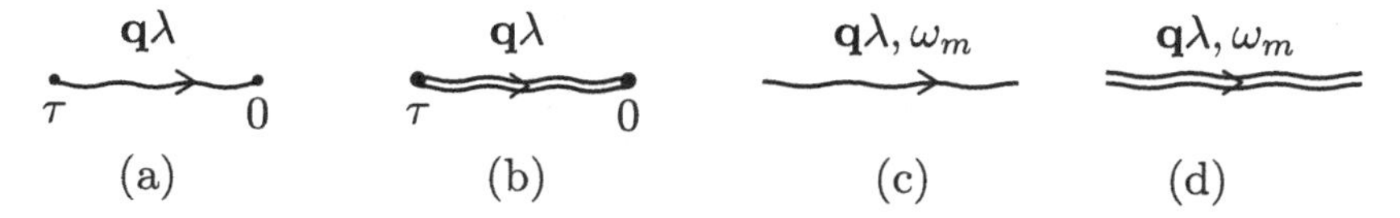

Figure 11.9 Pictorial representation of the phonon Green's function: (a) $d^0(\mathbf{q}\lambda, \tau)$, (b) $d(\mathbf{q}\lambda, \tau)$, (c) $d^0(\mathbf{q}\lambda, \omega_m)$, and (d) $d(\mathbf{q}\lambda, \omega_m)$. Here, d^0 is the noninteracting (free) phonon Green's function, while d is the interacting function.

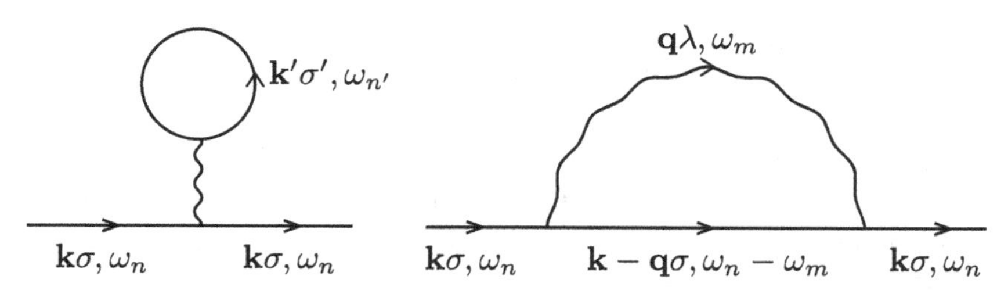

Figure 11.10 The two diagrams of second order in the electron–phonon coupling.

In drawing Feynman diagrams, the phonon Green's function is depicted as in Figure 11.9.

11.9 Feynman rules for the electron–phonon interaction

Treating the electron–phonon interaction as a perturbation, we can expand the electron Green's function to various orders in the perturbation. Since the thermal average of the product of an odd number of phonon field operators is zero, only even orders in the perturbation expansion will survive. The derivation of the Feynman rules from Wick's theorem proceeds in exactly the same way as in Chapter 9. Here, we simply write the rules for calculating the electron Green's function.

(1) At order $2n$ in the electron–phonon interaction (since only even orders survive), draw all topologically distinct diagrams with n phonon lines, two external electron lines, and $2n - 1$ internal electron lines.
(2) To each electron line of coordinates $(\mathbf{k}\sigma, \omega_n)$, assign $g^0(\mathbf{k}\sigma, \omega_n)$.
(3) To each phonon line of coordinates $(\mathbf{q}\lambda, \omega_m)$, assign $|M_{\mathbf{q}\lambda}|^2 d^0(\mathbf{q}\lambda, \omega_m)$.
(4) At each vertex, conserve wave vector, frequency, and spin.
(5) Sum over all internal coordinates.
(6) Multiply each electron loop by -1.
(7) Multiply by the factor $(1/\hbar)^{2n}(-1/\beta\hbar)^n$.

For example, consider the two diagrams that arise in second-order perturbation in the electron–phonon interaction (see Figure 11.10). In the first diagram,

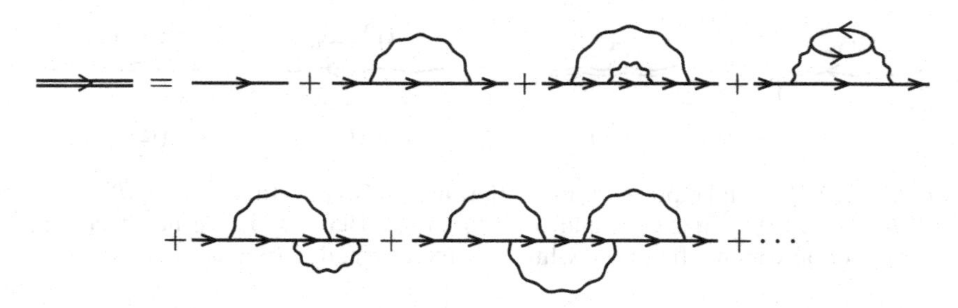

Figure 11.11 Some representative diagrams in the perturbation expansion of the electron Green's function. The perturbation is the electron–phonon interaction.

conservation of wave vector at the vertex implies that the phonon line has zero wave vector. However, the term $\mathbf{q} = \mathbf{0}$ is absent in the electron–phonon interaction, and this diagram should be excluded. Using the Feynman rules, the contribution of the second diagram in Figure 11.10 is

$$\delta g(\mathbf{k}\sigma, \omega_n) = -\frac{1}{\beta\hbar^3}\left[g^0(\mathbf{k}\sigma, \omega_n)\right]^2 \sum_{\mathbf{q}\lambda m} |M_{\mathbf{q}\lambda}|^2 g^0(\mathbf{k}-\mathbf{q}\sigma, \omega_n - \omega_m) d^0(\mathbf{q}\lambda, \omega_m).$$

11.10 Electron self energy

In a simple metal with one partially filled band, electrical conductivity is given by $ne^2\tau/m^*$, where n is the number of electrons per unit volume, m^* is the effective electron mass, and τ is the average lifetime of the electronic states near the Fermi surface (see, e.g., [Omar, 1993]). In a pure metal, the lifetime of an electronic state is determined by the electron–phonon interaction. Here, we calculate the electron self energy that is due to interaction with phonons; the imaginary part of the self energy is related to the lifetime of the electronic state.

We consider a system of electrons and phonons. The Hamiltonian is

$$\bar{H} = \sum_{\mathbf{k}\sigma} \bar{\epsilon}_{\mathbf{k}} c^\dagger_{\mathbf{k}\sigma} c_{\mathbf{k}\sigma} + \sum_{\mathbf{q}\lambda} \hbar\omega_{\mathbf{q}\lambda}(a^\dagger_{\mathbf{q}\lambda} a_{\mathbf{q}\lambda} + 1/2) + \sum_{\mathbf{k}\sigma} \sum_{\mathbf{q}\lambda}{}' M_{\mathbf{q}\lambda} c^\dagger_{\mathbf{k}+\mathbf{q}\sigma} c_{\mathbf{k}\sigma} \phi_{\mathbf{q}\lambda}. \tag{11.65}$$

The first term describes a collection of electrons in the conduction band of a metal; interactions among the electrons are taken in an average way, with the effect being simply a renormalization of the electron mass. The second term is the Hamiltonian for a system of noninteracting phonons, and the third term is the electron–phonon interaction, with the term $\mathbf{q} = \mathbf{0}$ excluded ($M_{\mathbf{q}=0,\lambda} = 0$).

The perturbation expansion of the electron Green's function is depicted in Figure 11.11, where some representative diagrams are shown. The last two diagrams in Figure 11.11 are, in fact, similar to the one-phonon diagram (the second

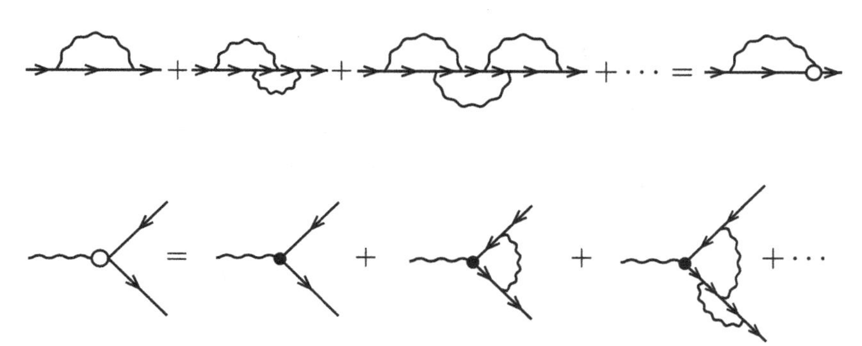

Figure 11.12 Diagrams that can be added to produce a single diagram with a corrected vertex.

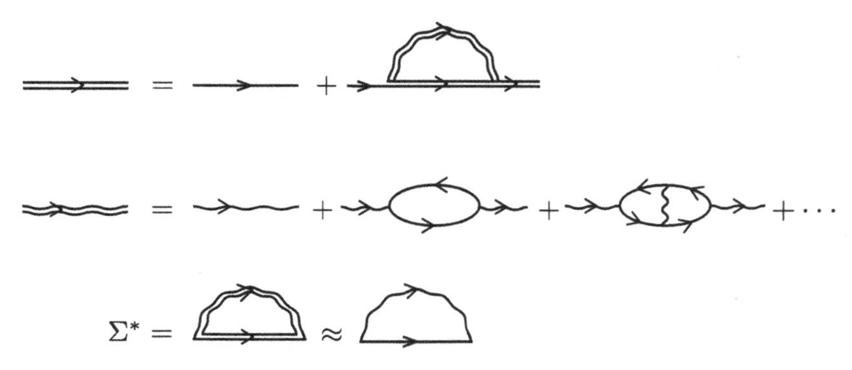

Figure 11.13 Electron and phonon Green's functions for a system of coupled electrons and phonons. Vertex corrections are ignored. The electron proper self energy Σ^* is approximated by replacing full (dressed, or interacting) electron and phonon propagators with bare (noninteracting) propagators.

one on the RHS in the figure), except for some vertex corrections, as shown in Figure 11.12. A remarkable theorem, due to Migdal (Migdal, 1958), states the following:

$$\circ = \bullet \left[1 + O(\sqrt{m^*/M}) \right] \tag{11.66}$$

where $\circ$ ($\bullet$) is the electron–phonon interaction matrix element in the presence (absence) of vertex corrections, m^* is the effective electron mass, and M is the ion mass. Thus, according to Migdal's theorem, vertex corrections may be ignored, since the error made is of the order of one percent ($\sqrt{m^*/M} \approx 0.01$). With that in mind, the electron Green's function may now be expanded as in Figure 11.13.

In calculating the electron self energy, we approximate the interacting electron and phonon Green's functions by using bare ones. The calculation can be carried out using the Feynman rules that were mentioned in the previous section. We

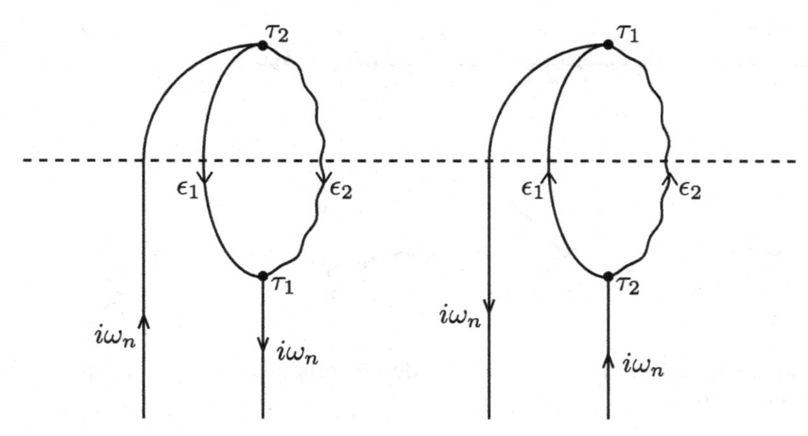

Figure 11.14 The two time-ordered diagrams that are used to calculate the electron self energy that is due to electron–phonon interaction. The external lines have coordinates $(\mathbf{k}\sigma, \omega_n)$. The internal electron line has coordinates $(\mathbf{k}-\mathbf{q}\sigma, \epsilon_1)$, while the internal phonon line has coordinates $(\mathbf{q}\lambda, \epsilon_2)$. The horizontal dashed line is a section.

relegate this approach to the Problems section. Here, we calculate the self energy using Dzyaloshinski's rules for time-ordered diagrams (see Section 9.8). There are two time-ordered diagrams (see Figure 11.14). The self energy is given by

$$\Sigma^*(\mathbf{k}\sigma, \omega_n) = -\frac{1}{\hbar^2}\sum_{\mathbf{q}\lambda}|M_{\mathbf{q}\lambda}|^2 \int_{-\infty}^{\infty}\frac{d\epsilon_1}{2\pi}\int_{-\infty}^{\infty}\frac{d\epsilon_2}{2\pi} \times \frac{P_g^{0<}(\mathbf{k}-\mathbf{q}\sigma, \epsilon_1)P_d^{0<}(\mathbf{q}\lambda, \epsilon_2) - P_g^{0>}(\mathbf{k}-\mathbf{q}\sigma, \epsilon_1)P_d^{0>}(\mathbf{q}\lambda, \epsilon_2)}{i\omega_n - \epsilon_1 - \epsilon_2}. \tag{11.67}$$

The electron spectral functions are

$$P_g^{0>}(\mathbf{k}-\mathbf{q}\sigma, \epsilon) = -2\pi(1-f_\epsilon)\delta(\epsilon - \bar{\epsilon}_{\mathbf{k}-\mathbf{q}}/\hbar)$$
$$P_g^{0<}(\mathbf{k}-\mathbf{q}\sigma, \epsilon) = 2\pi f_\epsilon \delta(\epsilon - \bar{\epsilon}_{\mathbf{k}-\mathbf{q}}/\hbar)$$

(see Eqs [6.55], [8.32], and [8.34]). The phonon spectral functions are given in Eq. (11.63). Inserting these into the expression for Σ^*, and noting that $n_{\omega_{\mathbf{q}\lambda}} = -1 - n_{-\omega_{\mathbf{q}\lambda}}$, we find that

$$\Sigma^*(\mathbf{k}\sigma, \omega_n) = \frac{1}{\hbar^2}\sum_{\mathbf{q}\lambda}|M_{\mathbf{q}\lambda}|^2\left[\frac{n_{\omega_{\mathbf{q}\lambda}} + f_{\mathbf{k}-\mathbf{q}}}{i\omega_n - \bar{\epsilon}_{\mathbf{k}-\mathbf{q}}/\hbar + \omega_{\mathbf{q}\lambda}} + \frac{1 + n_{\omega_{\mathbf{q}\lambda}} - f_{\mathbf{k}-\mathbf{q}}}{i\omega_n - \bar{\epsilon}_{\mathbf{k}-\mathbf{q}}/\hbar - \omega_{\mathbf{q}\lambda}}\right]. \tag{11.68}$$

The retarded self energy is obtained by replacing $i\omega_n$ with $\omega + i0^+$:

$$Im\,\Sigma_R^*(\mathbf{k}\sigma,\omega) = \frac{-\pi}{\hbar^2}\sum_{\mathbf{q}\lambda}|M_{\mathbf{q}\lambda}|^2\left[(n_{\omega_{\mathbf{q}\lambda}} + f_{\mathbf{k}-\mathbf{q}})\delta(\omega - \bar{\epsilon}_{\mathbf{k}-\mathbf{q}}/\hbar + \omega_{\mathbf{q}\lambda})\right.$$

$$\left. + (1 + n_{\omega_{\mathbf{q}\lambda}} - f_{\mathbf{k}-\mathbf{q}})\delta(\omega - \bar{\epsilon}_{\mathbf{k}-\mathbf{q}}/\hbar - \omega_{\mathbf{q}\lambda})\right]. \tag{11.69}$$

The first (second) term in the brackets corresponds to phonon absorption (emission). The lifetime of an electron in state $|\mathbf{k}\sigma\rangle$ is given by

$$\tau_{\mathbf{k}\sigma} = \frac{-1}{2Im\,\Sigma_R^*(\mathbf{k}\sigma, \bar{\epsilon}_{\mathbf{k}}/\hbar)}. \tag{11.70}$$

In writing Eq. (11.70), we have replaced ω in Σ_R^* with $\bar{\epsilon}_{\mathbf{k}}/\hbar$; this is an approximation. In fact, $\hbar\omega$ should be replaced with the shifted energy, which is obtained by solving the equation $\omega - \bar{\epsilon}_{\mathbf{k}}/\hbar + Re\,\Sigma_R^*(\mathbf{k}\sigma,\omega) = 0$.

11.11 The electromagnetic field

In free space, away from charge and current sources, the electromagnetic field is described by the following Maxwell's equations:

$$\nabla.\mathbf{E} = 0 \tag{11.71a}$$

$$\nabla.\mathbf{B} = 0 \tag{11.71b}$$

$$\nabla\times\mathbf{E} = -\frac{1}{c}\frac{\partial\mathbf{B}}{\partial t}\quad(\text{cgs}),\qquad \nabla\times\mathbf{E} = -\frac{\partial\mathbf{B}}{\partial t}\quad(SI) \tag{11.71c}$$

$$\nabla\times\mathbf{B} = \frac{1}{c}\frac{\partial\mathbf{E}}{\partial t}\quad(\text{cgs}),\qquad \nabla\times\mathbf{B} = \mu_0\epsilon_0\frac{\partial\mathbf{E}}{\partial t}\quad(SI). \tag{11.71d}$$

In the following treatment, we use the cgs system of units. The second and third Maxwell's equations are automatically satisfied if we express $\mathbf{E}$ and $\mathbf{B}$ in terms of a scalar potential $\Phi(\mathbf{r},t)$ and a vector potential $\mathbf{A}(\mathbf{r},t)$:

$$\mathbf{E} = -\nabla\Phi - \frac{1}{c}\frac{\partial\mathbf{A}}{\partial t},\qquad \mathbf{B} = \nabla\times\mathbf{A}. \tag{11.72}$$

This is because the divergence of a curl is zero ($\nabla.\nabla\times\mathbf{A}$) and the curl of a gradient is zero ($\nabla\times\nabla\Phi = 0$). The first and fourth Maxwell's equations are now written as

$$\nabla^2\Phi + \frac{1}{c}\frac{\partial}{\partial t}(\nabla.\mathbf{A}) = 0 \tag{11.73}$$

$$\nabla\times\nabla\times\mathbf{A} = -\frac{1}{c}\frac{\partial}{\partial t}\nabla\Phi - \frac{1}{c^2}\frac{\partial^2}{\partial t^2}\mathbf{A}. \tag{11.74}$$

Using the identity

$$\nabla \times \nabla \times \mathbf{A} = -\nabla^2 \mathbf{A} + \nabla(\nabla . \mathbf{A}), \tag{11.75}$$

we can rewrite Eq. (11.74) in the following form:

$$\nabla^2 \mathbf{A} - \frac{1}{c^2}\frac{\partial^2}{\partial t^2}\mathbf{A} = \nabla\left(\nabla . \mathbf{A} + \frac{1}{c}\frac{\partial \Phi}{\partial t}\right). \tag{11.76}$$

Simplification is achieved by exploiting a freedom in the choice of Φ and $\mathbf{A}$: under the gauge transformation,

$$\mathbf{A} \to \mathbf{A}' = \mathbf{A} + \nabla \Lambda, \quad \Phi \to \Phi' = \Phi - \frac{1}{c}\frac{\partial \Lambda}{\partial t},$$

where $\Lambda(\mathbf{r}, t)$ is any smooth function, both $\mathbf{E}$ and $\mathbf{B}$ remain unchanged. Choosing a particular function $\Lambda(\mathbf{r}, t)$ is called "fixing the gauge." In one gauge, called the radiation gauge, $\Lambda(\mathbf{r}, t)$ is chosen such that $\nabla^2 \Lambda(\mathbf{r}, t) = -\nabla . \mathbf{A}$ and $(1/c)(\partial \Lambda / \partial t) = \Phi$. Thus, in the radiation gauge, $\Phi' = 0$ and $\nabla . \mathbf{A}' = 0$. Relabeling $(\Phi', \mathbf{A}')$ as $(\Phi, \mathbf{A})$, we can write

$$\mathbf{E} = -\frac{1}{c}\frac{\partial \mathbf{A}}{\partial t}, \quad \mathbf{B} = \nabla \times \mathbf{A}, \tag{11.77}$$

where $\mathbf{A}$ satisfies the wave equation

$$\nabla^2 \mathbf{A} - \frac{1}{c^2}\frac{\partial^2 \mathbf{A}}{\partial t^2} = 0. \tag{11.78}$$

This equation is to be solved subject to the constraint $\nabla . \mathbf{A} = 0$. We assume that the electromagnetic field is enclosed in a large cube of volume $V = L^3$, and that it obeys periodic boundary conditions. Our approach is similar to the one we followed in studying atomic vibrations: we first find the normal modes and then write the general solution as a linear combination of these modes. The normal modes are given by

$$\mathbf{A}^{nor}_{\mathbf{q}\lambda}(\mathbf{r}, t) = \frac{1}{\sqrt{V}} \boldsymbol{\epsilon}_\lambda(\mathbf{q}) e^{i(\mathbf{q}.\mathbf{r} - \omega_{\mathbf{q}\lambda} t)}$$

where $\boldsymbol{\epsilon}_\lambda(\mathbf{q})$ is a unit polarization vector. The requirement $\nabla . \mathbf{A} = 0 \Rightarrow \mathbf{q}.\boldsymbol{\epsilon}_\lambda(\mathbf{q}) = 0$; the normal modes are transverse modes, so $\lambda = 1, 2$. Inserting the above expression into the wave equation, we find that the equation is satisfied if $\omega_{\mathbf{q}\lambda} = cq$, independent of λ; henceforth, we write $\omega_{\mathbf{q}}$ and drop the subscript λ. The periodic boundary conditions imply that the allowed values for $\mathbf{q}$ are

$$q_x, q_y, q_z = 0, \pm 2\pi/L, \pm 4\pi/L, \ldots.$$

The general solution of the wave equation is written as

$$\mathbf{A}(\mathbf{r}, t) = \left(\frac{4\pi c^2}{V}\right)^{1/2} \sum_{\mathbf{q}} \sum_{\lambda=1}^{2} A_{\mathbf{q}\lambda} \boldsymbol{\epsilon}_\lambda(\mathbf{q}) e^{i\mathbf{q}.\mathbf{r}}. \tag{11.79}$$

The factor $e^{-i\omega_{\mathbf{q}}t}$ is absorbed into $A_{\mathbf{q}\lambda}$, which satisfies the equation

$$\ddot{A}_{\mathbf{q}\lambda} = -\omega_{\mathbf{q}}^2 A_{\mathbf{q}\lambda}.$$

The expansion coefficients $A_{\mathbf{q}\lambda}$ are the normal coordinates of the electromagnetic field. They are similar to the $Q_{\mathbf{q}\lambda}$ coefficients that appear in the expansion of the ionic displacements in a crystal. The additional factor $(4\pi c^2)^{1/2}$ in Eq. (11.79) is inserted for later convenience. We note that, since $\mathbf{A}(\mathbf{r}, t)$ is real, $\mathbf{A}^*(\mathbf{r}, t) = \mathbf{A}(\mathbf{r}, t)$; it follows that $A^*_{\mathbf{q}\lambda} = A_{-\mathbf{q}\lambda}$ and $\boldsymbol{\epsilon}^*_\lambda(\mathbf{q}) = \boldsymbol{\epsilon}_\lambda(-\mathbf{q})$. Orthonormality of the normal modes implies that $\boldsymbol{\epsilon}^*_\lambda(\mathbf{q}).\boldsymbol{\epsilon}_{\lambda'}(\mathbf{q}) = \delta_{\lambda\lambda'}$.

The electric and magnetic fields are obtained by using Eq. (11.77):

$$\mathbf{E} = -\sqrt{4\pi/V} \sum_{\mathbf{q}\lambda} \dot{A}_{\mathbf{q}\lambda} \boldsymbol{\epsilon}_\lambda(\mathbf{q}) e^{i\mathbf{q}.\mathbf{r}}, \quad \mathbf{B} = i\sqrt{4\pi c^2/V} \sum_{\mathbf{q}\lambda} A_{\mathbf{q}\lambda} \mathbf{q} \times \boldsymbol{\epsilon}_\lambda(\mathbf{q}) e^{i\mathbf{q}.\mathbf{r}}. \tag{11.80}$$

The Lagrangian for the electromagnetic field (Jackson, 1999) is given by

$$L = \frac{1}{8\pi} \int (|\mathbf{E}|^2 - |\mathbf{B}|^2) d^3r \quad \text{(cgs)}. \tag{11.81}$$

Inserting the expressions for $\mathbf{E}$ and $\mathbf{B}$ from Eq. (11.80), and noting that $|\boldsymbol{\epsilon}_\lambda(\mathbf{q}) \times \mathbf{q}| = q$ (since $\boldsymbol{\epsilon}_\lambda(\mathbf{q}) \perp \mathbf{q}$) and $\omega_{\mathbf{q}} = cq$, we can show that

$$L = \frac{1}{2} \sum_{\mathbf{q}\lambda} \left(\dot{A}_{\mathbf{q}\lambda} \dot{A}_{-\mathbf{q}\lambda} - \omega_{\mathbf{q}}^2 A_{\mathbf{q}\lambda} A_{-\mathbf{q}\lambda}\right). \tag{11.82}$$

The momentum conjugate to $A_{\mathbf{q}\lambda}$ is $P_{\mathbf{q}\lambda} = \partial L/\partial \dot{A}_{\mathbf{q}\lambda} = \dot{A}_{-\mathbf{q}\lambda}$. The Hamiltonian is therefore given by

$$H = \frac{1}{2} \sum_{\mathbf{q}\lambda} \left(P_{\mathbf{q}\lambda} P_{-\mathbf{q}\lambda} + \omega_{\mathbf{q}}^2 A_{\mathbf{q}\lambda} A_{-\mathbf{q}\lambda}\right). \tag{11.83}$$

This is the Hamiltonian for a collection of harmonic oscillators. A quantum theory is obtained in a similar way as we did for phonons:

$$H = \sum_{\mathbf{q}\lambda} \hbar\omega_{\mathbf{q}\lambda} \left(b^\dagger_{\mathbf{q}\lambda} b_{\mathbf{q}\lambda} + 1/2\right), \tag{11.84}$$

where

$$A_{\mathbf{q}\lambda} = \sqrt{\frac{\hbar}{2\omega_{\mathbf{q}}}} \left(b_{\mathbf{q}\lambda} + b^\dagger_{-\mathbf{q}\lambda}\right), \quad P_{\mathbf{q}\lambda} = i\sqrt{\frac{\hbar\omega_{\mathbf{q}}}{2}} \left(b^\dagger_{\mathbf{q}\lambda} - b_{-\mathbf{q}\lambda}\right). \tag{11.85}$$

The quanta of the electromagnetic field are called photons. The operator $b^\dagger_{\mathbf{q}\lambda}(b_{\mathbf{q}\lambda})$ creates (annihilates) a photon of wave vector $\mathbf{q}$ and polarization λ.

11.12 Electron–photon interaction

In the presence of an electromagnetic field, described by the vector potential $\mathbf{A}$, the Hamiltonian for an electron in a crystal is

$$H = \frac{1}{2m}\left(\mathbf{p} + \frac{e}{c}\mathbf{A}\right)^2 + V(\mathbf{r}). \tag{11.86}$$

The electron charge is $-e$, and $V(\mathbf{r})$ is the periodic potential produced by the lattice of ions. We ignore the electron–phonon interaction for now. Expanding Eq. (11.86), we obtain

$$H = H_0 + H', \quad H_0 = p^2/2m + V(\mathbf{r}), \quad H' = \frac{e}{2mc}(\mathbf{p}.\mathbf{A} + \mathbf{A}.\mathbf{p}) + \frac{e^2}{2mc^2}A^2.$$

H_0 is the Hamiltonian for the electron in the absence of the electromagnetic field; its eigenstates $|s\mathbf{k}\sigma\rangle$ are characterized by a band index s, a wave vector $\mathbf{k}$, and a spin projection σ. H', when summed over all electrons, is the electron–photon interaction. The term in H' which is proportional to A^2 involves two-photon scattering processes. For weak fields, this term is generally far less important than the other term which involves single-photon processes; henceforth, the A^2 term will be ignored.

In the radiation gauge ($\mathbf{\nabla}.\mathbf{A} = 0$), the two terms $\mathbf{p}.\mathbf{A}$ and $\mathbf{A}.\mathbf{p}$ are equal. To see this, consider the action of $\mathbf{p}.\mathbf{A}$ on any function $f(\mathbf{r})$:

$$\begin{aligned}\mathbf{p}.\mathbf{A}f(\mathbf{r}) &= -i\hbar\mathbf{\nabla}.(\mathbf{A}f(\mathbf{r})) = -i\hbar(\mathbf{\nabla}.\mathbf{A})f(\mathbf{r}) - i\hbar\mathbf{A}.\mathbf{\nabla}f(\mathbf{r}) \\ &= 0 - i\hbar\mathbf{A}.\mathbf{\nabla}f(\mathbf{r}) = \mathbf{A}.\mathbf{p}f(\mathbf{r}).\end{aligned}$$

The electron–photon interaction Hamiltonian is obtained by summing H' over all electrons:

$$H_{e-\text{photon}} = \frac{e}{mc}\sum_j \mathbf{A}(\mathbf{r}_j, t).\mathbf{p}_j. \tag{11.87}$$

$\mathbf{p}_j$ is the momentum of the jth electron whose position is $\mathbf{r}_j$. Since this is a one-body operator, its second quantized form is

$$H_{e-\text{photon}} = \frac{e}{mc}\sum_{s\mathbf{k}\sigma}\sum_{s'\mathbf{k}'}\langle s'\mathbf{k}'\sigma|\mathbf{A}.\mathbf{p}|s\mathbf{k}\sigma\rangle c^\dagger_{s'\mathbf{k}'\sigma}c_{s\mathbf{k}\sigma}.$$

The matrix element is given by

$$\langle s'\mathbf{k}'\sigma|\mathbf{A}.\mathbf{p}|s\mathbf{k}\sigma\rangle = \left(\frac{4\pi c^2}{V\varepsilon(\infty)}\right)^{1/2}\sum_{\mathbf{q}\lambda} A_{\mathbf{q}\lambda}\,\boldsymbol{\epsilon}_\lambda(\mathbf{q}).\langle s'\mathbf{k}'\sigma|e^{i\mathbf{q}.\mathbf{r}}\mathbf{p}|s\mathbf{k}\sigma\rangle.$$

Note that, in expanding $\mathbf{A}(\mathbf{r}, t)$, we have modified Eq. (11.79), replacing c with $c/n = c/\sqrt{\varepsilon(\infty)}$, where $n = \sqrt{\varepsilon(\infty)}$ is the index of refraction of the medium (the crystal). This is because in $H_{e-\text{photon}}$, $\mathbf{A}(\mathbf{r}, t)$ is the vector potential in the medium, where the speed of light is c/n. The matrix element on the RHS of the above equation is evaluated using Bloch's theorem:

$$I = \langle s'\mathbf{k}'\sigma|e^{i\mathbf{q}.\mathbf{r}}\mathbf{p}|s\mathbf{k}\sigma\rangle = \int u^*_{s'\mathbf{k}'}(\mathbf{r})e^{-i\mathbf{k}'.\mathbf{r}}e^{i\mathbf{q}.\mathbf{r}}\mathbf{p}u_{s\mathbf{k}}(\mathbf{r})e^{i\mathbf{k}.\mathbf{r}}d^3r.$$

Here, $u_{s'\mathbf{k}'}(\mathbf{r})$ and $u_{s\mathbf{k}}(\mathbf{r})$ are periodic functions having the same periodicity as the lattice. Changing variables from $\mathbf{r}$ to $\mathbf{r} + \mathbf{R}_n$, where $\mathbf{R}_n$ is any lattice vector, and using the periodicity property of $u_{s'\mathbf{k}'}(\mathbf{r})$ and $u_{s\mathbf{k}}(\mathbf{r})$, we find

$$I = Ie^{-i(\mathbf{k}'-\mathbf{k}-\mathbf{q}).\mathbf{R}_n}.$$

This means that $\mathbf{k}' = \mathbf{k} + \mathbf{q} + \mathbf{G}$, where $\mathbf{G}$ is a reciprocal lattice vector. Since $\mathbf{k}, \mathbf{k}' \in$ FBZ, $\mathbf{G}$ must carry $\mathbf{k} + \mathbf{q}$ into the FBZ. For visible light, $q \approx 10^5$ cm^{-1}, which is too small compared to the width of the Brillouin zone ($\approx 10^8$ cm^{-1}). Thus, $\mathbf{G}$ is generally equal to zero unless $\mathbf{k}$ is extremely close to the Brillouin zone edge. The electron–photon interaction is therefore given by

$$H_{e-\text{photon}} = \sum_{ss'}\sum_{\mathbf{k}\sigma}\sum_{\mathbf{q}\lambda} P^{ss'}_{\mathbf{k},\mathbf{k}+\mathbf{q}}(\lambda)c^\dagger_{s'\mathbf{k}+\mathbf{q}\sigma}c_{s\mathbf{k}\sigma}(b_{\mathbf{q}\lambda} + b^\dagger_{-\mathbf{q}\lambda}) \tag{11.88}$$

where

$$P^{ss'}_{\mathbf{k},\mathbf{k}+\mathbf{q}}(\lambda) = \frac{e}{m}\left(\frac{2\pi\hbar}{V\omega_{\mathbf{q}}\varepsilon(\infty)}\right)^{1/2}\langle s'\mathbf{k}+\mathbf{q}\sigma|e^{i\mathbf{q}.\mathbf{r}}\boldsymbol{\epsilon}_\lambda(\mathbf{q}).\mathbf{p}|s\mathbf{k}\sigma\rangle. \tag{11.89}$$

The electron–photon interaction is thus seen to be a sum of terms, each of which represents a scattering process whereby an electron in state $|s\mathbf{k}\sigma\rangle$ is scattered into state $|s'\mathbf{k}+\mathbf{q}\sigma\rangle$ by the absorption (emission) of a photon of wave vector $\mathbf{q}\,(-\mathbf{q})$ and polarization λ.

11.13 Light scattering by crystals

In this section, we discuss the general theory of light scattering by crystals (Van Hove, 1954; Loudon, 1963). In the next section we will focus on the specific case of Raman scattering in insulators.

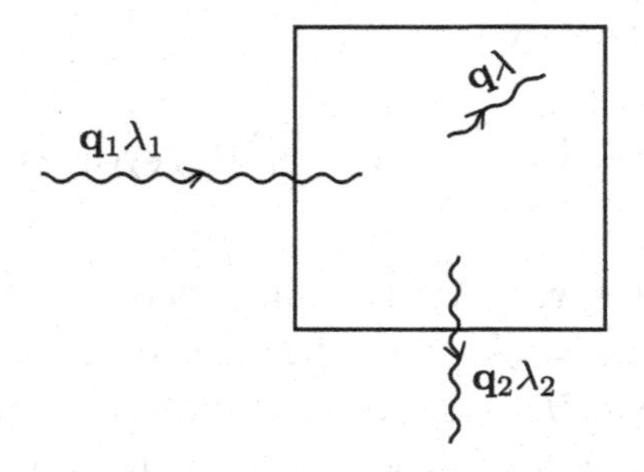

Figure 11.15 Light scattering by a crystal: an incident photon of frequency $\omega_{\mathbf{q}_1\lambda_1}$ is absorbed, and a photon of frequency $\omega_{\mathbf{q}_2\lambda_2}$ is emitted. In the process, a quantum of an excitation mode of the crystal is created or annihilated in order to conserve energy and momentum.

Consider a process in which a photon of wave vector $\mathbf{q}_1$ and polarization λ_1 is absorbed by a crystal, and a photon of coordinates $(\mathbf{q}_2\lambda_2)$ is created. The process is accompanied by the creation or annihilation of a quantum of an excitation mode of the crystal, having coordinates $(\mathbf{q}\lambda)$, such that momentum and energy are conserved. The process is depicted in Figure 11.15.

The scattering process is described by the Hamiltonian

$$H' = \sum_{\mathbf{q}\lambda} \sum_{\mathbf{q}'\lambda'} \sum_{\mathbf{q}''\lambda''} \Gamma(\mathbf{q}'\lambda', \mathbf{q}''\lambda'', \mathbf{q}\lambda)\phi^{\dagger}_{\mathbf{q}\lambda} b^{\dagger}_{\mathbf{q}''\lambda''} b_{\mathbf{q}'\lambda'} \tag{11.90}$$

where $b^{\dagger}_{\mathbf{q}\lambda}(b_{\mathbf{q}\lambda})$ creates (annihilates) a photon of coordinates $(\mathbf{q}\lambda)$, and

$$\phi^{\dagger}_{\mathbf{q}\lambda} = a^{\dagger}_{\mathbf{q}\lambda} + a_{-\mathbf{q}\lambda} \tag{11.91}$$

is the field operator for the excitation mode (phonon or plasmon, for example) of the crystal. Γ is the matrix element for the scattering process; it is determined by considering the detailed mechanism through which the process takes place. In the next section, we will calculate this quantity for the specific case of Raman scattering by an insulating crystal.

The initial and final states of the system, which consists of the photons and the crystal, are denoted by $|I\rangle$ and $|F\rangle$, respectively:

$$|I\rangle = |n_1\rangle|n_2\rangle|i\rangle, \quad |F\rangle = |n_1 - 1\rangle|n_2 + 1\rangle|f\rangle.$$

Here, n_1 is the number of incident photons of coordinates $(\mathbf{q}_1\lambda_1)$, n_2 is the number of scattered photons of coordinates $(\mathbf{q}_2\lambda_2)$, $|i\rangle$ is the initial state of the crystal, and $|f\rangle$ is its final state.

The probability per unit time (the transition rate) for scattering from the initial state is given by the Fermi golden rule,

$$W = \frac{2\pi}{\hbar} \sum_F |\langle F|H'|I\rangle|^2 \delta(E_F - E_I). \tag{11.92}$$

E_I and E_F are the energies of the initial and final states. Writing $E_I = \hbar\omega_I$, $E_F = \hbar\omega_F$, and using

$$b_{\mathbf{q}\lambda}|n_{\mathbf{q}\lambda}\rangle = \sqrt{n_{\mathbf{q}\lambda}}|n_{\mathbf{q}\lambda} - 1\rangle, \quad b^\dagger_{\mathbf{q}\lambda}|n_{\mathbf{q}\lambda}\rangle = \sqrt{n_{\mathbf{q}\lambda} + 1}|n_{\mathbf{q}\lambda} + 1\rangle,$$

we obtain

$$W = \frac{2\pi}{\hbar^2} \sum_{\mathbf{q}\lambda} \sum_{\mathbf{q}'\lambda'} \sum_f \Gamma(\mathbf{q}_1\lambda_1, \mathbf{q}_2\lambda_2, \mathbf{q}\lambda)\Gamma^*(\mathbf{q}_1\lambda_1, \mathbf{q}_2\lambda_2, \mathbf{q}'\lambda')$$
$$\times \langle i|\phi_{\mathbf{q}'\lambda'}|f\rangle\langle f|\phi^\dagger_{\mathbf{q}\lambda}|i\rangle n_1(n_2+1)\delta(\omega_f - \omega_i - \omega), \tag{11.93}$$

where $\hbar\omega_i$ ($\hbar\omega_f$) is the energy of the crystal's initial (final) state, and $\hbar\omega = \hbar\omega_1 - \hbar\omega_2$ is the energy transferred to the crystal. Noting that

$$\delta(\omega_f - \omega_i - \omega) = \frac{1}{2\pi} \int_{-\infty}^{\infty} e^{-i(\omega_f - \omega_i - \omega)t} dt,$$

the expression for W becomes

$$W = \frac{n_1(n_2+1)}{\hbar^2} \sum_{\mathbf{q}\lambda} \sum_{\mathbf{q}'\lambda'} \sum_f \Gamma(\mathbf{q}_1\lambda_1, \mathbf{q}_2\lambda_2, \mathbf{q}\lambda)\Gamma^*(\mathbf{q}_1\lambda_1, \mathbf{q}_2\lambda_2, \mathbf{q}'\lambda')$$
$$\times \int \langle i|\phi_{\mathbf{q}'\lambda'}|f\rangle\langle f|\phi^\dagger_{\mathbf{q}\lambda}|i\rangle e^{-i(\omega_f - \omega_i - \omega)t} dt$$
$$= \frac{n_1(n_2+1)}{\hbar^2} \sum_{\mathbf{q}\lambda} \sum_{\mathbf{q}'\lambda'} \sum_f \Gamma(\mathbf{q}_1\lambda_1, \mathbf{q}_2\lambda_2, \mathbf{q}\lambda)\Gamma^*(\mathbf{q}_1\lambda_1, \mathbf{q}_2\lambda_2, \mathbf{q}'\lambda')$$
$$\times \int \langle i|e^{iHt/\hbar}\phi_{\mathbf{q}'\lambda'}e^{-iHt/\hbar}|f\rangle\langle f|\phi^\dagger_{\mathbf{q}\lambda}|i\rangle e^{i\omega t} dt,$$

where H is the crystal Hamiltonian. The sum over the final states of the crystal is now carried out ($\sum_f |f\rangle\langle f| = 1$); we obtain

$$W = \frac{n_1(n_2+1)}{\hbar^2} \sum_{\mathbf{q}\lambda} |\Gamma(\mathbf{q}_1\lambda_1, \mathbf{q}_2\lambda_2, \mathbf{q}\lambda)|^2 \int_{-\infty}^{\infty} \langle i|\phi_{\mathbf{q}\lambda}(t)\phi^\dagger_{\mathbf{q}\lambda}(0)|i\rangle e^{i\omega t} dt. \tag{11.94}$$

The transition rate depends on the initial state of the crystal. At zero temperature, $|i\rangle$ is the ground state $|\Psi_0\rangle$ of the crystal, and its energy is E_0. At finite temperature, other states $|\Psi_n\rangle$ with energy E_n have a nonzero probability of being occupied. Hence, at finite temperature, the matrix element in Eq. (11.94) is replaced by the thermal average $\langle\phi_{\mathbf{q}\lambda}(t)\phi^\dagger_{\mathbf{q}\lambda}(0)\rangle$, which is the correlation function $C(\mathbf{q}\lambda, t)$ of the excitation mode in the crystal. The integral in Eq. (11.94) then becomes the Fourier transform of $C(\mathbf{q}\lambda, t)$,

$$W = \frac{n_1(n_2+1)}{\hbar^2}\sum_{\mathbf{q}\lambda}|\Gamma(\mathbf{q}_1\lambda_1, \mathbf{q}_2\lambda_2, \mathbf{q}\lambda)|^2 C(\mathbf{q}\lambda, \omega). \tag{11.95}$$

The differential scattering cross-section $d^2\sigma/d\omega d\Omega$ is the number of transitions per unit time per unit solid angle per unit frequency interval per unit incident flux. The number of transitions per unit time into a solid angle $d\Omega$ and a frequency interval $(\omega_2, \omega_2 + d\omega)$ is obtained by multiplying W by the number of photon states, of a given polarization, in the interval $d\omega d\Omega$. The number of such photon states is $(V/8\pi^3)q_2^2 dq d\Omega = (V/8\pi^3 c^3)\omega_2^2 d\omega d\Omega$. The incident flux is the number of photons striking a unit area of the crystal per unit time. During a time interval Δt, the photons in a volume $Ac\Delta t$ strike an area A of the crystal (the incident light is assumed to be normal to the surface of the crystal). Since the number of incident photons per unit volume is n_1/V, the incident flux is $n_1 c/V$. Therefore,

$$\frac{d^2\sigma}{d\omega d\Omega} = \frac{(n_2+1)V^2\omega_2^2}{8\pi^3\hbar^2c^4}\sum_{\mathbf{q}\lambda}|\Gamma(\mathbf{q}_1\lambda_1, \mathbf{q}_2\lambda_2, \mathbf{q}\lambda)|^2 C(\mathbf{q}\lambda, \omega).$$

Using the fluctuation–dissipation theorem (see Eq. [6.49]) which relates the correlation function to the imaginary part of the retarded Green's function, our final result is

$$\frac{d^2\sigma}{d\omega d\Omega} = -\frac{(n_\omega+1)(n_2+1)V^2\omega_2^2}{8\pi^3\hbar^2c^4}\sum_{\mathbf{q}\lambda}|\Gamma(\mathbf{q}_1\lambda_1, \mathbf{q}_2\lambda_2, \mathbf{q}\lambda)|^2 Im D^R(\mathbf{q}\lambda, \omega) \tag{11.96}$$

where $n_\omega = (e^{\beta\hbar\omega} - 1)^{-1}$ is the Bose–Einstein distribution function. We have managed to express the differential scattering cross-section for light scattering by a crystal in terms of the retarded function of the crystal excitation that participates in the scattering process. This function is obtained by analytic continuation of the corresponding imaginary-time function which, in turn, can be calculated using the Feynman diagram techniques developed in previous chapters.

11.14 Raman scattering in insulators

In a Raman scattering experiment, a photon of frequency ω_1, incident on a crystal, is absorbed, and a photon of frequency ω_2 is created. The process is accompanied by the emission (Stoke's scattering) or the absorption (anti-Stoke's scattering) of an optical phonon. At low temperatures, the optical phonon occupation number is small; hence, Raman scattering processes in which a phonon is created are generally more important than those in which a phonon is absorbed. Here, we consider Raman scattering with phonon emission.

Our system consists of a crystal (an insulator), the radiation (electromagnetic) field, and lattice vibrations (phonons). The Hamiltonian is

$$H = H_0 + H_{ER} + H_{EL} = H_0 + H'. \tag{11.97}$$

H_0 is the sum of the Hamiltonians for the electrons in the crystal, the radiation field, and the lattice vibrations. H_{ER} is the electron–photon interaction, and H_{EL} is the electron–phonon interaction:

$$H_{ER} = \sum_{ss'}\sum_{\mathbf{k}\sigma}\sum_{\mathbf{q}\lambda} P^{ss'}_{\mathbf{k},\mathbf{k}+\mathbf{q}}(\lambda) c^\dagger_{s'\mathbf{k}+\mathbf{q}\sigma} c_{s\mathbf{k}\sigma}(b_{\mathbf{q}\lambda} + b^\dagger_{-\mathbf{q}\lambda}) \tag{11.98}$$

$$H_{EL} = \sum_{ss'}\sum_{\mathbf{k}\sigma}\sum_{\mathbf{q}\lambda} M^{ss'}_{\mathbf{q}\lambda} c^\dagger_{s'\mathbf{k}+\mathbf{q}\sigma} c_{s\mathbf{k}\sigma}(a_{\mathbf{q}\lambda} + a^\dagger_{-\mathbf{q}\lambda}). \tag{11.99}$$

The process of interest involves the annihilation of a photon, the creation of another photon, and the creation of a phonon. This is a third-order process in which H_{ER} acts twice and H_{EL} acts once. We need to transform the Hamiltonian into a form that contains an effective photon–phonon interaction in which a photon is scattered and a phonon is created. The Hamiltonian, as given above, is written in terms of a certain basis set that spans the Hilbert space of the system (electrons, phonons, and photons). The basis consists of states $|m\rangle = |s\mathbf{k}\sigma\rangle|\mathbf{q}\lambda\rangle_{\text{phonon}}|\mathbf{q}'\lambda'\rangle_{\text{photon}}$. In this basis, H_0 is diagonal. We can transform to a new basis set of states $|\tilde{m}\rangle = U|m\rangle$, where $U^\dagger U = 1$. In the new basis, the Hamiltonian matrix elements are $\langle\tilde{m}|H|\tilde{n}\rangle = \langle m|U^\dagger H U|n\rangle = \langle m|\tilde{H}|n\rangle$, where $\tilde{H} = U^\dagger H U$. In other words, the change of basis is equivalent to applying a similarity transformation to the Hamiltonian. We thus consider the following similarity (canonical) transformation: ($U = e^{-S}$)

$$\tilde{H} = e^S H e^{-S} \tag{11.100}$$

where S is an operator such that $S^\dagger = -S$. $\tilde{H}$ has the same eigenvalues as H, and its eigenstates are obtained by the operator e^S acting on the corresponding eigenstates of H. Expanding $e^{\pm S}$, we obtain

$$\begin{aligned}\tilde{H} &= (1 + S + S^2/2! + S^3/3! + \cdots)H(1 - S + S^2/2! - S^3/3! + \cdots)\\ &= H + [S,H] + \frac{1}{2!}[S,[S,H]] + \frac{1}{3!}[S,[S,[S,H]]] + \cdots\\ &= H_0 + H' + [S,H_0] + [S,H'] + \frac{1}{2}[S,[S,H_0]] + \frac{1}{2}[S,[S,H']]\\ &\quad + \frac{1}{6}[S,[S,[S,H_0]]] + \cdots .\end{aligned}$$

The operator S is now chosen such that

$$[S, H_0] = -H'. \qquad (11.101)$$

With this choice,

$$\tilde{H} = H_0 + \frac{1}{2}[S, H'] + \frac{1}{3}[S, [S, H']] + \cdots$$
$$= H_0 + H_2 + H_3 + \cdots. \qquad (11.102)$$

Consider any two eigenstates $|I\rangle$ and $|F\rangle$ of H_0. Equation (11.101) gives

$$\langle F|SH_0|I\rangle - \langle F|H_0 S|I\rangle = -\langle F|H'|I\rangle \Rightarrow (E_F - E_I)\langle F|S|I\rangle = \langle F|H'|I\rangle$$
$$\Longrightarrow \langle F|S|I\rangle = \frac{\langle F|H'|I\rangle}{E_F - E_I}. \qquad (11.103)$$

This shows that S is proportional to H'. Since our interest is in third-order processes, we consider the third term on the RHS of Eq. (11.102):

$$H_3 = \frac{1}{3}[S, [S, H']] = \frac{1}{3}(S^2 H' - 2SH'S + H'S^2).$$

Using Eq. (11.103) and two resolutions of identity, we can write

$$\langle F|H_3|I\rangle = \frac{1}{3}\sum_{m,n} \frac{\langle F|H'|n\rangle\langle n|H'|m\rangle\langle m|H'|I\rangle}{(E_F - E_n)(E_n - E_m)(E_m - E_I)}[E_F - E_I + 3(E_m - E_n)].$$

In Fermi's golden rule, the initial and final states have the same energy; setting $E_F = E_I$, we obtain

$$\langle F|H_3|I\rangle = \sum_{m,n} \frac{\langle F|H'|n\rangle\langle n|H'|m\rangle\langle m|H'|I\rangle}{(E_I - E_n)(E_I - E_m)}. \qquad (11.104)$$

In Raman scattering with phonon emission, the initial state $|I\rangle$ consists of an incident photon of coordinates $(\mathbf{q}_1\lambda_1)$ and frequency ω_1, and a crystal (an insulator) in its electronic ground state (all of its valence bands are occupied and all of its conduction bands are empty). The final state $|F\rangle$ consists of a scattered photon of coordinates $(\mathbf{q}_2\lambda_2)$ and frequency ω_2 and a phonon of coordinates $(\mathbf{q}\lambda)$ and frequency $\omega = \omega_1 - \omega_2$. In state $|F\rangle$, the crystal is still in its electronic ground state. For these initial and final states, $\langle F|H_3|I\rangle$ corresponds to the matrix element $\Gamma(\mathbf{q}_1\lambda_1, \mathbf{q}_2\lambda_2, \mathbf{q}\lambda)$ in Eq. (11.90).

Replacing H' by $H_{\mathrm{ER}} + H_{\mathrm{EL}}$, the product of the three matrix elements on the RHS of Eq. (11.104) becomes a sum of eight terms, each of which is a product of three matrix elements. Of these eight terms, three are nonzero: $\langle F|H_{\mathrm{EL}}|n\rangle\langle n|H_{\mathrm{ER}}|m\rangle\langle m|H_{\mathrm{ER}}|I\rangle$, $\langle F|H_{\mathrm{ER}}|n\rangle\langle n|H_{\mathrm{EL}}|m\rangle\langle m|H_{\mathrm{ER}}|I\rangle$, and $\langle F|H_{\mathrm{ER}}|n\rangle\langle n|H_{\mathrm{ER}}|m\rangle\langle m|H_{\mathrm{EL}}|I\rangle$; this follows from the definition of $|I\rangle$ and $|F\rangle$.

Furthermore, in $H_{\rm ER}$, a photon is either emitted or absorbed, so that we can write $H_{\rm ER} = H_{\rm ER}^{em} + H_{\rm ER}^{ab}$, and each of the surviving three terms now becomes a sum of four terms, only two of which are nonzero (the number of photons in $|I\rangle$ and $|F\rangle$ is the same). Therefore, the matrix element for Stoke's Raman scattering (phonon emission) is

$$\begin{aligned}\langle F|H|I\rangle = \sum_{m,n} \frac{1}{(E_I - E_n)(E_I - E_m)} \big[&\langle F|H_{\rm EL}^{em}|n\rangle\langle n|H_{\rm ER}^{em}|m\rangle\langle m|H_{\rm ER}^{ab}|I\rangle \\ &+ \langle F|H_{\rm EL}^{em}|n\rangle\langle n|H_{\rm ER}^{ab}|m\rangle\langle m|H_{\rm ER}^{em}|I\rangle + \langle F|H_{\rm ER}^{em}|n\rangle\langle n|H_{\rm EL}^{em}|m\rangle\langle m|H_{\rm ER}^{ab}|I\rangle \\ &+ \langle F|H_{\rm ER}^{ab}|n\rangle\langle n|H_{\rm EL}^{em}|m\rangle\langle m|H_{\rm ER}^{em}|I\rangle + \langle F|H_{\rm ER}^{ab}|n\rangle\langle n|H_{\rm ER}^{em}|m\rangle\langle m|H_{\rm EL}^{em}|I\rangle \\ &+ \langle F|H_{\rm ER}^{em}|n\rangle\langle n|H_{\rm ER}^{ab}|m\rangle\langle m|H_{\rm EL}^{em}|I\rangle\big]. \qquad (11.105)\end{aligned}$$

The various coupling Hamiltonians appearing in the above equation are

$$H_{\rm EL}^{em} = \sum_{ss'}\sum_{\mathbf{k}\sigma} M_{-\mathbf{q}\lambda}^{ss'} c_{s'\mathbf{k}-\mathbf{q}\sigma}^{\dagger} c_{s\mathbf{k}\sigma} a_{\mathbf{q}\lambda}^{\dagger} \qquad (11.106a)$$

$$H_{\rm ER}^{em} = \sum_{ss'}\sum_{\mathbf{k}\sigma} P_{\mathbf{k},\mathbf{k}-\mathbf{q}_2}^{ss'}(\lambda_2) c_{s'\mathbf{k}-\mathbf{q}_2\sigma}^{\dagger} c_{s\mathbf{k}\sigma} b_{\mathbf{q}_2\lambda_2}^{\dagger} \qquad (11.106b)$$

$$H_{\rm ER}^{ab} = \sum_{ss'}\sum_{\mathbf{k}\sigma} P_{\mathbf{k},\mathbf{k}+\mathbf{q}_1}^{ss'}(\lambda_1) c_{s'\mathbf{k}+\mathbf{q}_1\sigma}^{\dagger} c_{s\mathbf{k}\sigma} b_{\mathbf{q}_1\lambda_1}. \qquad (11.106c)$$

We have not summed over the wave vectors and polarizations of the photons and phonons since we consider the absorption of a photon of specific coordinates $(\mathbf{q}_1\lambda_1)$, the emission of a photon of specific coordinates $(\mathbf{q}_2\lambda_2)$, and the emission of a phonon of specific coordinates $(\mathbf{q}\lambda)$. As for the insulator (the crystal), we only consider transitions between the highest occupied band and the lowest empty band, and we assume that the energy of the incident photon is lower than the energy gap, so that the transitions are virtual processes. Now we consider the various terms in Eq. (11.105).

The sequence of processes that occur in the first term on the RHS of Eq. (11.105) is illustrated in Figure 11.16. Here, $E_I = \hbar\omega_1$, $E_m = \epsilon_{c\mathbf{k}+\mathbf{q}_1} - \epsilon_{v\mathbf{k}} \simeq \epsilon_{c\mathbf{k}} - \epsilon_{v\mathbf{k}}$ (since q_1 is much smaller than the width of the Brillouin zone). We also assume that there is little dispersion in the bands such that $\epsilon_{c\mathbf{k}} - \epsilon_{v\mathbf{k}} \simeq E_g$, where E_g is the energy gap. The first term on the RHS of Eq. (11.105) is thus approximately equal to

$$\delta_{\mathbf{q},\mathbf{q}_1-\mathbf{q}_2} \sum_{\mathbf{k}\sigma} \frac{M_{\mathbf{q}\lambda}^{cv}\big[P_{\mathbf{k}+\mathbf{q}_1,\mathbf{k}+\mathbf{q}_1-\mathbf{q}_2}^{cc}(\lambda_2) + P_{\mathbf{k}+\mathbf{q}_2,\mathbf{k}}^{vv}(\lambda_2)\big]P_{\mathbf{k},\mathbf{k}+\mathbf{q}_1}^{vc}(\lambda_1)}{(\hbar\omega_1 - \hbar\omega_2 - E_g)(\hbar\omega_1 - E_g)}.$$

We can simplify the notation: since the photon wave vector is very small compared to the extent of the Brillouin zone, we can replace $P_{\mathbf{k}+\mathbf{q}_1,\mathbf{k}+\mathbf{q}_1-\mathbf{q}_2}^{cc}(\lambda_2)$ with $P_{\mathbf{k}}^{cc}(\lambda_2)$, i.e., we assume that the electron–photon matrix element depends only on $\mathbf{k}$. Similar

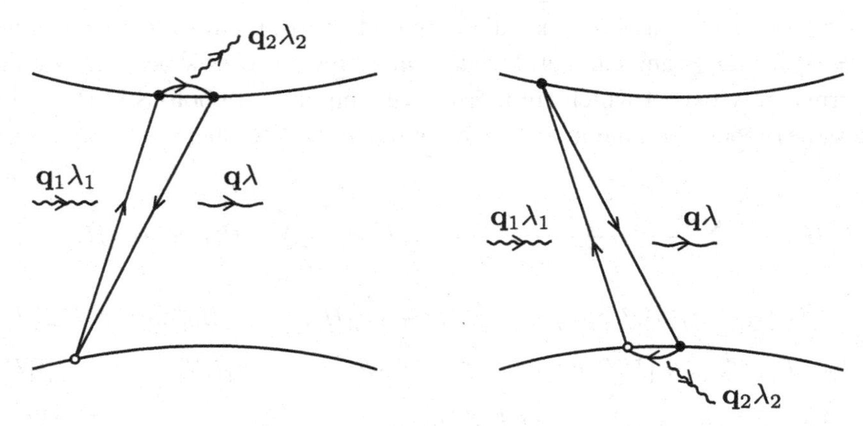

Figure 11.16 The sequence of processes in the first term of Eq. (11.105). A photon $(\mathbf{q}_1\lambda_1)$ is absorbed and an electron is promoted to the conduction band c (the upper band), leaving behind a hole in the valence band v. The next step can proceed in two different ways: in (a), the electron is scattered in the conduction band and a photon $(\mathbf{q}_2\lambda_2)$ is emitted, while in (b) an electron in the valence band is scattered and a photon $(\mathbf{q}_2\lambda_2)$ is emitted (we can also say that the hole is scattered). The third step, common to both, is an electron–hole recombination accompanied by the emission of a phonon $(\mathbf{q}\lambda)$. Note that for this sequence of processes, $\mathbf{q} = \mathbf{q}_1 - \mathbf{q}_2$, which is a statement of wave vector conservation.

replacements are made for other P-matrix elements. The above expression is now written as

$$\delta_{\mathbf{q},\mathbf{q}_1-\mathbf{q}_2} \sum_{\mathbf{k}\sigma} \frac{M^{cv}_{\mathbf{q}\lambda}\left[P^{cc}_{\mathbf{k}}(\lambda_2) + P^{vv}_{\mathbf{k}}(\lambda_2)\right] P^{vc}_{\mathbf{k}}(\lambda_1)}{(\hbar\omega_1 - \hbar\omega_2 - E_g)(\hbar\omega_1 - E_g)}.$$

The remaining five terms are evaluated in a similar way; we obtain

$$\begin{aligned}
\Gamma(\mathbf{q}_1\lambda_1, \mathbf{q}_2\lambda_2, \mathbf{q}\lambda) = \delta_{\mathbf{q},\mathbf{q}_1-\mathbf{q}_2} \sum_{\mathbf{k}\sigma} \Bigg[& \frac{M^{cv}_{\mathbf{q}\lambda}\left[P^{cc}_{\mathbf{k}}(\lambda_2) + P^{vv}_{\mathbf{k}}(\lambda_2)\right] P^{vc}_{\mathbf{k}}(\lambda_1)}{(\hbar\omega_1 - \hbar\omega_2 - E_g)(\hbar\omega_1 - E_g)} \\
& + \frac{M^{cv}_{\mathbf{q}\lambda}\left[P^{cc}_{\mathbf{k}}(\lambda_1) + P^{vv}_{\mathbf{k}}(\lambda_1)\right] P^{vc}_{\mathbf{k}}(\lambda_2)}{(\hbar\omega_1 - \hbar\omega_2 - E_g)(-\hbar\omega_2 - E_g)} + \frac{P^{cv}_{\mathbf{k}}(\lambda_2)\left[M^{cc}_{\mathbf{q}\lambda} + M^{vv}_{\mathbf{q}\lambda}\right] P^{vc}_{\mathbf{k}}(\lambda_1)}{(\hbar\omega_1 - \hbar\omega_{\mathbf{q}\lambda} - E_g)(\hbar\omega_1 - E_g)} \\
& + \frac{P^{cv}_{\mathbf{k}}(\lambda_1)\left[M^{cc}_{\mathbf{q}\lambda} + M^{vv}_{\mathbf{q}\lambda}\right] P^{vc}_{\mathbf{k}}(\lambda_2)}{(\hbar\omega_2 + \hbar\omega_{\mathbf{q}\lambda} + E_g)(\hbar\omega_2 + E_g)} + \frac{P^{cv}_{\mathbf{k}}(\lambda_1)\left[P^{cc}_{\mathbf{k}}(\lambda_2) + P^{vv}_{\mathbf{k}}(\lambda_2)\right] M^{vc}_{\mathbf{q}\lambda}}{(\hbar\omega_2 + \hbar\omega_{\mathbf{q}\lambda} + E_g)(\hbar\omega_{\mathbf{q}\lambda} + E_g)} \\
& + \frac{P^{cv}_{\mathbf{k}}(\lambda_2)\left[P^{cc}_{\mathbf{k}}(\lambda_1) + P^{vv}_{\mathbf{k}}(\lambda_1)\right] M^{vc}_{\mathbf{q}\lambda}}{(\hbar\omega_1 - \hbar\omega_{\mathbf{q}\lambda} - E_g)(-\hbar\omega_{\mathbf{q}\lambda} - E_g)} \Bigg].
\end{aligned} \tag{11.107}$$

The Kronecker delta ensures that momentum is conserved. If the energy of the incident photon is close to the energy gap ($\hbar\omega_1 \lesssim E_g$), then the third term on the

RHS of Eq. (11.107) dominates over the other five terms; this is called resonant Raman scattering.

In obtaining the above result for Γ, we have assumed that the intermediate states consist of an electron in the conduction band and a hole in the valence band. The electron and the hole were treated as independent particles. In insulators, the attraction between electrons and holes may be significant, leading to the formation of excitons, which are bound electron–hole pairs (Kittel, 2005). A description of Raman scattering in insulators which takes exciton formation into account can be formulated (Ganguly and Birman, 1967), but we will stop here and not take that road.

Further reading

Hayes, W. and Loudon, R. (2004). *Scattering of Light by Crystals*. New York: Dover.
Loudon, R. (2000). *The Quantum Theory of Light*, 3rd edn. Oxford: Oxford University Press.
Madelung, O. (1978). *Introduction to Solid State Theory*. Berlin: Springer.
Venkataraman, G., Feldkamp, L.A., and Sahni, V.C. (1975). *Dynamics of Perfect Crystals*. Cambridge, MA: MIT Press.
Ziman, J.M. (1960). *Electrons and Phonons*. Oxford: Oxford University Press.

Problems

11.1 *Lesser free-phonon Green's function*. Derive Eq. (11.61).

11.2 *Electron self energy*. Using the Feynman diagram rules, derive the expression for the electron self energy, given in Eq. (11.68), due to electron–phonon interaction.

11.3 *Phonon self energy*. Consider a system of electrons and phonons with a Hamiltonian

$$H = \sum_{\mathbf{k}\sigma} \epsilon_{\mathbf{k}} c^{\dagger}_{\mathbf{k}\sigma} c_{\mathbf{k}\sigma} + \sum_{\mathbf{q}\lambda} \hbar\omega_{\mathbf{q}\lambda}(a^{\dagger}_{\mathbf{q}\lambda} a_{\mathbf{q}\lambda} + 1/2) + \sum_{\mathbf{k}\sigma}\sum_{\mathbf{q}\lambda} M_{\mathbf{q}\lambda} c^{\dagger}_{\mathbf{k}+\mathbf{q}\sigma} c_{\mathbf{k}\sigma} \phi_{\mathbf{q}\lambda}.$$

(a) Write down the perturbation expansion for the phonon Green's function $d(\mathbf{q}\lambda, \tau)$.

(b) Using Wick's theorem, obtain $d(\mathbf{q}\lambda, \tau)$ to second order in the electron–phonon interaction.

(c) Show that, to second order in the electron–phonon interaction,

$$d(\mathbf{q}\lambda, \omega_m) = d^0(\mathbf{q}\lambda, \omega_m) + \frac{V}{\hbar}|M_{\mathbf{q}}\lambda|^2 d^0(\mathbf{q}\lambda, \omega_m)\Pi^0(\mathbf{q}, \omega_m) d^0(\mathbf{q}\lambda, \omega_m),$$

where $\Pi^0(\mathbf{q}, \omega_m)$ is the polarizability of noninteracting electrons.

Figure 11.17 Phonon Green's function in the random phase approximation. The shaded bubble is the polarizability of interacting electrons. The single wavy line represents the bare (noninteracting) phonon Green's function, while the double wavy line represents the dressed (interacting) phonon Green's function.

11.4 *Electron–phonon interaction in the jellium model.* In the jellium model of a metal, the positive ions are replaced by a positive background of constant charge density. The longitudinal phonon frequency reduces to the ionic plasma frequency $\Omega_q = (4\pi Z^2 e^2 n_i/M)^{1/2}$, where M is the ionic mass, Ze is the ionic charge, and $n_i = N/V$ is the number of ions per unit volume. The longitudinal modes are assumed to be dispersionless. In reality, the phonon frequency approaches zero as $q \to 0$. The inclusion of electron–phonon interaction is necessary to produce the correct behavior.

In the jellium model, the electron–phonon interaction is obtained from Eq. (11.40) by the replacements: $\omega_{\mathbf{q}\lambda} \to \Omega$, $\mathbf{G} \to \mathbf{0}$, $V_{\mathbf{q}} \to -4\pi Ze^2/q^2$. Thus,

$$H_{e\text{-phonon}} = \sum_{\mathbf{k}\sigma\mathbf{q}} M_{\mathbf{q}} c^{\dagger}_{\mathbf{k}+\mathbf{q}\sigma} c_{\mathbf{k}\sigma} \phi_{\mathbf{q}\lambda}, \qquad M_{\mathbf{q}} = i\frac{4\pi Ze^2}{Vq}\sqrt{\frac{N\hbar}{2M\Omega}}\,\hat{\mathbf{q}}.\boldsymbol{\epsilon}_L(\mathbf{q}).$$

In the random phase approximation, the dressed phonon Green's function is depicted in Figure 11.17. It is given by

$$d(\mathbf{q},\omega_m) = d^0(\mathbf{q},\omega_m) + \frac{V}{\hbar}|M_{\mathbf{q}}|^2 d^0(\mathbf{q},\omega_m)\Pi(\mathbf{q},\omega_m) d(\mathbf{q},\omega_m),$$

where $\Pi(\mathbf{q},\omega_m)$ is the polarizability of interacting electrons. Since the phonon frequency is much smaller than the electron plasmon frequency, we are justified in replacing $\Pi(\mathbf{q},\omega_m)$ with $\Pi(\mathbf{q},0)$.

Show that, as $q \to 0$ the renormalized phonon frequency is given by $\omega = vq$. What is the value of v?

11.5 *Electromagnetic field Lagrangian.* Define the 4-vectors

$$\partial^\mu = \left(\frac{1}{c}\frac{\partial}{\partial t}, -\nabla\right), \quad \partial_\mu = \left(\frac{1}{c}\frac{\partial}{\partial t}, \nabla\right), \quad A^\mu = (\phi, \mathbf{A}), \quad A_\mu = (\phi, -\mathbf{A})$$

and the tensors

$$F^{\mu\nu} = \partial^\mu A^\nu - \partial^\nu A^\mu, \quad F_{\mu\nu} = \partial_\mu A_\nu - \partial_\nu A_\mu.$$

Here, the indices μ and ν take the values 0, 1, 2, and 3. The Lagrangian is given by $\int \mathcal{L} d^3r$ where $\mathcal{L}$ is the Lagrangian density. The Euler–Lagrange

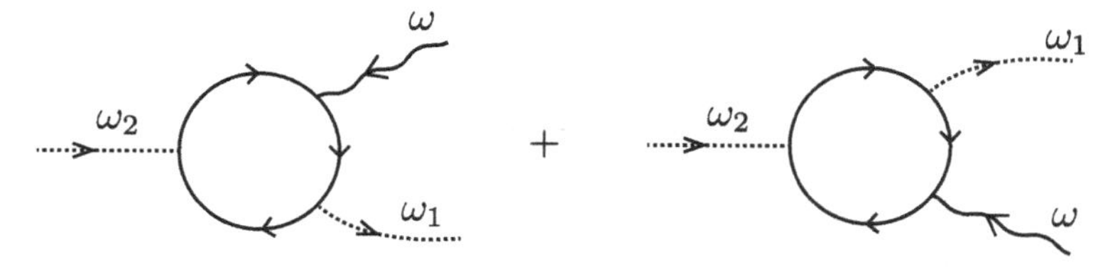

Figure 11.18 The two diagrams that describe Raman scattering in insulators. Solid lines are electron lines, dotted lines are photon lines, and wavy lines are phonon lines.

equations (one for each ν) are

$$\partial^\mu \frac{\partial \mathcal{L}}{\partial(\partial^\mu A_\nu)} = \frac{\partial \mathcal{L}}{\partial A_\nu},$$

where a repeated index (μ in the above equation) is summed over.

(a) Show that the Euler–Lagrange equations yield Maxwell's equations in free space if

$$\mathcal{L} = -\frac{1}{16\pi} F_{\mu\nu} F^{\mu\nu}.$$

(b) Show that

$$\mathcal{L} = \frac{1}{8\pi}(E^2 - B^2).$$

(c) Derive Eq. (11.82).

11.6 *Raman tensor.* Raman scattering in insulators is described by the two Feynman diagrams shown in Figure 11.18. Each diagram represents a process in which a photon of coordinates $(\mathbf{q}_1\lambda_1)$ is annihilated, a photon of coordinates $(\mathbf{q}_2\lambda_2)$ is created, and a phonon of coordinates $(\mathbf{q}\lambda)$ is created. In each diagram there are three interactions (one electron–phonon interaction and two electron–photon interactions) occurring at three different times. Use Dzyaloshinski's rules for time-ordered diagrams (there are six time-ordered diagrams corresponding to each of the two Feynman diagrams) to calculate the Raman tensor $\Gamma(\mathbf{q}_1\lambda_1, \mathbf{q}_2\lambda_2, \mathbf{q}\lambda)$.

12

Superconductivity

False friends are common. Yes, but where
True nature links a friendly pair,
The blessing is as rich as rare.

–From the Panchatantra
Translated by Arthur W. Ryder

The magnet of their course is gone, or only points in vain
The shore to which their shiver'd sail shall never stretch again.

–Lord Byron, Youth and Age

Superconductivity was discovered in 1911 by H. Kamerlingh Onnes soon after he succeeded in liquefying helium (Onnes, 1911). He observed that the resistivity of mercury dropped suddenly as its temperature was lowered below a certain critical value T_C (for Hg, $T_C = 4.2$ K). Over the years, it was found that many additional elements and compounds similarly transition to a superconducting state. In this state, materials exhibit properties that are strikingly different from the normal state. Below we discuss the most important features of superconductors.

12.1 Properties of superconductors

The first important property of a material that undergoes a superconducting transition is that its resistivity drops to zero below a critical temperature (see Figure 12.1). In a superconducting ring, a persistent electric current flows without any observable attenuation for as long as one is willing to watch.

The application of a sufficiently strong magnetic field destroys superconductivity and returns a material to its normal state. The value of the critical magnetic field is denoted in the literature by H_C, and it is a function of temperature. $H_C(T)$ is largest at $T = 0$, dropping to zero at the transition temperature T_C, as shown in Figure 12.2. The temperature dependence of H_C is approximated by

$$H_C(T) = H_C(0)(1 - T^2/T_C^2). \tag{12.1}$$

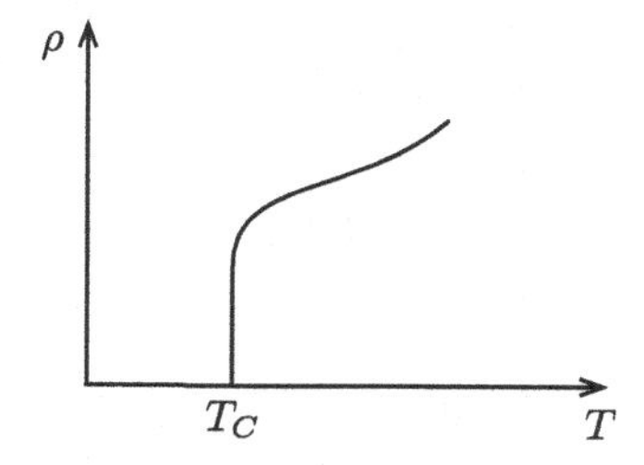

Figure 12.1 At the superconducting critical temperature T_C, the resistivity of a material drops to zero.

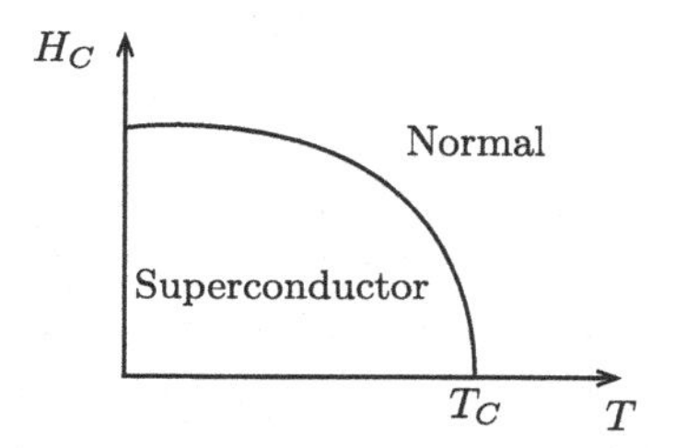

Figure 12.2 The critical magnetic field that destroys superconductivity varies with temperature.

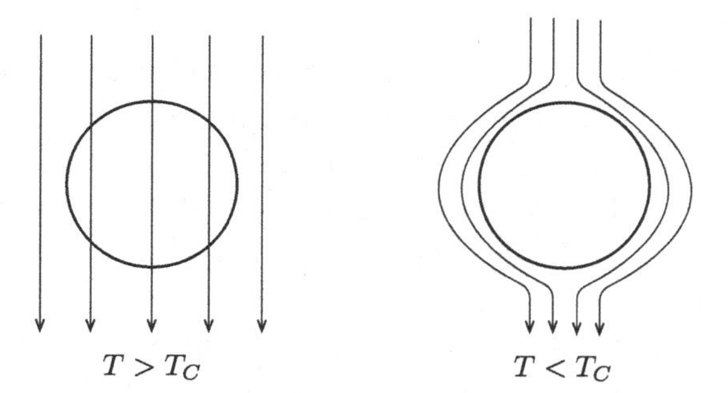

Figure 12.3 The Meissner effect. In the normal state ($T > T_C$), a magnetic field penetrates the material. In the superconducting state ($T < T_C$), the magnetic field is expelled from the bulk of the material.

Another crucial property of the superconducting state is perfect diamagnetism: when a material is cooled in the presence of a magnetic field to below T_C, the magnetic flux is expelled from the inside of the superconductor, as illustrated in Figure 12.3. This is known as the Meissner effect (Meissner and Ochsenfeld, 1933). The flux expulsion occurs due to the appearance of surface currents. These produce a magnetic field which cancels out the applied one within the sample. The

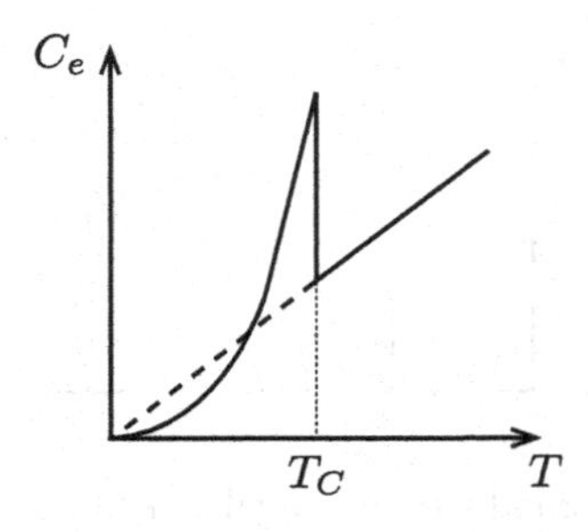

Figure 12.4 Electronic specific heat as a function of temperature in a superconductor. The dashed line is the would-be specific heat $C_e(T)$ had the metal remained in the normal state (it is obtained experimentally by measuring $C_e(T)$ in the presence of a magnetic field larger than the critical field).

Meissner effect is not a consequence of the vanishing of the resistivity; rather, it is an independent property of the superconductor. Ohm's law $\mathbf{E} = \rho \mathbf{J}$, together with Maxwell's equation $\nabla \times \mathbf{E} = -(1/c)\partial \mathbf{B}/\partial t$, imply that if $\rho = 0$, the magnetic field remains constant over time. Hence, if a magnetic field penetrates a sample and temperature is lowered to below T_C, the vanishing of its resistivity implies that **B** remains frozen within the sample (the argument is a bit subtle and is developed further in the next section). However, this is not how a superconductor behaves. Zero resistivity and perfect diamagnetism are two *independent* properties of a superconductor.

In the presence of an applied magnetic field, superconductors exhibit one of two types of behavior. Type-I superconductors have only one critical field $H_C(T)$; fields below H_C are excluded from the bulk of the superconductor. By contrast, a type-II superconductor has two critical magnetic fields. For an applied field below its lower critical field $H_{C_1}(T)$, flux expulsion is complete, similar to the type-I case. For fields larger than its upper critical field $H_{C_2}(T)$, superconductivity is destroyed, and the applied field penetrates the sample completely. However, for fields in between the two critical fields, $H_{C_1}(T) < H < H_{C_2}(T)$, there is partial penetration by the magnetic flux, and the sample contains both normal and superconducting regions.

Specific heat also behaves anomalously in superconductors. In normal metals at low temperature, electronic specific heat varies linearly with temperature: $C_e = \alpha T$. In superconductors, as the temperature drops, electronic specific heat suddenly jumps to a higher value at T_C, then decreases, eventually falling below values expected for a normal metal, as illustrated in Figure 12.4. Detailed analysis of experimental data indicates that, in the superconducting state, $C_e \propto \exp(-\Delta/k_B T)$. This behavior is characteristic of a system whose excited states are separated from the ground state by an energy gap of 2Δ. More evidence for the existence of a gap in the energy spectrum of a superconductor is provided by tunneling experiments. When two metals are brought into close contact, separated by only a thin insulating layer, electrons tunnel from one metal to the other; equilibrium is established when the chemical potentials of both metals become equal. An applied voltage raises the chemical potential of one metal relative to the other, leading to a flow of electrons from the metal with the higher chemical potential to the other metal. If one of the metals is a superconductor and the system is cooled to below T_C, no flow of electrons occurs until the applied voltage exceeds a threshold value given by $\mathrm{eV} = \Delta$. This indicates that, in the superconducting state, chemical potential sits in the middle of an energy gap of size 2Δ.

The last property of superconductors we discuss is the isotope effect. Accurate measurements reveal that, in most superconductors, a slight shift in the value of T_C occurs as ionic mass changes through the use of different isotopes. This effect indicates that electron–phonon interaction plays an important role in the mechanism of superconductivity. It is the only plausible conclusion we can draw, since changing the isotopes should have no effect on the energy bands or on the Coulomb interaction between electrons.

It is worth noting that the superconducting transition temperature is generally very low. Until 1986, the highest recorded T_C was 23.3 K, and it belonged to Nb_3Ge. However, toward the end of 1986, a new era was ushered in with the discovery of the high-T_C copper oxide family of superconductors (Bednorz and Müller, 1986). These compounds contain copper oxide planes separated by insulating layers. The compound discovered by Bednorz and Müller belongs to the family of La-based superconductors that are obtained by doping the insulating parent compound La_2CuO_4, whose crystal structure is shown in Figure 12.5.

In $La_2\,CuO_4$, the CuO_2 planes are separated by two LaO layers, and each Cu ion is surrounded by an elongated octahedron of oxygen ions. The configurations of the valence electrons in the atoms of La_2CuO_4 are as follows: La: $5d^16s^2$, Cu: $3d^{10}4s^1$, O: $2s^22p^4$. An oxygen atom needs two electrons to fill its outer shell; to fulfill this need, every La atom loses its three valence electrons, and every Cu atom loses two valence electrons. The compound is thus more appropriately represented as $La_2^{3+}Cu^{2+}O_4^{2-}$. With the loss of two electrons, the Cu^{2+} ion has the configuration $Cu^{2+}: 3d^9$. There is a hole on each Cu site. In an independent-electron model, the compound would be metallic, since each hole could hop from one Cu site to another. However, Coulomb repulsion between two holes on the same Cu site tends to prevent such hopping from taking place, so the holes remain localized on the Cu sites. The magnetic moments on neighboring Cu sites are aligned in opposite directions, the result of a mechanism called superexchange that occurs due to intervening oxygen ions. La_2CuO_4 is thus an antiferromagnetic insulator, known as a Hubbard–Mott insulator.

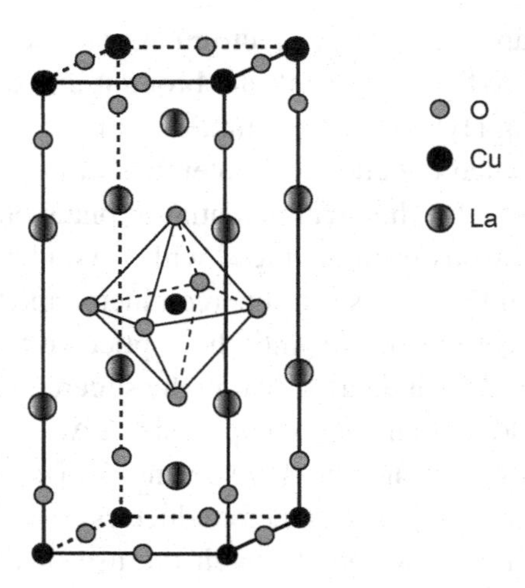

Figure 12.5 Crystal structure of La_2CuO_4. The separation between the CuO_2 planes is 6.6 Å, Cu-O separation in the plane is 1.9 Å, while it is 2.4 Å perpendicular to the plane.

Upon doping, which involves the replacement of a certain percentage of La^{3+} ions by Ba^{2+} or Sr^{2+} ions, fewer electrons are donated to the CuO_2 planes, and some mobile holes are produced on the oxygen sites. The resulting compound becomes metallic, and for optimal doping, it is superconducting at a critical temperature of 36 K.

Shortly after Bednorz and Müller's discovery, many compounds containing copper oxide planes were synthesized and found to be superconducting at temperatures exceeding 77 K, the temperature at which nitrogen is liquefied. Since liquid nitrogen is much less expensive than liquid helium, this was a very important achievement. However, the ultimate goal, room temperature superconductivity, remains elusive.

More recently, another class of layered, iron-based, high-temperature superconductors has been discovered (Kamihara et al., 2008). When compounds with the general formula LnOFeAs, where Ln is a lanthanide (Ln = La, Ce, Pr, ...) are doped with fluorine, they become superconductive at a critical temperature ranging from 25 K to 55 K. The parent compounds, LnOFeAs, consist of stacks of alternating LnO and FeAs layers. Neutron scattering measurements (De la Cruz et al., 2008) as well as numerical calculations (Yildrim, 2008; Alyahyaei and Jishi, 2009) reveal that, in the ground state, the magnetic moments of the iron ions adopt an antiferromagnetic order. As in the copper oxide family, the parent compounds are antiferromagnetic insulators, becoming superconductive only upon doping.

As we will show later, superconductivity arises because of the existence of an effective attraction between electrons in a thin shell near the Fermi surface. The effective attraction between electrons in conventional (pre-1986) superconductors is mediated by phonons. In high-T_C superconductors, the pairing mechanism may be different. Nevertheless, the framework presented by the theory of conventional superconductors is essential for understanding all classes of superconductors. This theory is discussed later in this chapter. We begin, however, by considering a phenomenological model of the magnetic properties of superconductors, since the notions and ideas introduced in this model are relevant to the treatment (within microscopic theory) of the response of a superconductor to an applied magnetic field.

12.2 The London equation

We saw that a magnetic flux is expelled from the bulk of a superconductor (the Meissner effect). In what way should the electrodynamics of a superconductor differ from that of a normal metal in order to account for the Meissner effect? This question was examined by the brothers F. London and H. London (London and London, 1935) two years after the discovery of the Meissner effect.

In order to clearly elucidate the distinction between a perfect conductor and a superconductor, let us begin by considering a normal metal containing n conduction electrons per unit volume. In the presence of a static (time-independent) electric field $\mathbf{E}$, an electron is accelerated, but it is also scattered by phonons and impurities. These scattering processes cause a damping of the electron's motion. Taking the damping force to be proportional to the electron's velocity, Newton's second law gives

$$m\dot{\mathbf{v}} = -e\mathbf{E} - m\mathbf{v}/\tau \tag{12.2}$$

where τ, the relaxation time, is the average time between scattering events. The current density is $\mathbf{J} = -ne\mathbf{v}$. Under steady-state conditions, $\mathbf{J}$ is constant: $\dot{\mathbf{v}} = 0$, and $\mathbf{v} = -e\mathbf{E}\tau/m$; hence $\mathbf{J} = (ne^2\tau/m)\mathbf{E}$. Ohm's law, $\mathbf{E} = \rho\mathbf{J}$, then implies that the resistivity $\rho = m/(ne^2\tau)$.

In a perfect conductor, $\rho = 0$; the relaxation time τ is thus infinite, and Eq. (12.2) becomes

$$m\dot{\mathbf{v}} = -e\mathbf{E} \Longrightarrow \dot{\mathbf{J}} = (ne^2/m)\mathbf{E}. \tag{12.3}$$

Taking the curl on both sides of Eq. (12.3), and using Maxwell's equation $\nabla \times \mathbf{E} = (-1/c)\partial\mathbf{B}/\partial t$ (cgs), we find

$$\frac{\partial}{\partial t}\left(\nabla \times \mathbf{J} + \frac{ne^2}{mc}\mathbf{B}\right) = 0. \tag{12.4}$$

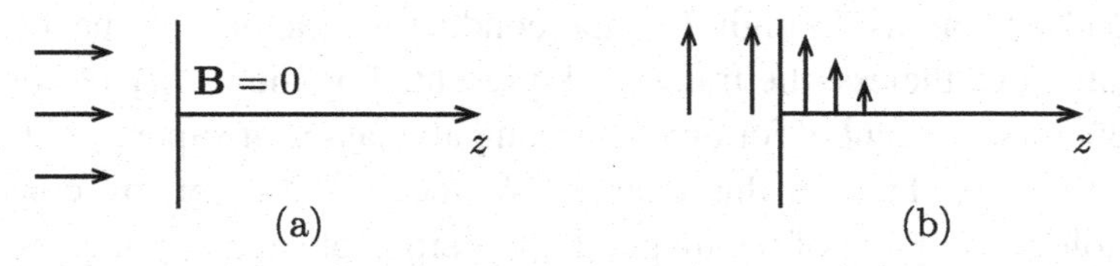

Figure 12.6 (a) A magnetic field is perpendicular to the surface of a semi-infinite superconducting slab. No field penetration takes place. (b) The field is parallel to the slab surface. Inside the superconductor, the field decays exponentially.

This relation, along with the Maxwell equation (for a static **E**-field)

$$\nabla \times \mathbf{B} = \frac{4\pi}{c}\mathbf{J}, \tag{12.5}$$

determines the magnetic field and the current density in a perfect conductor. Note that any static **B**-field determines, through Eq. (12.5), a static **J**, and therefore Eq. (12.4) will be automatically satisfied. Equations (12.4) and (12.5) are thus consistent with the existence of an arbitrary static magnetic field inside a perfect conductor. We pointed this out in the previous section. This behavior is, however, incompatible with the observed Meissner effect in superconductors; hence, zero resistivity is a necessary, but not sufficient, condition for superconductivity.

It was conjectured by the London brothers that the magnetic field and current density in a superconductor satisfy the relation

$$\nabla \times \mathbf{J} + \frac{ne^2}{mc}\mathbf{B} = 0. \tag{12.6}$$

This is known as the London equation. Whereas for a perfect conductor the LHS of Eq. (12.6) is only required to be time-independent (see Eq. [12.4]), it is identically equal to zero for a superconductor.

Taking the curl on both sides of Eq. (12.5) and using the vector identity

$$\nabla \times \nabla \times \mathbf{B} = \nabla(\nabla \cdot \mathbf{B}) - \nabla^2\mathbf{B}, \tag{12.7}$$

along with Maxwell's equation $\nabla \cdot \mathbf{B} = 0$ and Eq. (12.6), we obtain

$$\nabla^2\mathbf{B} = \frac{4\pi ne^2}{mc^2}\mathbf{B}. \tag{12.8}$$

Similarly, taking the curl on both sides of Eq. (12.6) and using Eq. (12.5), we find

$$\nabla^2\mathbf{J} = \frac{4\pi ne^2}{mc^2}\mathbf{J}. \tag{12.9}$$

We solve Eq. (12.8) for **B** inside a superconducting semi-infinite ($z \geq 0$) slab for the following two cases, which are illustrated in Figure 12.6.

- **B** is parallel to the z-axis and varies only along the z-direction, i.e., $\mathbf{B} = (0, 0, B(z))$. In this case, $\nabla \cdot \mathbf{B} = 0 \Rightarrow \partial B(z)/\partial z = 0 \Rightarrow B(z)$ is constant, independent of z. Equation (12.8) then implies that $B(z) = 0$ inside the superconductor.
- **B** is parallel to the x-axis and varies along the z-direction: $\mathbf{B} = (B(z), 0, 0)$. In this case, Eq. (12.8) becomes

$$\frac{\partial^2 B(z)}{\partial z^2} = \frac{4\pi n e^2}{mc^2} B(z), \quad z \geq 0.$$

Its solution is

$$B(z) = B(0) \exp(-z/\lambda_L), \quad z \geq 0. \tag{12.10}$$

The parameter λ_L, known as the London penetration depth, is given by

$$\lambda_L = \left(\frac{mc^2}{4\pi n e^2} \right)^{1/2}. \tag{12.11}$$

In most superconductors, $\lambda_L = 10^2 - 10^3$ Å. The magnetic field decays exponentially inside the superconductor, and it only penetrates a small distance, of the order of λ_L, into the superconductor.

12.3 Effective electron–electron interaction

In addition to the weak, screened Coulomb interaction between electrons in a metal, there is an attractive interaction which results from their coupling to lattice vibrations. The existence of such an attraction and its possible role in superconductivity was first noted by H. Frohlich (Frohlich, 1950). As an electron moves within a crystal, it pulls the positive ions in its vicinity. The ions respond by moving toward the electron. However, by the time the ions have been maximally displaced, the electron, due to its much higher speed, is long gone. The region into which the ions move, however, now has excess positive charge; a second electron that happens to pass by is attracted by this excess positive charge. This state of affairs is illustrated in Figure 12.7. What we end up with, in effect, is an attractive interaction between two electrons. In contrast to the instantaneous Coulomb repulsion between electrons, the attractive interaction is a retarded one. The time it takes for the ions to achieve maximal displacement is of the order of $1/\omega_D$, where ω_D (Debye frequency) is a typical phonon frequency: $\omega_D \approx 10^{13}\ \text{s}^{-1}$. In metals, the typical electron velocity is the Fermi velocity $v_F \approx 10^6$ m/s. Thus, by the time the ions are maximally displaced, the first electron is ~ 1000 Å away; the attractive interaction can operate between electrons that are

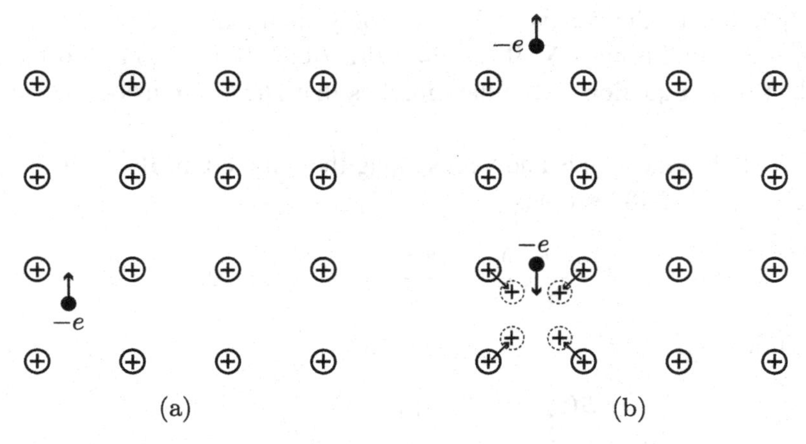

Figure 12.7 Effective electron–electron interaction. (a) An electron attracts the positive ions in its vicinity. (b) By the time the ions are maximally displaced, the electron is very far away, but a second electron is attracted by the resulting excess positive charge. In effect, there is an attraction between the two electrons.

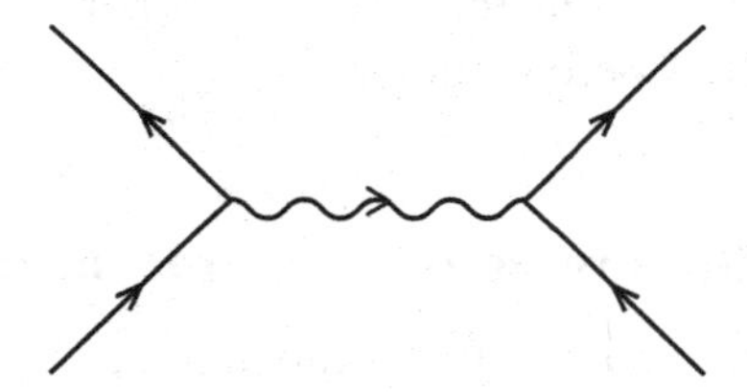

Figure 12.8 A virtual process in which a phonon is exchanged between two electrons, giving rise to an effective electron–electron interaction.

very far apart. At such distances the screened Coulomb repulsion is completely negligible.

An alternative description of the lattice-mediated interaction between electrons can be formulated using the language of phonons. The interaction between the electrons and the ions may be viewed in terms of the electrons' emission and absorption of phonons. We can consider a virtual process whereby an electron emits a phonon, which, in turn, is absorbed by another electron (see Figure 12.8). Since the phonon energy is typically $\hbar\omega_D$, the uncertainty principle, $\Delta E \Delta t \sim \hbar$, implies that the phonon will live for a time $\tau \sim 1/\omega_D \approx 10^{-13}$ s. Since the typical phonon velocity is $\sim 10^3$ m/s (the speed of sound in solids), the electron that absorbs the phonon should be very close ($\sim$ 1 Å) to the location where the phonon is emitted.

The Hamiltonian for the electron–phonon system is

$$H = H_0 + H' \tag{12.12a}$$

$$H_0 = \sum_{\mathbf{k}\sigma} \epsilon_{\mathbf{k}} c^\dagger_{\mathbf{k}\sigma} c_{\mathbf{k}\sigma} + \sum_{\mathbf{q}\lambda} \hbar\omega_{\mathbf{q}\lambda}(a^\dagger_{\mathbf{q}\lambda} a_{\mathbf{q}\lambda} + 1/2) \tag{12.12b}$$

$$H' = \sum_{\mathbf{k}\sigma} \sum_{\mathbf{q}\lambda} M_{\mathbf{q}\lambda} c^\dagger_{\mathbf{k}+\mathbf{q}\sigma} c_{\mathbf{k}\sigma} (a_{\mathbf{q}\lambda} + a^\dagger_{-\mathbf{q}\lambda}). \tag{12.12c}$$

H_0 is the Hamiltonian that describes the conduction electrons and the free phonons, and H' is the electron–phonon interaction. Here, $c^\dagger_{\mathbf{k}\sigma}$ ($c_{\mathbf{k}\sigma}$) creates (annihilates) an electron in a state specified by the wave vector $\mathbf{k}$ and spin projection σ, and $a^\dagger_{\mathbf{q}\lambda}$ ($a_{\mathbf{q}\lambda}$) creates (annihilates) a phonon of wave vector $\mathbf{q}$ and branch index λ. We have assumed that the metal has only one partially filled band, and that electrons scatter within this band; hence, the band index has been dropped.

We can obtain an expression for the electron–electron interaction mediated by phonons by carrying out a change of basis, as we did in Section 11.14. The second quantized form of the Hamiltonian, as given in Eq. (12.12), is obtained by using the basis set of states $|n\rangle = |\mathbf{k}\sigma\rangle|\mathbf{q}\lambda\rangle$. We transform to a new basis set of states $|\tilde{n}\rangle = U|n\rangle$, where $U^\dagger U = 1$. In the new basis, the matrix elements of the Hamiltonian are given by

$$\langle\tilde{m}|H|\tilde{n}\rangle = \langle m|U^\dagger H U|n\rangle = \langle m|\tilde{H}|n\rangle. \tag{12.13}$$

Thus, the change of basis is equivalent to applying a similarity transformation to the Hamiltonian: $H \to \tilde{H} = U^\dagger H U$. Let $U = e^S$, where $S^\dagger = -S$ (in order for U to be unitary). Then,

$$\tilde{H} = e^{-S} H e^S. \tag{12.14}$$

The operator S will be chosen so as to eliminate the electron–phonon interaction in first order. Alternatively, we may define new electron and phonon operators through a canonical transformation

$$\tilde{c}_{\mathbf{k}\sigma} = e^{-S} c_{\mathbf{k}\sigma} e^S, \quad \tilde{a}_{\mathbf{q}\lambda} = e^{-S} a_{\mathbf{q}\lambda} e^S,$$

rewrite the Hamiltonian in terms of the new operators, and choose S so as to eliminate the electron–phonon interaction in first order.

Expanding the exponential operators in Eq. (12.14),

$$\tilde{H} = (1 - S + S^2/2! + \cdots)(H_0 + H')(1 + S + S^2/2! + \cdots)$$

and choosing S such that

$$[S, H_0] = H', \tag{12.15}$$

we obtain

$$\tilde{H} = H_0 + \frac{1}{2}\left[H', S\right] + \cdots . \tag{12.16}$$

Note that, because H_0 and H' are hermitian operators, the operator S, defined by Eq. (12.15), does indeed satisfy the requirement that $S^\dagger = -S$. Considering any two eigenkets, $|m\rangle$ and $|n\rangle$, of H_0, with corresponding eigenvalues E_m and E_n, Eq. (12.15) gives

$$\langle m|S|n\rangle = \frac{\langle m|H'|n\rangle}{E_n - E_m}. \tag{12.17}$$

Since the effect of H' is to scatter an electron from a state with energy $\epsilon_{\mathbf{k}}$ into a state with energy $\epsilon_{\mathbf{k}+\mathbf{q}}$ either by the absorption of a phonon $(\mathbf{q}\lambda)$ or by the emission of a phonon $(-\mathbf{q}\lambda)$, the energy difference $E_n - E_m$ is either $\epsilon_{\mathbf{k}} + \hbar\omega_{\mathbf{q}\lambda} - \epsilon_{\mathbf{k}+\mathbf{q}}$ (corresponding to phonon absorption) or $\epsilon_{\mathbf{k}} - \hbar\omega_{\mathbf{q}\lambda} - \epsilon_{\mathbf{k}+\mathbf{q}}$ (corresponding to phonon emission). Therefore, S is given by

$$S = \sum_{\mathbf{k}\sigma}\sum_{\mathbf{q}\lambda} M_{\mathbf{q}\lambda} c^\dagger_{\mathbf{k}+\mathbf{q}\sigma} c_{\mathbf{k}\sigma}\left(\frac{a_{\mathbf{q}\lambda}}{\epsilon_{\mathbf{k}} - \epsilon_{\mathbf{k}+\mathbf{q}} + \hbar\omega_{\mathbf{q}\lambda}} + \frac{a^\dagger_{-\mathbf{q}\lambda}}{\epsilon_{\mathbf{k}} - \epsilon_{\mathbf{k}+\mathbf{q}} - \hbar\omega_{\mathbf{q}\lambda}}\right). \tag{12.18}$$

One can check that S, as given above, satisfies Eq. (12.15). The Hamiltonian in Eq. (12.16) becomes

$$\tilde{H} = H_0 + \frac{1}{2}\sum_{\mathbf{k}\sigma}\sum_{\mathbf{q}\lambda}\sum_{\mathbf{k}'\sigma'}\sum_{\mathbf{q}'\lambda'} M_{\mathbf{q}\lambda} M_{\mathbf{q}'\lambda'}\left[c^\dagger_{\mathbf{k}+\mathbf{q}\sigma} c_{\mathbf{k}\sigma}\left(a_{\mathbf{q}\lambda} + a^\dagger_{-\mathbf{q}\lambda}\right),\right.$$

$$\left. c^\dagger_{\mathbf{k}'+\mathbf{q}'\sigma'} c_{\mathbf{k}'\sigma'}\left(\frac{a_{\mathbf{q}'\lambda'}}{\epsilon_{\mathbf{k}'} - \epsilon_{\mathbf{k}'+\mathbf{q}'} + \hbar\omega_{\mathbf{q}'\lambda'}} + \frac{a^\dagger_{-\mathbf{q}'\lambda'}}{\epsilon_{\mathbf{k}'} - \epsilon_{\mathbf{k}'+\mathbf{q}'} - \hbar\omega_{\mathbf{q}'\lambda'}}\right)\right].$$

Out of the many terms in the commutator, there are two terms that contain four electron operators; they arise from commuting $a_{\mathbf{q}\lambda}$ with $a^\dagger_{-\mathbf{q}'\lambda'}$, and $a^\dagger_{-\mathbf{q}\lambda}$ with $a_{\mathbf{q}'\lambda'}$. The other terms all contain two electron and two phonon operators. We thus write

$$\tilde{H} = H_0 + \frac{1}{2}\sum_{\mathbf{k}\sigma}\sum_{\mathbf{q}\lambda}\sum_{\mathbf{k}'\sigma'} M_{\mathbf{q}\lambda} M_{-\mathbf{q}\lambda} c^\dagger_{\mathbf{k}+\mathbf{q}\sigma} c_{\mathbf{k}\sigma} c^\dagger_{\mathbf{k}'-\mathbf{q}\sigma'} c_{\mathbf{k}'\sigma'}$$

$$\times\left(\frac{1}{\epsilon_{\mathbf{k}'} - \epsilon_{\mathbf{k}'-\mathbf{q}} - \hbar\omega_{\mathbf{q}\lambda}} - \frac{1}{\epsilon_{\mathbf{k}'} - \epsilon_{\mathbf{k}'-\mathbf{q}} + \hbar\omega_{\mathbf{q}\lambda}}\right)$$

$$+ \text{ (terms containing two electron and two phonon operators).}$$

From the commutation relations of the electron operators, it follows that

$$c_{\mathbf{k}\sigma} c^\dagger_{\mathbf{k}'-\mathbf{q}\sigma'} = \delta_{\mathbf{k}',\mathbf{k}+\mathbf{q}}\delta_{\sigma\sigma'} - c^\dagger_{\mathbf{k}'-\mathbf{q}\sigma'} c_{\mathbf{k}\sigma} \tag{12.19a}$$

$$c_{\mathbf{k}\sigma} c_{\mathbf{k}'\sigma'} = -c_{\mathbf{k}'\sigma'} c_{\mathbf{k}\sigma}. \tag{12.19b}$$

Using these, along with $M_{-\mathbf{q}\lambda} = M^*_{\mathbf{q}\lambda}$, we can write

$$\tilde{H} = H_0 + H_1 + H_2 + \text{“others.”}$$

where “others” are terms containing two electron and two phonon operators, and

$$H_1 = \sum_{\mathbf{k}\sigma} \sum_{\mathbf{q}\lambda} |M_{\mathbf{q}\lambda}|^2 c^\dagger_{\mathbf{k}+\mathbf{q}\sigma} c_{\mathbf{k}+\mathbf{q}\sigma} \left[\frac{\hbar\omega_{\mathbf{q}\lambda}}{\left(\epsilon_{\mathbf{k}+\mathbf{q}} - \epsilon_{\mathbf{k}}\right)^2 - \left(\hbar\omega_{\mathbf{q}\lambda}\right)^2} \right]$$

$$= \sum_{\mathbf{k}\sigma} \sum_{\mathbf{q}\lambda} |M_{\mathbf{q}\lambda}|^2 \left[\frac{\hbar\omega_{\mathbf{q}\lambda}}{\left(\epsilon_{\mathbf{k}} - \epsilon_{\mathbf{k}-\mathbf{q}}\right)^2 - \left(\hbar\omega_{\mathbf{q}\lambda}\right)^2} \right] c^\dagger_{\mathbf{k}\sigma} c_{\mathbf{k}\sigma} \tag{12.20}$$

$$H_2 = \sum_{\mathbf{k}\sigma} \sum_{\mathbf{k}'\sigma'} \sum_{\mathbf{q}\lambda} |M_{\mathbf{q}\lambda}|^2 \left[\frac{\hbar\omega_{\mathbf{q}\lambda}}{\left(\epsilon_{\mathbf{k}'} - \epsilon_{\mathbf{k}'-\mathbf{q}}\right)^2 - \left(\hbar\omega_{\mathbf{q}\lambda}\right)^2} \right] c^\dagger_{\mathbf{k}+\mathbf{q}\sigma} c^\dagger_{\mathbf{k}'-\mathbf{q}\sigma'} c_{\mathbf{k}'\sigma'} c_{\mathbf{k}\sigma}. \tag{12.21}$$

The term H_1 can be absorbed into H_0; it simply leads to a renormalization of the single particle energy. On the other hand, the term H_2 represents an interaction between electrons. The electron–electron interaction, mediated by phonons, is thus given by

$$H'_{int} = \sum_{\mathbf{k}\sigma} \sum_{\mathbf{k}'\sigma'} \sum_{\mathbf{q}} V_{\mathbf{k}'\mathbf{q}} c^\dagger_{\mathbf{k}+\mathbf{q}\sigma} c^\dagger_{\mathbf{k}'-\mathbf{q}\sigma'} c_{\mathbf{k}'\sigma'} c_{\mathbf{k}\sigma} \tag{12.22}$$

$$V_{\mathbf{k}'\mathbf{q}} = \sum_{\lambda} |M_{\mathbf{q}\lambda}|^2 \left[\frac{\hbar\omega_{\mathbf{q}\lambda}}{\left(\epsilon_{\mathbf{k}'} - \epsilon_{\mathbf{k}'-\mathbf{q}}\right)^2 - \left(\hbar\omega_{\mathbf{q}\lambda}\right)^2} \right]. \tag{12.23}$$

Consider a shell surrounding the Fermi surface. The inner and outer surfaces of the shell are constant energy surfaces with energies $E_F - \hbar\omega_D$ and $E_F + \hbar\omega_D$, respectively, where E_F is the Fermi energy and $\hbar\omega_D$ is a typical phonon energy. Equation (12.23) tells us that if two electrons remain in states that lie within this shell, the phonon-mediated interaction between them is attractive.

12.4 Cooper pairs

At $T = 0$, the ground state of an electron gas is obtained by filling all states up to the Fermi wave vector k_F. Let us imagine adding two extra electrons to the system and turning on an attractive interaction between them. We assume that the attractive interaction between the two extra electrons exists only when the two electrons occupy states in a shell of energy width $\hbar\omega_D$ (the typical phonon energy) that surrounds the Fermi sphere (see Figure 12.9). We also assume that the two added electrons interact with other electrons only through the Pauli exclusion principle:

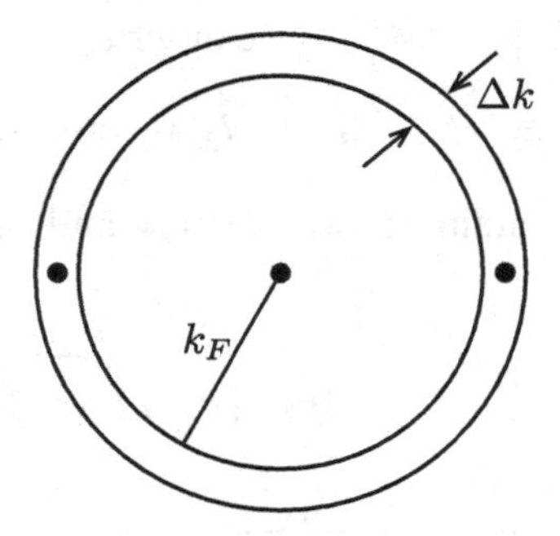

Figure 12.9 Two extra electrons are added to the Fermi sphere of radius k_F. If the added electrons are in a shell around the Fermi sphere of width Δk, the interaction between the two electrons is attractive; otherwise, the interaction is zero. Here, $(\hbar^2 k_F/m)\Delta k = \hbar\omega_D$.

the role of the Fermi sea of electrons is simply to prevent the two added electrons from occupying any state below the Fermi surface. Absent an attractive interaction, the ground state of the two added electrons is obtained if each has energy E_F. In the presence of the attractive interaction, what is the ground state of the two added electrons? The answer to this question was provided in a seminal paper by Cooper (Cooper, 1956). The two extra electrons can scatter off each other from states $|\mathbf{k}_1\sigma_1, \mathbf{k}_2\sigma_2\rangle$ into states $|\mathbf{k}_1 + \mathbf{q}\sigma_1, \mathbf{k}_2 - \mathbf{q}\sigma_2\rangle$. Conservation of momentum dictates that $\mathbf{k}_1 + \mathbf{k}_2 = \mathbf{K}$ must remain unchanged. Since the two electrons are constrained to remain within a shell of energy width $\hbar\omega_D$ surrounding the Fermi sphere, the conservation of momentum means that for a given $\mathbf{K}$, the wave vectors $\mathbf{k}_1$ and $\mathbf{k}_2$ will be restricted to the region of intersection of the two shells in $\mathbf{k}$-space centered on $\mathbf{0}$ and $\mathbf{K}$ (see Figure 12.10). Since we are interested in the lowest energy state, it is sufficient for us to consider the case when the region of attractive interaction is maximal; this occurs when $\mathbf{K} = 0$, for then the shaded region in Figure 12.10 coincides with the whole shell. Henceforth, we assume that the two added electrons have wave vectors $\mathbf{k}$ and $-\mathbf{k}$. Denoting the positions of the two added electrons by $\mathbf{r}_1$ and $\mathbf{r}_2$, and their wave function by $\psi(\mathbf{r}_1\sigma_1, \mathbf{r}_2\sigma_2)$, the Schrödinger equation reads

$$\left[p_1^2/2m + p_2^2/2m + U(\mathbf{r}_1 - \mathbf{r}_2)\right]\psi(\mathbf{r}_1\sigma_1, \mathbf{r}_2\sigma_2) = E\psi(\mathbf{r}_1\sigma_1, \mathbf{r}_2\sigma_2) \qquad (12.24)$$

where $U(\mathbf{r}_1 - \mathbf{r}_2)$ is the interaction energy of the two electrons; it depends on $\mathbf{r}_1 - \mathbf{r}_2$ due to the translational invariance of the system. Since the Hamiltonian is spin-independent, the stationary states can be written as the product of a spatial function and a spin function. The two electrons are continually scattered from states $|\mathbf{k}\sigma_1, -\mathbf{k}\sigma_2\rangle$ into states $|\mathbf{k}'\sigma_1, -\mathbf{k}'\sigma_2\rangle$; hence, we consider a solution to the

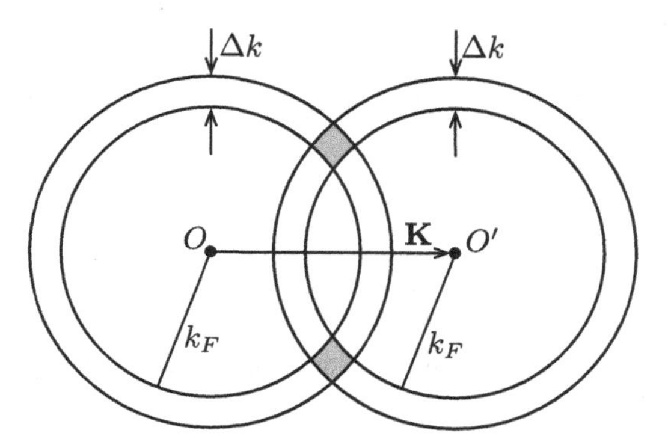

Figure 12.10 $\mathbf{k}_1$ and $\mathbf{k}_2$ are restricted to a shell of width Δk surrounding a Fermi sphere of radius k_F. For a given $\mathbf{K}$, the requirement that $\mathbf{k}_1 + \mathbf{k}_2 = \mathbf{K}$ is satisfied only if $\mathbf{k}_1$ and $\mathbf{k}_2$ are restricted to the region where the two shells centered at O and O′ intersect. O and O′ are two points in $\mathbf{k}$-space that are separated by the vector $\mathbf{K}$. The region of intersection is the volume obtained by rotating the shaded area in the figure around the OO′ axis.

Schrödinger equation of the form:

$$\psi(\mathbf{r}_1\sigma_1, \mathbf{r}_2\sigma_2) = \sum_{\mathbf{k}} g(\mathbf{k}) \frac{1}{V} e^{i\mathbf{k}.(\mathbf{r}_1 - \mathbf{r}_2)} \chi(\sigma_1, \sigma_2) \tag{12.25}$$

where V is the system's volume, $(1/V)e^{i\mathbf{k}.(\mathbf{r}_1-\mathbf{r}_2)}$ is the spatial part of the wave function corresponding to the ket $|\mathbf{k}\sigma_1, -\mathbf{k}\sigma_2\rangle$, i.e., it is $\langle \mathbf{r}_1, \mathbf{r}_2 | \mathbf{k}, -\mathbf{k}\rangle$, and the expansion coefficients $g(\mathbf{k})$ are to be determined. The spin function $\chi(\sigma_1, \sigma_2)$ can be chosen to be antisymmetric (singlet) or symmetric (triplet). For the singlet state, the antisymmetry of $\psi(\mathbf{r}_1\sigma_1, \mathbf{r}_2\sigma_2)$ under the interchange $(\mathbf{r}_1\sigma_1) \leftrightarrow (\mathbf{r}_2\sigma_2)$ requires that the spatial part be symmetric, i.e., $g(-\mathbf{k}) = g(\mathbf{k})$. For the triplet state, the spatial part of the wave function is antisymmetric, i.e., $g(-\mathbf{k}) = -g(\mathbf{k})$. Furthermore, the restriction of the states to a shell of energy width $\hbar\omega_D$ around the Fermi sphere implies that $g(\mathbf{k})$ is nonvanishing only for $E_F < \epsilon_{\mathbf{k}} < E_F + \hbar\omega_D$. Substituting the wave function, as given in Eq. (12.25), into the Schrödinger equation, we obtain

$$\sum_{\mathbf{k}'} (2\epsilon_{\mathbf{k}'} - E) g(\mathbf{k}') e^{i\mathbf{k}'.\mathbf{r}} + \sum_{\mathbf{k}'} g(\mathbf{k}') U(\mathbf{r}) e^{i\mathbf{k}'.\mathbf{r}} = 0 \tag{12.26}$$

where $\mathbf{r} = \mathbf{r}_1 - \mathbf{r}_2$. Multiplying by $(1/V)e^{-i\mathbf{k}.\mathbf{r}}$, integrating over the system's volume, and using

$$\int_V e^{i(\mathbf{k}'-\mathbf{k}).\mathbf{r}} d^3r = V\delta_{\mathbf{k}\mathbf{k}'}, \tag{12.27}$$

we obtain the following equation:

$$(2\epsilon_{\mathbf{k}} - E)g(\mathbf{k}) + \frac{1}{V}\sum_{\mathbf{k}'} U_{\mathbf{k}\mathbf{k}'}g(\mathbf{k}') = 0, \quad E_F < \epsilon_{\mathbf{k}}, \epsilon_{\mathbf{k}'} < E_F + \hbar\omega_D. \quad (12.28)$$

$U_{\mathbf{k}\mathbf{k}'}$ is the Fourier transform of the attractive interaction,

$$U_{\mathbf{k}\mathbf{k}'} = \int_V e^{-i(\mathbf{k}-\mathbf{k}').\mathbf{r}}U(\mathbf{r})d^3r. \quad (12.29)$$

Note that, since $U(\mathbf{r})$ is real, it follows that $U^*_{\mathbf{k}\mathbf{k}'} = U_{\mathbf{k}'\mathbf{k}}$. Moreover, since $U(\mathbf{r}) = U(-\mathbf{r})$, $U_{\mathbf{k}\mathbf{k}'}$ must be real.

We solve for $g(\mathbf{k})$ by considering a simple model for which

$$U_{\mathbf{k}\mathbf{k}'} = -U_0, \quad E_F < \epsilon_{\mathbf{k}}, \epsilon_{\mathbf{k}'} < E_F + \hbar\omega_D$$

where $U_0 > 0$. For values of $\mathbf{k}$ and $\mathbf{k}'$ such that $\epsilon_{\mathbf{k}}$ or $\epsilon_{\mathbf{k}'}$ lies outside the range indicated above, $U_{\mathbf{k}\mathbf{k}'} = 0$. The fact that $U_{\mathbf{k}\mathbf{k}'}$ is negative reflects the assumption that the interaction between the added electrons is attractive. Equation (12.28) now reduces to

$$(2\epsilon_{\mathbf{k}} - E)g(\mathbf{k}) = \frac{U_0}{V}\sum_{\mathbf{k}'} g(\mathbf{k}') \quad E_F < \epsilon_{\mathbf{k}}, \epsilon_{\mathbf{k}'} < E_F + \hbar\omega_D. \quad (12.30)$$

For a triplet state, $g(-\mathbf{k}') = -g(\mathbf{k}')$, and the RHS of Eq. (12.30) vanishes. Thus, for the triplet state, $E = 2\epsilon_{\mathbf{k}}$; the attractive interaction has no effect on the energy of the two added electrons. For the singlet state, on the other hand, $g(-\mathbf{k}') = g(\mathbf{k}')$, and the RHS of Eq. (12.30) does not vanish. Further analysis is now restricted to the singlet state, in which the two electrons have opposite spins.

Dividing both sides of Eq. (12.30) by $(2\epsilon_{\mathbf{k}} - E)$, then summing over $\mathbf{k}$, we obtain

$$1 = \frac{U_0}{V}\sum_{\mathbf{k}} \frac{1}{2\epsilon_{\mathbf{k}} - E}, \quad E_F < \epsilon_{\mathbf{k}} < E_F + \hbar\omega_D.$$

The sum over $\mathbf{k}$ is a sum over states of one spin projection. Since the number of such states in the energy range $(\epsilon, \epsilon + d\epsilon)$ is $D_\sigma(\epsilon)d\epsilon$, where $D_\sigma(\epsilon)$ is the density of states per spin, the above equation may be written as

$$1 = \frac{U_0}{V}\int_{E_F}^{E_F+\hbar\omega_D} \frac{D_\sigma(\epsilon)}{2\epsilon - E}d\epsilon.$$

Since it is generally true that in metals $\hbar\omega_D \ll E_F$ ($\hbar\omega_D \approx 20$ meV, $E_F \approx 5$ eV), we may assume that $D_\sigma(\epsilon)$ is equal to its value at the Fermi energy,

$$1 = \frac{U_0 D_\sigma(E_F)}{V}\int_{E_F}^{E_F+\hbar\omega_D} \frac{d\epsilon}{2\epsilon - E} = \frac{1}{2}U_0 d_\sigma(E_F)\ln\left(\frac{2E_F + 2\hbar\omega_D - E}{2E_F - E}\right)$$

where $d_\sigma(E_F) = D_\sigma(E_F)/V$ is the density of states per unit volume per spin. This equation is easily solved for E,

$$E = 2E_F - \frac{2\hbar\omega_D \exp\{-2/[U_0 d_\sigma(E_F)]\}}{1 - \exp\{-2/[U_0 d_\sigma(E_F)]\}}.$$

In the weak coupling limit ($U_0 d_\sigma(E_F) \ll 1$),

$$E \simeq 2E_F - 2\hbar\omega_D \exp\{-2/[U_0 d_\sigma(E_F)]\}. \tag{12.31}$$

The following remarks are in order:

1. No matter how weak the attractive interaction is, the two electrons form a bound state, known as a Cooper pair, whose energy is lower than $2E_F$.
2. The energy E of the bound state is not an analytic function of U_0 as $U_0 \to 0$, i.e., E cannot be expanded in powers of U_0. Thus, the result for E cannot be obtained by a perturbation expansion in powers of U_0.
3. The binding energy of the Cooper pair increases as U_0 increases; the stronger the electron–phonon interaction, the larger the binding energy.
4. The binding energy increases as the density of states at the Fermi energy increases.

12.5 BCS theory of superconductivity

A microscopic theory of superconductivity was presented in 1957 by Bardeen, Cooper, and Schrieffer (BCS) (Bardeen et al., 1957). The idea that a weakly attractive interaction between two electrons leads to the formation of a Cooper pair was a major clue that led to a fuller description of the superconducting ground state. The attractive interaction scatters a pair of electrons from states $|\mathbf{k}\uparrow, -\mathbf{k}\downarrow\rangle$ into states $|\mathbf{k}'\uparrow, -\mathbf{k}'\downarrow\rangle$ (see Figure 12.11). BCS considered the following model Hamiltonian

$$H_{\mathrm{BCS}} = \sum_{\mathbf{k}\sigma} \epsilon_{\mathbf{k}} c^\dagger_{\mathbf{k}\sigma} c_{\mathbf{k}\sigma} + \sum_{\mathbf{k}\mathbf{k}'} U_{\mathbf{k}\mathbf{k}'} c^\dagger_{\mathbf{k}'\uparrow} c^\dagger_{-\mathbf{k}'\downarrow} c_{-\mathbf{k}\downarrow} c_{\mathbf{k}\uparrow} \tag{12.32}$$

which describes the scattering processes mentioned above. In order to determine the ground state, a variational approach is adopted, with a trial wave function proposed and the corresponding energy minimized. The BCS trial state is taken as

$$|\Psi\rangle = \prod_{\mathbf{k}} \left(u_{\mathbf{k}} + v_{\mathbf{k}} c^\dagger_{\mathbf{k}\uparrow} c^\dagger_{-\mathbf{k}\downarrow} \right) |0\rangle \tag{12.33}$$

where $|0\rangle$ is the vacuum state, $u_{\mathbf{k}}$ and $v_{\mathbf{k}}$ are parameters to be determined, and they are assumed to be real. The state is normalized if

$$u^2_{\mathbf{k}} + v^2_{\mathbf{k}} = 1. \tag{12.34}$$

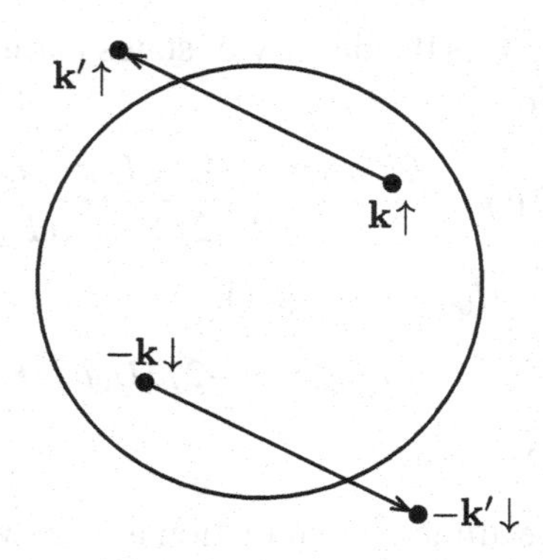

Figure 12.11 Scattering processes that contribute to the BCS Hamiltonian. Two electrons in states $|\mathbf{k}\uparrow\rangle$ and $|-\mathbf{k}\downarrow\rangle$ are scattered into states $|\mathbf{k}'\uparrow\rangle$ and $|-\mathbf{k}'\downarrow\rangle$. The matrix element for this scattering process is $U_{\mathbf{kk}'}$.

The form of the wave function implies that $v_{\mathbf{k}}^2$ is the probability that the pair state $|\mathbf{k}\uparrow, -\mathbf{k}\downarrow\rangle$ is occupied, and $u_{\mathbf{k}}^2$ is the probability for it to be empty. Note that $|\Psi\rangle$ would describe the normal ground state if $u_{\mathbf{k}} = 0,\ v_{\mathbf{k}} = 1$ for $k < k_F$, and $u_{\mathbf{k}} = 1,\ v_{\mathbf{k}} = 0$ for $k > k_F$.

If the state $|\Psi\rangle$ is expanded, we see that it is a linear combination of states with varying numbers of pairs, i.e., $|\Psi\rangle$ is not an eigenstate of the number operator N_{op} given by

$$N_{\text{op}} = \sum_{\mathbf{k}\sigma} c_{\mathbf{k}\sigma}^{\dagger} c_{\mathbf{k}\sigma}. \tag{12.35}$$

This should not cause any alarm; the system is considered to be in contact with a particle reservoir at $T = 0$. In other words, the system is assumed to be a member of a grand canonical ensemble; as we know from Chapter 5, the properties of a system may be obtained using any of various ensembles. What we require here is that the average number of electrons, $\langle\Psi|N_{\text{op}}|\Psi\rangle$, be equal to N, the actual number of conduction electrons in the crystal.

Our problem is thus to minimize the energy $\langle\Psi|H|\Psi\rangle$ subject to the constraint that $\langle\Psi|N_{\text{op}}|\Psi\rangle = N$. This is achieved by introducing a Lagrange multiplier μ and minimizing $\langle\Psi|H|\Psi\rangle - \mu\langle\Psi|N_{\text{op}}|\Psi\rangle = \langle\Psi|H - \mu N_{\text{op}}|\Psi\rangle$ without any conditions. The Lagrange multiplier μ is then determined by the requirement that $\langle\Psi|N_{\text{op}}|\Psi\rangle = N$. The parameter μ turns out to be nothing but the Fermi energy E_F. Defining $\bar{\epsilon}_{\mathbf{k}} = \epsilon_{\mathbf{k}} - \mu$, we can write

$$\bar{H} = H - \mu N_{\text{op}} = \sum_{\mathbf{k}} \bar{\epsilon}_{\mathbf{k}} \left(c_{\mathbf{k}\uparrow}^{\dagger} c_{\mathbf{k}\uparrow} + c_{-\mathbf{k}\downarrow}^{\dagger} c_{-\mathbf{k}\downarrow} \right) + \sum_{\mathbf{kk}'} U_{\mathbf{kk}'} c_{\mathbf{k}'\uparrow}^{\dagger} c_{-\mathbf{k}'\downarrow}^{\dagger} c_{-\mathbf{k}\downarrow} c_{\mathbf{k}\uparrow}. \tag{12.36}$$

Using the commutation properties of the creation and annihilation operators, it is not difficult to show that

$$\bar{E} = \langle\Psi|\bar{H}|\Psi\rangle = 2\sum_{\mathbf{k}} v_{\mathbf{k}}^2\bar{\epsilon}_{\mathbf{k}} + \sum_{\mathbf{k}\mathbf{k}'} U_{\mathbf{k}\mathbf{k}'}u_{\mathbf{k}}v_{\mathbf{k}}u_{\mathbf{k}'}v_{\mathbf{k}'}. \tag{12.37}$$

$\bar{E}$ is viewed as a function of $u_{\mathbf{k}}$ and $v_{\mathbf{k}}$, and we seek the values of $u_{\mathbf{k}}$ and $v_{\mathbf{k}}$ that minimize $\bar{E}$. Since $u_{\mathbf{k}}^2 + v_{\mathbf{k}}^2 = 1$, there exists an angle $\theta_{\mathbf{k}}$ such that

$$u_{\mathbf{k}} = \cos\theta_{\mathbf{k}}, \qquad v_{\mathbf{k}} = \sin\theta_{\mathbf{k}}. \tag{12.38}$$

The expression for $\bar{E}$ becomes

$$\bar{E} = 2\sum_{\mathbf{k}} \bar{\epsilon}_{\mathbf{k}}\sin^2\theta_{\mathbf{k}} + \frac{1}{4}\sum_{\mathbf{k}\mathbf{k}'} U_{\mathbf{k}\mathbf{k}'}\sin(2\theta_{\mathbf{k}})\sin(2\theta_{\mathbf{k}'}). \tag{12.39}$$

The minimization condition $\partial\bar{E}/\partial\theta_{\mathbf{k}} = 0$ yields

$$2\bar{\epsilon}_{\mathbf{k}}\sin(2\theta_{\mathbf{k}}) + \cos(2\theta_{\mathbf{k}})\sum_{\mathbf{k}'} U_{\mathbf{k}\mathbf{k}'}\sin(2\theta_{\mathbf{k}'}) = 0. \tag{12.40}$$

Reintroducing $u_{\mathbf{k}}$ and $v_{\mathbf{k}}$: $\sin(2\theta_{\mathbf{k}}) = 2u_{\mathbf{k}}v_{\mathbf{k}}$, $\cos(2\theta_{\mathbf{k}}) = u_{\mathbf{k}}^2 - v_{\mathbf{k}}^2$, the above equation becomes

$$2\bar{\epsilon}_{\mathbf{k}}u_{\mathbf{k}}v_{\mathbf{k}} + (u_{\mathbf{k}}^2 - v_{\mathbf{k}}^2)\sum_{\mathbf{k}'} U_{\mathbf{k}\mathbf{k}'}u_{\mathbf{k}'}v_{\mathbf{k}'} = 0. \tag{12.41}$$

We now define the energy gap parameter by

$$\Delta_{\mathbf{k}} = -\sum_{\mathbf{k}'} U_{\mathbf{k}\mathbf{k}'}u_{\mathbf{k}'}v_{\mathbf{k}'} \tag{12.42}$$

and thus obtain

$$2\bar{\epsilon}_{\mathbf{k}}u_{\mathbf{k}}v_{\mathbf{k}} - \Delta_{\mathbf{k}}\left(u_{\mathbf{k}}^2 - v_{\mathbf{k}}^2\right) = 0. \tag{12.43}$$

Keeping in mind that $u_{\mathbf{k}}^2 + v_{\mathbf{k}}^2 = 1$, the following solutions are obtained

$$u_{\mathbf{k}}^2 = \frac{1}{2}\left[1 + \frac{\bar{\epsilon}_{\mathbf{k}}}{\sqrt{\bar{\epsilon}_{\mathbf{k}}^2 + \Delta_{\mathbf{k}}^2}}\right], \qquad v_{\mathbf{k}}^2 = \frac{1}{2}\left[1 - \frac{\bar{\epsilon}_{\mathbf{k}}}{\sqrt{\bar{\epsilon}_{\mathbf{k}}^2 + \Delta_{\mathbf{k}}^2}}\right]. \tag{12.44}$$

Note that if $U_{\mathbf{k}\mathbf{k}'} = 0$, $\Delta_{\mathbf{k}}$ vanishes, and $v_{\mathbf{k}}^2 = 1$ for $\bar{\epsilon}_{\mathbf{k}} < 0$, while $v_{\mathbf{k}}^2 = 0$ for $\bar{\epsilon}_{\mathbf{k}} > 0$. This is the situation in a normal metal where $v_{\mathbf{k}} = 1$ for $\epsilon_{\mathbf{k}} < E_F$ and $v_{\mathbf{k}} = 0$ for $\epsilon_{\mathbf{k}} > E_F$. Since $\bar{\epsilon}_{\mathbf{k}} = \epsilon_{\mathbf{k}} - \mu$, it follows that μ is simply E_F. A plot of $v_{\mathbf{k}}^2$ vs. $\bar{\epsilon}_{\mathbf{k}}$ is shown in Figure 12.12. Using the above expressions for $u_{\mathbf{k}}^2$ and $v_{\mathbf{k}}^2$, Eq. (12.42) becomes

$$\Delta_{\mathbf{k}} = -\frac{1}{2}\sum_{\mathbf{k}'} U_{\mathbf{k}\mathbf{k}'}\frac{\Delta_{\mathbf{k}'}}{\sqrt{\bar{\epsilon}_{\mathbf{k}'}^2 + \Delta_{\mathbf{k}'}^2}}. \tag{12.45}$$

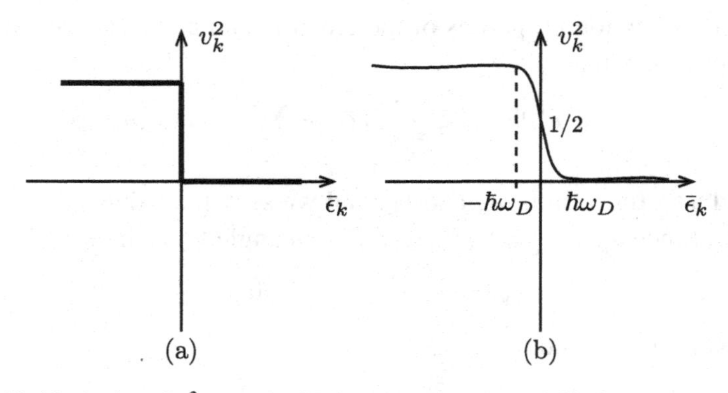

Figure 12.12 A plot of $v_{\mathbf{k}}^2$ vs. $\bar{\epsilon}_{\mathbf{k}}$ in (a) the normal state, and (b) the superconducting state.

In principle, this equation determines the gap parameter. In general, a solution is difficult to come by, but a simple solution is obtained if we adopt the following model for the attractive interaction:

$$U_{\mathbf{kk'}} = \begin{cases} -U_0 & -\hbar\omega_D < \bar{\epsilon}_{\mathbf{k}},\ \bar{\epsilon}_{\mathbf{k'}} < \hbar\omega_D \\ 0 & \text{otherwise} \end{cases}. \tag{12.46}$$

It follows from Eq. (12.42) that $\Delta_{\mathbf{k}}$ is constant, independent of $\mathbf{k}$, for $-\hbar\omega_D < \bar{\epsilon}_{\mathbf{k}} < \hbar\omega_D$, and zero otherwise. Writing the constant as Δ_0, Eq. (12.45) becomes

$$\frac{U_0}{2}\sum_{\mathbf{k}} \frac{1}{\sqrt{\bar{\epsilon}_{\mathbf{k}}^2 + \Delta_0^2}} = 1, \qquad -\hbar\omega_D < \bar{\epsilon}_{\mathbf{k}} < \hbar\omega_D.$$

Converting the sum over $\mathbf{k}$ into an integral over energy, we obtain

$$1 = \frac{U_0 D_\sigma(E_F)}{2}\int_{-\hbar\omega_D}^{\hbar\omega_D} \frac{d\epsilon}{\sqrt{\epsilon^2 + \Delta_0^2}} = U_0 D_\sigma(E_F)\sinh^{-1}\left(\frac{\hbar\omega_D}{\Delta_0}\right)$$

$$\Longrightarrow \Delta_0 = \frac{\hbar\omega_D}{\sinh\left[\frac{1}{U_0 D_\sigma(E_F)}\right]}. \tag{12.47}$$

$D_\sigma(E_F)$ is the density of states for one spin projection at the Fermi energy. In the weak coupling limit ($U_0 D_\sigma(E_F) \ll 1$), the gap parameter is given by

$$\Delta_0 \simeq 2\hbar\omega_D \exp\left[\frac{-1}{U_0 D_\sigma(E_F)}\right]. \tag{12.48}$$

The energy of the BCS ground state is

$$\bar{E}_S = \sum_{\mathbf{k}} \left(2v_{\mathbf{k}}^2 \bar{\epsilon}_{\mathbf{k}} - \Delta_{\mathbf{k}} u_{\mathbf{k}} v_{\mathbf{k}}\right) = \sum_{\mathbf{k}} \left[\bar{\epsilon}_{\mathbf{k}} - \frac{2\bar{\epsilon}_{\mathbf{k}}^2 + \Delta_{\mathbf{k}}^2}{2\sqrt{\bar{\epsilon}_{\mathbf{k}}^2 + \Delta_{\mathbf{k}}^2}} \right] \tag{12.49}$$

(see Eqs [12.37], [12.42], and [12.44]). Assuming, as before, that $\Delta_{\mathbf{k}} = \Delta_0$ for $-\hbar\omega_D < \bar{\epsilon}_{\mathbf{k}} < \hbar\omega_D$, and is zero otherwise, we see that the summand in the above equation is equal to $2\bar{\epsilon}_{\mathbf{k}}$ for $\bar{\epsilon}_{\mathbf{k}} < -\hbar\omega_D$ and is equal to zero for $\bar{\epsilon}_{\mathbf{k}} > \hbar\omega_D$. The energy $\bar{E}_N$ of the normal ground state is the sum over **k** of $2\bar{\epsilon}_{\mathbf{k}}$ up to $\bar{\epsilon}_{\mathbf{k}} = 0$. Hence

$$\bar{E}_S - \bar{E}_N = \sum_{\mathbf{k}}{}' \left[\bar{\epsilon}_{\mathbf{k}} - \frac{2\bar{\epsilon}_{\mathbf{k}}^2 + \Delta_{\mathbf{k}}^2}{2\sqrt{\bar{\epsilon}_{\mathbf{k}}^2 + \Delta_{\mathbf{k}}^2}} \right] - \sum_{\mathbf{k}}{}'' 2\bar{\epsilon}_{\mathbf{k}}.$$

The prime over the sum means that the sum is restricted to values of **k** such that $-\hbar\omega_D < \bar{\epsilon}_{\mathbf{k}} < \hbar\omega_D$, while the double prime indicates that the sum over **k** is restricted so that $-\hbar\omega_D < \bar{\epsilon}_{\mathbf{k}} < 0$. Writing the sum over **k** as an integral over energy, we find

$$\begin{aligned} \bar{E}_S - \bar{E}_N &= D_\sigma(E_F) \int_{-\hbar\omega_D}^{\hbar\omega_D} \left[\epsilon - \frac{2\epsilon^2 + \Delta_0^2}{2\sqrt{\epsilon^2 + \Delta_0^2}} \right] d\epsilon - D_\sigma(E_F) \int_{-\hbar\omega_D}^{0} 2\epsilon d\epsilon \\ &= D_\sigma(E_F)\hbar\omega_D \left[\hbar\omega_D - \sqrt{(\hbar\omega_D)^2 + \Delta_0^2} \right]. \end{aligned}$$

For weak coupling ($\Delta_0 \ll \hbar\omega_D$), the above equation, upon expansion of the square root, reduces to

$$\bar{E}_S - \bar{E}_N \simeq -\frac{1}{2} D_\sigma(E_F)\Delta_0^2. \tag{12.50}$$

The superconducting state is lower in energy than the normal state; hence, in the presence of an attractive interaction between electrons near the Fermi surface, the normal state becomes unstable, and the system undergoes a transition to a superconducting state.

We note that the BCS theory, by replacing $U_{\mathbf{kk}'}$ by $-U_0$, it neglects the fact that the attractive interaction between electrons (mediated by phonons) is retarded. This is a good approximation in superconductors where the electron–phonon interaction is weak, but it does not provide an accurate description of strong-coupling superconductors, where the electron–phonon interaction is strong. For a review of strong-coupling theory of superconductivity, the reader is referred to the article by Scalapino (1969).

Finally, we briefly touch upon a certain feature, mentioned earlier, of the superconducting ground state. The BCS Hamiltonian commutes with the number of particles operator, $[H_{\text{BCS}}, N_{\text{op}}] = 0$, but the number of particles in the ground state wave function is not constant. Stated differently, the Hamiltonian possesses a certain symmetry which the ground state lacks. The superconducting state is thus characterized by a broken symmetry. To elaborate this point further, we note that the BCS Hamiltonian given in Eq. (12.32) is invariant under the global transformation

$$c_{\mathbf{k}\sigma} \to e^{-i\phi} c_{\mathbf{k}\sigma}, \quad c^{\dagger}_{\mathbf{k}\sigma} \to e^{i\phi} c^{\dagger}_{\mathbf{k}\sigma}.$$

Under this transformation, the normal state $|F\rangle = \prod'_{\mathbf{k}\sigma} c^{\dagger}_{\mathbf{k}\sigma}|0\rangle$ remains invariant; it simply aquires a constant phase (the prime on the product sign indicates that $k < k_F$). However, the BCS ground state, given in Eq. (12.33), is not invariant under this transformation.

12.6 Mean field approach

The superconducting ground state may also be obtained using a mean field approach. An additional benefit of this approach is the elucidation of the nature of excited states. Our starting point is again the BCS Hamiltonian

$$\bar{H}_{BCS} = \sum_{\mathbf{k}\sigma} \bar{\epsilon}_{\mathbf{k}} c^{\dagger}_{\mathbf{k}\sigma} c_{\mathbf{k}\sigma} + \sum_{\mathbf{k}\mathbf{k}'} U_{\mathbf{k}\mathbf{k}'} c^{\dagger}_{\mathbf{k}'\uparrow} c^{\dagger}_{-\mathbf{k}'\downarrow} c_{-\mathbf{k}\downarrow} c_{\mathbf{k}\uparrow} = \bar{H}_0 + H'.$$

We define a fluctuation operator $d_{\mathbf{k}}$ that represents the deviation of $c_{-\mathbf{k}\downarrow} c_{\mathbf{k}\uparrow}$ from its average in the ground state,

$$d_{\mathbf{k}} = c_{-\mathbf{k}\downarrow} c_{\mathbf{k}\uparrow} - \langle c_{-\mathbf{k}\downarrow} c_{\mathbf{k}\uparrow} \rangle. \tag{12.51a}$$

Similarly,

$$d^{\dagger}_{\mathbf{k}} = c^{\dagger}_{\mathbf{k}\uparrow} c^{\dagger}_{-\mathbf{k}\downarrow} - \langle c^{\dagger}_{\mathbf{k}\uparrow} c^{\dagger}_{-\mathbf{k}\downarrow} \rangle. \tag{12.51b}$$

In a normal metal, the quantities $\langle c_{-\mathbf{k}\downarrow} c_{\mathbf{k}\uparrow} \rangle$ and $\langle c^{\dagger}_{\mathbf{k}\uparrow} c^{\dagger}_{-\mathbf{k}\downarrow} \rangle$ vanish, but this is not the case in a superconductor, where the ground state is not an eigenstate of the number of particles operator. In terms of the fluctuation operators, the interaction Hamiltonian is given by

$$H' = \sum_{\mathbf{k}\mathbf{k}'} U_{\mathbf{k}\mathbf{k}'} \left\{ \langle c^{\dagger}_{\mathbf{k}'\uparrow} c^{\dagger}_{-\mathbf{k}'\downarrow} \rangle d_{\mathbf{k}} + \langle c_{-\mathbf{k}\downarrow} c_{\mathbf{k}\uparrow} \rangle d^{\dagger}_{\mathbf{k}'} + \langle c^{\dagger}_{\mathbf{k}'\uparrow} c^{\dagger}_{-\mathbf{k}'\downarrow} \rangle \langle c_{-\mathbf{k}\downarrow} c_{\mathbf{k}\uparrow} \rangle + d^{\dagger}_{\mathbf{k}'} d_{\mathbf{k}} \right\}.$$

In the mean field approximation, the last term in the above expression, which is bilinear in fluctuation operators, is ignored. The assumption made is that the fluctuations of $c_{-\mathbf{k}\downarrow} c_{\mathbf{k}\uparrow}$ and $c^{\dagger}_{\mathbf{k}\uparrow} c^{\dagger}_{-\mathbf{k}\downarrow}$ about their average values are small.

Defining the gap parameter by

$$\Delta_{\mathbf{k}} = -\sum_{\mathbf{k}'} U_{\mathbf{k}'\mathbf{k}} \langle c_{-\mathbf{k}'\downarrow} c_{\mathbf{k}'\uparrow} \rangle, \tag{12.52}$$

the mean field Hamiltonian may be written as

$$H_{MF} = \sum_{\mathbf{k}\sigma} \bar{\epsilon}_{\mathbf{k}} c^{\dagger}_{\mathbf{k}\sigma} c_{\mathbf{k}\sigma} - \sum_{\mathbf{k}} \left(\Delta^*_{\mathbf{k}} d_{\mathbf{k}} + \Delta_{\mathbf{k}} d^{\dagger}_{\mathbf{k}} \right) - \sum_{\mathbf{k}} \Delta^*_{\mathbf{k}} \langle c_{-\mathbf{k}\downarrow} c_{\mathbf{k}\uparrow} \rangle$$
$$= \sum_{\mathbf{k}\sigma} \bar{\epsilon}_{\mathbf{k}} c^{\dagger}_{\mathbf{k}\sigma} c_{\mathbf{k}\sigma} - \sum_{\mathbf{k}} \Delta^*_{\mathbf{k}} c_{-\mathbf{k}\downarrow} c_{\mathbf{k}\uparrow} - \sum_{\mathbf{k}} \Delta_{\mathbf{k}} c^{\dagger}_{\mathbf{k}\uparrow} c^{\dagger}_{-\mathbf{k}\downarrow} + \sum_{\mathbf{k}} \Delta_{\mathbf{k}} \langle c^{\dagger}_{\mathbf{k}\uparrow} c^{\dagger}_{-\mathbf{k}\downarrow} \rangle. \tag{12.53}$$

H_{MF} can be diagonalized by means of a canonical transformation known as the Bogoliubov–Valatin transformation (Bogoliubov, 1958; Valatin, 1958):

$$\gamma_{\mathbf{k}\uparrow} = u_{\mathbf{k}} c_{\mathbf{k}\uparrow} - v_{\mathbf{k}} c^{\dagger}_{-\mathbf{k}\downarrow}, \quad \gamma_{-\mathbf{k}\downarrow} = u_{\mathbf{k}} c_{-\mathbf{k}\downarrow} + v_{\mathbf{k}} c^{\dagger}_{\mathbf{k}\uparrow} \tag{12.54a}$$

$$\gamma^{\dagger}_{\mathbf{k}\uparrow} = u^*_{\mathbf{k}} c^{\dagger}_{\mathbf{k}\uparrow} - v^*_{\mathbf{k}} c_{-\mathbf{k}\downarrow}, \quad \gamma^{\dagger}_{-\mathbf{k}\downarrow} = u^*_{\mathbf{k}} c^{\dagger}_{-\mathbf{k}\downarrow} + v^*_{\mathbf{k}} c_{\mathbf{k}\uparrow}. \tag{12.54b}$$

The new operators must satisfy the same commutation relations as the original ones; we thus require that

$$\{\gamma_{\mathbf{k}\sigma}, \gamma_{\mathbf{k}'\sigma'}\} = \left\{\gamma^{\dagger}_{\mathbf{k}\sigma}, \gamma^{\dagger}_{\mathbf{k}'\sigma'}\right\} = 0, \quad \left\{\gamma_{\mathbf{k}\sigma}, \gamma^{\dagger}_{\mathbf{k}'\sigma'}\right\} = \delta_{\mathbf{k}\mathbf{k}'} \delta_{\sigma\sigma'}. \tag{12.55}$$

These are satisfied provided that

$$|u_{\mathbf{k}}|^2 + |v_{\mathbf{k}}|^2 = 1. \tag{12.56}$$

Using Eq. (12.54), we solve for the c-operators in terms of the γ-operators,

$$c_{\mathbf{k}\uparrow} = u^*_{\mathbf{k}} \gamma_{\mathbf{k}\uparrow} + v_{\mathbf{k}} \gamma^{\dagger}_{-\mathbf{k}\downarrow}, \quad c_{-\mathbf{k}\downarrow} = u^*_{\mathbf{k}} \gamma_{-\mathbf{k}\downarrow} - v_{\mathbf{k}} \gamma^{\dagger}_{\mathbf{k}\uparrow} \tag{12.57a}$$

$$c^{\dagger}_{\mathbf{k}\uparrow} = u_{\mathbf{k}} \gamma^{\dagger}_{\mathbf{k}\uparrow} + v^*_{\mathbf{k}} \gamma_{-\mathbf{k}\downarrow}, \quad c^{\dagger}_{-\mathbf{k}\downarrow} = u_{\mathbf{k}} \gamma^{\dagger}_{-\mathbf{k}\downarrow} - v^*_{\mathbf{k}} \gamma_{\mathbf{k}\uparrow}. \tag{12.57b}$$

Inserting these terms into Eq. (12.53), and then laboring through some tedious calculations, we find

$$H_{MF} = \sum_{\mathbf{k}} \left[\bar{\epsilon}_{\mathbf{k}} \left(|u_{\mathbf{k}}|^2 - |v_{\mathbf{k}}|^2 \right) + \Delta_{\mathbf{k}} u_{\mathbf{k}} v^*_{\mathbf{k}} + \Delta^*_{\mathbf{k}} u^*_{\mathbf{k}} v_{\mathbf{k}} \right] \left(\gamma^{\dagger}_{\mathbf{k}\uparrow} \gamma_{\mathbf{k}\uparrow} + \gamma^{\dagger}_{-\mathbf{k}\downarrow} \gamma_{-\mathbf{k}\downarrow} \right)$$
$$+ \sum_{\mathbf{k}} \left(2\bar{\epsilon}_{\mathbf{k}} u_{\mathbf{k}} v_{\mathbf{k}} + \Delta^*_{\mathbf{k}} v^2_{\mathbf{k}} - \Delta_{\mathbf{k}} u^2_{\mathbf{k}} \right) \gamma^{\dagger}_{\mathbf{k}\uparrow} \gamma^{\dagger}_{-\mathbf{k}\downarrow}$$
$$+ \sum_{\mathbf{k}} \left(2\bar{\epsilon}_{\mathbf{k}} u^*_{\mathbf{k}} v^*_{\mathbf{k}} + \Delta_{\mathbf{k}} v^{*2}_{\mathbf{k}} - \Delta^*_{\mathbf{k}} u^{*2}_{\mathbf{k}} \right) \gamma_{-\mathbf{k}\downarrow} \gamma_{\mathbf{k}\uparrow}$$
$$+ \sum_{\mathbf{k}} \left[2\bar{\epsilon}_{\mathbf{k}} |v_{\mathbf{k}}|^2 - \Delta_{\mathbf{k}} u_{\mathbf{k}} v^*_{\mathbf{k}} - \Delta^*_{\mathbf{k}} u^*_{\mathbf{k}} v_{\mathbf{k}} + \Delta_{\mathbf{k}} \langle c^{\dagger}_{\mathbf{k}\uparrow} c^{\dagger}_{-\mathbf{k}\downarrow} \rangle \right]. \tag{12.58}$$

The first term describes single-particle excitations, while the last term is a constant that represents the energy of the system in the absence of single-particle excitations – the ground state energy. The troublesome terms are the third and fourth ones; they are not diagonal. However, the only condition imposed on $u_{\mathbf{k}}$ and $v_{\mathbf{k}}$ so far is Eq. (12.56). We can take advantage of the available freedom regarding the choice of $u_{\mathbf{k}}$ and $v_{\mathbf{k}}$ by demanding that the troublesome terms vanish. We thus impose the condition

$$2\bar{\epsilon}_{\mathbf{k}} u_{\mathbf{k}} v_{\mathbf{k}} + \Delta_{\mathbf{k}}^* v_{\mathbf{k}}^2 - \Delta_{\mathbf{k}} u_{\mathbf{k}}^2 = 0. \tag{12.59}$$

To solve for $u_{\mathbf{k}}$ and $v_{\mathbf{k}}$, we set

$$u_{\mathbf{k}} = |u_{\mathbf{k}}| e^{i\theta_{\mathbf{k}}}, \quad v_{\mathbf{k}} = |v_{\mathbf{k}}| e^{i\phi_{\mathbf{k}}}, \quad \Delta_{\mathbf{k}} = |\Delta_{\mathbf{k}}| e^{2i\delta_{\mathbf{k}}}.$$

The condition on $u_{\mathbf{k}}$ and $v_{\mathbf{k}}$ becomes

$$2\bar{\epsilon}_{\mathbf{k}} |u_{\mathbf{k}} v_{\mathbf{k}}| e^{i(\theta_{\mathbf{k}}+\phi_{\mathbf{k}})} + |\Delta_{\mathbf{k}}| |v_{\mathbf{k}}|^2 e^{2i(\phi_{\mathbf{k}}-\delta_{\mathbf{k}})} - |\Delta_{\mathbf{k}}| |u_{\mathbf{k}}|^2 e^{2i(\theta_{\mathbf{k}}+\delta_{\mathbf{k}})} = 0.$$

Choosing $\theta_{\mathbf{k}}$, $\phi_{\mathbf{k}}$, and $\delta_{\mathbf{k}}$ such that $\theta_{\mathbf{k}} = -\phi_{\mathbf{k}} = -\delta_{\mathbf{k}}$, we obtain

$$2\bar{\epsilon}_{\mathbf{k}} |u_{\mathbf{k}}| |v_{\mathbf{k}}| + |\Delta_{\mathbf{k}}| \left(|v_{\mathbf{k}}|^2 - |u_{\mathbf{k}}|^2 \right) = 0.$$

This is to be solved along with the constraint $|u_{\mathbf{k}}|^2 + |v_{\mathbf{k}}|^2 = 1$; we find

$$|u_{\mathbf{k}}|^2 = \frac{1}{2}\left[1 + \frac{\bar{\epsilon}_{\mathbf{k}}}{\sqrt{\bar{\epsilon}_{\mathbf{k}}^2 + |\Delta_{\mathbf{k}}|^2}}\right], \quad |v_{\mathbf{k}}|^2 = \frac{1}{2}\left[1 - \frac{\bar{\epsilon}_{\mathbf{k}}}{\sqrt{\bar{\epsilon}_{\mathbf{k}}^2 + |\Delta_{\mathbf{k}}|^2}}\right]. \tag{12.60}$$

The phase of $u_{\mathbf{k}}$ is not determined; it can be chosen arbitrarily. Setting $\theta_{\mathbf{k}} = \phi_{\mathbf{k}} = \delta_{\mathbf{k}} = 0$ is tantamount to choosing $u_{\mathbf{k}}$, $v_{\mathbf{k}}$, and Δ_k to be real. When the values given above for $u_{\mathbf{k}}$ and $v_{\mathbf{k}}$ are inserted into Eq. (12.58), the first term in the Hamiltonian takes a particularly simple form:

$$H_{MF} = \sum_{\mathbf{k}} E_{\mathbf{k}} \left(\gamma_{\mathbf{k}\uparrow}^\dagger \gamma_{\mathbf{k}\uparrow} + \gamma_{-\mathbf{k}\downarrow}^\dagger \gamma_{-\mathbf{k}\downarrow} \right) + \sum_{\mathbf{k}} \left[2\bar{\epsilon}_{\mathbf{k}} v_{\mathbf{k}}^2 - 2\Delta_{\mathbf{k}} u_{\mathbf{k}} v_{\mathbf{k}} + \Delta_{\mathbf{k}} \langle c_{\mathbf{k}\uparrow}^\dagger c_{-\mathbf{k}\downarrow}^\dagger \rangle \right] \tag{12.61}$$

where $E_{\mathbf{k}} = \sqrt{\bar{\epsilon}_{\mathbf{k}}^2 + \Delta_{\mathbf{k}}^2}$. The second term is the ground state energy, while the first term describes excitations above the ground state.

The ground state $|\Psi_0\rangle$ is the state with no excitations; it is defined by the requirement that

$$\gamma_{\mathbf{k}\uparrow} |\Psi_0\rangle = \gamma_{-\mathbf{k}\downarrow} |\Psi_0\rangle = 0, \quad \forall \mathbf{k} \in \text{FBZ}. \tag{12.62}$$

The solution of the above equation is given by

$$|\Psi_0\rangle = \prod_{\mathbf{k}} \gamma_{\mathbf{k}\uparrow} \gamma_{-\mathbf{k}\downarrow} |0\rangle$$

where $|0\rangle$ is the vacuum state. That $|\Psi_0\rangle$ satisfies Eq. (12.62) follows from the commutation relations of the γ-operators. Using Eq. (12.54),

$$\begin{aligned}
|\Psi_0\rangle &= \prod_{\mathbf{k}} \left(u_{\mathbf{k}} c_{\mathbf{k}\uparrow} - v_{\mathbf{k}} c^{\dagger}_{-\mathbf{k}\downarrow}\right)\left(u_{\mathbf{k}} c_{-\mathbf{k}\downarrow} + v_{\mathbf{k}} c^{\dagger}_{\mathbf{k}\uparrow}\right)|0\rangle \\
&= \prod_{\mathbf{k}} \left(u_{\mathbf{k}} v_{\mathbf{k}} - v^2_{\mathbf{k}} c^{\dagger}_{-\mathbf{k}\downarrow} c^{\dagger}_{\mathbf{k}\uparrow}\right)|0\rangle \\
&= \left(\prod_{\mathbf{k}} v_{\mathbf{k}}\right) \prod_{\mathbf{k}} \left(u_{\mathbf{k}} + v_{\mathbf{k}} c^{\dagger}_{\mathbf{k}\uparrow} c^{\dagger}_{-\mathbf{k}\downarrow}\right)|0\rangle .
\end{aligned}$$

Since $u^2_{\mathbf{k}} + v^2_{\mathbf{k}} = 1$, the normalized ground state is

$$|\Psi_0\rangle = \prod_{\mathbf{k}} \left(u_{\mathbf{k}} + v_{\mathbf{k}} c^{\dagger}_{\mathbf{k}\uparrow} c^{\dagger}_{-\mathbf{k}\downarrow}\right)|0\rangle . \tag{12.63}$$

This is the same state we saw earlier using the variational method. Again, $v^2_{\mathbf{k}}$ is the probability that the pair $|\mathbf{k}\uparrow, -\mathbf{k}\downarrow\rangle$ is occupied, and $u^2_{\mathbf{k}}$ is the probability that it is empty.

The ground state energy is the second term of the Hamiltonian given in Eq. (12.61). At $T = 0$, $\langle c^{\dagger}_{\mathbf{k}\uparrow} c^{\dagger}_{-\mathbf{k}\downarrow}\rangle = \langle\Psi_0| c^{\dagger}_{\mathbf{k}\uparrow} c^{\dagger}_{-\mathbf{k}\downarrow}|\Psi_0\rangle$. We can evaluate this directly by using the expression given for $|\Psi_0\rangle$ in Eq. (12.63); alternatively, we can use Eq. (12.57) to write

$$c^{\dagger}_{\mathbf{k}\uparrow} c^{\dagger}_{-\mathbf{k}\downarrow} = u^2_{\mathbf{k}} \gamma^{\dagger}_{\mathbf{k}\uparrow} \gamma^{\dagger}_{-\mathbf{k}\downarrow} - v^2_{\mathbf{k}} \gamma_{-\mathbf{k}\downarrow} \gamma_{\mathbf{k}\uparrow} - u_{\mathbf{k}} v_{\mathbf{k}} \gamma^{\dagger}_{\mathbf{k}\uparrow} \gamma_{\mathbf{k}\uparrow} + u_{\mathbf{k}} v_{\mathbf{k}} \gamma_{-\mathbf{k}\downarrow} \gamma^{\dagger}_{-\mathbf{k}\downarrow} .$$

Since $\gamma_{\mathbf{k}\uparrow}|\Psi_0\rangle = \langle\Psi_0|\gamma^{\dagger}_{\mathbf{k}\uparrow} = 0$, $\gamma_{-\mathbf{k}\downarrow}\gamma^{\dagger}_{-\mathbf{k}\downarrow} = 1 - \gamma^{\dagger}_{-\mathbf{k}\downarrow}\gamma_{-\mathbf{k}\downarrow}$, and $\gamma_{-\mathbf{k}\downarrow}|\Psi_0\rangle = 0$, we obtain

$$\langle\Psi_0| c^{\dagger}_{\mathbf{k}\uparrow} c^{\dagger}_{-\mathbf{k}\downarrow}|\Psi_0\rangle = u_{\mathbf{k}} v_{\mathbf{k}} . \tag{12.64}$$

The ground state energy is thus given by

$$\bar{E}_S = \sum_{\mathbf{k}} \left(2\bar{\epsilon}_{\mathbf{k}} v^2_{\mathbf{k}} - \Delta_{\mathbf{k}} u_{\mathbf{k}} v_{\mathbf{k}}\right) . \tag{12.65}$$

This is exactly the same expression obtained earlier using a variational approach (see Eq. [12.49]).

An alternative expression for $\bar{E}_S$ can be written. From Eq. (12.60),

$$\bar{\epsilon}_{\mathbf{k}} = E_{\mathbf{k}}\left(1 - 2v^2_{\mathbf{k}}\right), \quad u^2_{\mathbf{k}} v^2_{\mathbf{k}} = \frac{\Delta^2_{\mathbf{k}}}{4E^2_{\mathbf{k}}} . \tag{12.66}$$

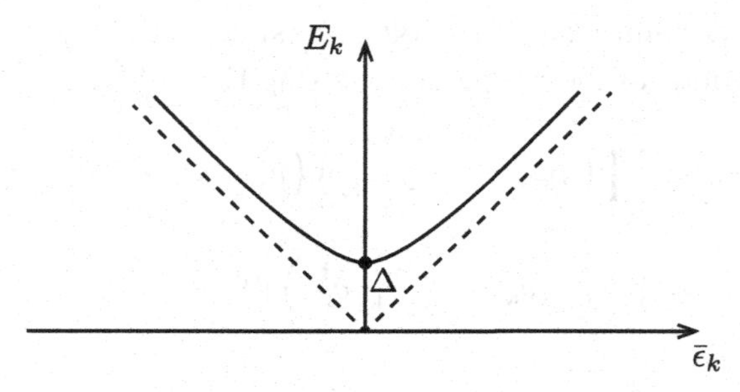

Figure 12.13 Single-particle excitation energy as a function of $\bar{\epsilon}_{\mathbf{k}} = \epsilon_{\mathbf{k}} - E_F$. The excitation energy has a minimum value equal to Δ. The function $E_{\mathbf{k}}$ aymptotically approaches the two dashed lines with slopes of ± 1.

It follows that

$$\bar{E}_S = -\sum_{\mathbf{k}} 4E_{\mathbf{k}} v_{\mathbf{k}}^4 + \sum_{\mathbf{k}} \left(2E_{\mathbf{k}} v_{\mathbf{k}}^2 - \frac{\Delta_{\mathbf{k}}^2}{2E_{\mathbf{k}}} \right).$$

The last term may be written as

$$\sum_{\mathbf{k}} \left(2E_{\mathbf{k}} v_{\mathbf{k}}^2 - \frac{\Delta_{\mathbf{k}}^2}{2E_{\mathbf{k}}} \right) = \sum_{\mathbf{k}} 2E_{\mathbf{k}} \left(v_{\mathbf{k}}^2 - \frac{\Delta_{\mathbf{k}}^2}{4E_{\mathbf{k}}^2} \right) = \sum_{\mathbf{k}} 2E_{\mathbf{k}} \left(v_{\mathbf{k}}^2 - u_{\mathbf{k}}^2 v_{\mathbf{k}}^2 \right)$$
$$= \sum_{\mathbf{k}} 2E_{\mathbf{k}} v_{\mathbf{k}}^2 \left(1 - u_{\mathbf{k}}^2 \right) = \sum_{\mathbf{k}} 2E_{\mathbf{k}} v_{\mathbf{k}}^4.$$

The ground state energy reduces to

$$\bar{E}_S = -2 \sum_{\mathbf{k}} E_{\mathbf{k}} v_{\mathbf{k}}^4. \tag{12.67}$$

Going back to the mean field Hamiltonian of Eq. (12.61), the first term describes excitations above the ground state. The single-particle excitation has energy $E_{\mathbf{k}} = \sqrt{\bar{\epsilon}_{\mathbf{k}}^2 + \Delta_{\mathbf{k}}^2}$. If we adopt the approximation that $\Delta_{\mathbf{k}}$ is independent of $\mathbf{k}$, as we did in the previous section, we see that the minimum energy for a single-particle excitation is equal to Δ, which corresponds to particles at the Fermi surface ($\bar{\epsilon}_{\mathbf{k}} = \epsilon_{\mathbf{k}} - E_F = 0$). The single-particle excitation energy is plotted in Figure 12.13 as a function of $\bar{\epsilon}_{\mathbf{k}}$. Note that, in a normal metal, it is possible to excite an electron from a state just below the Fermi surface to a state just above the Fermi surface by adding an infinitesimal amount of energy. This is not the case for a superconductor.

Finally, we note that even though the minimum single-particle excitation energy is equal to Δ, the lowest excited state has energy 2Δ above the ground state energy. This is because the lowest excited state involves breaking up a Cooper pair: an electron is scattered out of the state $|\mathbf{k}\uparrow\rangle$, leaving behind an unpaired electron in the state $|-\mathbf{k}\downarrow\rangle$. If the pair state $|\mathbf{k}\uparrow, -\mathbf{k}\downarrow\rangle$ is occupied in the ground state ($v_{\mathbf{k}}^2 = 1$), then it is unoccupied in the excited state ($v_{\mathbf{k}}^2 = 0$). Using Eq. (12.67), the change in energy is $2E_{\mathbf{k}}$, which has a minimum value of 2Δ. Another way (Taylor and Heinonen, 2002) to arrive at this result is by realizing that in a superconductor, single-particle excitations are always created in pairs (never singly). Any perturbation will scatter electrons between states; thus, any perturbation Hamiltonian will contain an equal number of electron annihilation and creation operators (the minimum is one of each kind). For example, consider a perturbation of the form:

$$H' = \sum_{\mathbf{k}\neq\mathbf{k}'\sigma} V_{\mathbf{k}\mathbf{k}'} c_{\mathbf{k}'\sigma}^{\dagger} c_{\mathbf{k}\sigma} = \sum_{\mathbf{k}\neq\mathbf{k}'} V_{\mathbf{k}\mathbf{k}'} c_{\mathbf{k}'\uparrow}^{\dagger} c_{\mathbf{k}\uparrow} + \sum_{\mathbf{k}\neq\mathbf{k}'} V_{\mathbf{k}\mathbf{k}'} c_{\mathbf{k}'\downarrow}^{\dagger} c_{\mathbf{k}\downarrow}.$$

When this acts on the ground state, the first term in H' gives

$$\begin{aligned}\sum_{\mathbf{k}\neq\mathbf{k}'} V_{\mathbf{k}\mathbf{k}'} c_{\mathbf{k}'\uparrow}^{\dagger} c_{\mathbf{k}\uparrow} |\Psi_0\rangle &= \sum_{\mathbf{k}\neq\mathbf{k}'} V_{\mathbf{k}\mathbf{k}'} \left(u_{\mathbf{k}'} \gamma_{\mathbf{k}'\uparrow}^{\dagger} + v_{\mathbf{k}'} \gamma_{-\mathbf{k}'\downarrow}\right)\left(u_{\mathbf{k}} \gamma_{\mathbf{k}\uparrow} + v_{\mathbf{k}} \gamma_{-\mathbf{k}\downarrow}^{\dagger}\right) |\Psi_0\rangle \\ &= \sum_{\mathbf{k}\neq\mathbf{k}'} V_{\mathbf{k}\mathbf{k}'} u_{\mathbf{k}'} v_{\mathbf{k}} \gamma_{\mathbf{k}'\uparrow}^{\dagger} \gamma_{-\mathbf{k}\downarrow}^{\dagger} |\Psi_0\rangle,\end{aligned}$$

the other terms being zero. A similar expression is obtained if the second term in H' acts on $|\Psi_0\rangle$. Therefore, only pairs of particles are excited, and the minimum excitation energy, equal to 2Δ, is obtained if $\bar{\epsilon}_{\mathbf{k}} = \bar{\epsilon}_{\mathbf{k}'} = 0$.

12.7 Green's function approach to superconductivity

We now turn to Green's function as a method for studying superconductivity. The relevant Hamiltonian for describing superconductivity is the BCS Hamiltonian

$$\bar{H} = \sum_{\mathbf{k}\sigma} \bar{\epsilon}_{\mathbf{k}} c_{\mathbf{k}\sigma}^{\dagger} c_{\mathbf{k}\sigma} + \sum_{\mathbf{k}\mathbf{k}'} U_{\mathbf{k}\mathbf{k}'} c_{\mathbf{k}'\uparrow}^{\dagger} c_{-\mathbf{k}'\downarrow}^{\dagger} c_{-\mathbf{k}\downarrow} c_{\mathbf{k}\uparrow} = \bar{H}_0 + H'. \tag{12.68}$$

Here, $\bar{\epsilon}_{\mathbf{k}}$ is the single-particle energy measured from the chemical potential, and the sum over $\mathbf{k}$ and $\mathbf{k}'$ is restricted to values that satisfy $-\hbar\omega_D < \bar{\epsilon}_{\mathbf{k}}, \bar{\epsilon}_{\mathbf{k}'} < \hbar\omega_D$. The imaginary-time Green's function for spin-up electrons is

$$g(\mathbf{k}\uparrow, \tau) = -\langle T c_{\mathbf{k}\uparrow}(\tau) c_{\mathbf{k}\uparrow}^{\dagger}(0)\rangle = -\theta(\tau)\langle c_{\mathbf{k}\uparrow}(\tau) c_{\mathbf{k}\uparrow}^{\dagger}(0)\rangle + \theta(-\tau)\langle c_{\mathbf{k}\uparrow}^{\dagger}(0) c_{\mathbf{k}\uparrow}(\tau)\rangle. \tag{12.69}$$

The modified Heisenberg operator $c_{\mathbf{k}\uparrow}(\tau)$ is given by

$$c_{\mathbf{k}\uparrow}(\tau) = e^{\bar{H}\tau/\hbar} c_{\mathbf{k}\uparrow} e^{-\bar{H}\tau/\hbar} \tag{12.70}$$

where $c_{\mathbf{k}\uparrow} = c_{\mathbf{k}\uparrow}(0)$. The equation of motion for Green's function is

$$\frac{\partial}{\partial \tau} g(\mathbf{k}\uparrow, \tau) = -\delta(\tau)\langle c_{\mathbf{k}\uparrow} c^{\dagger}_{\mathbf{k}\uparrow}\rangle - \delta(\tau)\langle c^{\dagger}_{\mathbf{k}\uparrow} c_{\mathbf{k}\uparrow}\rangle - \theta(\tau)\left\langle \frac{\partial}{\partial \tau} c_{\mathbf{k}\uparrow}(\tau) c^{\dagger}_{\mathbf{k}\uparrow}(0)\right\rangle$$
$$+ \theta(-\tau)\left\langle c^{\dagger}_{\mathbf{k}\uparrow}(0)\frac{\partial}{\partial \tau} c_{\mathbf{k}\uparrow}(\tau)\right\rangle.$$

Using $\{c_{\mathbf{k}\uparrow}, c^{\dagger}_{\mathbf{k}\uparrow}\} = 1$, the above equation reduces to

$$\frac{\partial}{\partial \tau} g(\mathbf{k}\uparrow, \tau) = -\delta(\tau) - \left\langle T\frac{\partial}{\partial \tau} c_{\mathbf{k}\uparrow}(\tau) c^{\dagger}_{\mathbf{k}\uparrow}(0)\right\rangle. \tag{12.71}$$

It follows from Eq. (12.70) that

$$\frac{\partial}{\partial \tau} c_{\mathbf{k}\uparrow}(\tau) = \frac{1}{\hbar}\left[\bar{H}, c_{\mathbf{k}\uparrow}(\tau)\right] = \frac{1}{\hbar}\left[\bar{H}_0, c_{\mathbf{k}\uparrow}(\tau)\right] + \frac{1}{\hbar}\left[H', c_{\mathbf{k}\uparrow}(\tau)\right].$$

Note that

$$\bar{H} = e^{\bar{H}\tau/\hbar}\bar{H}e^{-\bar{H}\tau/\hbar} = \bar{H}(\tau) = \bar{H}_0(\tau) + H'(\tau).$$

The commutators are evaluated using the relation

$$[AB, C] = A[B, C] + [A, C]B = A\{B, C\} - \{A, C\}B.$$

We find

$$[\bar{H}_0, c_{\mathbf{k}\uparrow}] = \sum_{\mathbf{k}'\sigma'} \bar{\epsilon}_{\mathbf{k}'}[c^{\dagger}_{\mathbf{k}'\sigma'} c_{\mathbf{k}'\sigma'}, c_{\mathbf{k}\uparrow}] = -\bar{\epsilon}_{\mathbf{k}} c_{\mathbf{k}\uparrow}$$

$$[H', c_{\mathbf{k}\uparrow}] = \sum_{\mathbf{k}_1\mathbf{k}_2} U_{\mathbf{k}_1\mathbf{k}_2}[c^{\dagger}_{\mathbf{k}_2\uparrow} c^{\dagger}_{-\mathbf{k}_2\downarrow} c_{-\mathbf{k}_1\downarrow} c_{\mathbf{k}_1\uparrow}, c_{\mathbf{k}\uparrow}]$$
$$= \sum_{\mathbf{k}_1\mathbf{k}_2} U_{\mathbf{k}_1\mathbf{k}_2}[c^{\dagger}_{\mathbf{k}_2\uparrow} c^{\dagger}_{-\mathbf{k}_2\downarrow}, c_{\mathbf{k}\uparrow}] c_{-\mathbf{k}_1\downarrow} c_{\mathbf{k}_1\uparrow} = -\sum_{\mathbf{k}'} U_{\mathbf{k}'\mathbf{k}} c^{\dagger}_{-\mathbf{k}\downarrow} c_{-\mathbf{k}'\downarrow} c_{\mathbf{k}'\uparrow}.$$

The equation of motion for Green's function becomes

$$\left(\frac{\partial}{\partial \tau} + \frac{\bar{\epsilon}_{\mathbf{k}}}{\hbar}\right) g(\mathbf{k}\uparrow, \tau) = -\delta(\tau) + \frac{1}{\hbar}\sum_{\mathbf{k}'} U_{\mathbf{k}'\mathbf{k}}\left\langle T c^{\dagger}_{-\mathbf{k}\downarrow}(\tau) c_{-\mathbf{k}'\downarrow}(\tau) c_{\mathbf{k}'\uparrow}(\tau) c^{\dagger}_{\mathbf{k}\uparrow}(0)\right\rangle. \tag{12.72}$$

As is usually the case, the equation of motion of the one-particle Green's function contains a two-particle Green's function (the second term on the RHS of Eq. [12.72]). Ideally, we would construct the equation of motion for *this* function as well, but then a three-particle Green's function would appear, and so on; the system of equations never closes on itself. This, of course, reflects the fact that the problem is not exactly solvable; we need to resort to some approximation scheme.

We assume that the particles are weakly interacting; the effect of the interaction is considered only to the extent that it leads to the formation of Cooper pairs whose number is not constant. In other words, we evaluate the average of the time-ordered product in Eq. (12.72) for a noninteracting system, one whose energy eigenstates are not eigenstates of the number operator. We may then apply Wick's theorem,

$$\begin{aligned}\left\langle Tc^{\dagger}_{-\mathbf{k}\downarrow}(\tau)c_{-\mathbf{k}'\downarrow}(\tau)c_{\mathbf{k}'\uparrow}(\tau)c^{\dagger}_{\mathbf{k}\uparrow}(0)\right\rangle &= -\left\langle Tc_{-\mathbf{k}'\downarrow}(\tau)c^{\dagger}_{-\mathbf{k}\downarrow}(\tau)\right\rangle\left\langle Tc_{\mathbf{k}'\uparrow}(\tau)c^{\dagger}_{\mathbf{k}\uparrow}(0)\right\rangle\delta_{\mathbf{k}\mathbf{k}'}\\ &\quad -\left\langle Tc_{\mathbf{k}'\uparrow}(\tau)c_{-\mathbf{k}'\downarrow}(\tau)\right\rangle\left\langle Tc^{\dagger}_{-\mathbf{k}\downarrow}(\tau)c^{\dagger}_{\mathbf{k}\uparrow}(0)\right\rangle.\end{aligned}$$

In a normal metal, where all the stationary states can be chosen to be simultaneous eigenstates of H and N_{op}, only the first term on the RHS of the above equation survives, and the approximation is the Hartree–Fock approximation. This term is simply $g(\mathbf{k}\uparrow, \tau)$ multiplied by a time-independent function; it leads to a renormalization of the single-particle energy, and it will be dropped in what follows. The second term vanishes in a normal metal but does not vanish in a superconductor, where states are not eigenstates of the number of particles operator. We thus define two new "anomalous" Green's functions,

$$F(\mathbf{k}, \tau) = -\langle Tc_{\mathbf{k}\uparrow}(\tau)c_{-\mathbf{k}\downarrow}(0)\rangle, \quad F^{\dagger}(\mathbf{k}, \tau) = -\left\langle Tc^{\dagger}_{-\mathbf{k}\downarrow}(\tau)c^{\dagger}_{\mathbf{k}\uparrow}(0)\right\rangle. \tag{12.73}$$

The equation of motion for $g(\mathbf{k}\uparrow, \tau)$ is now written as

$$\begin{aligned}\left(\frac{\partial}{\partial\tau} + \bar{\epsilon}_{\mathbf{k}}/\hbar\right) g(\mathbf{k}\uparrow, \tau) &= -\delta(\tau) - \frac{1}{\hbar}\sum_{\mathbf{k}'} U_{\mathbf{k}'\mathbf{k}} F(\mathbf{k}', 0)F^{\dagger}(\mathbf{k}, \tau)\\ &= -\delta(\tau) + \frac{1}{\hbar}\Delta_{\mathbf{k}} F^{\dagger}(\mathbf{k}, \tau).\end{aligned} \tag{12.74}$$

We have introduced the gap parameter $\Delta_{\mathbf{k}}$ defined by

$$\Delta_{\mathbf{k}} = -\sum_{\mathbf{k}'} U_{\mathbf{k}'\mathbf{k}} F(\mathbf{k}', 0) = -\sum_{\mathbf{k}'} U_{\mathbf{k}'\mathbf{k}} \left\langle c_{-\mathbf{k}'\downarrow}c_{\mathbf{k}'\uparrow}\right\rangle. \tag{12.75}$$

To solve for $g(\mathbf{k}\uparrow, \tau)$, we write the equation of motion for $F^{\dagger}(\mathbf{k}, \tau)$,

$$\begin{aligned}\frac{\partial}{\partial\tau}F^{\dagger}(\mathbf{k}, \tau) &= \frac{\partial}{\partial\tau}\left[-\theta(\tau)\left\langle c^{\dagger}_{-\mathbf{k}\downarrow}(\tau)c^{\dagger}_{\mathbf{k}\uparrow}(0)\right\rangle + \theta(-\tau)\left\langle c^{\dagger}_{\mathbf{k}\uparrow}(0)c^{\dagger}_{-\mathbf{k}\downarrow}(\tau)\right\rangle\right]\\ &= -\delta(\tau)\left\langle c^{\dagger}_{-\mathbf{k}\downarrow}c^{\dagger}_{\mathbf{k}\uparrow} + c^{\dagger}_{\mathbf{k}\uparrow}c^{\dagger}_{-\mathbf{k}\downarrow}\right\rangle - \left\langle T\frac{\partial}{\partial\tau}c^{\dagger}_{-\mathbf{k}\downarrow}(\tau)c^{\dagger}_{\mathbf{k}\uparrow}(0)\right\rangle.\end{aligned}$$

Since $\{c^{\dagger}_{-\mathbf{k}\downarrow}, c^{\dagger}_{\mathbf{k}\uparrow}\} = 0$, the first term on the RHS vanishes. The second term is obtained by evaluating the commutator $[H, c^{\dagger}_{-\mathbf{k}\downarrow}]$. We end up with

$$\left(\frac{\partial}{\partial\tau} - \bar{\epsilon}_{\mathbf{k}}/\hbar\right) F^{\dagger}(\mathbf{k}, \tau) = \left(\frac{\Delta^{*}_{\mathbf{k}}}{\hbar}\right) g(\mathbf{k}\uparrow, \tau). \tag{12.76}$$

The coupled equations for $g(\mathbf{k}\uparrow, \tau)$ and $F^\dagger(\mathbf{k}, \tau)$ (Eqs (12.74) and (12.76)) are solved by Fourier expanding

$$g(\mathbf{k}\uparrow, \tau) = \frac{1}{\beta\hbar} \sum_{n=-\infty}^{\infty} g(\mathbf{k}\uparrow, \omega_n) e^{-i\omega_n \tau}, \quad F^\dagger(\mathbf{k}, \tau) = \frac{1}{\beta\hbar} \sum_{n=-\infty}^{\infty} F^\dagger(\mathbf{k}, \omega_n) e^{-i\omega_n \tau} \tag{12.77}$$

where $\omega_n = (2n+1)\pi/\beta\hbar$ and n is an integer. The coupled equations become

$$(-i\omega_n + \bar{\epsilon}_\mathbf{k}/\hbar) g(\mathbf{k}\uparrow, \omega_n) = -1 + (\Delta_\mathbf{k}/\hbar) F^\dagger(\mathbf{k}, \omega_n) \tag{12.78}$$

$$(-i\omega_n - \bar{\epsilon}_\mathbf{k}/\hbar) F^\dagger(\mathbf{k}, \omega_n) = (\Delta_\mathbf{k}^*/\hbar) g(\mathbf{k}\uparrow, \omega_n). \tag{12.79}$$

These are Gorkov equations in momentum-frequency space (Gorkov, 1958). Their solution is straightforward:

$$g(\mathbf{k}\uparrow, \omega_n) = \frac{i\omega_n + \bar{\epsilon}_\mathbf{k}/\hbar}{(i\omega_n)^2 - \left(\bar{\epsilon}_\mathbf{k}^2 + |\Delta_\mathbf{k}|^2\right)/\hbar^2} \tag{12.80}$$

$$F^\dagger(\mathbf{k}, \omega_n) = \frac{-\Delta_\mathbf{k}^*/\hbar}{(i\omega_n)^2 - \left(\bar{\epsilon}_\mathbf{k}^2 + |\Delta_\mathbf{k}|^2\right)/\hbar^2}. \tag{12.81}$$

Green's function can also be expressed another way. Using the expressions for $u_\mathbf{k}^2$ and $v_\mathbf{k}^2$ given in Eq. (12.60), we can show that

$$g(\mathbf{k}\uparrow, \omega_n) = \frac{u_\mathbf{k}^2}{i\omega_n - E_\mathbf{k}/\hbar} + \frac{v_\mathbf{k}^2}{i\omega_n + E_\mathbf{k}/\hbar} \tag{12.82}$$

where $E_\mathbf{k} = \sqrt{\bar{\epsilon}_\mathbf{k}^2 + \Delta_\mathbf{k}^2}$. The retarded Green's function $G^R(\mathbf{k}\uparrow, \omega)$ is obtained from $g(\mathbf{k}\uparrow, \omega_n)$ by replacing $i\omega_n$ with $\omega + i0^+$. The spectral density function $A(\mathbf{k}\uparrow, \omega)$ is equal to $-2\, Im\, G^R(\mathbf{k}\uparrow, \omega)$; hence

$$A(\mathbf{k}\uparrow, \omega) = 2\pi\hbar \left[u_\mathbf{k}^2 \delta(\hbar\omega - E_\mathbf{k}) + v_\mathbf{k}^2 \delta(\hbar\omega + E_\mathbf{k})\right]. \tag{12.83}$$

The spectral density function consists of two delta-function peaks. The first peak, at $E_\mathbf{k}$, corresponds to the energy of an electron added to the system in state $|\mathbf{k}\uparrow\rangle$. The second peak, at $-E_\mathbf{k}$, is the energy of an electron removed from state $|\mathbf{k}\uparrow\rangle$. To add an electron into state $|\mathbf{k}\uparrow\rangle$, the pair state $|\mathbf{k}\uparrow, -\mathbf{k}\downarrow\rangle$ needs to be unoccupied; the probability of that is $u_\mathbf{k}^2$. To remove an electron from state $|\mathbf{k}\uparrow\rangle$, the pair state $|\mathbf{k}\uparrow, -\mathbf{k}\downarrow\rangle$ must be occupied; the probability of that is $v_\mathbf{k}^2$.

If the pair state $|\mathbf{k}\uparrow, -\mathbf{k}\downarrow\rangle$ is occupied, then $u_\mathbf{k}^2 = 0$ and $v_\mathbf{k}^2 = 1$, the spectral density function has one peak at $\hbar\omega = -E_\mathbf{k}$, and the energy of the electron in state $|\mathbf{k}\uparrow\rangle$ is $-E_\mathbf{k}$. If, on the other hand, the pair state $|\mathbf{k}\uparrow, -\mathbf{k}\downarrow\rangle$ is empty, then $u_\mathbf{k}^2 = 1$ and $v_\mathbf{k}^2 = 0$, the spectral function has one peak at $\omega = E_\mathbf{k}/\hbar$, and the energy of the electron in state $|\mathbf{k}\uparrow\rangle$ is $E_\mathbf{k}$. If $\mathbf{k}$ is such that $\epsilon_\mathbf{k} = \mu$, then the energy of the electron in state $|\mathbf{k}\uparrow\rangle$ is either $-\Delta_\mathbf{k}$ or $+\Delta_\mathbf{k}$, depending on whether the pair state

Figure 12.14 Pictorial representation of the anomalous Green's functions (a) $F(\mathbf{k}, \tau)$ and (b) $F^{\dagger}(\mathbf{k}, \tau)$.

$|\mathbf{k}\uparrow, -\mathbf{k}\downarrow\rangle$ is occupied or empty. In a model where $\Delta_{\mathbf{k}} = \Delta$, independent of $\mathbf{k}$, there is a gap in the single-particle energy spectrum of 2Δ.

In obtaining the above results, we used the Hartree–Fock (mean field) approximation. Alternatively, we could have started with the mean field Hamiltonian (see Eq. [12.53]) and proceeded to calculate $g(\mathbf{k}\uparrow, \omega_n)$ and $F^{\dagger}(\mathbf{k}, \omega_n)$. The results obtained would be identical.

We have obtained Green's function using the equation of motion approach. Another way to obtain the same results is by means of a diagrammatic expansion. Using this approach, we construct Dyson's equation for $g(\mathbf{k}\uparrow, \omega_n)$,

$$g(\mathbf{k}\uparrow, \omega_n) = g^0(\mathbf{k}\uparrow, \omega_n) + g^0(\mathbf{k}\uparrow, \omega_n)\Sigma^*(\mathbf{k}\uparrow, \omega_n)g(\mathbf{k}\uparrow, \omega_n) \tag{12.84}$$

where $\Sigma^*(\mathbf{k}\uparrow, \omega_n)$ is the proper (irreducible) self energy. In the normal state, the self energy consists of diagrams containing only Green's function and interaction lines, because when we apply Wick's theorem in this situation, only contractions that involve one annihilation and one creation operator are nonvanishing. By contrast, in the superconducting state, anomalous Green's functions appear, so we need to expand our store of Green's functions. Diagrammatically, the anomalous Green's functions are represented as in Figure 12.14. These functions have vanishing zero-order values, i.e., they vanish in the noninteracting system. This is clearly so, since the Hamiltonian is H_0 in the absence of interactions and the system is in the normal state.

Every Green's function diagram consists of a series of self energy diagrams connected by the zeroth-order Green's function. In the normal state, a single line enters the self energy part and another single line leaves it, as shown in Figure 12.15a. The corresponding diagrammatic expansion for $g(\mathbf{k}\uparrow, \omega_n)$ is shown in Figure 12.15b. In the superconducting state, two new types of self-energy diagrams become possible: two single lines enter or leave the self-energy part, as in Figure 12.15c. Ignoring normal-state corrections, the diagrammatic expansion for $g(\mathbf{k}\uparrow, \omega_n)$ is shown in Figure 12.15d. An examination of this figure reveals that structures occurring after the first self energy part correspond to a new function. Graphically, this new function is characterized by two external lines pointing outward; it is the function $F^{\dagger}(\mathbf{k}, \omega_n)$. The analogue of Dyson's equation is shown in Figure 12.15e.

We can now write Dyson's equation for a superconductor. Ignoring normal-state corrections, Dyson's equation (in the Hartree–Fock, or mean field approximation)

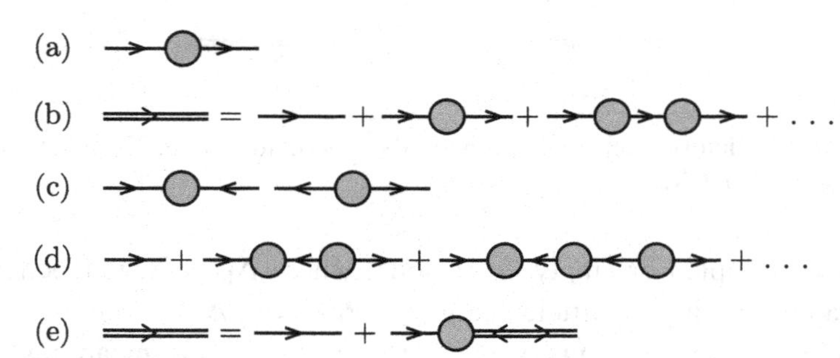

Figure 12.15 (a) In the normal state diagram, a single line enters the self energy part and a single line leaves it. (b) The expansion of Green's function in the normal state. (c) In the superconducting state, additional diagrams appear in which two single lines enter or leave the self energy part. (d) Diagrams that appear in the expansion of Green's function in a superconductor. (e) The Dyson-like equation for Green's function in a superconductor.

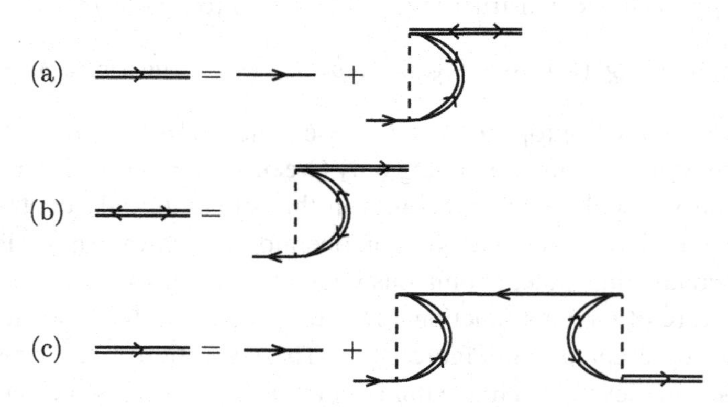

Figure 12.16 Diagrams of a superconductor in the Hartree–Fock approximation, ignoring normal-state corrections. (a) The Dyson-like equation for $g(\mathbf{k}\uparrow, \tau)$. (b) The equation for $F^\dagger(\mathbf{k}, \tau)$. (c) Dyson's equation for $g(\mathbf{k}\uparrow, \tau)$, obtained by combining (a) and (b).

is depicted graphically in Figure 12.16. The algebraic expressions corresponding to diagrams 12.16a and 12.16b are, respectively,

$$g(\mathbf{k}\uparrow, \tau) = g^0(\mathbf{k}\uparrow, \tau) + \frac{1}{\hbar}\int_0^{\beta\hbar} d\tau_1 g^0(\mathbf{k}\uparrow, \tau - \tau_1)F^\dagger(\mathbf{k}, \tau_1)\sum_{\mathbf{k}'} U_{\mathbf{k}'\mathbf{k}} F(\mathbf{k}', 0) \tag{12.85}$$

$$F^\dagger(\mathbf{k}, \tau) = -\frac{1}{\hbar}\int_0^{\beta\hbar} d\tau_1 g^0(-\mathbf{k}\downarrow, \tau_1 - \tau) g(\mathbf{k}\uparrow, \tau_1)\sum_{\mathbf{k}'} U_{\mathbf{k}\mathbf{k}'} F^\dagger(\mathbf{k}', 0). \tag{12.86}$$

The signs before the integrals may be checked by writing the first order perturbation term and applying Wick's theorem. Using the definition of the gap parameter

$\Delta_{\mathbf{k}}$ (see Eq. [12.75]), with $F^{\dagger}(\mathbf{k}, \tau = 0) = F^{*}(\mathbf{k}, \tau = 0)$, and Fourier-expanding $g^{0}(\mathbf{k}\uparrow, \tau)$, $g(\mathbf{k}\uparrow, \tau)$, and $F^{\dagger}(\mathbf{k}, \tau)$, we obtain

$$g(\mathbf{k}\uparrow, \omega_n) = g^{0}(\mathbf{k}\uparrow, \omega_n) - (\Delta_{\mathbf{k}}/\hbar) g^{0}(\mathbf{k}\uparrow, \omega_n) F^{\dagger}(\mathbf{k}, \omega_n) \tag{12.87}$$

$$F^{\dagger}(\mathbf{k}, \omega_n) = (\Delta_{\mathbf{k}}^{*}/\hbar) g^{0}(-\mathbf{k}\downarrow, -\omega_n) g(\mathbf{k}\uparrow, \omega_n). \tag{12.88}$$

The solution of Eqs (12.87) and (12.88) is identical to the one shown in Eqs (12.80) and (12.81).

To determine the gap consistency condition, we adopt the following simple model, which was considered earlier:

$$U_{\mathbf{k}\mathbf{k}'} = \begin{cases} -U_0 & -\hbar\omega_D < \bar{\epsilon}_{\mathbf{k}}, \bar{\epsilon_{\mathbf{k}'}} < \hbar\omega_D \\ 0 & \text{otherwise} \end{cases}. \tag{12.89}$$

Within this model, the gap parameter is independent of $\mathbf{k}$ and is written as Δ; it is given by

$$\Delta = U_0 {\sum_{\mathbf{k}}}' F(\mathbf{k}, \tau = 0) \Rightarrow \Delta^{*} = U_0 {\sum_{\mathbf{k}}}' F^{*}(\mathbf{k}, \tau = 0) = U_0 {\sum_{\mathbf{k}}}' F^{\dagger}(\mathbf{k}, \tau = 0).$$

The prime on the summation means that the sum is restricted to values of $\mathbf{k}$ such that $|\bar{\epsilon}_{\mathbf{k}}| < \hbar\omega_D$. Using

$$F^{\dagger}(\mathbf{k}, \tau = 0) = (\beta\hbar)^{-1} \sum_{n=-\infty}^{\infty} F^{\dagger}(\mathbf{k}, \omega_n)$$

along with Eq. (12.81), we obtain

$$-\frac{U_0}{\beta\hbar^2} {\sum_{\mathbf{k}}}' \sum_{n=-\infty}^{\infty} \frac{1}{(i\omega_n)^2 - \left(\bar{\epsilon}_{\mathbf{k}}^2 + |\Delta|^2\right)/\hbar^2} = 1. \tag{12.90}$$

Since $\omega_n = (2n+1)\pi/\beta\hbar$, the summand reduces to $-1/\omega_n^2$ as $n \to \pm\infty$; the series is convergent. We may thus introduce the redundant convergence factor $e^{i\omega_n 0^+}$, which allows us to evaluate the sum over n using the method of partial fractions,

$$\begin{aligned}\sum_{n=-\infty}^{\infty} \frac{1}{(i\omega_n)^2 - \left(\bar{\epsilon}_{\mathbf{k}}^2 + |\Delta|^2\right)/\hbar^2} &= \frac{\hbar}{2E_{\mathbf{k}}} \sum_{n=-\infty}^{\infty} e^{i\omega_n 0^+} \left[\frac{1}{i\omega_n - E_{\mathbf{k}}/\hbar} - \frac{1}{i\omega_n + E_{\mathbf{k}}/\hbar}\right] \\ &= \frac{\beta\hbar^2}{2E_{\mathbf{k}}} \left(f_{E_{\mathbf{k}}} - f_{-E_{\mathbf{k}}}\right)\end{aligned} \tag{12.91}$$

where $f_{E_{\mathbf{k}}} = \left(1 + e^{\beta E_{\mathbf{k}}}\right)^{-1}$ is the Fermi distribution function. In evaluating the frequency sum we have made use of Eq. (9.14). Since $f_{-E} = 1 - f_E$, Eq. (12.90) may be written as

$$\frac{U_0}{2}\sum_{\mathbf{k}}{}' \frac{1-2f_{E_{\mathbf{k}}}}{E_{\mathbf{k}}} = 1. \tag{12.92}$$

Replacing the sum over **k** by an integral over the energy, we obtain

$$\frac{1}{2}U_0 D_\sigma(E_F)\int_{-\hbar\omega_D}^{\hbar\omega_D} \frac{\tanh\left(\beta\sqrt{\epsilon^2+|\Delta|^2}/2\right)}{\sqrt{\epsilon^2+|\Delta|^2}}d\epsilon = 1. \tag{12.93}$$

This is the condition that the gap parameter must satisfy.

12.8 Determination of the transition temperature

The gap consistency condition, specified in Eq. (12.93), can be used to determine T_C, the transition temperature to the superconducting state. At $T = 0$, the gap condition reduces to Eq. (12.47). As T increases, the numerator of the integrand in Eq. (12.93) decreases, and in order to maintain the validity of the equation, the denominator must also decrease. Hence, $\Delta(T)$ is a decreasing function of temperature. At $T = T_C$, the system reverts to the normal state, where the gap parameter vanishes and Eq. (12.93) reduces to

$$1 = U_0 D_\sigma(E_F)\int_0^{\hbar\omega_D} \epsilon^{-1}\tanh\left(\epsilon/2k_B T_C\right)d\epsilon = U_0 D_\sigma(E_F)\int_0^{\theta} x^{-1}\tanh x\, dx$$

where $\theta = \hbar\omega_D/2k_B T_C$. Integrating by parts,

$$\frac{1}{U_0 D_\sigma(E_F)} = \ln x\, \tanh x|_0^\theta - \int_0^\theta \mathrm{sech}^2 x\, \ln x\, dx.$$

For weak coupling, $\theta \gg 1$; we can then replace $\tanh\theta$ by 1 and extend the upper limit of integration to infinity (this is possible because $\mathrm{sech}^2 x$ is a rapidly decreasing function of x for large x):

$$\frac{1}{U_0 D_\sigma(E_F)} = \ln\left(\frac{\hbar\omega_D}{2k_B T_C}\right) - \int_0^\infty \mathrm{sech}^2 x\, \ln x\, dx.$$

The integral on the RHS is tabulated; it is equal to $\ln(\pi/4) - \gamma$, where $\gamma \simeq 0.577$ is Euler's constant. Therefore

$$\frac{1}{U_0 D_\sigma(E_F)} = \ln\left(\frac{2\hbar\omega_D}{\pi k_B T_C}\right) + \gamma.$$

Rearranging terms, we find

$$k_B T_C = \frac{2}{\pi} e^{\gamma} \hbar\omega_D \exp\left[\frac{-1}{U_0 D_\sigma(E_F)}\right]$$
$$\simeq 1.14\hbar\omega_D \exp\left[\frac{-1}{U_0 D_\sigma(E_F)}\right]. \tag{12.94}$$

12.9 The Nambu formalism

We now discuss another formalism, introduced by Nambu (Nambu, 1960), that will be useful when we study the response of a superconductor to a weak magnetic field. Since we have been using mean field theory, and will continue to do so, it is convenient to start our analysis from the mean field Hamiltonian given in Eq. (12.53):

$$H_{MF} = \sum_{\mathbf{k}\sigma} \bar{\epsilon}_{\mathbf{k}} c^{\dagger}_{\mathbf{k}\sigma} c_{\mathbf{k}\sigma} - \sum_{\mathbf{k}} \Delta^{*}_{\mathbf{k}} c_{-\mathbf{k}\downarrow} c_{\mathbf{k}\uparrow} - \sum_{\mathbf{k}} \Delta_{\mathbf{k}} c^{\dagger}_{\mathbf{k}\uparrow} c^{\dagger}_{-\mathbf{k}\downarrow} + \sum_{\mathbf{k}} \Delta_{\mathbf{k}} \langle c^{\dagger}_{\mathbf{k}\uparrow} c^{\dagger}_{-\mathbf{k}\downarrow} \rangle.$$

We define two new operators

$$\alpha_{\mathbf{k}} = \begin{pmatrix} c_{\mathbf{k}\uparrow} \\ c^{\dagger}_{-\mathbf{k}\downarrow} \end{pmatrix}, \quad \alpha^{\dagger}_{\mathbf{k}} = \begin{pmatrix} c^{\dagger}_{\mathbf{k}\uparrow} & c_{-\mathbf{k}\downarrow} \end{pmatrix}. \tag{12.95}$$

The Nambu Green's function is defined by

$$\tilde{g}(\mathbf{k}, \tau) = -\left\langle T \alpha_{\mathbf{k}}(\tau) \alpha^{\dagger}_{\mathbf{k}}(0) \right\rangle. \tag{12.96}$$

This is a matrix Green's function,

$$\tilde{g}(\mathbf{k}, \tau) = -\left\langle T \begin{pmatrix} c_{\mathbf{k}\uparrow}(\tau) \\ c^{\dagger}_{-\mathbf{k}\downarrow}(\tau) \end{pmatrix} \begin{pmatrix} c^{\dagger}_{\mathbf{k}\uparrow}(0) & c_{-\mathbf{k}\downarrow}(0) \end{pmatrix} \right\rangle$$
$$= -\begin{pmatrix} \langle T c_{\mathbf{k}\uparrow}(\tau) c^{\dagger}_{\mathbf{k}\uparrow}(0) \rangle & \langle T c_{\mathbf{k}\uparrow}(\tau) c_{-\mathbf{k}\downarrow}(0) \rangle \\ \langle T c^{\dagger}_{-\mathbf{k}\downarrow}(\tau) c^{\dagger}_{\mathbf{k}\uparrow}(0) \rangle & \langle T c^{\dagger}_{-\mathbf{k}\downarrow}(\tau) c_{-\mathbf{k}\downarrow}(0) \rangle \end{pmatrix}$$
$$= \begin{pmatrix} g(\mathbf{k}\uparrow, \tau) & F(\mathbf{k}, \tau) \\ F^{\dagger}(\mathbf{k}, \tau) & -g(-\mathbf{k}\downarrow, -\tau) \end{pmatrix}. \tag{12.97}$$

The equation of motion is

$$\frac{\partial}{\partial \tau} \tilde{g}(\mathbf{k}, \tau) = -\delta(\tau) - \left\langle T \frac{\partial}{\partial \tau} \alpha_{\mathbf{k}}(\tau) \alpha^{\dagger}_{\mathbf{k}}(0) \right\rangle. \tag{12.98}$$

The evaluation of the time derivative of the α-operator proceeds as follows:

$$\hbar\frac{\partial}{\partial\tau}c_{\mathbf{k}\uparrow}(\tau) = [H, c_{\mathbf{k}\uparrow}] = -\bar{\epsilon}_{\mathbf{k}}c_{\mathbf{k}\uparrow}(\tau) + \Delta_{\mathbf{k}}c^{\dagger}_{-\mathbf{k}\downarrow}(\tau) \tag{12.99}$$

$$\hbar\frac{\partial}{\partial\tau}c^{\dagger}_{-\mathbf{k}\downarrow}(\tau) = [H, c^{\dagger}_{-\mathbf{k}\downarrow}] = \bar{\epsilon}_{\mathbf{k}}c^{\dagger}_{-\mathbf{k}\downarrow}(\tau) + \Delta^*_{\mathbf{k}}c_{\mathbf{k}\uparrow}(\tau). \tag{12.100}$$

In matrix form, these equations are written as

$$\hbar\frac{\partial}{\partial\tau}\begin{pmatrix} c_{\mathbf{k}\uparrow}(\tau) \\ c^{\dagger}_{-\mathbf{k}\downarrow}(\tau)\end{pmatrix} = \begin{pmatrix} -\bar{\epsilon}_{\mathbf{k}} & \Delta_{\mathbf{k}} \\ \Delta^*_{\mathbf{k}} & \bar{\epsilon}_{\mathbf{k}}\end{pmatrix}\begin{pmatrix} c_{\mathbf{k}\uparrow}(\tau) \\ c^{\dagger}_{-\mathbf{k}\downarrow}(\tau)\end{pmatrix}. \tag{12.101}$$

Introducing the matrices

$$\sigma_3 = \begin{pmatrix} 1 & 0 \\ 0 & -1\end{pmatrix}, \quad \sigma_+ = \begin{pmatrix} 0 & 1 \\ 0 & 0\end{pmatrix}, \quad \sigma_- = \begin{pmatrix} 0 & 0 \\ 1 & 0\end{pmatrix} \tag{12.102}$$

we can recast Eq. (12.101) into the following form:

$$\hbar\frac{\partial}{\partial\tau}\alpha_{\mathbf{k}}(\tau) = -\bar{\epsilon}_{\mathbf{k}}\sigma_3\alpha_{\mathbf{k}}(\tau) + (\Delta_{\mathbf{k}}\sigma_+ + \Delta^*_{\mathbf{k}}\sigma_-)\alpha_{\mathbf{k}}(\tau). \tag{12.103}$$

The equation of motion for the Nambu Green's function now becomes

$$\left(\hbar\frac{\partial}{\partial\tau} + \bar{\epsilon}_{\mathbf{k}}\sigma_3 - \Delta_{\mathbf{k}}\sigma_+ - \Delta^*_{\mathbf{k}}\sigma_-\right)\tilde{g}(\mathbf{k}, \tau) = -\hbar\delta(\tau). \tag{12.104}$$

Fourier expanding $\tilde{g}(\mathbf{k}, \tau) = (\beta\hbar)^{-1}\sum_n \tilde{g}(\mathbf{k}, \omega_n)e^{-i\omega_n\tau}$, we obtain

$$\tilde{g}(\mathbf{k}, \omega_n) = \left[i\omega_n - (\bar{\epsilon}_{\mathbf{k}}/\hbar)\sigma_3 + (\Delta_{\mathbf{k}}/\hbar)\sigma_+ + (\Delta^*_{\mathbf{k}}/\hbar)\sigma_-\right]^{-1}. \tag{12.105}$$

The matrix inversion is straightforward; the result is

$$\tilde{g}(\mathbf{k}, \omega_n) = \frac{i\omega_n + (\bar{\epsilon}_{\mathbf{k}}/\hbar)\sigma_3 - (\Delta_{\mathbf{k}}/\hbar)\sigma_+ - (\Delta^*_{\mathbf{k}}/\hbar)\sigma_-}{(i\omega_n)^2 - \left(\bar{\epsilon}^2_{\mathbf{k}} + |\Delta_{\mathbf{k}}|^2\right)/\hbar^2}. \tag{12.106}$$

The Green's function $g(\mathbf{k}\uparrow, \omega_n)$ is simply $\tilde{g}_{11}(\mathbf{k}, \omega_n)$; hence

$$g(\mathbf{k}\uparrow, \omega_n) = \frac{i\omega_n + \bar{\epsilon}_{\mathbf{k}}/\hbar}{(i\omega_n)^2 - E^2_{\mathbf{k}}/\hbar^2}. \tag{12.107}$$

This is the same expression obtained earlier (see Eq. [12.80]).

It is possible to expand the Nambu Green's function in a perturbation series and apply Wick's theorem. The resulting Feynman diagrams obey essentially the same rules as do those for Matsubara Green's function, except that:

1. A single electron line stands for the diagonal Green's function whose entries are $g^0(\mathbf{k}\uparrow, \omega_n)$ and $-g^0(-\mathbf{k}\downarrow, -\omega_n)$.
2. The electron–electron Coulomb matrix element carries an extra factor σ_3, as does the electron–phonon interaction matrix element.

3. For a closed electron loop, the trace is taken over the matrix product of the matrices that represent the lines that make up the loop.

In the normal state, there is no advantage whatsoever to using the Nambu formalism instead of the Matsubara method. In the superconducting state, however, there is an advantage to using the Nambu Green's function: its perturbation series is the same as that of the normal state. As a result, the Feynman diagrams that appear in the expansion of the Nambu propagator are exactly the same diagrams as those seen in the normal state.

12.10 Response to a weak magnetic field

In this section, we calculate the current that results in a superconductor from the presence of a weak magnetic field **B** using linear response theory. The field is represented by a vector potential **A**, where $\mathbf{B} = \nabla \times \mathbf{A}$. In the presence of **A**, the current-density operator is (per Problem 3.7)

$$\mathbf{J}(\mathbf{r}) = \mathbf{j}^D(\mathbf{r}) + \mathbf{j}^P(\mathbf{r}). \tag{12.108}$$

The first term is the diamagnetic current-density operator, while the second term is the paramagnetic current-density operator,

$$\mathbf{j}^D(\mathbf{r}, t) = -\frac{e^2}{mc}\mathbf{A}(\mathbf{r}, t)n(\mathbf{r}), \quad \mathbf{j}^P(\mathbf{r}) = \frac{ie\hbar}{2m}\sum_\sigma \left[\Psi_\sigma^\dagger \nabla \Psi_\sigma - \left(\nabla \Psi_\sigma^\dagger\right)\Psi_\sigma\right] \tag{12.109}$$

where $n(\mathbf{r})$ is the number-density operator and $\Psi_\sigma(\mathbf{r})$ and $\Psi_\sigma^\dagger(\mathbf{r})$ are field operators. Within linear response theory (first order in **A**),

$$\langle \mathbf{j}^D \rangle(\mathbf{r}, t) = -\frac{e^2}{mc}\langle n(\mathbf{r}, t)\rangle_0 \mathbf{A}(\mathbf{r}, t) = -\frac{ne^2}{mc}\mathbf{A}(\mathbf{r}, t) \tag{12.110}$$

where n is the number of electrons per unit volume. Note that since $\mathbf{j}^D$ is already proportional to **A**, the ensemble average of $n(\mathbf{r}, t)$ is taken over the unperturbed system, i.e., over the system in the absence of the vector potential.

To determine $\langle \mathbf{j}^P \rangle$ in the presence of **A**, we need to determine H^{ext}, the perturbation which arises from the presence of **A**; this is given by

$$H^{\text{ext}}(t) = -\frac{1}{c}\int \mathbf{j}^P(\mathbf{r}).\mathbf{A}(\mathbf{r}, t)d^3r + \frac{e^2}{2mc^2}\int A^2(\mathbf{r}, t)n(\mathbf{r})d^3r \tag{12.111}$$

(see Problem 12.6). In the absence of the vector potential, $\langle \mathbf{j}^P(\mathbf{r}, t)\rangle_0 = 0$. To first order in $\mathbf{A}$, $\langle \mathbf{j}^P \rangle$ is given by Kubo's formula (see Eq. [6.74])

$$\begin{aligned}\langle j^P_\alpha\rangle(\mathbf{r}, t) &= -\frac{i}{\hbar}\int_{-\infty}^{t} dt' \langle [j^P_{\alpha,H}(\mathbf{r}, t), H^{\text{ext}}_H(t')]\rangle \\ &= \frac{i}{\hbar c}\int_{-\infty}^{t} dt' \int d^3r' \sum_\beta \langle [j^P_{\alpha,H}(\mathbf{r}, t), j^P_{\beta,H}(\mathbf{r}', t')]\rangle A_\beta(\mathbf{r}', t') \qquad (12.112)\end{aligned}$$

where $\alpha, \beta = x, y, z$ and $\mathbf{j}^P_H$ is the paramagnetic current-density operator in the Heisenberg picture. Since both $\mathbf{j}^P$ and H commute with the number of particles operator N, $\mathbf{j}^P_H = \mathbf{j}^P_{\bar{H}}$, where $\bar{H} = H - \mu N$. Fourier transforming $\langle \mathbf{J}(\mathbf{r}, t)\rangle = \langle \mathbf{j}^D\rangle(\mathbf{r}, t) + \langle \mathbf{j}^P\rangle(\mathbf{r}, t)$, we find

$$\langle J_\alpha(\mathbf{q}, \omega)\rangle = -\frac{ne^2}{mc}A_\alpha(\mathbf{q}, \omega) - \frac{1}{\hbar c}\sum_\beta D^R_{\alpha\beta}(\mathbf{q}\omega)A_\beta(\mathbf{q}\omega). \qquad (12.113)$$

$D^R_{\alpha\beta}(\mathbf{q}, \omega)$ is the Fourier transform of the retarded current–current correlation function $D^R_{\alpha\beta}(\mathbf{q}, t)$,

$$D^R_{\alpha\beta}(\mathbf{q}, t) = -i\theta(t)(1/V)\langle [j^P_\alpha(\mathbf{q}, t), j^P_\beta(-\mathbf{q}, 0)]\rangle \qquad (12.114)$$

where $\theta(t)$ is the step function and V is the system's volume. The operators inside the commutator are modified Heisenberg picture operators.

The retarded function $D^R_{\alpha\beta}(\mathbf{q}, \omega)$ is obtained by analytic continuation from the corresponding imaginary-time correlation function $D_{\alpha\beta}(\mathbf{q}, \omega)$, which is the Fourier transform of

$$D_{\alpha\beta}(\mathbf{q}, \tau) = -(1/V)\langle T j_{P\alpha}(\mathbf{q}, \tau) j_{P\beta}(-\mathbf{q}, 0)\rangle. \qquad (12.115)$$

We evaluate the correlation function using Nambu's formalism. We first need to express the current-density operator in terms of the Nambu creation and annihilation operators (the α-operators). The paramagnetic current-density operator is

$$\begin{aligned}\mathbf{j}^P(\mathbf{q}) &= -\frac{e\hbar}{2m}\sum_{\mathbf{k}\sigma}(2\mathbf{k} + \mathbf{q})c^\dagger_{\mathbf{k}\sigma}c_{\mathbf{k}+\mathbf{q}\sigma} \\ &= -\frac{e\hbar}{2m}\sum_{\mathbf{k}}(2\mathbf{k} + \mathbf{q})\left[c^\dagger_{\mathbf{k}\uparrow}c_{\mathbf{k}+\mathbf{q}\uparrow} + c^\dagger_{\mathbf{k}\downarrow}c_{\mathbf{k}+\mathbf{q}\downarrow}\right] \qquad (12.116)\end{aligned}$$

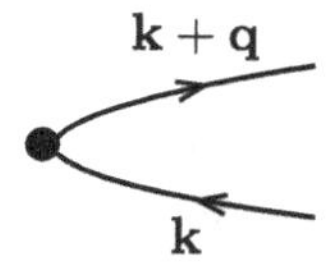

Figure 12.17 Graphical representation of the paramagnetic current-density operator $j_\alpha^P(\mathbf{q})$. The vertex $\bullet = -\frac{e\hbar}{2m}(2k_\alpha + q_\alpha)$.

(see Problem 3.7). In the last term on the RHS, we make a change of variable: $\mathbf{k} \to -\mathbf{k} - \mathbf{q}$,

$$\mathbf{j}^P(\mathbf{q}) = -\frac{e\hbar}{2m}\sum_{\mathbf{k}}(2\mathbf{k}+\mathbf{q})\left[c^\dagger_{\mathbf{k}\uparrow}c_{\mathbf{k}+\mathbf{q}\uparrow} - c^\dagger_{-\mathbf{k}-\mathbf{q}\downarrow}c_{-\mathbf{k}\downarrow}\right]$$

$$= -\frac{e\hbar}{2m}\sum_{\mathbf{k}}(2\mathbf{k}+\mathbf{q})\left[c^\dagger_{\mathbf{k}\uparrow}c_{\mathbf{k}+\mathbf{q}\uparrow} + c_{-\mathbf{k}\downarrow}c^\dagger_{-\mathbf{k}-\mathbf{q}\downarrow} - \delta_{\mathbf{q},0}\right].$$

Since $\sum_{\mathbf{k}}\mathbf{k} = 0$, the term containing the Kronecker delta yields zero. The remaining two terms inside the brackets add up to $\alpha^\dagger_{\mathbf{k}}\alpha_{\mathbf{k}+\mathbf{q}}$. Therefore,

$$\mathbf{j}^P(\mathbf{q}) = -\frac{e\hbar}{2m}\sum_{\mathbf{k}}(2\mathbf{k}+\mathbf{q})\alpha^\dagger_{\mathbf{k}}\alpha_{\mathbf{k}+\mathbf{q}}. \tag{12.117}$$

The α-component of the paramagnetic current, $j_\alpha^P(\mathbf{q})$, is represented graphically in Figure 12.17. Inserting the above expression into Eq. (12.115) we obtain

$$D_{\alpha\beta}(\mathbf{q},\tau) = \frac{-e^2\hbar^2}{4m^2V}\sum_{\mathbf{k}\mathbf{k}'}(2k_\alpha+q_\alpha)(2k'_\beta-q_\beta)\left\langle T\alpha^\dagger_{\mathbf{k}}(\tau)\alpha_{\mathbf{k}+\mathbf{q}}(\tau)\alpha^\dagger_{\mathbf{k}'}(0)\alpha_{\mathbf{k}'-\mathbf{q}}(0)\right\rangle. \tag{12.118}$$

We can evaluate $D_{\alpha\beta}(\mathbf{q},\tau)$ by means of a perturbation expansion followed by the use of Wick's theorem. $D_{\alpha\beta}(\mathbf{q},\tau)$ is similar to the dressed pair bubble in the interacting electron gas which we studied in Chapter 10. Graphically, $D_{\alpha\beta}(\mathbf{q},\omega_m)$ is given in Figure 12.18. In evaluating $D_{\alpha\beta}(\mathbf{q},\omega_m)$ we keep only one pair bubble, as indicated in Figure 12.18. In this case, $\mathbf{k}' = \mathbf{k}+\mathbf{q}$; it follows that $2k'_\beta - q_\beta = 2k_\beta + q_\beta$. Since we have a closed electron loop, there is an additional factor of -1, and the trace must be taken over the matrix product. The Feynman rules thus yield the following expression,

$$D_{\alpha\beta}(\mathbf{q},\omega_m) = \frac{\hbar e^2}{4m^2\beta V}\sum_{\mathbf{k},n}(2k_\alpha+q_\alpha)(2k_\beta+q_\beta)Tr\left[\tilde{g}(\mathbf{k},\omega_n)\tilde{g}(\mathbf{k}+\mathbf{q},\omega_m+\omega_n)\right]. \tag{12.119}$$

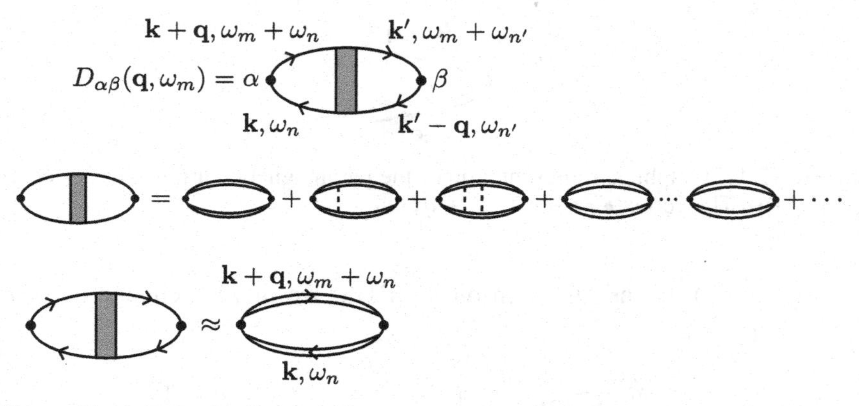

Figure 12.18 Perturbation expansion of $D_{\alpha\beta}(\mathbf{q}, \omega_m)$. In the lowest order, only one pair bubble is retained. The double lines represent the Nambu matrix Green's function.

The frequency summation is best carried out by using the spectral representation of the matrix Green's function,

$$\tilde{g}(\mathbf{k}, \omega_n) = \int_{-\infty}^{\infty} \frac{\tilde{A}(\mathbf{k}, \epsilon)}{i\omega_n - \epsilon} \frac{d\epsilon}{2\pi} = -\int_{-\infty}^{\infty} \frac{Im\, \tilde{G}^R(\mathbf{k}, \epsilon)}{i\omega_n - \epsilon} \frac{d\epsilon}{\pi} \tag{12.120}$$

where $\tilde{A}(\mathbf{k}, \epsilon) = -2Im\, \tilde{G}^R(\mathbf{k}, \epsilon)$ is the spectral density function and $\tilde{G}^R(\mathbf{k}, \epsilon)$ is the retarded matrix Green's function. Thus

$$D_{\alpha\beta}(\mathbf{q}, \omega_m) = \frac{\hbar e^2}{4m^2\beta V} \sum_{\mathbf{k},n} (2k_\alpha + q_\alpha)(2k_\beta + q_\beta) \int_{-\infty}^{\infty} \frac{d\epsilon_1}{\pi} \int_{-\infty}^{\infty} \frac{d\epsilon_2}{\pi} \times \frac{Tr\left[Im\, \tilde{G}^R(\mathbf{k}, \epsilon_1)\, Im\, \tilde{G}^R(\mathbf{k}+\mathbf{q}, \epsilon_2)\right]}{(i\omega_n - \epsilon_1)(i\omega_n + i\omega_m - \epsilon_2)}. \tag{12.121}$$

We now carry out the summation over n. Since the series is convergent (as $n \to \infty$, the summand $\to -1/\omega_n^2$), we introduce the (redundant) $e^{i\omega_n 0^+}$ factor,

$$\sum_{n=-\infty}^{\infty} \frac{1}{(i\omega_n - \epsilon_1)(i\omega_n + i\omega_m - \epsilon_2)} = \sum_{n=-\infty}^{\infty} \frac{e^{i\omega_n 0^+}}{i\omega_m + \epsilon_1 - \epsilon_2} \left(\frac{1}{i\omega_n - \epsilon_1} - \frac{1}{i\omega_n + i\omega_m - \epsilon_2} \right) = \frac{\beta\hbar\left(f_{\epsilon_1} - f_{\epsilon_2}\right)}{i\omega_m + \epsilon_1 - \epsilon_2} \tag{12.122}$$

where we used the frequency summation formula (see Eq. [9.14]). Here, f_ϵ is the Fermi distribution function,

$$f_\epsilon = (e^{\beta\hbar\epsilon} + 1)^{-1}. \tag{12.123}$$

We can thus write

$$D_{\alpha\beta}(\mathbf{q},\omega_m) = \frac{\hbar^2 e^2}{4m^2 V}\sum_{\mathbf{k},n}(2k_\alpha + q_\alpha)(2k_\beta + q_\beta)\int_{-\infty}^{\infty}\frac{d\epsilon_1}{\pi}\int_{-\infty}^{\infty}\frac{d\epsilon_2}{\pi}$$
$$\times \frac{f_{\epsilon_1} - f_{\epsilon_2}}{i\omega_m + \epsilon_1 - \epsilon_2}\, Tr\left[Im\ \tilde{G}^R(\mathbf{k},\epsilon_1)\ Im\ \tilde{G}^R(\mathbf{k}+\mathbf{q},\epsilon_2)\right]. \quad (12.124)$$

The retarded function $\tilde{G}^R(\mathbf{k},\epsilon) = \tilde{g}(\mathbf{k},\omega_n)|_{i\omega_n\to\epsilon+i0^+}$. The matrix function $\tilde{g}(\mathbf{k},\omega_n)$, given in Eq. (12.107), can be written as

$$\tilde{g}(\mathbf{k},\omega_n) = \frac{\hbar}{2E_\mathbf{k}}\left(\frac{1}{i\omega_n - E_\mathbf{k}/\hbar} - \frac{1}{i\omega_n + E_\mathbf{k}/\hbar}\right)\begin{bmatrix} i\omega_n + \bar{\epsilon}_\mathbf{k}/\hbar & -\Delta_\mathbf{k}/\hbar \\ -\Delta^*_\mathbf{k}/\hbar & i\omega_n - \bar{\epsilon}_\mathbf{k}/\hbar \end{bmatrix}. \quad (12.125)$$

Replacing $i\omega_n$ with $\epsilon + i0^+$ and taking the imaginary part, we find

$$Im\ \tilde{G}^R(\mathbf{k},\epsilon) = -\frac{\pi\hbar}{2E_\mathbf{k}}[\delta(\epsilon - E_\mathbf{k}/\hbar) - \delta(\epsilon + E_\mathbf{k}/\hbar)]\begin{bmatrix} \epsilon + \bar{\epsilon}_\mathbf{k}/\hbar & -\Delta_\mathbf{k}/\hbar \\ -\Delta^*_\mathbf{k}/\hbar & \epsilon - \bar{\epsilon}_\mathbf{k}/\hbar \end{bmatrix}. \quad (12.126)$$

To simplify the calculations, we adopt the BCS model, where the electron–electron interaction is a negative constant in a shell of energy width $2\hbar\omega_D$ that encloses the Fermi surface. In this model, $\Delta_\mathbf{k}$ is a real constant, independent of $\mathbf{k}$. In this case, a straightforward calculation yields

$$Tr\left[Im\,\tilde{G}^R(\mathbf{k},\epsilon_1)\,Im\,\tilde{G}^R(\mathbf{k}+\mathbf{q},\epsilon_2)\right] = \pi^2\hbar^2\,[\delta(\epsilon_1 - E_\mathbf{k}/\hbar) - \delta(\epsilon_1 + E_\mathbf{k}/\hbar)]$$
$$\times\left[\delta(\epsilon_2 - E_{\mathbf{k}+\mathbf{q}}/\hbar) - \delta(\epsilon_2 + E_{\mathbf{k}+\mathbf{q}}/\hbar)\right]\frac{\epsilon_1\epsilon_2 + \bar{\epsilon}_\mathbf{k}\bar{\epsilon}_{\mathbf{k}+\mathbf{q}}/\hbar^2 + \Delta^2/\hbar^2}{2E_\mathbf{k}E_{\mathbf{k}+\mathbf{q}}}.$$

We now specify to the case of a static magnetic field: $\omega_m = 0$. The Dirac-delta functions in the above expression for the trace of the matrix product make it possible to carry out the integrations over ϵ_1 and ϵ_2 in Eq. (12.124). Noting that $f_{-\epsilon} = 1 - f_\epsilon$, it is not difficult to show that

$$D_{\alpha\beta}(\mathbf{q},0) = \frac{\hbar^3 e^2}{4m^2 V}\sum_{\mathbf{k}}(2k_\alpha + q_\alpha)(2k_\beta + q_\beta)\left[\left(1 + \frac{\bar{\epsilon}_\mathbf{k}\bar{\epsilon}_{\mathbf{k}+\mathbf{q}} + \Delta^2}{E_\mathbf{k}E_{\mathbf{k}+\mathbf{q}}}\right)\frac{f_{E_\mathbf{k}} - f_{E_{\mathbf{k}+\mathbf{q}}}}{E_\mathbf{k} - E_{\mathbf{k}+\mathbf{q}}}\right.$$
$$\left. + \left(1 - \frac{\bar{\epsilon}_\mathbf{k}\bar{\epsilon}_{\mathbf{k}+\mathbf{q}} + \Delta^2}{E_\mathbf{k}E_{\mathbf{k}+\mathbf{q}}}\right)\frac{f_{E_\mathbf{k}} + f_{E_{\mathbf{k}+\mathbf{q}}} - 1}{E_\mathbf{k} + E_{\mathbf{k}+\mathbf{q}}}\right]. \quad (12.127)$$

This is still a complicated expression. We restrict ourselves further to the case of a uniform magnetic field: $\mathbf{q} \to 0$. In this case,

$$\bar{\epsilon}_{\mathbf{k}+\mathbf{q}} \to \bar{\epsilon}_\mathbf{k}\,,\quad E_{\mathbf{k}+\mathbf{q}} \to E_\mathbf{k},\quad \bar{\epsilon}_\mathbf{k}\bar{\epsilon}_{\mathbf{k}+\mathbf{q}} + \Delta^2 \to E_\mathbf{k}^2,\quad \frac{f_{E_\mathbf{k}} - f_{E_{\mathbf{k}+\mathbf{q}}}}{E_\mathbf{k} - E_{\mathbf{k}+\mathbf{q}}} \to \frac{\partial f_{E_\mathbf{k}}}{\partial E_\mathbf{k}}.$$

The correlation function reduces to

$$D_{\alpha\beta}(\mathbf{q} \to 0,\ \omega_m = 0) \to \frac{\hbar^3 e^2}{4m^2 V} \sum_{\mathbf{k}} 4k_\alpha k_\beta \left[(1+1)\frac{\partial f}{\partial E_{\mathbf{k}}} + 0 \right]$$

$$= \frac{2\hbar^3 e^2}{m^2 V} \sum_{\mathbf{k}} k_\alpha k_\beta \frac{\partial f_{E_{\mathbf{k}}}}{\partial E_{\mathbf{k}}}.$$

For $\alpha \neq \beta$, the sum over $\mathbf{k}$ yields zero since $E_{-\mathbf{k}} = E_{\mathbf{k}}$ and $f_{E_{-\mathbf{k}}} = f_{E_{\mathbf{k}}}$. Therefore,

$$D^R_{\alpha\beta}(\mathbf{q} = 0,\ \omega_m = 0) = D_{\alpha\beta}(\mathbf{q} = 0,\ \omega_m = 0) = \frac{2\hbar^3 e^2}{m^2 V} \delta_{\alpha\beta} \sum_{\mathbf{k}} k_\alpha^2 \frac{\partial f_{E_{\mathbf{k}}}}{\partial E_{\mathbf{k}}}$$

$$= \frac{2\hbar^3 e^2}{3m^2 V} \delta_{\alpha\beta} \sum_{\mathbf{k}} k^2 \frac{\partial f_{E_{\mathbf{k}}}}{\partial E_{\mathbf{k}}}.$$

The current density is thus given by

$$\langle J_\alpha \rangle = -\frac{ne^2}{mc} A_\alpha - \frac{2\hbar^2 e^2}{3m^2 c V} \sum_{\mathbf{k}} k^2 \frac{\partial f_{E_{\mathbf{k}}}}{\partial E_{\mathbf{k}}} A_\alpha.$$

We can rewrite this expression as follows:

$$\langle \mathbf{J} \rangle = -\frac{n_s e^2}{mc} \mathbf{A} \tag{12.128}$$

where

$$n_s = n + \frac{2\hbar^2}{3mV} \sum_{\mathbf{k}} k^2 \frac{\partial f_{E_{\mathbf{k}}}}{\partial E_{\mathbf{k}}} \tag{12.129}$$

is interpreted as the density of superconducting electrons. Taking the curl of both sides of Eq. (12.128) yields the London equation, which results in a Meissner effect as long as $n_s \neq 0$, as shown in Section 12.2.

At $T = 0$, $\partial f_{E_{\mathbf{k}}}/\partial E_{\mathbf{k}} = -\delta(E_{\mathbf{k}}) = -\delta\left(\sqrt{\bar{\epsilon}_{\mathbf{k}}^2 + \Delta^2}\right) = 0$ since $\Delta^2 > 0$. In this case, $n_s = n$ and the Meissner effect exists. As T increases from zero, n_s decreases, and at $T = T_C$, $\Delta = 0$ and $E_{\mathbf{k}} = \bar{\epsilon}_{\mathbf{k}}$. Since T_C is small, we assume that at T_C the electrons occupy all states below the Fermi energy E_F; in this case, $\partial f_{E_{\mathbf{k}}}/\partial E_{\mathbf{k}} = \partial f/\partial \bar{\epsilon}_{\mathbf{k}} = -\delta(\bar{\epsilon}_{\mathbf{k}}) = -\delta(\epsilon_{\mathbf{k}} - E_F)$. Assuming that $\epsilon_{\mathbf{k}} = \hbar^2 k^2/2m$, then at $T = T_C$

$$n_s = n - \frac{4}{3V} \sum_{\mathbf{k}} \epsilon_{\mathbf{k}}\, \delta(\epsilon_{\mathbf{k}} - E_F) = n - \frac{2}{3V} \sum_{\mathbf{k}\sigma} E_F\, \delta(\epsilon_{\mathbf{k}} - E_F)$$

$$= n - \frac{2}{3} E_F\, d(E_F) \tag{12.130}$$

where $d(E_F)$ is the density of states, per unit volume, at the Fermi energy. A simple calculation of $d(E_F)$ shows that $d(E_F) = 3n/2E_F$; hence at $T = T_C$, $n_s = 0$, and the Meissner effect disappears.

12.11 Infinite conductivity

If a superconductor is part of an electric circuit through which a current flows, the voltage drop across the superconductor will be zero because of its infinite conductivity (or zero resistivity). This means that the average electric field inside a superconductor must be zero. Alternatively, we can say that the electric field in a perfect conductor produces a current that increases with time; the circuit will achieve a steady state only when the electric field inside the perfect conductor is zero.

To test whether the BCS theory predicts infinite conductivity, we consider the response of the superconductor to a uniform (constant in space) steady (constant in time) electric field. The simplest approach is to consider a sinusoidal field and take the limit as the frequency tends to zero. We thus consider a field $\mathbf{E}(\mathbf{q} = 0, \omega) = \mathbf{E}(\omega)$. Since $\mathbf{E}(t) = -(1/c)\partial\mathbf{A}/\partial t$, the corresponding vector potential is such that $\mathbf{E}(\mathbf{q} = 0, \omega) = (i\omega/c)\mathbf{A}(\mathbf{q} = 0, \omega)$. In evaluating the current (see Eq. [12.113]), we should first take the limit $\mathbf{q} \to 0$ followed by the limit $\omega \to 0$,

$$\lim_{\omega\to 0}\langle J_\alpha(0,\omega)\rangle = -\frac{ne^2}{mc}\lim_{\omega\to 0} A_\alpha(0,\omega) - \frac{1}{\hbar c}\sum_\beta \lim_{\omega\to 0}\lim_{\mathbf{q}\to 0} D^R_{\alpha\beta}(\mathbf{q},\omega)A_\beta(0,\omega). \tag{12.131}$$

In studying the Meissner effect, the limits were taken in reverse order: first $\omega \to 0$, then $\mathbf{q} \to 0$. It turns out that $D^R_{\alpha\beta}(0, 0)$ does not depend on the order in which the limits are taken, and we arrive at essentially the same result as in the previous section:

$$\lim_{\omega\to 0}\langle J_\alpha(\mathbf{q} = 0,\omega)\rangle = -\frac{n_s e^2}{mc}\lim_{\omega\to 0} A_\alpha(0,\omega) = \lim_{\omega\to 0}\frac{in_s e^2}{m\omega}E_\alpha(0,\omega). \tag{12.132}$$

This means that for a slowly varying electric field ($\omega \to 0$)

$$\frac{\partial\mathbf{J}}{\partial t} = \frac{n_s e^2}{m}\mathbf{E}(t). \tag{12.133}$$

This is precisely the equation for current density in a system containing n_s free electrons that are not subjected to any damping ($\mathbf{J} = -n_s e\mathbf{v} \Rightarrow \partial\mathbf{J}/\partial t = -n_s e\mathbf{a} = n_s e^2\mathbf{E}/m$). The equation clearly shows that as long as $n_s \neq 0$, a steady uniform electric field produces a current that increases with time; this is the signature of a perfect conductor.

Further reading

Abrikosov, A.A., Gorkov, L.P., and Dzyaloshinski, I.E. (1963). *Methods of Quantum Field Theory in Statistical Physics*. New York: Dover Publications.
Fetter, A.L. and Walecka, J.D. (1971). *Quantum Theory of Many-Particle Systems*. New York: McGraw-Hill.
Rickayzen, G. (1965). *Theory of Superconductivity*. New York: Wiley.
Schrieffer, J.R. (1964). *Theory of Superconductivity*. New York: W.A. Benjamin, Inc.
Taylor, P.L. and Heinonen, O. (2002). *A Quantum Approach to Condensed Matter Physics*. Cambridge: Cambridge University Press.
Tinkham, M. (2004). *Superconductivity*, 2nd edn. New York: Dover Publications.

Problems

12.1 *The operator S.* Show that S, given in Eq. (12.18), satisfies Eq. (12.15).

12.2 *Ground state energy.* Verify Eq. (12.37) for the ground state energy of a superconductor.

12.3 *The anomalous Green's function.* Derive Eq. (12.76), the equation of motion for $F^\dagger(\mathbf{k}, \tau)$.

12.4 *Dirac-delta function.* In writing (12.78), we used the following equation:

$$\delta(\tau) = \frac{1}{\beta\hbar} \sum_{n=-\infty}^{n=\infty} e^{-i\omega_n \tau} \qquad \omega_n = (2n+1)\pi/\beta\hbar.$$

Verify this equation. Hint: use the frequency sum formula (Eq. [9.14]) and its complex conjugate. Also note that $f_{\bar{\epsilon}_\mathbf{k}=0} = 1/2$.

12.5 *Equation of motion.* Starting from the mean field Hamiltonian, as given in Eq. (12.53), show that the equations of motion for g and $F^\dagger$ are given by Eqs (12.74) and (12.76).

12.6 *Perturbation due to an electromagnetic field.* In the presence of a vector potential $\mathbf{A}$, the kinetic energy portion of the Hamiltonian is obtained by the replacement $\mathbf{p} \to \mathbf{p} + e\mathbf{A}/c$. Hence,

$$T = \sum_\sigma \int \Psi_\sigma^\dagger(\mathbf{r})(-i\hbar\nabla + e\mathbf{A}/c)^2 \Psi_\sigma(\mathbf{r}) d^3r.$$

Show that $T = T_{\mathbf{A}=\mathbf{0}} + H^{\text{ext}}$, where H^{ext} is given by Eq. (12.111).

12.7 $D_{\alpha\beta}(\mathbf{q}, 0)$. Verify the validity of Eq. (12.127).

12.8 *Pair fluctuations in the ground state.* Define the operator χ by

$$\chi = \frac{1}{V}\sum_{\mathbf{k}} c_{-\mathbf{k}\downarrow}c_{\mathbf{k}\uparrow}$$

where V is the volume of the superconducting system. Define $\langle\chi\rangle$ to be equal to $\langle\Psi_0|\chi|\Psi_0\rangle$, where $|\Psi_0\rangle$ is the BCS ground state. Show that, as $V \to \infty$, $\langle\chi^2\rangle - \langle\chi\rangle^2$ vanishes.

12.9 *Superconductor in a magnetic field.* In the presence of a magnetic field described by the vector potential $\mathbf{A}(\mathbf{r})$, the superconducting system is not translationally invariant. The various Green's functions of the superconductor are

$$g(\mathbf{r}\uparrow\tau,\mathbf{r}'\uparrow\tau') = -\langle T\Psi_\uparrow(\mathbf{r}\tau)\Psi_\uparrow^\dagger(\mathbf{r}'\tau')\rangle,\; F(\mathbf{r}\tau,\mathbf{r}'\tau') = -\langle T\Psi_\uparrow(\mathbf{r}\tau)\Psi_\downarrow(\mathbf{r}'\tau')\rangle,$$

and $F^\dagger(\mathbf{r}\tau,\mathbf{r}'\tau') = -\langle T\Psi_\downarrow^\dagger(\mathbf{r}\tau)\Psi_\uparrow^\dagger(\mathbf{r}'\tau')\rangle$. The Hamiltonian is

$$\bar{H} = \sum_\sigma \int d^3r\Psi_\sigma^\dagger(\mathbf{r})\left\{\frac{1}{2m}\left[-i\hbar\nabla + \frac{e\mathbf{A}(\mathbf{r})}{c}\right]^2 - \mu\right\}\Psi_\sigma(\mathbf{r})$$
$$- U_0\int d^3r\Psi_\uparrow^\dagger(\mathbf{r})\Psi_\downarrow^\dagger(\mathbf{r})\Psi_\downarrow(\mathbf{r})\Psi_\uparrow(\mathbf{r}).$$

The gap function is given by $\Delta(\mathbf{r}) = U_0F(\mathbf{r}\tau^+,\mathbf{r}\tau)$.

(a) Show that

$$\left\{i\hbar\omega_n + \frac{\hbar^2}{2m}\left[\nabla + \frac{ie\mathbf{A}(\mathbf{r})}{\hbar c}\right]^2 + \mu\right\}g(\mathbf{r},\mathbf{r}',\omega_n) + \Delta(\mathbf{r})F^\dagger(\mathbf{r},\mathbf{r}',\omega_n)$$
$$= \hbar\delta(\mathbf{r}-\mathbf{r}')$$

$$\left\{-i\hbar\omega_n + \frac{\hbar^2}{2m}\left[\nabla - \frac{ie\mathbf{A}(\mathbf{r})}{\hbar c}\right]^2 + \mu\right\}F^\dagger(\mathbf{r},\mathbf{r}',\omega_n) - \Delta^*(\mathbf{r})g(\mathbf{r},\mathbf{r}',\omega_n) = 0$$

where $\Delta^*(\mathbf{r}) = U_0F^\dagger(\mathbf{r}\tau^+,\mathbf{r}\tau) = (U_0/\beta\hbar)\sum_n F^\dagger(\mathbf{r},\mathbf{r},\omega_n)$.

(b) Let $\tilde{g}^0(\mathbf{r},\mathbf{r}',\omega_n)$ be the temperature Green's function of the metal in the normal state in the presence of the vector potential $\mathbf{A}(\mathbf{r})$. The equation for $\tilde{g}^0(\mathbf{r},\mathbf{r}',\omega_n)$ is obtained from the above equation for $g(\mathbf{r},\mathbf{r}',\omega_n)$ by setting $\Delta(\mathbf{r})$ equal to zero. Show that

$$g(\mathbf{r},\mathbf{r}',\omega_n) = \tilde{g}^0(\mathbf{r},\mathbf{r}',\omega_n) - \frac{1}{\hbar}\int d^3l\,\tilde{g}^0(\mathbf{r},\mathbf{l},\omega_n)\Delta(\mathbf{l})F^\dagger(\mathbf{l},\mathbf{r}',\omega_n)$$

$$F^\dagger(\mathbf{r},\mathbf{r}',\omega_n) = \frac{1}{\hbar}\int d^3l\,\tilde{g}^0(\mathbf{l},\mathbf{r},-\omega_n)g(\mathbf{l},\mathbf{r}',\omega_n)\Delta^*(\mathbf{l}).$$

(c) As the magnetic field approaches the critical field, $\Delta \to 0$ and $g \to \tilde{g}^0$. Show that, in this limit,

$$\Delta^*(\mathbf{r}) = \frac{U_0}{\beta\hbar^2}\sum_n \int d^3l\, g^0(\mathbf{l},\mathbf{r},-\omega_n) g^0(\mathbf{l},\mathbf{r},\omega_n) \exp\left(\frac{2ie}{\hbar c}\int_{\mathbf{l}}^{\mathbf{r}} \mathbf{A}(\mathbf{s}).d\mathbf{s}\right)\Delta^*(\mathbf{l})$$

where $g^0(\mathbf{l},\mathbf{r},\omega_n) = g^0(\mathbf{l}-\mathbf{r},\omega_n)$ is the temperature Green's function in the normal state in the absence of a magnetic field.

12.10 *Two-band model of superconductivity.* Consider a metal where two different energy bands cross the Fermi surface. For example, when graphite is intercalated with alkali atoms, such as K or Rb, partial charge transfer from the alkali atoms to the graphite planes takes place. This results in a Fermi surface which has two components: an almost two-dimensional graphite π band at the zone edge, and an approximately spherical s band (associated with alkali-metal-derived orbitals) centered at the Brillouin zone center (Dresselhaus and Dresselhaus, 1981). Let us consider a model in which superconductivity in such compounds is due to a coupling between the s and π bands. The model Hamiltonian is

$$\bar{H} = \sum_{\mathbf{k}\sigma} \bar{\epsilon}_{\mathbf{k}} c^\dagger_{\mathbf{k}\sigma} c_{\mathbf{k}\sigma} + \sum_{\mathbf{p}\sigma} \bar{\xi}_{\mathbf{p}} b^\dagger_{\mathbf{p}\sigma} b_{\mathbf{p}\sigma} - U_0 \sum_{\mathbf{kp}} \left(b^\dagger_{\mathbf{p}\uparrow} b^\dagger_{-\mathbf{p}\downarrow} c_{-\mathbf{k}\downarrow} c_{\mathbf{k}\uparrow} + \text{H.C.}\right)$$

where $c^\dagger_{\mathbf{k}\sigma}$ ($c_{\mathbf{k}\sigma}$) creates (annihilates) an electron, in the s band, of wave vector $\mathbf{k}$ and spin projection σ, $b^\dagger_{\mathbf{p}\sigma}$ ($b_{\mathbf{p}\sigma}$) creates (annihilates) an electron, in the π band, of wave vector $\mathbf{p}$ and spin projection σ, $\bar{\epsilon}_{\mathbf{k}} = \epsilon_{\mathbf{k}} - \mu$ is the energy of an electron in the s band, measured from the chemical potential, $\bar{\xi}_{\mathbf{p}}$ is the corresponding energy of an electron in the π band, and H.C. stands for "hermitian conjugate." The constant U_0 is nonvanishing only if $-\hbar\omega_D < \bar{\epsilon}_{\mathbf{k}}, \bar{\xi}_{\mathbf{p}} < \hbar\omega_D$, where $\hbar\omega_D$ is a cutoff energy. For the s band, the relevant Green's functions are

$$g_s(\mathbf{k}\uparrow,\tau) = -\langle T c_{\mathbf{k}\uparrow}(\tau) c^\dagger_{\mathbf{k}\uparrow}(0)\rangle, \quad F^\dagger_s(\mathbf{k},\tau) = -\langle T c^\dagger_{-\mathbf{k}\downarrow}(\tau) c^\dagger_{\mathbf{k}\uparrow}(0)\rangle$$

and for the π band

$$g_\pi(\mathbf{p}\uparrow,\tau) = -\langle T b_{\mathbf{p}\uparrow}(\tau) b^\dagger_{\mathbf{p}\uparrow}(0)\rangle, \quad F^\dagger_\pi(\mathbf{p},\tau) = -\langle T b^\dagger_{-\mathbf{p}\downarrow}(\tau) b^\dagger_{\mathbf{p}\uparrow}(0)\rangle.$$

The gap parameters are given by

$$\Delta^*_s = U_0 \sum_{\mathbf{k}} F^\dagger_s(\mathbf{k},0^-), \quad \Delta^*_\pi = U_0 \sum_{\mathbf{p}} F^\dagger_\pi(\mathbf{p},0^-).$$

(a) Show that

$$g_s(\mathbf{k}\uparrow,\omega_n)=\frac{-\hbar(i\hbar\omega_n+\bar{\epsilon}_{\mathbf{k}})}{\hbar^2\omega_n^2+\bar{\epsilon}_{\mathbf{k}}^2+|\Delta_\pi|^2},\quad g_\pi(\mathbf{p}\uparrow,\omega_n)=\frac{-\hbar(i\hbar\omega_n+\bar{\xi}_{\mathbf{p}})}{\hbar^2\omega_n^2+\bar{\xi}_{\mathbf{p}}^2+|\Delta_s|^2}.$$

(b) Show that

$$\Delta_s^*=U_0k_BT\sum_{\mathbf{k}n}\frac{\Delta_\pi^*}{\hbar^2\omega_n^2+\bar{\epsilon}_{\mathbf{k}}^2+|\Delta_\pi|^2}$$

$$\Delta_\pi^*=U_0k_BT\sum_{\mathbf{p}n}\frac{\Delta_s^*}{\hbar^2\omega_n^2+\bar{\xi}_{\mathbf{p}}^2+|\Delta_\pi|^2}.$$

(c) Show that the superconducting critical temperature is given by

$$k_BT_C\simeq 1.14\hbar\omega_D\exp\left(\frac{-1}{U_0\sqrt{D_{\sigma s}(0)D_{\sigma\pi}(0)}}\right)$$

where $D_{\sigma s}(0)$ and $D_{\sigma\pi}(0)$ are, respectively, the densities of states per spin orientation, at the Fermi energy, of the s and π bands.

(d) In the presence of a magnetic field described by the vector potential $\mathbf{A}(\mathbf{r})$,

$$\begin{aligned}H=&\sum_\sigma\int d^3r\Psi_{s\sigma}^\dagger(\mathbf{r})\left[\frac{1}{2m_s}\left(-i\hbar\nabla+\frac{e\mathbf{A}(\mathbf{r})}{c}\right)^2-\mu\right]\Psi_{s\sigma}(\mathbf{r})\\&+\sum_\sigma\int d^3r\Psi_{\pi\sigma}^\dagger(\mathbf{r})\left[\frac{1}{2m_\pi}\left(-i\hbar\nabla+\frac{e\mathbf{A}(\mathbf{r})}{c}\right)^2-\mu\right]\Psi_{\pi\sigma}(\mathbf{r})\\&-U_0\left[\int d^3r\Psi_{\pi\uparrow}^\dagger(\mathbf{r})\Psi_{\pi\downarrow}^\dagger(\mathbf{r})\Psi_{s\downarrow}(\mathbf{r})\Psi_{s\uparrow}(\mathbf{r})+\text{H.C.}\right]\end{aligned}$$

where m_s (m_π) is the effective mass of an electron in the s (π) band. g_s and $F_s^\dagger$ are now given by

$$g_s(\mathbf{r}\uparrow\tau,\mathbf{r}'\uparrow\tau')=-\langle T\Psi_{s\uparrow}(\mathbf{r}\tau)\Psi_{s\uparrow}^\dagger(\mathbf{r}'\tau')\rangle$$

$$F_s^\dagger(\mathbf{r}\tau,\mathbf{r}'\tau')=-\langle T\Psi_{s\downarrow}^\dagger(\mathbf{r}\tau)\Psi_{s\uparrow}^\dagger(\mathbf{r}'\tau')\rangle.$$

g_π and $F_\pi^\dagger$ are similarly defined. The gap functions are now given by

$$\Delta_s^*(\mathbf{r})=\frac{U_0}{\beta\hbar}\sum_n e^{i\omega_n0^+}F_s^\dagger(\mathbf{r},\mathbf{r},\omega_n),\quad \Delta_\pi^*(\mathbf{r})=\frac{U_0}{\beta\hbar}\sum_n e^{i\omega_n0^+}F_\pi^\dagger(\mathbf{r},\mathbf{r},\omega_n).$$

Show that, as the magnetic field approaches the critical field, the gap functions satisfy the following equations:

$$\Delta_s^*(\mathbf{r}) = \frac{U_0}{\beta\hbar^2}\sum_n \int d^3l\, g_s^0(\mathbf{l}, \mathbf{r}, -\omega_n) g_s^0(\mathbf{l}, \mathbf{r}, \omega_n) \exp\left(\frac{2ie}{\hbar c}\int_{\mathbf{l}}^{\mathbf{r}} \mathbf{A}(\mathbf{s}).d\mathbf{s}\right) \Delta_\pi^*(\mathbf{l})$$

$$\Delta_\pi^*(\mathbf{r}) = \frac{U_0}{\beta\hbar^2}\sum_n \int d^3l\, g_\pi^0(\mathbf{l}, \mathbf{r}, -\omega_n) g_\pi^0(\mathbf{l}, \mathbf{r}, \omega_n) \exp\left(\frac{2ie}{\hbar c}\int_{\mathbf{l}}^{\mathbf{r}} \mathbf{A}(\mathbf{s}).d\mathbf{s}\right) \Delta_s^*(\mathbf{l})$$

where g_i^0 $(i = s, \pi)$ is Green's function of an electron in band i in the normal state in the absence of a magnetic field. These equations can be used to determine the upper critical field H_{c2} in a type-II superconductor that is described by a two-band model (Jishi et al., 1991; Jishi and Dresselhaus, 1992).

13

Nonequilibrium Green's function

Back and forth, without a moment's rest,
An endless flow
To where the field shall lead:
No more, no less

13.1 Introduction

Thus far, we have studied systems in equilibrium. In Chapter 8 we developed a perturbation expansion for the imaginary-time Green's function which was made possible by the similarity between $\exp(-\beta\bar{H})$, which occurs in the statistical operator, and the time evolution operator $\exp(-i\bar{H}t/\hbar)$. Such similarities do not always fortuitously occur, however. A perturbation expansion is possible for the real-time causal Green's function at zero temperature, but not at finite temperature.

What approach can we use when a system is driven out of equilibrium by, for example, a time-dependent perturbation that is switched on at time t_0? In Chapter 6, we developed a method that gave a system's response to first order in the perturbation (linear response). That method, however, is incapable of dealing with the general case of nonlinear response. Moreover, when the Hamiltonian is time-dependent, the time evolution operator is no longer $\exp(-iHt/\hbar)$, and the Matsubara technique becomes inadequate.

We should point out that it is not necessary for a perturbation to be time-dependent to drive a system out of equilibrium. Consider the following example, depicted in Figure 13.1. Two metallic leads and a quantum dot (a nanostructure, for example) are initially separated. The left lead, the right lead, and the dot are initially in equilibrium, with each part having its own chemical potential. Assume that $\mu_L > \mu_R$. The Hamiltonian for the system is the sum of the Hamiltonians for the leads and the dot. At time t_0, the dot and the leads are brought into contact, and a coupling between the dot and the leads is established. As a result, current

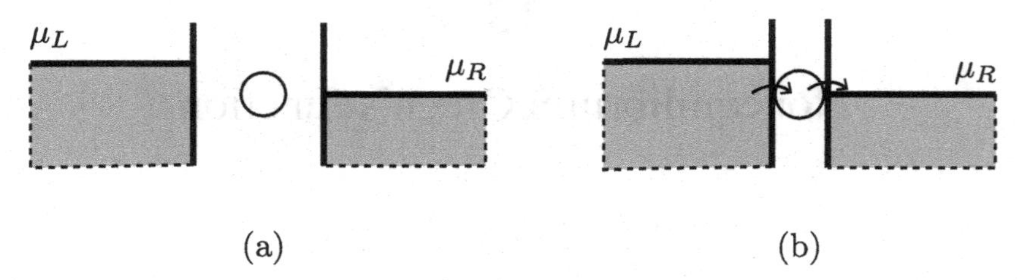

Figure 13.1 A system driven out of equilibrium. (a) A three-component system consisting of two metallic leads (left and right) and a central quantum dot. The components are separated and each is in equilibrium. (b) The leads and the dot are brought into contact, and a coupling is established between the leads and the dot. The coupling causes a current to flow, driving the system out of equilibrium.

begins to flow, resulting in the dot now being out of equilibrium. In this case, the perturbation is the coupling between the dot and the leads; apart from being switched on at time t_0, the perturbation is time-independent. The techniques we used earlier employing Green's function for systems in equilibrium cannot deal with this situation; for example, the equilibrium methods cannot give the current through the quantum dot.

To develop a method applicable to systems out of equilibrium, it is helpful to understand why the equilibrium methods fail. Toward this end, we take a closer look at the real-time causal Green's function. Before doing so, however, we discuss the Schrödinger, Heisenberg, and interaction pictures of quantum mechanics, in the general case where the Hamiltonian is time-dependent. These pictures were described earlier for a time-independent Hamiltonian.

13.2 Schrödinger, Heisenberg, and interaction pictures

We consider a many-particle system whose Hamiltonian is

$$\mathcal{H}(t) = H_0 + V + H_{\text{ext}}(t) = H_0 + H'(t). \tag{13.1}$$

H_0 is the Hamiltonian for the noninteracting system, V is the interaction among the particles, and $H_{\text{ext}}(t)$ is a (possibly) time-dependent potential.

13.2.1 The Schrödinger picture

In the Schrödinger picture, the usual picture of quantum mechanics, time dependence resides in the state $|\psi_S(t)\rangle$, which evolves in time according to the Schrödinger equation

$$i\hbar\frac{\partial}{\partial t}|\psi_S(t)\rangle = \mathcal{H}(t)|\psi_S(t)\rangle. \tag{13.2}$$

On the other hand, dynamical variables are represented by hermitian operators that have no explicit time dependence. Given the state $|\psi_S(t_0)\rangle$ at some initial time t_0, the state at time t is given by

$$|\psi_S(t)\rangle = U(t, t_0)|\psi_S(t_0)\rangle \tag{13.3}$$

where $U(t, t_0)$ is an evolution operator; it satisfies the equation

$$i\hbar\frac{\partial}{\partial t}U(t, t_0) = \mathcal{H}(t)U(t, t_0). \tag{13.4}$$

This differential equation for $U(t, t_0)$, along with the boundary condition $U(t_0, t_0) = 1$, can be converted into an integral equation,

$$U(t, t_0) = 1 - \frac{i}{\hbar}\int_{t_0}^{t} dt_1 \mathcal{H}(t_1)U(t_1, t_0). \tag{13.5}$$

The integral equation is solved by iteration. Exactly as we did in Chapter 8, we can write for $t > t_0$,

$$\begin{aligned} U(t, t_0) &= \sum_{n=0}^{\infty}\left(\frac{-i}{\hbar}\right)^n \frac{1}{n!}\int_{t_0}^{t} dt_1 \ldots \int_{t_0}^{t} dt_n T\left[\mathcal{H}(t_1)\ldots\mathcal{H}(t_n)\right] \\ &\equiv T\exp\left[\frac{-i}{\hbar}\int_{t_0}^{t} dt'\mathcal{H}(t')\right], \qquad t > t_0 \end{aligned} \tag{13.6}$$

where T is the time-ordering operator which orders operators with increasing time arguments from right to left. For $t < t_0$, it is possible to show that

$$U(t, t_0) = \tilde{T}\exp\left[\frac{-i}{\hbar}\int_{t_0}^{t} dt'\mathcal{H}(t')\right] \qquad t < t_0. \tag{13.7}$$

$\tilde{T}$ is the antitime-ordering operator: it orders operators with increasing time arguments from left to right. It is not difficult to prove the following:

$$U(t, t) = 1 \tag{13.8a}$$

$$U^\dagger(t, t_0) = U^{-1}(t, t_0) = U(t_0, t) \tag{13.8b}$$

$$U(t, t'')U(t'', t') = U(t, t'). \tag{13.8c}$$

The average value in a pure quantum state of an observable represented by the operator A varies with time according to

$$\langle A\rangle(t) = \langle\psi_S(t)|A_S|\psi_S(t)\rangle \tag{13.9}$$

where A_S is the operator A in the Schrödinger picture. More generally, the system may be in a statistical mixture of states, where we may have only limited information about the system; for example, we may know only its volume, temperature, and chemical potential. Then

$$\langle A\rangle(t) = \sum_n p_n \langle \psi_{n_S}(t)|A_S|\psi_{n_S}(t)\rangle \tag{13.10}$$

where p_n is the probability of state $|\psi_n\rangle$ occurring in the ensemble. An equivalent expression for the ensemble average is

$$\langle A\rangle(t) = Tr\,[\rho_S(t)A_S] \tag{13.11}$$

where $\rho_S(t)$ is the statistical operator in the Schrödinger picture,

$$\rho_S(t) = \sum_n p_n |\psi_{n_S}(t)\rangle\langle\psi_{n_S}(t)|. \tag{13.12}$$

The statistical operator is time-dependent, due to the explicit time dependence of the states. From the Schrödinger equation and its complex conjugate, it follows that

$$i\hbar\frac{\partial}{\partial t}\rho_S(t) = [\mathcal{H}(t), \rho_S(t)]. \tag{13.13}$$

This is the quantum Liouville equation; its solution, easily verified, is

$$\rho_S(t) = U(t, t_0)\rho_S(t_0)U^\dagger(t, t_0). \tag{13.14}$$

13.2.2 The Heisenberg picture

In contrast to the Schrödinger picture, states in the Heisenberg picture are time-independent, while operators are time-dependent. By definition, the two pictures agree at some time t_0 that can be chosen at will,

$$|\psi_{\mathcal{H}}\rangle = |\psi_S(t_0)\rangle, \quad A_{\mathcal{H}}(t_0) = A_S(t_0). \tag{13.15}$$

The expectation value of an operator, a measurable quantity, must be the same in both pictures:

$$\begin{aligned}\langle\psi_{\mathcal{H}}|A_{\mathcal{H}}(t)|\psi_{\mathcal{H}}\rangle &= \langle\psi_S(t)|A_S|\psi_S(t)\rangle = \langle\psi_S(t_0)|U^\dagger(t, t_0)A_S U(t, t_0)|\psi_S(t_0)\rangle \\ &= \langle\psi_{\mathcal{H}}|U^\dagger(t, t_0)A_S U(t, t_0)|\psi_{\mathcal{H}}\rangle.\end{aligned}$$

We thus conclude that

$$A_{\mathcal{H}}(t) = U^\dagger(t, t_0)A_S U(t, t_0). \tag{13.16}$$

The equation of motion of the Heisenberg operator is obtained by differentiating both sides of the above equation,

$$i\hbar \frac{d}{dt} A_{\mathcal{H}}(t) = [A_{\mathcal{H}}(t), \mathcal{H}_{\mathcal{H}}(t)] \tag{13.17}$$

where

$$\mathcal{H}_{\mathcal{H}}(t) = U^{\dagger}(t, t_0)\mathcal{H}(t)U(t, t_0). \tag{13.18}$$

The statistical operator in the Heisenberg picture is given by

$$\rho_{\mathcal{H}} = U^{\dagger}(t, t_0)\rho_S(t)U(t, t_0) = U^{\dagger}(t, t_0)U(t, t_0)\rho(t_0)U^{\dagger}(t, t_0)U(t, t_0) = \rho(t_0). \tag{13.19}$$

The statistical operator is time-independent in the Heisenberg picture. In a way, the statistical operator is unusual: as opposed to other operators, it is time-dependent in the Schrödinger picture and time-independent in the Heisenberg picture. In a system with a statistical mixture of states,

$$\langle A(t)\rangle = Tr[\rho_{\mathcal{H}} A_{\mathcal{H}}(t)] = Tr[\rho(t_0)A_{\mathcal{H}}(t)]. \tag{13.20}$$

13.2.3 The interaction picture

The interaction picture is intermediate between the Schrödinger and Heisenberg pictures. In the interaction picture, states and operators are related to those in the Schrödinger picture as follows:

$$|\psi_I(t)\rangle = e^{iH_0(t-t_0)/\hbar}|\psi_S(t)\rangle, \quad \hat{A}(t) = e^{iH_0(t-t_0)/\hbar} A_S e^{-iH_0(t-t_0)/\hbar}. \tag{13.21}$$

The caret or "hat" above an operator identifies it as an interaction picture operator. The three pictures coincide at $t = t_0$,

$$|\psi_S(t_0)\rangle = |\psi_{\mathcal{H}}\rangle = |\psi_I(t_0)\rangle, \quad A_S = A_{\mathcal{H}}(t_0) = \hat{A}(t_0). \tag{13.22}$$

It is straightforward to show that

$$i\hbar \frac{d}{dt}\hat{A}(t) = [\hat{A}(t), H_0], \quad i\hbar \frac{\partial}{\partial t}|\psi_I(t)\rangle = \hat{H}'(t)|\psi_I(t)\rangle. \tag{13.23}$$

The time evolution of the state $|\psi_I(t)\rangle$ is similar to that of $|\psi_S(t)\rangle$, except that $\hat{H}'(t)$ appears in place of $\mathcal{H}(t)$; hence,

$$|\psi_I(t)\rangle = S(t, t_0)|\psi_I(t_0)\rangle \tag{13.24}$$

$$S(t, t_0) = \begin{cases} T\exp\left[\frac{-i}{\hbar}\int_{t_0}^{t} \hat{H}'(t')dt'\right] & t > t_0 \\ \tilde{T}\exp\left[\frac{-i}{\hbar}\int_{t_0}^{t} \hat{H}'(t')dt'\right] & t < t_0. \end{cases} \tag{13.25}$$

Table 13.1 *Relations between the three pictures of quantum mechanics. The Hamiltonian is $\mathcal{H}(t) = H_0 + H'(t)$. At $t = t_0$, the three pictures coincide.*

	Schrödinger	Heisenberg	Interaction
$\vert\psi\rangle$	$U(t,t_0)\vert\psi_S(t_0)\rangle$	$\vert\psi_{\mathcal{H}}(t_0)\rangle$	$S(t,t_0)\vert\psi_I(t_0)\rangle$
A	A_S	$U^\dagger(t,t_0)A_S U(t,t_0)$	$e^{iH_0(t-t_0)/\hbar}A_S e^{-iH_0(t-t_0)/\hbar}$
ρ	$U(t,t_0)\rho(t_0)U^\dagger(t,t_0)$	$\rho(t_0)$	$S(t,t_0)\rho(t_0)S^\dagger(t,t_0)$

The S-matrix, or scattering matrix, satisfies the following relations

$$S(t,t) = 1, \quad S^\dagger(t,t_0) = S^{-1}(t,t_0) = S(t_0,t), \quad S(t,t'')S(t'',t') = S(t,t'). \tag{13.26}$$

A relation between $S(t,t_0)$ and $U(t,t_0)$ can be derived. For an arbitrary $|\psi_I\rangle$,

$$\begin{aligned} S(t,t_0)|\psi_I(t_0)\rangle &= |\psi_I(t)\rangle = e^{iH_0(t-t_0)/\hbar}|\psi_S(t)\rangle = e^{iH_0(t-t_0)/\hbar}U(t,t_0)|\psi_S(t_0)\rangle \\ &= e^{iH_0(t-t_0)/\hbar}U(t,t_0)|\psi_I(t_0)\rangle \\ &\Longrightarrow S(t,t_0) = e^{iH_0(t-t_0)/\hbar}U(t,t_0). \end{aligned} \tag{13.27}$$

In the interaction picture, the statistical operator is given by

$$\begin{aligned} \hat{\rho}(t) &= \sum_n p_n|\psi_{n_I}(t)\rangle\langle\psi_{n_I}(t)| = \sum_n p_n e^{iH_0(t-t_0)/\hbar}|\psi_{n_S}(t)\rangle\langle\psi_{n_S}(t)|e^{-iH_0(t-t_0)/\hbar} \\ &= \sum_n p_n e^{iH_0(t-t_0)/\hbar}U(t,t_0)|\psi_{n_S}(t_0)\rangle\langle\psi_{n_S}(t_0)|U^\dagger(t,t_0)e^{-iH_0(t-t_0)/\hbar}. \end{aligned}$$

We thus find

$$\hat{\rho}(t) = S(t,t_0)\hat{\rho}(t_0)S^\dagger(t,t_0). \tag{13.28}$$

Finally, we relate operators in the Heisenberg and interaction pictures:

$$A_{\mathcal{H}}(t) = U^\dagger(t,t_0)A_S U(t,t_0) = U^\dagger(t,t_0)e^{-iH_0(t-t_0)/\hbar}\hat{A}(t)e^{iH_0(t-t_0)/\hbar}U(t,t_0).$$

Using Eq. (13.27), we find

$$A_{\mathcal{H}}(t) = S^\dagger(t,t_0)\hat{A}(t)S(t,t_0). \tag{13.29}$$

Table 13.1 provides a summary of the three pictures of quantum mechanics.

13.3 The malady and the remedy

We have stated that a perturbation expansion for the real-time Green's function is not feasible for a system at finite temperature, or for a system that is not in equilibrium. It is instructive to see exactly why such an expansion fails; this will point the way to the construction of a Green's function that is more appropriate for the study of systems that are not in equilibrium.

Let us consider a system of interacting particles with a Hamiltonian

$$H = H_0 + V.$$

At time $t = t_0$, a (possibly) time-dependent perturbation $H_{\mathrm{ext}}(t)$ is applied. The Hamiltonian is now given by

$$\mathcal{H} = H + H_{\mathrm{ext}}(t) = H_0 + V + H_{\mathrm{ext}}(t) \equiv H_0 + H'(t)$$

where $H_{\mathrm{ext}}(t < t_0) = 0$.

We now take a closer look at the real-time causal Green's function $G(1, 1') \equiv G(\mathbf{r}\sigma t, \mathbf{r}'\sigma' t')$ defined by

$$G(1, 1') = -i\langle T[\psi_{\mathcal{H}}(1)\psi^{\dagger}_{\mathcal{H}}(1')]\rangle \tag{13.30}$$

where T is the time ordering operator; $\psi_{\mathcal{H}}(1)$ is the field operator (in the Heisenberg picture) which annihilates a particle of spin projection σ at position $\mathbf{r}$ and time t; $\psi^{\dagger}_{\mathcal{H}}(1')$ creates a particle, at time t', of coordinates $(\mathbf{r}'\sigma')$; and $\langle \ldots \rangle$ stands for a grand canonical ensemble average,

$$\langle \cdots \rangle = \frac{Tr[\rho(t_0)\cdots]}{Tr[\rho(t_0)]}.$$

In writing the ensemble average, we used the fact that the statistical operator is time-independent in the Heisenberg picture. Using Eqs (13.26) and (13.29), we can write

$$\begin{aligned} T\left[\psi_{\mathcal{H}}(1)\psi^{\dagger}_{\mathcal{H}}(1')\right] &= \theta(t - t')S^{\dagger}(t, t_0)\hat{\psi}(1)S(t, t')\hat{\psi}^{\dagger}(1')S(t', t_0) \\ &\quad \pm \theta(t' - t)S^{\dagger}(t', t_0)\hat{\psi}^{\dagger}(1')S(t', t)\hat{\psi}(1)S(t, t_0). \end{aligned}$$

The lower (upper) sign refers to fermions (bosons), and $\theta(t - t')$ is the step function: $\theta(t - t') = 1\,(0)$ if $t > t'$ $(t < t')$. Using $S^{\dagger}(t, t_0) = S(t_0, t)$, the above relation may be written in a more compact form:

$$T\left[\psi_{\mathcal{H}}(1)\psi^{\dagger}_{\mathcal{H}}(1')\right] = S(t_0, t_m)T[S(t_m, t_0)\hat{\psi}(1)\hat{\psi}^{\dagger}(1')]$$

where $t_m = \max(t, t')$. We have used the fact that, under time ordering, S-operators commute with field operators, since fermion operators come in pairs in the S-operator. Green's function is thus given by

$$iG(1, 1') = \left\langle S(t_0, t_m)T[S(t_m, t_0)\hat{\psi}(1)\hat{\psi}^{\dagger}(1')]\right\rangle. \tag{13.31}$$

Multiplying by $S(t_m, \infty)S(\infty, t_m) = 1$ after $S(t_0, t_m)$ in Eq. (13.31), and moving $S(\infty, t_m)$ inside the time-ordered product (this is allowed because $S(\infty, t_m)$ is an expansion in time-ordered products of operators and all the times in $S(\infty, t_m)$ occur later than the times inside the T-product), we obtain

$$iG(1, 1') = \left\langle S(t_0, \infty) T[S(\infty, t_0)\hat{\psi}(1)\hat{\psi}^\dagger(1')] \right\rangle. \tag{13.32}$$

Thus far, our treatment is applicable whether or not the system is in equilibrium, and whether its temperature is zero or finite. We now consider a system in equilibrium ($H_{\text{ext}} = 0$, $\mathcal{H} = H_0 + V$) at zero temperature. The ensemble average reduces to an average over the interacting ground state (in the Heisenberg picture):

$$iG(1, 1') = \left\langle \psi_{0_H} \middle| S(t_0, \infty) T[S(\infty, t_0)\hat{\psi}(1)\hat{\psi}^\dagger(1')] \middle| \psi_{0_H} \right\rangle. \tag{13.33}$$

Since the pictures coincide at $t = t_0$, we may replace $|\psi_{0_H}\rangle$ with $|\psi_{0_I}(t_0)\rangle$. The difficulty in the above expression arises because the ground state of the interacting system is unknown. Is there a way to express Green's function in terms of the noninteracting ground state? To accomplish this, we invoke the mathematical trick of switching the interaction on and off adiabatically: we assume that V is turned on and off with infinite slowness:

$$V(t) = e^{-\epsilon|t-t_0|} V \tag{13.34}$$

where ϵ is a small positive number that is eventually set equal to zero. In the remote past and in the distant future, the particles are noninteracting. At $t = -\infty$ the interaction is slowly turned on, and it attains its full strength at $t = t_0$. The noninteracting ground state $|\psi_{0_I}(-\infty)\rangle$ evolves adiabatically to the interacting ground state at $t = t_0$:

$$|\psi_{0_H}\rangle = |\psi_{0_I}(t_0)\rangle = S_\epsilon(t_0, -\infty)|\psi_{0_I}(-\infty)\rangle. \tag{13.35}$$

Here, S_ϵ is the evolution operator determined by $V(t)$ in Eq. (13.34). Putting this into Eq. (13.33) and using Eq. (13.26), we obtain

$$iG(1, 1') = \left\langle \psi_{0_I}(\infty) | T[S_\epsilon(\infty, -\infty)\hat{\psi}(1)\hat{\psi}^\dagger(1')] | \psi_{0_I}(-\infty) \right\rangle. \tag{13.36}$$

The state $|\psi_{0_I}(\infty)\rangle = S_\epsilon(\infty, -\infty)|\psi_{0_I}(-\infty)\rangle$ is the state obtained from the noninteracting ground state in the remote past by adiabatic evolution to the distant future, where the system is also noninteracting; hence, both $|\psi_{0_I}(-\infty)\rangle$ and $|\psi_{0_I}(\infty)\rangle$ are ground states of H_0. Since the ground state is nondegenerate, these two states can only differ by a phase factor,

$$|\psi_{0_I}(\infty)\rangle = e^{i\phi}|\psi_{0_I}(-\infty)\rangle$$

$$e^{i\phi} = \langle\psi_{0_I}(-\infty)|\psi_{0_I}(\infty)\rangle = \langle\psi_{0_I}(-\infty)|S_\epsilon(\infty, -\infty)|\psi_{0_I}(-\infty)\rangle.$$

Writing $|\psi_{0_I}(-\infty)\rangle \equiv |\Phi_0\rangle$, Green's function may be written as

$$iG(1,1') = \frac{\langle\Phi_0|T[S_\epsilon(\infty,-\infty)\hat{\psi}(1)\hat{\psi}^\dagger(1')]|\Phi_0\rangle}{\langle\Phi_0|S_\epsilon(\infty,-\infty)|\Phi_0\rangle}.$$

Finally, we take the limit $\epsilon \to 0$. The existence of the above expression in this limit is assured by a theorem to that effect (Gell-Mann and Low, 1951). The final expression for the causal Green's function for a system in equilibrium at zero temperature is thus

$$iG(1,1') = \frac{\langle\Phi_0|T[S(\infty,-\infty)\hat{\psi}(1)\hat{\psi}^\dagger(1')]|\Phi_0\rangle}{\langle\Phi_0|S(\infty,-\infty)|\Phi_0\rangle}. \tag{13.37}$$

This form of Green's function allows for a perturbation expansion, which in turn gives rise, through the application of Wick's theorem (applicable because the average is over the noninteracting system), to a series of connected Feynman diagrams.

We have shown that all is well in equilibrium at zero temperature. The crucial property used in arriving at the above expression is that the ground state is nondegenerate; consequently, states in the remote past and distant future coincide. At finite temperature, however, an ensemble average is taken over *all* states. The excited states of a many-particle system are generally degenerate, and the argument we developed above will break down. In nonequilibrium, even at zero temperature, the state at $t = \infty$ is not simply related to the ground state at $t = -\infty$. For example, a time-dependent perturbation would pump energy into the system, causing transitions to excited states; even after the perturbation was turned off, the system would not necessarily revert to the ground state. As another example, if we were to couple two different metals by bringing them into contact with a thin insulating layer, electrons would flow from the metal with the higher chemical potential to the other metal. If the coupling was then turned off (by separating the metals), the new ground state would not be the same as the initial one; the two metals would no longer be charge-neutral. We are stuck with the term $S(t_0,\infty)$ outside the time-ordered product in Eq. (13.32). Thus, a perturbation expansion for the real-time causal Green's function is not valid if the system is at finite temperature and/or out of equilibrium. In equilibrium at finite temperature, going to imaginary time produces a Green's function (Matsubara function) that admits a perturbation expansion, but this approach is futile in the case of nonequilibrium.

How should we proceed when a system is out of equilibrium? For a clue, we go back to Eqs (13.31) and (13.32); these equations hold in a general nonequilibrium setting. Using Eq. (13.25), we can write

$$iG(1,1') = \left\langle \tilde{T}\left[e^{-\frac{i}{\hbar}\int_{t_m}^{t_0}\hat{H}'(t_1)dt_1}\right] T\left[e^{-\frac{i}{\hbar}\int_{t_0}^{t_m}\hat{H}'(t_1)dt_1}\hat{\psi}(1)\hat{\psi}^\dagger(1')\right]\right\rangle. \tag{13.38}$$

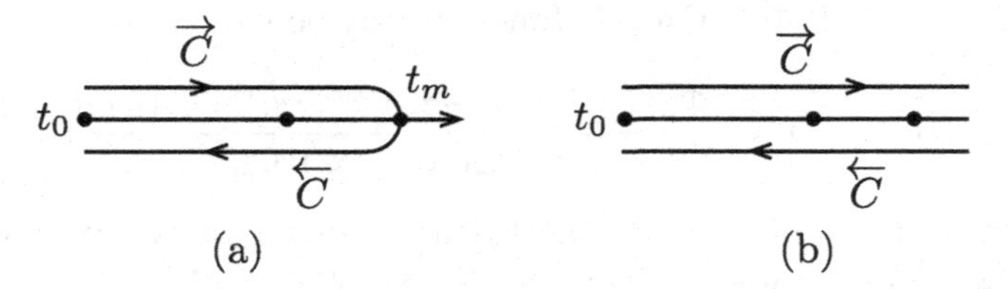

Figure 13.2 $C = \overrightarrow{C} \cup \overleftarrow{C}$. The contour runs along the real-time axis from t_0 to $max(t, t')$, as in (a), or to ∞, as in (b), and back along the real-time axis to t_0. For the sake of clarity, $\overrightarrow{C}$ and $\overleftarrow{C}$ are drawn above and below the real-time axis.

We can bring this expression into some formal similarity with the expression used for states in equilibrium by proceeding as follows. We introduce a contour-ordering operator T_C, along a contour C which consists of two parts: $\overrightarrow{C}$ from t_0 to t_m (or ∞, if we start from Eq. [3.32]) and $\overleftarrow{C}$ from t_m (or ∞) to t_0, as depicted in Figure 13.2. The contour-ordering operator is defined by

$$T_C[A(\tau)B(\tau')] = \begin{cases} A(\tau)B(\tau') & \tau \overset{C}{>} \tau' \\ \pm B(\tau')A(\tau) & \tau' \overset{C}{>} \tau \end{cases} \tag{13.39}$$

where the lower (upper) sign refers to fermions (bosons), and the time along the contour is denoted by τ. The statement $\tau \overset{C}{>} \tau'$ means that τ lies further along the contour than τ', regardless of the numerical values of τ and τ'. Thus, ordering along $\overrightarrow{C}$ corresponds to normal-time ordering, whereas ordering along $\overleftarrow{C}$ corresponds to antitime ordering:

$$T_{\overrightarrow{C}} = T, \quad T_{\overleftarrow{C}} = \tilde{T}.$$

We may rewrite Eq. (13.38) as

$$\begin{aligned} iG(1, 1') &= \left\langle T_{\overleftarrow{C}} \left[e^{-\frac{i}{\hbar}\int_{\overleftarrow{C}} \hat{H}'(\tau_1)d\tau_1} \right] T_{\overrightarrow{C}} \left[e^{-\frac{i}{\hbar}\int_{\overrightarrow{C}} \hat{H}'(\tau_1)d\tau_1} \hat{\psi}(1)\hat{\psi}^\dagger(1') \right] \right\rangle \\ &= \left\langle T_C \left[e^{-\frac{i}{\hbar}\int_C \hat{H}'(\tau_1)d\tau_1} \hat{\psi}(\mathbf{r}\sigma\tau)\hat{\psi}^\dagger(\mathbf{r}'\sigma'\tau') \right] \right\rangle, \quad \tau, \tau' \in \overrightarrow{C}. \end{aligned} \tag{13.40}$$

All operators are now under contour ordering. The above expression is simply Eq. (13.32) rewritten. Equation (13.32) does not give a perturbation expansion for $G(1, 1')$, so Eq. (13.40) does not either. However, the form of $G(1, 1')$ in Eq. (13.40) suggests a generalization: rather than restricting τ and τ' and tying them to $\overrightarrow{C}$, they can be freed so as to lie anywhere on the contour C. This step, as it turns out, produces a Green's function that admits a perturbation expansion, one that is relevant for systems that are out of equilibrium.

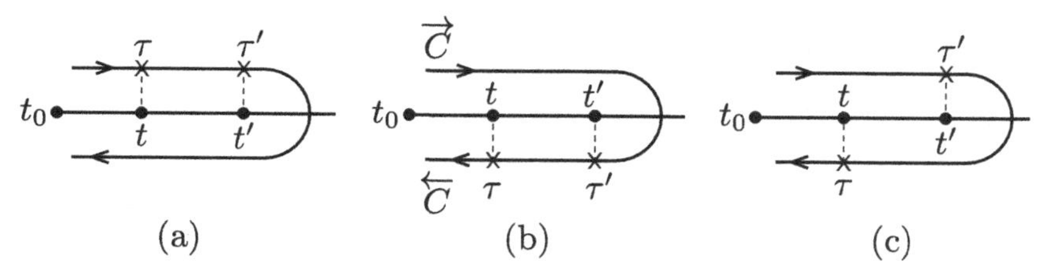

Figure 13.3 The contour C consists of a forward part ($\overrightarrow{C}$) and a backward part ($\overleftarrow{C}$). Both lie along the real-time axis but are shown displaced from it for the sake of clarity. τ and τ' are the locations of t and t' on the contour. In (a) both t and t' are on the forward part $\overrightarrow{C}$, while in (b) they are both on the backward part $\overleftarrow{C}$. In (c), $t \in \overleftarrow{C}$, $t' \in \overrightarrow{C}$. Note that while $t < t'$ on the real-time axis in this figure, as contour times $\tau \overset{C}{>} \tau'$ in (b) and (c), while $\tau \overset{C}{<} \tau'$ in (a).

13.4 Contour-ordered Green's function

We define the contour-ordered Green's function by

$$G_c(\mathbf{r}\sigma\tau, \mathbf{r}'\sigma'\tau') = -i\langle T_C\, \psi_{\mathcal{H}}(\mathbf{r}\sigma\tau)\psi^\dagger_{\mathcal{H}}(\mathbf{r}'\sigma'\tau')\rangle, \quad \tau, \tau' \in C. \tag{13.41}$$

T_C is the contour-time-ordering operator which places operators with time arguments that are further along the contour on the left. The contour C starts at t_0, goes to $t_m = \max(t, t')$ or to any point beyond t_m on the real-time axis, and goes back to t_0, passing through t and t' exactly once, as shown in Figure 13.3. The contour may pass through t along the forward path or along the backward path; ditto for t'. The ensemble average is over operators in the Heisenberg picture; here the statistical operator is time-independent: $\rho(t) = \rho(t_0)$, where t_0 is the time when the external field is switched on. For times prior to t_0, the system is assumed to be in equilibrium; hence, the contour-ordered Green's function is given by

$$G_c(1, 1') = \frac{-i\,Tr\left[e^{-\beta(H-\mu N)} T_C\, \psi_{\mathcal{H}}(1)\psi^\dagger_{\mathcal{H}}(1')\right]}{Tr[e^{-\beta(H-\mu N)}]} \tag{13.42}$$

where $H = H_0 + V$ is the Hamiltonian for the interacting system in the absence of an external field, $(1) = (\mathbf{r}\sigma\tau)$, and $(1') = (\mathbf{r}'\sigma'\tau')$. Since the contour C is the union of two segments, $\overrightarrow{C}$ and $\overleftarrow{C}$, there are four possible outcomes depending on the locations of t and t' on the contour:

1. $t, t' \in \overrightarrow{C}$; we saw in the previous section that in this case $G_c(1, 1')$ coincides with the causal (time-ordered) Green's function $G^T(1, 1')$.
2. $t \in \overrightarrow{C}, t' \in \overleftarrow{C}$; in this case $\tau' \overset{C}{>} \tau$ and

$$G_c(1, 1') = \mp i\langle\psi^\dagger_{\mathcal{H}}(\mathbf{r}'\sigma't')\psi_{\mathcal{H}}(\mathbf{r}\sigma t)\rangle = G^<(1, 1').$$

3. $t \in \overleftarrow{C}, t' \in \overrightarrow{C}$; then $\tau \overset{C}{>} \tau'$ and

$$G_c(1, 1') = -i\langle \psi_{\mathcal{H}}(\mathbf{r}\sigma t)\psi^\dagger_{\mathcal{H}}(\mathbf{r}'\sigma' t')\rangle = G^>(1, 1').$$

4. $t, t' \in \overleftarrow{C}$; in this case $T_C = T_{\overleftarrow{C}} = \tilde{T}$ and

$$G_c(1, 1') = -i\langle \tilde{T}\psi_{\mathcal{H}}(1)\psi^\dagger_{\mathcal{H}}(1')\rangle = G^{\tilde{T}}(1, 1')$$

where $G^{\tilde{T}}(1, 1')$ is the antitime-ordered (anticausal) Green's function. We can summarize the above results as:

$$G_c(1, 1') = \begin{cases} G^T(1, 1') & t,\ t' \in \overrightarrow{C} \\ G^<(1, 1') & t \in \overrightarrow{C},\ t' \in \overleftarrow{C} \\ G^>(1, 1') & t \in \overleftarrow{C},\ t' \in \overrightarrow{C} \\ G^{\tilde{T}}(1, 1') & t,\ t' \in \overleftarrow{C}. \end{cases} \tag{13.43}$$

Since

$$G^T(1, 1') = -i\theta(t - t')\langle \psi_{\mathcal{H}}(1)\psi^\dagger_{\mathcal{H}}(1')\rangle \mp i\theta(t' - t)\langle \psi^\dagger_{\mathcal{H}}(1')\psi_{\mathcal{H}}(1)\rangle,$$
$$G^{\tilde{T}}(1, 1') = -i\theta(t' - t)\langle \psi_{\mathcal{H}}(1)\psi^\dagger_{\mathcal{H}}(1')\rangle \mp i\theta(t - t')\langle \psi^\dagger_{\mathcal{H}}(1')\psi_{\mathcal{H}}(1)\rangle,$$

it follows that

$$\begin{aligned} G^T(1, 1') + G^{\tilde{T}}(1, 1') &= -i\langle \psi_{\mathcal{H}}(1)\psi^\dagger_{\mathcal{H}}(1')\rangle \mp i\langle \psi^\dagger_{\mathcal{H}}(1')\psi_{\mathcal{H}}(1)\rangle \\ &= G^>(1, 1') + G^<(1, 1'). \end{aligned} \tag{13.44}$$

Therefore, three of the four functions contained in $G_c(1, 1')$ are independent. The retarded and advanced Green's functions, G^R and G^A, respectively, are given by

$$\begin{aligned} G^R(1, 1') &= -i\theta(t - t')\langle [\psi_{\mathcal{H}}(1),\ \psi^\dagger_H(1')]_\mp\rangle \\ &= -i\theta(t - t')\langle \psi_{\mathcal{H}}(1)\psi^\dagger_{\mathcal{H}}(1') \mp \psi^\dagger_{\mathcal{H}}(1')\psi_{\mathcal{H}}(1)\rangle \\ &= \theta(t - t')\left[G^>(1, 1') - G^<(1, 1')\right] \end{aligned} \tag{13.45}$$

$$\begin{aligned} G^A(1, 1') &= i\theta(t' - t)\langle [\psi_{\mathcal{H}}(1),\ \psi^\dagger_{\mathcal{H}}(1')]_\mp\rangle \\ &= i\theta(t' - t)\langle \psi_{\mathcal{H}}(1)\psi^\dagger_{\mathcal{H}}(1') \mp \psi^\dagger_{\mathcal{H}}(1')\psi_{\mathcal{H}}(1)\rangle \\ &= \theta(t' - t)\left[-G^>(1, 1') + G^<(1, 1')\right]. \end{aligned} \tag{13.46}$$

Finally, the following relations among the various functions can be verified:

$$G^T = G^< + G^R = G^> + G^A, \quad G^{\tilde{T}} = G^< - G^A = G^> - G^R. \tag{13.47}$$

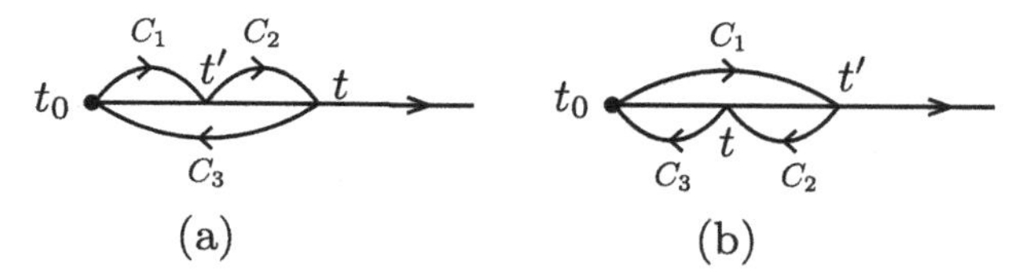

Figure 13.4 $C = C_1 \cup C_2 \cup C_3$. The contour segments actually lie along the time axis, but are shown displaced from it for the sake of clarity. In both (a) and (b), $\tau \overset{C}{>} \tau'$.

13.5 Kadanoff–Baym and Keldysh contours

In order to develop a perturbation expansion for $G_c(1, 1')$, we need to express it in terms of interaction-picture operators. The Hamiltonian is

$$\mathcal{H}(t) = H + H_{\text{ext}}(t) = H_0 + V + H_{\text{ext}}(t) = H_0 + H'(t). \tag{13.48}$$

The contour Green's function, in terms of interaction-picture operators (identified by hats), is given by

$$G_c(1, 1') = -i \left\langle T_C \left[e^{-\frac{i}{\hbar} \int_C \hat{H}'(\tau_1) d\tau_1} \hat{\psi}(1) \hat{\psi}^\dagger(1') \right] \right\rangle \tag{13.49}$$

where C is the contour depicted in Figure 13.2 or Figure 13.3. A proof of this result is presented below.

We prove Eq. (13.49) for the case when t lies further along the contour than t'; the opposite case is proved in a similar way. For $\tau \overset{C}{>} \tau'$,

$$iG(1, 1') = \langle \psi_{\mathcal{H}}(1) \psi^\dagger_{\mathcal{H}}(1') \rangle = \langle S(t_0, t) \hat{\psi}(1) S(t, t') \hat{\psi}^\dagger(1') S(t', t_0) \rangle.$$

In writing this, we have used Eqs (13.26) and (13.29). To calculate the RHS of Eq. (13.49), we divide the contour into three segments C_1, C_2, and C_3 (see Figure 13.4). The contour $C = C_1 \cup C_2 \cup C_3$, and $\int_C = \int_{C_1} + \int_{C_2} + \int_{C_3}$. The contour-ordered product on the RHS of Eq. (13.49), denoted by B, can be written as

$$B = T_C \left[e^{-\frac{i}{\hbar} \int_C \hat{H}'(\tau_1) d\tau_1} \hat{\psi}(1) \hat{\psi}^\dagger(1') \right]$$
$$= \sum_{n=0}^{\infty} \frac{1}{n!} \left(\frac{-i}{\hbar} \right)^n \int_C d\tau_1 \ldots \int_C d\tau_n T_C \left[\hat{H}'(\tau_1) \ldots \hat{H}'(\tau_n) \hat{\psi}(1) \hat{\psi}^\dagger(1') \right].$$

Let us consider the term of order n. Each integral over C is a sum of three integrals, and the term of order n is thus the sum of 3^n terms. Consider one such term that has k integrals along C_3 ($k = 0, 1, \ldots, n$), l integrals along C_2 ($l = 0, 1, \ldots, n - k$), and $n - k - l$ integrals along C_1. There is a total of $n!/[k!l!(n - k - l)!]$ such terms that differ only by a relabeling of their time indices; since the times are integrated over, these terms are equal. Therefore,

$$B = \sum_{n=0}^{\infty} \frac{1}{n!}\left(\frac{-i}{\hbar}\right)^n \sum_{k=0}^{n}\sum_{l=0}^{n-k} \frac{n!}{k!\,l!\,(n-k-l)!}$$
$$\times \int_{C_3} d\tau_1 \cdots \int_{C_3} d\tau_k\, T_{C_3}[\hat{H}'(\tau_1)\cdots\hat{H}'(\tau_k)]\hat{\psi}(1)$$
$$\times \int_{C_2} d\tau_{k+1} \cdots \int_{C_2} d\tau_{k+l}\, T_{C_2}[\hat{H}'(\tau_{k+1})\cdots\hat{H}'(\tau_{k+l})]\hat{\psi}^{\dagger}(1')$$
$$\times \int_{C_1} d\tau_{k+l+1} \cdots \int_{C_1} d\tau_n\, T_{C_1}[\hat{H}'(\tau_{k+l+1})\cdots\hat{H}'(\tau_n)].$$

Noting that

$$\sum_{k=0}^{n}\sum_{l=0}^{n-k} \frac{n!}{k!\,l!\,(n-k-l)!}\cdots = \sum_{k=0}^{\infty}\sum_{l=0}^{\infty}\sum_{m=0}^{\infty} \frac{n!}{k!\,l!\,m!}\delta_{n,k+l+m}\cdots$$

the following expression for B is obtained by summing over n first,

$$B = \sum_{k=0}^{\infty}\left(\frac{-i}{\hbar}\right)^k \frac{1}{k!}\int_{C_3} d\tau_1 \cdots \int_{C_3} d\tau_k\, T_{C_3}[\hat{H}'(\tau_1)\cdots\hat{H}'(\tau_k)]\hat{\psi}(1)$$
$$\times \sum_{l=0}^{\infty}\left(\frac{-i}{\hbar}\right)^l \frac{1}{l!}\int_{C_2} d\tau_1 \cdots \int_{C_2} d\tau_l\, T_{C_2}[\hat{H}'(\tau_1)\cdots\hat{H}'(\tau_l)]\hat{\psi}^{\dagger}(1')$$
$$\times \sum_{m=0}^{\infty}\left(\frac{-i}{\hbar}\right)^m \frac{1}{m!}\int_{C_1} d\tau_1 \cdots \int_{C_1} d\tau_m\, T_{C_1}[\hat{H}'(\tau_1)\ldots\hat{H}'(\tau_m)]$$
$$\equiv P\hat{\psi}(1)Q\hat{\psi}^{\dagger}(1')R.$$

C_1 extends from t_0 to $t' > t_0 \Rightarrow T_{C_1} = T \Rightarrow R = S(t', t_0)$. C_3 extends from t to $t_0 < t \Rightarrow T_{C_3} = \tilde{T} \Rightarrow P = S(t_0, t)$. C_2 extends from t' to t. If $t > t'$ then $T_{C_2} = T$ and $Q = S(t, t')$; if $t < t'$ then $T_{C_2} = \tilde{T}$ and again (see Eq. [13.25]) $Q = S(t, t')$. This ends the proof of Eq. (13.49) for the case $\tau \overset{C}{>} \tau'$.

Returning to Eq. (13.49), we replace $\hat{H}'$ with $\hat{V} + \hat{H}_{\text{ext}}$,

$$iG_c(1,1') = \left\langle T_C\left[e^{-\frac{i}{\hbar}\int_C(\hat{V}(\tau_1)+\hat{H}_{\text{ext}}(\tau_1))d\tau_1}\hat{\psi}(1)\hat{\psi}^{\dagger}(1')\right]\right\rangle$$
$$= \left\langle T_C\left[e^{-\frac{i}{\hbar}\int_C \hat{V}(\tau_1)d\tau_1}e^{-\frac{i}{\hbar}\int_C \hat{H}_{\text{ext}}(\tau_1)d\tau_1}\hat{\psi}(1)\hat{\psi}^{\dagger}(1')\right]\right\rangle. \qquad (13.50)$$

The last equality in the above equation is valid since $\hat{V}$ and $\hat{H}_{\text{ext}}$ contain an even number of fermion operators, so they commute under contour ordering. Defining the operators

$$S_C^V = \exp\left[-\frac{i}{\hbar}\int_C \hat{V}(\tau)d\tau\right], \quad S_C^{\text{ext}} = \exp\left[-\frac{i}{\hbar}\int_C \hat{H}_{\text{ext}}(\tau)d\tau\right], \tag{13.51}$$

we can write

$$iG_c(1,1') = \left\langle T_C\left[S_C^V S_C^{\text{ext}}\hat{\psi}(1)\hat{\psi}^\dagger(1')\right]\right\rangle. \tag{13.52}$$

From the definition of $iG_c(1,1')$, if $\hat{\psi}$ is replaced by 1 and $\hat{\psi}^\dagger$ is also replaced by 1, then $iG_c(1,1')$ reduces to 1. Hence, we deduce from Eq. (13.52) that $T_C\left(S_C^V S_C^{\text{ext}}\right) = 1$.

Setting $H_{\text{ext}}(t)$ equal to zero in Eq. (13.27), which leads to $U(t,t_0)$ becoming equal to $\exp[-iH(t-t_0)/\hbar]$, and using Eq. (13.25), we can write

$$\bar{S}^V(t,t_0) = e^{iH_0(t-t_0)/\hbar}e^{-iH(t-t_0)/\hbar} \tag{13.53}$$

where $\bar{S}^V(t,t_0)$ is defined by

$$\bar{S}^V(t,t_0) = T\left[\exp\left(-\frac{i}{\hbar}\int_{t_0}^{t}\hat{V}(t')dt'\right)\right]. \tag{13.54}$$

Comparing Eq. (13.53) with the following equation,

$$e^{-\beta(H-\mu N)} = e^{-\beta(H_0-\mu N)}e^{\beta H_0}e^{-\beta H},$$

which is valid since N commutes with H_0 and H, we can write

$$e^{-\beta(H-\mu N)} = e^{-\beta(H_0-\mu N)}\bar{S}^V(t_0 - i\beta\hbar, t_0). \tag{13.55}$$

The contour Green's function (see Eq. [13.52]) may now be expressed as

$$iG_c(1,1') = \frac{Tr\left[e^{-\beta(H_0-\mu N)}\bar{S}^V(t_0-i\beta\hbar,t_0)T_C[S_C^V S_C^{\text{ext}}\hat{\psi}(1)\hat{\psi}^\dagger(1')]\right]}{Tr\left[e^{-\beta(H_0-\mu N)}\bar{S}^V(t_0-i\beta\hbar,t_0)\right]}. \tag{13.56}$$

We move $\bar{S}^V(t_0-i\beta\hbar,t_0)$ through T_C and combine it with S_C^V,

$$iG_c(1,1') = \frac{Tr\left[e^{-\beta(H_0-\mu N)}T_{C'}[S_{C'}^V S_C^{\text{ext}}\hat{\psi}(1)\hat{\psi}^\dagger(1')]\right]}{Tr\left[e^{-\beta(H_0-\mu N)}\bar{S}^V(t_0-i\beta\hbar,t_0)\right]} \tag{13.57}$$

where

$$S_{C'}^V = \exp\left[-\frac{i}{\hbar}\int_{C'}\hat{V}(\tau)d\tau\right]. \tag{13.58}$$

Here $C' = C \cup [t_0, t_0 - i\beta\hbar]$ is the Kadanoff–Baym three-branch contour shown in Figure 13.5 (Kadanoff and Baym, 1962). The contour starts at t_0, goes to

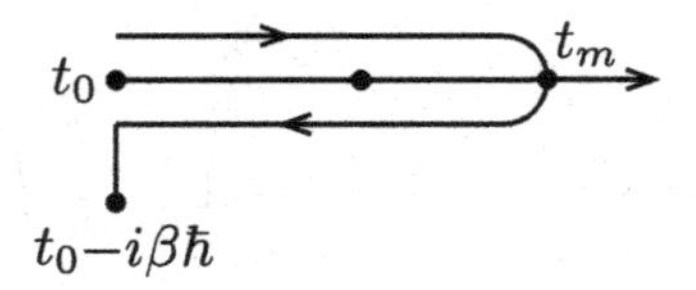

Figure 13.5 Kadanoff–Baym three-branch contour C'; it starts at t_0 and stretches to $\max(t, t')$, then returns to t_0, and down to $t_0 - i\beta\hbar$.

$t_m = \max(t, t')$ or any time beyond t_m, returns to t_0 on the backward path, and goes down to $t_0 - i\beta\hbar$. The operator $T_{C'}$ is the contour-time-ordering operator along C'.

We may, if we choose, insert $T_C\left(S_C^V S_C^{\text{ext}}\right) = 1$ after $\bar{S}^V(t_0 - i\beta\hbar, t_0)$ in the denominator of Eq. (13.57), and then divide the numerator and denominator by $Tr\left[e^{-\beta(H_0-\mu N)}\right]$; the result is

$$iG_c(1, 1') = \frac{\left\langle \left[T_{C'}[S_{C'}^V S_C^{\text{ext}} \hat{\psi}(1)\hat{\psi}^\dagger(1')\right]\right\rangle_0}{\left\langle T_{C'}\left(S_{C'}^V S_C^{\text{ext}}\right)\right\rangle_0} \tag{13.59}$$

where the ensemble average is now over the noninteracting system. Substituting the perturbation expansion for $S_{C'}^V$ and S_C^{ext} into the above expression, Wick's theorem (applicable here since the ensemble average is over the noninteracting system), yields a perturbation series for $G_c(1, 1')$.

The above expression for $G_c(1, 1')$ can be used to study the behavior of a system out of equilibrium at times $t > t_0$, after an external perturbation has been switched on at time t_0, while taking into account the initial correlations at $t = t_0$. Indeed, we have used $\rho(t_0) = e^{-\beta(H-\mu N)}/Tr[e^{-\beta(H-\mu N)}]$, which includes interactions among the particles. In many cases, however, we are only interested in studying the behavior of a system for times $t \gg t_0$. For example, regarding the system depicted in Figure 13.1, we may be interested in its steady state after all transients have died off. The steady state, if it develops at times $t \gg t_0$, will not depend on the initial state at time t_0. In such a case, we may use the statistical operator of the noninteracting system instead of that of the interacting system. Alternatively, if we are only interested in the behavior of the system for times $t \gg t_0$, we may take $t_0 = -\infty$ and assume that interactions are turned on adiabatically (but not turned off). In this case, the statistical operator $\rho(t_0 = -\infty)$ is that of a noninteracting system, and the branch of the contour C' that extends from t_0 to $t_0 - i\beta\hbar$ may be dropped, i.e., C' coincides with C. The contour C now extends from $-\infty$ to $\max(t, t')$ and back to $-\infty$. We might as well extend the contour to $+\infty$ so that it runs from $-\infty$ to $+\infty$ and back to $-\infty$; this is the Keldysh contour (Keldysh, 1965), depicted in Figure 13.6. The expression for the contour Green's function in

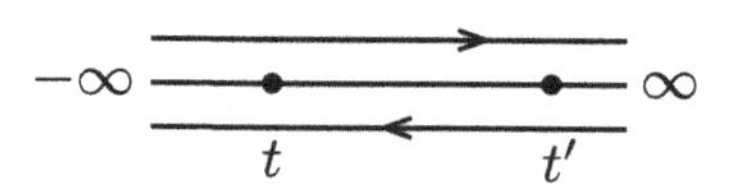

Figure 13.6 Keldysh contour C: it runs along the real time axis from $-\infty$ to $+\infty$ and back to $-\infty$, passing through t and t' exactly once.

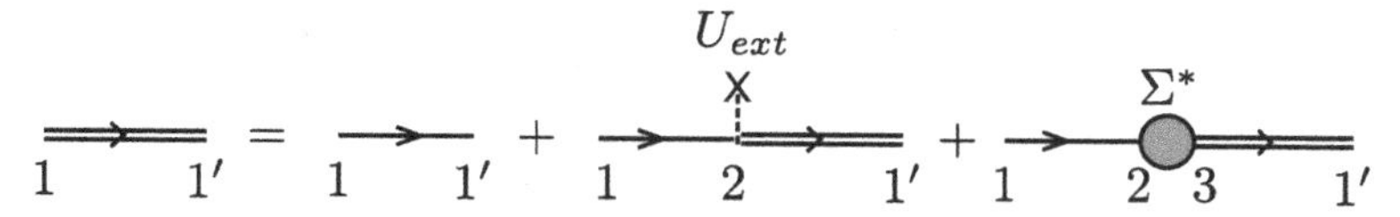

Figure 13.7 Dyson's equation. A double line represents G_c, the contour Green's function for the interacting system, while a single line represents G_c^0, the corresponding contour Green's function for the noninteracting system. U_{ext} is a one-body external potential and Σ^* is the irreducible self energy arising from interparticle interactions. In the figure, $1 = (\mathbf{r}\sigma\tau)$.

the Keldysh formalism becomes

$$G_c(1, 1') = -i \left\langle T_C \left[S_C^V S_C^{\text{ext}} \hat{\psi}(1) \hat{\psi}^\dagger(1') \right] \right\rangle_0 . \tag{13.60}$$

The above expression may also be written as

$$G_c(1, 1') = \frac{-i \left\langle T_C \left[S_C^V S_C^{\text{ext}} \hat{\psi}(1) \hat{\psi}^\dagger(1') \right] \right\rangle_0}{\left\langle T_C \, S_C^V \, S_C^{\text{ext}} \right\rangle_0} . \tag{13.61}$$

since the denominator in the above equation is equal to 1.

We may now expand S_C^V and S_C^{ext} in a power series in $\hat{V}$ and $\hat{H}_{\text{ext}}$. In the second quantized form, $\hat{V}$ and $\hat{H}_{\text{ext}}$ are written in terms of field operators in the interaction picture. Since the ensemble average is over the noninteracting system, Wick's theorem applies, and we end up with a perturbation expansion similar to the one for Matsubara Green's function. The only difference is that contour time ordering replaces time ordering, so that in the resulting Feynman diagrams, the Green's functions that appear are contour Green's functions. As before, all disconnected diagrams cancel out, and $G_c(1, 1')$ is a sum over connected diagrams.

13.6 Dyson's equation

The perturbation expansion of $G_c(1, 1')$ can be expressed in the form of a Dyson's equation, much like Matsubara Green's function. The interaction consists of two parts: a perturbation H_{ext} (due to an external field), which we take to be a one-body operator, and the interparticle interaction V, which is a two-body operator. A graphical representation of Dyson's equation is depicted in Figure 13.7. Thus, the

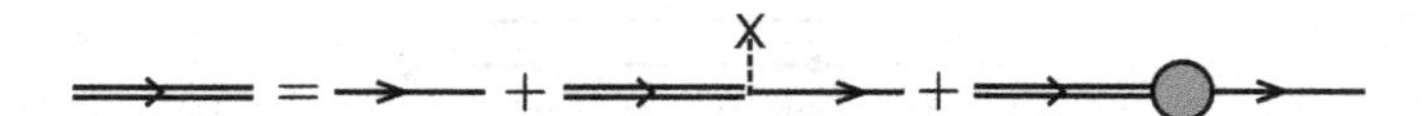

Figure 13.8 An alternative form of Dyson's equation.

expression for $G_c(1, 1')$ is as follows:

$$G_c(1, 1') = G_c^0(1, 1') + \int_C d2\, G_C^0(1, 2) U_{\text{ext}}(2) G_c(2, 1')$$
$$+ \int_C d2 \int_C d3\, G_c^0(1, 2) \Sigma^*(2, 3) G_c(3, 1') \tag{13.62}$$

where U_{ext} is the external potential giving rise to the external perturbation,

$$\hat{H}_{\text{ext}}(t) = \sum_\sigma \int \hat{\psi}^\dagger(\mathbf{r}\sigma t) U_{\text{ext}}(\mathbf{r}\sigma t) \hat{\psi}(\mathbf{r}\sigma t) d^3r,$$

and Σ^* is the irreducible self energy resulting from the pairwise interaction V among the particles of the system. In Eq. (13.62),

$$\int_C d2 = \sum_{\sigma_2} \int d^3r_2 \int_C d\tau_2.$$

We can adopt a compact matrix notation and write Dyson's equation as

$$G_c = G_c^0 + G_c^0 U G_c + G_c^0 \Sigma^* G_c \tag{13.63}$$

where $G_c^0 U G_c$ and $G_c^0 \Sigma^* G_c$ stand for the second and third terms, respectively, in Eq. (13.62).

We can also write Dyson's equation in an alternative form. Noting that

$$G_c = G_c^0(1 + UG_c + \Sigma^* G_c) \Rightarrow 1 + UG_c + \Sigma^* G_c = G_c^{0^{-1}} G_c$$
$$\Rightarrow G_c^{-1} + U + \Sigma^* = G_c^{0^{-1}} \Rightarrow 1 + G_c U + G_c \Sigma^* = G_c G_c^{0^{-1}},$$

we can write the following:

$$G_c = G_c^0 + G_c U G_c^0 + G_c \Sigma^* G_c^0. \tag{13.64}$$

This form of Dyson's equation is depicted in Figure 13.8. It is clear that Figures 13.7 and 13.8 produce identical perturbation series for G_c.

In practice, carrying out calculations with contour integrals is not convenient, and they should be reexpressed in terms of real-time integrals. The procedure for converting contour-time integrals into real-time integrals is known as analytic continuation (a misnomer, since the contour is attached to the real-time axis and no continuation from the complex plane takes place). The rules for this procedure are taken up next.

13.7 Langreth rules

The quantities directly related to observables are the lesser, greater, retarded, and advanced functions $G^<$, $G^>$, G^R, and G^A, respectively; these are functions of real times rather than contour times. We were forced to resort to the contour Green's function, not because it is directly related to observables, but because it can be expanded in a perturbation series, whereas no such expansion exists for $G^<$, $G^>$, G^R, and G^A. To make contact with physical quantities, these functions must be extracted from the contour Green's function; Langreth rules (Langreth, 1977) provide the vehicle for doing that.

We note that Dyson's equation contains terms that involve one or two contour-time integrals. In order to keep the discussion as general as possible, we introduce a general function $A(\tau, \tau')$, $\tau, \tau' \in C$, and the corresponding real-time functions,

$$A(\tau, \tau') = \begin{cases} A^T(t, t') & \tau, \tau' \in \overrightarrow{C} \\ A^<(t, t') & \tau \in \overrightarrow{C},\ \tau' \in \overleftarrow{C} \\ A^>(t, t') & \tau \in \overleftarrow{C},\ \tau' \in \overrightarrow{C} \\ A^{\tilde{T}}(t, t') & \tau, \tau' \in \overleftarrow{C}. \end{cases} \tag{13.65}$$

Properties of the function $A(\tau, \tau')$ resemble those of the contour Green's function $G_c(1, 1')$. Analogous to Eqs (13.45) and (13.46), we define $A^r(t, t')$ and $A^a(t, t')$ by the following relations

$$A^r(t, t') = \theta(t - t')\left[A^>(t, t') - A^<(t, t')\right] \tag{13.66}$$

$$A^a(t, t') = \theta(t' - t)\left[A^<(t, t') - A^>(t, t')\right]. \tag{13.67}$$

Now consider the contour integral of the form

$$C(\tau, \tau') = \int_C A(\tau, \tau_1) B(\tau_1, \tau') d\tau_1 \tag{13.68}$$

where B is another function dependent on two contour times and C is the contour shown in Figure 13.6. The lesser function $C^<(t, t')$ is given by

$$\begin{aligned} C^<(t, t') &= C(t \in \overrightarrow{C}, t' \in \overleftarrow{C}) \equiv C(t^{\rightarrow}, t'^{\leftarrow}) = \int_C d\tau_1 A(t^{\rightarrow}, \tau_1) B(\tau_1, t'^{\leftarrow}) \\ &= \int_{-\infty}^{\infty} dt_1 A(t^{\rightarrow}, t_1^{\rightarrow}) B(t_1^{\rightarrow}, t'^{\leftarrow}) + \int_{\infty}^{-\infty} dt_1 A(t^{\rightarrow}, t_1^{\leftarrow}) B(t_1^{\leftarrow}, t'^{\leftarrow}) \\ &= \int_{-\infty}^{\infty} dt_1 \left[A(t^{\rightarrow}, t_1^{\rightarrow}) B(t_1^{\rightarrow}, t'^{\leftarrow}) - A(t^{\rightarrow}, t_1^{\leftarrow}) B(t_1^{\leftarrow}, t'^{\leftarrow})\right] \\ &= \int_{-\infty}^{\infty} dt_1 \left[A^T(t, t_1) B^<(t_1, t') - A^<(t, t_1) B^{\tilde{T}}(t_1, t')\right]. \end{aligned}$$

From Eq. (13.47), we can write $A^{T} = A^{<} + A^{r}$, $B^{\tilde{T}} = B^{<} - B^{a}$. Thus,

$$C^{<}(t,t') = \int_{-\infty}^{\infty} dt_1 \left[A^{r}(t,t_1)B^{<}(t_1,t') + A^{<}(t,t_1)B^{a}(t_1,t')\right]. \tag{13.69}$$

This provides an expression for the lesser function in terms of functions of real time integrated over the real-time axis.

Similarly, we can obtain an expression for $C^{>}(t,t')$,

$$\begin{aligned}
C^{>}(t,t') &= C(t \in \overleftarrow{C}, t' \in \overrightarrow{C}) \equiv C(t^{\leftarrow}, t'^{\rightarrow}) = \int_C d\tau_1 A(t^{\leftarrow},\tau_1)B(\tau_1,t'^{\rightarrow}) \\
&= \int_{-\infty}^{\infty} dt_1 A(t^{\leftarrow}, t_1^{\rightarrow})B(t_1^{\rightarrow}, t'^{\rightarrow}) + \int_{\infty}^{-\infty} dt_1 A(t^{\leftarrow}, t_1^{\leftarrow})B(t_1^{\leftarrow}, t'^{\rightarrow}) \\
&= \int_{-\infty}^{\infty} dt_1 \left[A^{>}(t,t_1)B^{T}(t_1,t') - A^{\tilde{T}}(t,t_1)B^{>}(t_1,t')\right].
\end{aligned}$$

It follows from Eq. (13.47) that $B^{T} = B^{>} + B^{a}$ and $A^{\tilde{T}} = A^{>} - A^{r}$. Hence,

$$C^{>}(t,t') = \int_{-\infty}^{\infty} dt_1 \left[A^{r}(t,t_1)B^{>}(t_1,t') + A^{>}(t,t_1)B^{a}(t_1,t')\right]. \tag{13.70}$$

From Eq. (13.66), $C^{r}(t,t') = \theta(t-t')\left[C^{>}(t,t') - C^{<}(t,t')\right]$. Using Eqs (13.69) and (13.70), we obtain

$$\begin{aligned}
C^{r}(t,t') = \theta(t-t') \int_{-\infty}^{\infty} dt_1 \big\{ &\left[A^{>}(t,t_1) - A^{<}(t,t_1)\right] B^{a}(t_1,t') \\
&+ A^{r}(t,t_1)\left[B^{>}(t_1,t') - B^{<}(t_1,t')\right]\big\}.
\end{aligned}$$

Using Eqs (13.66) and (13.67), we can write

$$\begin{aligned}
C^{r}(t,t') &= \theta(t-t') \left\{ \int_{-\infty}^{t'} dt_1 \left[A^{>}(t,t_1) - A^{<}(t,t_1)\right]\left[B^{<}(t_1,t') - B^{>}(t_1,t')\right] \right. \\
&\quad \left. + \int_{-\infty}^{t} dt_1 \left[A^{>}(t,t_1) - A^{<}(t,t_1)\right]\left[B^{>}(t_1,t') - B^{<}(t_1,t')\right] \right\} \\
&= \theta(t-t') \int_{t'}^{t} dt_1 \left[A^{>}(t,t_1) - A^{<}(t,t_1)\right]\left[B^{>}(t_1,t') - B^{<}(t_1,t')\right].
\end{aligned}$$

Since $t > t'$, due to the step function $\theta(t-t')$, it follows that, in the above integrand, $t > t_1 > t'$; hence,

$$\begin{aligned}
C^{r}(t,t') = \int_{-\infty}^{\infty} & dt_1 \theta(t-t_1)\left[A^{>}(t,t_1) - A^{<}(t,t_1)\right] \\
&\times \theta(t_1-t')\left[B^{>}(t_1,t') - B^{<}(t_1,t')\right].
\end{aligned}$$

With the help of Eq. (13.66), the above relation reduces to

$$C^r(t,t') = \int_{-\infty}^{\infty} dt_1 A^r(t,t_1) B^r(t_1,t'). \tag{13.71}$$

A similar calculation for $C^a(t,t')$ yields

$$C^a(t,t') = \int_{-\infty}^{\infty} dt_1 A^a(t,t_1) B^a(t_1,t'). \tag{13.72}$$

We can adopt a simplified matrix notation to summarize our results:

$$C = AB \tag{13.73}$$

$$C^< = A^r B^< + A^< B^a \tag{13.74}$$

$$C^> = A^r B^> + A^> B^a \tag{13.75}$$

$$C^r = A^r B^r, \quad C^a = A^a B^a. \tag{13.76}$$

We now consider an expression with two integrations over contour time,

$$D(\tau,\tau') = \int_C d\tau_1 \int_C d\tau_2\, A(\tau,\tau_1) B(\tau_1,\tau_2) C(\tau_2,\tau'). \tag{13.77}$$

In matrix notation, $D = ABC$. Using Eqs (13.74) and (13.76),

$$\begin{aligned} D^< &= A^r(BC)^< + A^<(BC)^a \\ &= A^r B^r C^< + A^r B^< C^a + A^< B^a C^a. \end{aligned} \tag{13.78}$$

Similarly, we derive the following equations:

$$D^> = A^r B^r C^> + A^r B^> C^a + A^> B^a C^a \tag{13.79}$$

$$D^r = A^r B^r C^r, \quad D^a = A^a B^a C^a. \tag{13.80}$$

13.8 Keldysh equations

Applying Langreth rules from Eqs (13.76) and (13.80) to the two forms of Dyson's equation, (13.63) and (13.64), we obtain

$$G^{R,A} = G^{0\,R,A}\left(1 + UG^{R,A} + \Sigma^{*R,A} G^{R,A}\right) \tag{13.81a}$$

$$G^{R,A} = \left(1 + G^{R,A} U + G^{R,A} \Sigma^{*R,A}\right) G^{0\,R,A}. \tag{13.81b}$$

Note that since U depends on only one time, it is neither retarded nor advanced. Although the above equations look like Dyson's equation for the equilibrium Green's function, there is a subtle distinction: $\Sigma^{*R,A}$ depends not only on $G^{0\,R,A}$, but also on $G^<$ and $G^>$. We rearrange the first of these equations as follows:

$$G^{0R,A} = \left(1 - G^{0R,A}U - G^{0R,A}\Sigma^{*R,A}\right)G^{R,A}.$$

Putting this expression for $G^{0R,A}$ into Eq. (13.81b), we obtain

$$\left(1 + G^{R,A}U + G^{R,A}\Sigma^{*R,A}\right)\left(1 - G^{0R,A}U - G^{0R,A}\Sigma^{*R,A}\right) = 1. \tag{13.82}$$

If the rules presented in Eqs (13.74) and (13.78) are now applied to the first form of Dyson's equation, (13.63), the result is

$$G^< = G^{0<}\left(1 + UG^A + \Sigma^{*A}G^A\right) + G^{0R}\Sigma^{*<}G^A + \left(G^{0R}U + G^{0R}\Sigma^{*R}\right)G^<.$$

The same expression holds for $G^>$ if we replace $<$ with $>$ everywhere. Rearranging terms, the above expression for $G^<$ is written as

$$\left(1 - G^{0R}U - G^{0R}\Sigma^{*R}\right)G^< = G^{0<}\left(1 + UG^A + \Sigma^{*A}G^A\right) + G^{0R}\Sigma^{*<}G^A.$$

Multiplying by $\left(1 + G^RU + G^R\Sigma^{*R}\right)$ on the left, and using Eqs (13.81b) and (13.82), we obtain

$$G^< = \left(1 + G^RU + G^R\Sigma^{*R}\right)G^{0<}\left(1 + UG^A + \Sigma^{*A}G^A\right) + G^R\Sigma^{*<}G^A. \tag{13.83}$$

Similarly, we find, for the greater function,

$$G^> = \left(1 + G^RU + G^R\Sigma^{*R}\right)G^{0>}\left(1 + UG^A + \Sigma^{*A}G^A\right) + G^R\Sigma^{*>}G^A. \tag{13.84}$$

Equations (13.83) and (13.84) are the Keldysh equations for the lesser and greater functions.

13.9 Steady-state transport

We now turn our attention to the application of the nonequilibrium Green's function to transport in a system consisting of a small structure, such as a quantum dot, connected to two metallic leads (see Figure 13.1). The nonequilibrium problem is formulated as follows. Initially, the left lead, the dot, and the right lead are separated, and each is in equilibrium at its own chemical potential. Without any loss of generality, we assume that the chemical potential in the left lead is larger than that in the right lead: $\mu_L > \mu_R$. The statistical operator for the system is simply the direct product of the equilibrium statistical operators of the system's three separate components:

$$\rho = \rho_L^{eq} \otimes \rho_D^{eq} \otimes \rho_R^{eq}. \tag{13.85}$$

At time t_0, the components are brought into contact, and a coupling between the dot and the two leads is established, allowing electrons to tunnel from the leads to the dot and vice versa. The perturbation that drives the system out of equilibrium is the coupling between the dot and the two leads.

We choose to analyze the problem in the setting described above, although other choices are possible. For example, the three components may be initially in contact and in equilibrium at a common chemical potential. In this approach, the initial statistical operator is $e^{-\beta(H-\mu N)}/Tr\, e^{-\beta(H-\mu N)}$, where H is the Hamiltonian that includes the coupling between the dot and the leads, μ is the common chemical potential, and N is the number of particles operator for the whole system. The perturbation that drives the system out of equilibrium is the increase in the chemical potential of the left lead due to an applied bias voltage (Cini, 1980; Stefanucci and Almbladh, 2004). An increase in the chemical potential by Δ means an increase, in the amount of Δ, of the energy of each single-particle state in the left lead.The first approach is simpler, since our purpose is to study steady-state transport across the quantum dot.

13.9.1 Model Hamiltonian

The Hamiltonian for the system, consisting of the left lead, the right lead, and the dot, is written as

$$H = H_L + H_R + H_D + H_T. \tag{13.86}$$

H_L and H_R are the Hamiltonians for the left and right leads, respectively,

$$H_\alpha = \sum_{k\sigma} \epsilon_{k\alpha} c^\dagger_{k\sigma\alpha} c_{k\sigma\alpha}, \qquad \alpha = L, R \tag{13.87}$$

where σ is the spin projection, and k is a collective index representing the spatial quantum numbers of the electronic states in the leads. In writing H_α, we have assumed that the electrons in the leads are noninteracting, except for a possible average interaction which can be taken into account through a renormalization of the single-particle state energies ϵ_k. Neglecting correlations in metals generally yields a good approximation, especially for simple metals. The term H_D is the Hamiltonian for the dot,

$$H_D = H_D(\{d^\dagger_{n\sigma}\}, \{d_{n\sigma}\}). \tag{13.88}$$

H_D is expressed in terms of creation ($d^\dagger_{n\sigma}$) and annihilation ($d_{n\sigma}$) operators associated with single-particle states in the dot. These states are characterized by n and σ. Again, n is a collective index that stands for the spatial quantum numbers of the electronic states in the dot. Various model Hamiltonians for the dot may be chosen.

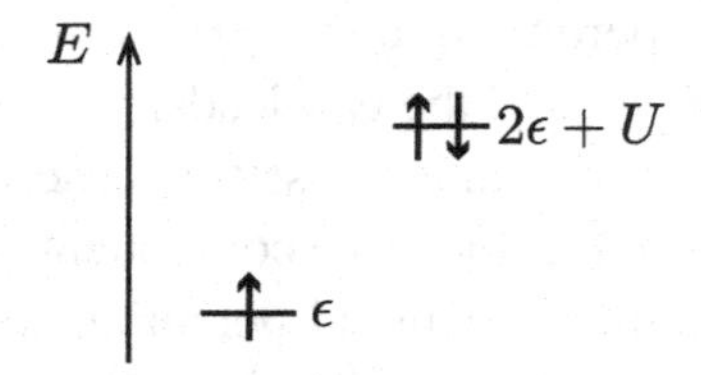

Figure 13.9 In the Anderson impurity model, the dot has only one energy level, and it can be occupied by up to two electrons. If there is only one electron, the energy of the dot is ϵ; if the level is doubly occupied, the energy of the dot is $2\epsilon + U$. The two electrons that occupy the level must necessarily have opposite spin projections.

The simplest model describes the dot in terms of noninteracting electrons,

$$H_D = \sum_{n\sigma} \epsilon_n d^\dagger_{n\sigma} d_{n\sigma}. \tag{13.89}$$

This model is used to describe resonant tunneling through a quantum dot. Another model is the Anderson impurity model (Anderson, 1961); here, it is assumed that the dot has only one level of energy ϵ such that

$$H_D = \epsilon \sum_{\sigma} d^\dagger_\sigma d_\sigma + U n_\uparrow n_\downarrow \tag{13.90}$$

where n_σ ($\sigma = \uparrow, \downarrow$) is the operator that represents the number of electrons in the level with spin projection σ. If one electron occupies this level, the energy of the dot is ϵ. However, if two electrons occupy the level, one with spin up and the other with spin down, the energy of the dot is $2\epsilon + U$, where $U > 0$ is the Coulomb repulsion energy of the two electrons (see Figure 13.9). Other model Hamiltonians for the dot may be considered; e.g., one may be formulated that includes the interaction between electrons and atomic vibrations in the dot.

The coupling between the dot and the leads is given by the last term in the Hamiltonian, H_T. The coupling is represented by terms that describe tunneling of electrons from the dot to the leads, and vice versa:

$$H_T = \sum_{k\sigma n} \sum_{\alpha=L,R} \left(V_{k\sigma\alpha,n\sigma} c^\dagger_{k\sigma\alpha} d_{n\sigma} + V^*_{k\sigma\alpha,n\sigma} d^\dagger_{n\sigma} c_{k\sigma\alpha} \right). \tag{13.91}$$

$V_{k\sigma\alpha,n\sigma}$ is the matrix element for the tunneling of an electron from state $|n\sigma\rangle$ in the dot into state $|k\sigma\rangle$ in lead α; it is determined by first-principles calculations, but here we take it as a known quantity. The second term in H_T is the hermitian conjugate of the first term, and it describes tunneling from the leads into the dot. It is assumed that in tunneling between the leads and the dot, an electron maintains its spin orientation. The equilibrium Hamiltonian is $H_L + H_R + H_D$, while H_T is the perturbation that drives the system out of equilibrium.

In considering the model Hamiltonian for the three-component system, we assume that no direct coupling exists between the left and right leads. Furthermore, we assume that the creation and annihilation operators in the Fock space of one component anticommute with the operators in the Fock space of another component.

13.9.2 Expression for the current

The electron current from the left lead into the dot is determined by the rate of change in the number of electrons in the left lead:

$$I_L(t) = -e\langle dN_L/dt\rangle \equiv -e\langle \dot{N}_L\rangle. \tag{13.92}$$

The operator N_L represents the number of electrons in the left lead,

$$N_L = \sum_{k\sigma} c^\dagger_{k\sigma L} c_{k\sigma L}. \tag{13.93}$$

Since $i\hbar\dot{N}_L = [N_L, H]$, and N_L commutes with H_L, H_R, and H_D,

$$I_L = (ie/\hbar)\langle [N_L, H_T]\rangle. \tag{13.94}$$

We can calculate the commutator

$$[N_L, H_T] = \left[\sum_{k'\sigma'} c^\dagger_{k'\sigma' L} c_{k'\sigma' L}, \sum_{k\sigma n}\sum_{\alpha=L,R}\left(V_{k\sigma\alpha,n\sigma} c^\dagger_{k\sigma\alpha} d_{n\sigma} + V^*_{k\sigma\alpha,n\sigma} d^\dagger_{n\sigma} c_{k\sigma\alpha}\right)\right]$$

by using $[AB, CD] = A\{B, C\}D - AC\{B, D\} + \{A, C\}DB - C\{A, D\}B$, $\{c^\dagger_{k'\sigma'\alpha'}, c^\dagger_{k\sigma\alpha}\} = \{c_{k'\sigma'\alpha'}, c_{k\sigma\alpha}\} = 0$, and $\{c_{k'\sigma'\alpha'}, c^\dagger_{k\sigma\alpha}\} = \delta_{kk'}\delta_{\sigma\sigma'}\delta_{\alpha\alpha'}$. The result is

$$I_L(t) = (ie/\hbar)\sum_{k\sigma n}\left\{V_{k\sigma L,n\sigma}\left\langle c^\dagger_{k\sigma L}(t) d_{n\sigma}(t)\right\rangle - V^*_{k\sigma L,n\sigma}\left\langle d^\dagger_{n\sigma}(t) c_{k\sigma L}(t)\right\rangle\right\}.$$

This expression motivates the definition of the mixed lesser functions

$$G^<_{n,kL}(t, t'; \sigma) = i\left\langle c^\dagger_{k\sigma L}(t') d_{n\sigma}(t)\right\rangle \tag{13.95}$$

$$G^<_{kL,n}(t, t'; \sigma) = i\left\langle d^\dagger_{n\sigma}(t') c_{k\sigma L}(t)\right\rangle. \tag{13.96}$$

In terms of these functions, the expression for the current is

$$I_L(t) = (e/\hbar)\sum_{k\sigma n}\left\{V_{k\sigma L,n\sigma} G^<_{n,kL}(t, t; \sigma) - V^*_{k\sigma L,n\sigma} G^<_{kL,n}(t, t; \sigma)\right\}.$$

For any two operators A and B, $\langle AB\rangle^* = \langle (AB)^\dagger\rangle = \langle B^\dagger A^\dagger\rangle$. Therefore,

$$G^<_{kL,n}(t, t; \sigma) = -\left[G^<_{n,kL}(t, t; \sigma)\right]^*. \tag{13.97}$$

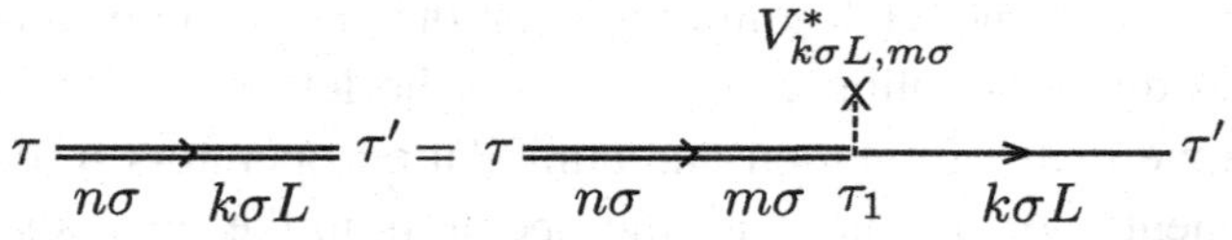

$$\tau \Longrightarrow \tau' = \tau \Longrightarrow \overset{V^*_{k\sigma L,m\sigma}}{\times} \longrightarrow \tau'$$
$$n\sigma \quad k\sigma L \qquad n\sigma \quad m\sigma \; \tau_1 \quad k\sigma L$$

Figure 13.10 Graphical representation of the Dyson-like equation for the contour Green's function $G_{n,kL}(\tau, \tau'; \sigma)$.

We can thus write

$$I_L(t) = (2e/\hbar)\mathrm{Re}\left[\sum_{k\sigma n} V_{k\sigma L,n\sigma} G^<_{n,kL}(t, t; \sigma)\right]. \tag{13.98}$$

To determine $G^<_{n,kL}(t, t; \sigma)$, we first calculate the contour Green's function $G_{n,kL}(\tau, \tau'; \sigma) = -i\langle T_C d_{n\sigma}(\tau)c^\dagger_{k\sigma L}(\tau')\rangle$, then apply Langreth rules. Recall that the unperturbed Hamiltonian is $H_L + H_R + H_D$, while the perturbation is H_T. In the absence of H_T, the contour Green's function $G^0_{n,kL}(\tau, \tau'; \sigma)$ vanishes, i.e., $\langle T_C d_{n\sigma}(\tau)c^\dagger_{k\sigma L}(\tau')\rangle_0 = 0$. A Dyson-like equation for the contour Green's function is shown in Figure 13.10.

The term $\sum_{k\sigma n} V_{k\sigma L,n\sigma} c^\dagger_{k\sigma L} d_{n\sigma}$ in H_T does not contribute to the contour function; $G_{n,kL}(\tau, \tau'; \sigma)$ contains $c^\dagger_{k\sigma L}$, whose contraction with $c^\dagger_{k'\sigma' L}$ is equal to zero (unless the left lead is a superconductor), and the contraction of $c^\dagger_{k\sigma L}$ with $d_{n\sigma}$ also gives zero: $\langle T_C d_{n\sigma} c^\dagger_{k\sigma L}\rangle_0 = 0$. The algebraic expression for the mixed contour function can be read off Figure 13.10:

$$G_{n,kL}(\tau, \tau'; \sigma) = \frac{1}{\hbar}\sum_m \int_C d\tau_1 G_{nm}(\tau, \tau_1; \sigma) V^*_{k\sigma L,m\sigma} G^0_{kL}(\tau_1, \tau'; \sigma) \tag{13.99}$$

where $G^0_{kL}(\tau_1, \tau'; \sigma)$ is the noninteracting contour Green's function of the electrons in the left lead, and $G_{nm}(\tau, \tau_1; \sigma)$ is the contour Green's function of the electrons in the dot:

$$G^0_{kL}(\tau_1, \tau'; \sigma) = -i\left\langle T_C\; c_{k\sigma L}(\tau)c^\dagger_{k\sigma L}(\tau')\right\rangle_0 \tag{13.100}$$

$$G_{nm}(\tau, \tau_1; \sigma) = -i\left\langle T_C\; d_{n\sigma}(\tau)d^\dagger_{m\sigma}(\tau')\right\rangle. \tag{13.101}$$

Alternatively, Eq. (13.99) can be derived by expanding $G_{n,kL}(\tau, \tau'; \sigma)$ in a perturbation series and applying Wick's theorem to contract the creation and annihilation operators of the electrons in the left lead (this is made possible by assuming that the leads contain noninteracting electrons).

We now apply the Langreth rule, Eq. (13.69), to obtain the lesser function

$$G^{<}_{n,kL}(t,t';\sigma) = \frac{1}{\hbar}\sum_m \int_{-\infty}^{\infty} dt_1 \left[G^R_{nm}(t,t_1;\sigma)G^{0\,<}_{kL}(t_1,t';\sigma) + G^{<}_{nm}(t,t_1;\sigma)G^{0\,A}_{kL}(t_1,t';\sigma)\right] V^*_{k\sigma L,m\sigma}. \tag{13.102}$$

As indicated earlier, our interest is in studying the system in the steady state, i.e., at times long after the moment when the perturbation is switched on. In this state, the current is independent of time, and all Green's functions depend on the difference between their time arguments. We can then Fourier transform the various functions that appear in Eq. (13.102):

$$G^R_{nm}(t,t_1;\sigma) = G^R_{nm}(t-t_1;\sigma) = \frac{1}{2\pi}\int_{-\infty}^{\infty} d\omega G^R_{nm}(\omega;\sigma)e^{-i\omega(t-t_1)}. \tag{13.103}$$

Similar expressions are written for the other functions that appear on the RHS of Eq. (13.102). Using the relation

$$\int_{-\infty}^{\infty} dt e^{i(\omega-\omega')t} = 2\pi\delta(\omega-\omega'),$$

we obtain

$$G^{<}_{n,kL}(t,t';\sigma) = \frac{1}{2\pi\hbar}\sum_m \int_{-\infty}^{\infty} d\omega e^{-i\omega(t-t')} V^*_{k\sigma L,m\sigma} \times \left[G^R_{nm}(\omega;\sigma)G^{0\,<}_{kL}(\omega;\sigma) + G^{<}_{nm}(\omega;\sigma)G^{0\,A}_{kL}(\omega;\sigma)\right]. \tag{13.104}$$

Setting $t'=t$ gives us $G^{<}_{n,kL}(t,t;\sigma)$. Inserting this into Eq. (13.98), we find

$$I_L = \frac{e}{\pi\hbar^2}\int_{-\infty}^{\infty} d\omega \mathrm{Re} \sum_{nm}\sum_{k\sigma} V_{k\sigma L,n\sigma}V^*_{k\sigma L,m\sigma} \times \left[G^R_{nm}(\omega;\sigma)G^{0\,<}_{kL}(\omega;\sigma) + G^{<}_{nm}(\omega;\sigma)G^{0\,A}_{kL}(\omega;\sigma)\right]. \tag{13.105}$$

For the left lead

$$G^{0\,<}_{kL}(\omega;\sigma) = 2\pi i f_L(\omega)\delta(\omega - \epsilon_{k\sigma L}/\hbar) \tag{13.106}$$

$$G^{0\,A}_{kL}(\omega;\sigma) = (\omega - \epsilon_{k\sigma L}/\hbar - i0^+)^{-1} \tag{13.107}$$

where $f_L(\omega)$ is the Fermi function in the left lead (see Problem 6.6 and Eq. [6.56]). We also note that

$$G^{<}_{nm}(t-t';\sigma) = i\langle d^\dagger_m(t')d_n(t)\rangle = \frac{1}{2\pi}\int_{-\infty}^{\infty} d\omega e^{-i\omega(t-t')}G^{<}_{nm}(\omega;\sigma)$$

and that

$$\frac{1}{2\pi}\int_{-\infty}^{\infty} d\omega e^{-i\omega(t'-t)} G^{<}_{mn}(\omega;\sigma) = G^{<}_{mn}(t'-t;\sigma) = i\langle d^{\dagger}_{n\sigma}(t) d_{m\sigma}(t')\rangle$$

$$= -\left[i\langle d^{\dagger}_{m\sigma}(t') d_{n\sigma}(t)\rangle\right]^{*} = -\left[G^{<}_{nm}(t-t';\sigma)\right]^{*}$$

$$= -\frac{1}{2\pi}\left[\int_{-\infty}^{\infty} d\omega e^{-i\omega(t-t')} G^{<}_{nm}(\omega;\sigma)\right]^{*} = -\frac{1}{2\pi}\int_{-\infty}^{\infty} d\omega e^{-i\omega(t'-t)} \left[G^{<}_{nm}(\omega;\sigma)\right]^{*}.$$

We thus conclude that

$$G^{<}_{mn}(\omega;\sigma) = -\left[G^{<}_{nm}(\omega;\sigma)\right]^{*}. \tag{13.108}$$

A similar calculation yields

$$G^{R}_{mn}(\omega;\sigma) = \left[G^{A}_{nm}(\omega;\sigma)\right]^{*}. \tag{13.109}$$

Using Eqs (13.106–13.109), the following expression for the current is derived:

$$I_L = \frac{ie}{2\pi\hbar}\sum_{\sigma}\int_{-\infty}^{\infty} d\omega Tr\{\Gamma^{L}(\omega;\sigma)[f_L(\omega)(G^{R}(\omega;\sigma) - G^{A}(\omega;\sigma)) + G^{<}(\omega;\sigma)]\} \tag{13.110}$$

where the level-width function $\Gamma^{L}(\omega;\sigma)$ is given by

$$\Gamma^{L}_{mn}(\omega;\sigma) = \frac{2\pi}{\hbar}\sum_{k} V^{*}_{k\sigma L,m\sigma} V_{k\sigma L,n\sigma}\delta(\omega - \epsilon_{k\sigma L}/\hbar). \tag{13.111}$$

In Eq. (13.110), a matrix notation is adopted: Γ^{L}, G^{R}, G^{A}, and $G^{<}$ are matrices with matrix elements Γ^{L}_{nm}, G^{R}_{nm}, G^{A}_{nm}, and $G^{<}_{nm}$, respectively. The product is a matrix product, and the trace of the resulting matrix is taken. Below, a derivation of the current formula is given.

We rewrite Eq. (13.105) as follows:

$$I_L = \frac{e}{\pi\hbar^2}\int_{-\infty}^{\infty} d\omega X(\omega)$$

$$X = \mathrm{Re}\sum_{nm}\sum_{k\sigma} V_{k\sigma L,n\sigma} V^{*}_{k\sigma L,m\sigma}\left[G^{R}_{nm}(\omega;\sigma) G^{0<}_{kL}(\omega;\sigma) + G^{<}_{nm}(\omega;\sigma) G^{0A}_{kL}(\omega;\sigma)\right]$$

The real part of a complex number z is $(z+z^{*})/2$; hence,

$$X = \frac{1}{2}\sum_{nm}\sum_{k\sigma}\{V_{k\sigma L,n\sigma} V^{*}_{k\sigma L,m\sigma}\left(G^{R}_{nm} G^{0<}_{kL} + G^{<}_{nm} G^{0A}_{kL}\right)$$

$$+ V^{*}_{k\sigma L,n\sigma} V_{k\sigma L,m\sigma}\left(G^{R*}_{nm} G^{0<*}_{kL} + G^{<*}_{nm} G^{0A*}_{kL}\right)\}.$$

The arguments of the functions have been suppressed for now. Using Eqs (13.108) and (13.109),

$$X = \frac{1}{2}\sum_{nm}\sum_{k\sigma}\left\{V_{k\sigma L,n\sigma}V^*_{k\sigma L,m\sigma}\left(G^R_{nm}G^{0<}_{kL} + G^<_{nm}G^{0A}_{kL}\right)\right.$$
$$\left. - V^*_{k\sigma L,n\sigma}V_{k\sigma L,m\sigma}\left(G^A_{mn}G^{0<}_{kL} + G^<_{mn}G^{0R}_{kL}\right)\right\}.$$

Interchanging n and m in the second term,

$$X = \frac{1}{2}\sum_{nm}\sum_{k\sigma}V_{k\sigma L,n\sigma}V^*_{k\sigma L,m\sigma}\left[\left(G^R_{nm} - G^A_{nm}\right)G^{0<}_{kL} + G^<_{nm}\left(G^{0A}_{kL} - G^{0R}_{kL}\right)\right].$$

The zeroth-order functions that appear in the above expression are known,

$$G^{0A}_{kL} - G^{0R}_{kL} = \left(\omega - \epsilon_{k\sigma L}/\hbar - i0^+\right)^{-1} - \left(\omega - \epsilon_{k\sigma L}/\hbar + i0^+\right)^{-1}$$
$$= 2\pi i\delta(\omega - \epsilon_{k\sigma L}/\hbar)$$
$$G^{0<}_{kL} = 2\pi i f_L(\omega)\delta(\omega - \epsilon_{k\sigma L}/\hbar).$$

Inserting these into the expression for X,

$$X(\omega) = \frac{i}{2}\sum_{nm}\sum_{k\sigma}V_{k\sigma L,n\sigma}V^*_{k\sigma L,m\sigma}2\pi\delta(\omega - \epsilon_{k\sigma L}/\hbar)$$
$$\times\left\{f_L(\omega)\left[G^R_{nm}(\omega;\sigma) - G^A_{nm}(\omega;\sigma)\right] + G^<_{nm}(\omega;\sigma)\right\}.$$

We now introduce the level-width function (a matrix),

$$\Gamma^L_{mn}(\omega;\sigma) = \frac{1}{\hbar}\sum_k V^*_{k\sigma L,m\sigma}V_{k\sigma L,n\sigma}2\pi\delta(\omega - \epsilon_{k\sigma L}/\hbar).$$

The expression for $X(\omega)$ reduces to

$$X(\omega) = \frac{i\hbar}{2}\sum_{\sigma nm}\Gamma^L_{mn}(\omega;\sigma)\left\{f_L(\omega)\left[G^R_{nm}(\omega;\sigma) - G^A_{nm}(\omega;\sigma)\right] + G^<_{nm}(\omega;\sigma)\right\}$$
$$= \frac{i\hbar}{2}\sum_\sigma Tr\left\{\Gamma^L(\omega;\sigma)\left[f_L(\omega)\left(G^R(\omega;\sigma) - G^A(\omega;\sigma)\right) + G^<(\omega;\sigma)\right]\right\}.$$

The expression for the current, Eq. (13.110), immediately follows.

The current from the right lead to the central dot has exactly the same expression as I_L except that R replaces L. In the steady state, $I = I_L = -I_R$. A symmetrical expression for the current is obtained by writing $I = (I_L - I_R)/2$; it is given by

$$I = \frac{ie}{4\pi\hbar}\sum_\sigma\int_{-\infty}^{\infty}d\omega Tr\left\{\left[f_L(\omega)\Gamma^L(\omega;\sigma) - f_R(\omega)\Gamma^R(\omega;\sigma)\right]\left[G^R(\omega;\sigma) - G^A(\omega;\sigma)\right]\right.$$
$$\left. + \left[\Gamma^L(\omega;\sigma) - \Gamma^R(\omega;\sigma)\right]G^<(\omega;\sigma)\right\}. \qquad (13.112)$$

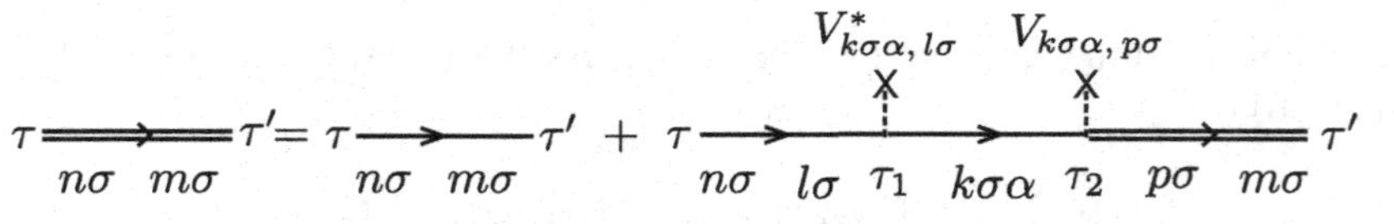

$V^*_{k\sigma\alpha,\,l\sigma}$ $V_{k\sigma\alpha,\,p\sigma}$

$\tau \Longrightarrow \tau' = \tau \longrightarrow \tau' + \tau \longrightarrow \tau_1 \longrightarrow \tau_2 \Longrightarrow \tau'$

$n\sigma\ \ m\sigma \qquad n\sigma\ \ m\sigma \qquad n\sigma\ \ l\sigma\ \ \tau_1\ \ k\sigma\alpha\ \ \tau_2\ \ p\sigma\ \ m\sigma$

Figure 13.11 Graphical representation of Dyson's equation for the contour Green's function $G_{nm}(\tau, \tau'; \sigma)$ of a quantum dot in contact with two metal leads.

This is the Meir–Wingreen formula (Meir and Wingreen, 1992) for the current in the steady state; it expresses the current in terms of the Green's functions of the dot that is in the central region, between the two leads. In general, the calculation of these functions is highly nontrivial.

A simplification is possible in the special case of proportional coupling, when the left and right level-width functions are proportional: $\Gamma^L(\omega) = \lambda\Gamma^R(\omega)$. Since $I = I_L = -I_R$, we can write $I = xI_L - (1-x)I_R$ for an arbitrary x. The current is then given by

$$I = \frac{ie}{2\pi\hbar}\sum_\sigma \int_{-\infty}^{\infty} d\omega$$
$$\times\, Tr\left[\Gamma^R\left\{[\lambda x f_L(\omega) - (1-x)f_R(\omega)]\left[G^R - G^A\right] + [\lambda x - (1-x)]\,G^<\right\}\right].$$

The arbitrary parameter x is now fixed so as to eliminate the term multiplying $G^<: x = 1/(1+\lambda)$; then,

$$I = \frac{ie}{2\pi\hbar}\sum_\sigma \int_{-\infty}^{\infty} d\omega\,[f_L(\omega) - f_R(\omega)]$$
$$\times\, Tr\left\{\frac{\Gamma^L(\omega;\sigma)\Gamma^R(\omega;\sigma)}{\Gamma^L(\omega;\sigma) + \Gamma^R(\omega;\sigma)}\left[G^R(\omega;\sigma) - G^A(\omega;\sigma)\right]\right\}. \tag{13.113}$$

The ratio of the coupling matrices is well defined, since Γ^L and Γ^R are proportional. The condition of proportional coupling does not generally hold, but it may be a reasonable approximation if the level-width functions have a weak dependence on energy.

13.10 Noninteracting quantum dot

In the case of a noninteracting quantum dot, the dot Hamiltonian is

$$H_D = \sum_{n\sigma} \epsilon_n d^\dagger_{n\sigma} d_{n\sigma}. \tag{13.114}$$

To evaluate the current, given by Eq. (13.110) or Eq. (13.112), we need $G^R - G^A$ and $G^<$. These are obtained from the contour Green's function, which, in turn, is determined by Dyson's equation; this is shown graphically in Figure (13.11).

Algebraically, Dyson's equation reads

$$G_{nm}(\omega;\sigma) = G^0_{nm}(\omega;\sigma) + \sum_{l,p} G^0_{nl}(\omega;\sigma)\Sigma_{lp}(\omega;\sigma)G_{pm}(\omega;\sigma) \tag{13.115}$$

where the proper (irreducible) self energy is given by

$$\Sigma_{lp}(\omega;\sigma) = \frac{1}{\hbar^2}\sum_{k,\alpha=L,R} V^*_{k\sigma\alpha,l\sigma} G^0_{k\alpha}(\omega;\sigma)V_{k\sigma\alpha,p\sigma}. \tag{13.116}$$

The superscript "*" on Σ is dropped in this section and in the next one. We note that $G^0_{nm}(\omega;\sigma) = G^0_{nn}(\omega;\sigma)\delta_{nm}$. In matrix notation, Dyson's equation is written as

$$G = G^0 + G^0\Sigma G \tag{13.117}$$

where G^0 is a diagonal matrix. Applying the Langreth rule, Eq. (13.80),

$$G^R = G^{0R} + G^{0R}\Sigma^R G^R, \quad G^A = G^{0A} + G^{0A}\Sigma^A G^A \tag{13.118}$$

where G^{0R} and G^{0A} are diagonal matrices, and

$$\hbar^2\Sigma^{R,A}_{lp}(\omega;\sigma) = \sum_{k\alpha} V^*_{k\sigma\alpha,l\sigma} G^{0R,A}_{k\alpha}(\omega;\sigma)V_{k\sigma\alpha,p\sigma}. \tag{13.119}$$

The self energy and the level-width function are related,

$$\begin{aligned}
\hbar\left[\Sigma^R_{lp}(\omega;\sigma) - \Sigma^A_{lp}(\omega;\sigma)\right] &= \frac{1}{\hbar}\sum_{k\alpha} V^*_{k\sigma\alpha,l\sigma}\left[G^{0R}_{k\alpha}(\omega;\sigma) - G^{0A}_{k\alpha}(\omega;\sigma)\right]V_{k\sigma\alpha,p\sigma} \\
&= \frac{1}{\hbar}\sum_{k\alpha}\left[-2\pi i\delta(\omega - \epsilon_{k\sigma\alpha}/\hbar)\right]V^*_{k\sigma\alpha,l\sigma}V_{k\sigma\alpha,p\sigma} \\
&= -i\left[\Gamma^L_{lp}(\omega;\sigma) + \Gamma^R_{lp}(\omega;\sigma)\right].
\end{aligned} \tag{13.120}$$

From Eq. (13.118) for G^R and G^A, the following expression is derived

$$G^R - G^A = -(i/\hbar)G^R\left(\Gamma^L + \Gamma^R\right)G^A. \tag{13.121}$$

A proof of this statement is provided in the shaded area below.

Consider Eq. (13.118):

$$G^R = G^{0R} + G^{0R}\Sigma^R G^R.$$

Multiply on the left by $\left(G^{0R}\right)^{-1}$ and on the right by $\left(G^R\right)^{-1}$; the result is

$$\left(G^{0R}\right)^{-1} = \left(G^R\right)^{-1} + \Sigma^R.$$

Similarly,

$$\left(G^{0A}\right)^{-1} = \left(G^A\right)^{-1} + \Sigma^A.$$

Since the matrices G^{0R} and G^{0A} are diagonal, $\left(G^{0R}\right)^{-1}$ and $\left(G^{0A}\right)^{-1}$ are also diagonal. Furthermore,

$$\left(G^{0R}\right)^{-1}_{nn} = \omega - \epsilon_n/\hbar + i0^+, \quad \left(G^{0A}\right)^{-1}_{nn} = \omega - \epsilon_n/\hbar - i0^+.$$

It follows that $\left(G^{0R}\right)^{-1} = \left(G^{0A}\right)^{-1}$. Therefore,

$$\left(G^R\right)^{-1} + \Sigma^R = \left(G^A\right)^{-1} + \Sigma^A.$$

Multiplying by G^R on the left and by G^A on the right, we obtain

$$G^A + G^R\Sigma^R G^A = G^R + G^R\Sigma^A G^A \Longrightarrow G^R - G^A = G^R\left(\Sigma^R - \Sigma^A\right)G^A.$$

Using Eq. (13.120) to replace $\left(\Sigma^R - \Sigma^A\right)$ by $(-i/\hbar)\left(\Gamma^L + \Gamma^R\right)$, Eq. (13.121) is obtained.

In order to obtain $G^<$, we use the Keldysh equation (13.83). Here, $U = 0$ (see Eq. [3.117]), and the Keldysh equation reduces to

$$G^< = \left(1 + G^R\Sigma^R\right)G^{0<}\left(1 + \Sigma^A G^A\right) + G^R\Sigma^< G^A. \tag{13.122}$$

From Eq. (13.81) we deduce that $1 + \Sigma^A G^A = \left(G^{0A}\right)^{-1} G^A$. Thus,

$$G^< = \left(1 + G^R\Sigma^R\right)G^{0<}\left(G^{0A}\right)^{-1} G^A + G^R\Sigma^< G^A. \tag{13.123}$$

Since $G^{0<}_{nm} = 2\pi i f(\omega)\delta(\omega - \epsilon_n/\hbar)\delta_{nm}$ and $\left(G^{0A}\right)^{-1}_{nm} = (\omega - \epsilon_n/\hbar - i0^+)\delta_{nm}$, it follows that $G^{0<}\left(G^{0A}\right)^{-1} = 0$. The Keldysh equation thus takes the simple form

$$G^< = G^R\Sigma^< G^A. \tag{13.124}$$

From the expression for Σ, Eq. (13.116), we find

$$\Sigma^<_{lp} = \frac{1}{\hbar^2}\sum_{k\alpha} V^*_{k\sigma\alpha,l\sigma} G^{0<}_{k\alpha} V_{k\sigma\alpha,p\sigma} = \frac{2\pi i}{\hbar^2}\sum_{\alpha} f_\alpha \sum_k V^*_{k\sigma\alpha,l\sigma}\delta(\omega - \epsilon_{k\sigma\alpha}/\hbar)V_{k\sigma\alpha,p\sigma}$$

$$= (i/\hbar)(f_L\Gamma^L + f_R\Gamma^R)_{lp}. \tag{13.125}$$

Therefore,

$$G^< = (i/\hbar)G^R\left(f_L\Gamma^L + f_R\Gamma^R\right)G^A. \tag{13.126}$$

Putting Eqs (13.121) and (13.126) into the equation for the current, Eq. (13.110), we obtain

$$I = \frac{e}{2\pi\hbar^2}\int_{-\infty}^{\infty} d\omega\,[f_L(\omega) - f_R(\omega)]\,T(\omega), \tag{13.127}$$

where

$$T(\omega) = \sum_{\sigma} Tr\left[\Gamma^L(\omega;\sigma)G^R(\omega;\sigma)\Gamma^R(\omega;\sigma)G^A(\omega;\sigma)\right]. \quad (13.128)$$

$(1/\hbar^2)T(\omega)$ is interpreted as a transmission probability. Equation (13.127) for the current is known as the Landauer formula (Landauer, 1957, 1970). We note that it is applicable only if the quantum dot is modeled as a noninteracting system.

13.11 Coulomb blockade in the Anderson model

We now consider the quantum dot to have a single energy level and onsite Coulomb repulsion, which is the Anderson impurity model. The dot Hamiltonian is

$$H_D = \sum_{\sigma} \epsilon d_\sigma^\dagger d_\sigma + U n_\uparrow n_\downarrow \quad (13.129)$$

where ϵ is the energy of the level, $U > 0$ is the onsite Coulomb repulsion, and $n_\uparrow (n_\downarrow)$ is the operator representing the number of electrons in the level with spin up (down). For simplicity, we assume that $\Gamma^L = \lambda\Gamma^R$, so that we can use Eq. (13.113) for the current. In this case, where there is only a single level in the quantum dot, the quantities Γ^L, Γ^R, G^R, and G^A are all scalars. From the spectral representations of G^R and G^A (see Eqs (6.36) and (6.37)) we deduce that $Re\, G^R = \mathrm{Re}\, G^A$ and $Im\, G^R = -Im\, G^A$; it follows that $G^R - G^A = 2i\, Im\, G^R$. The expression for the current thus takes the form:

$$I = \frac{-e}{\pi\hbar} \sum_{\sigma} \int_{-\infty}^{\infty} d\omega \left[f_L(\omega) - f_R(\omega)\right] \times$$

$$Tr\left\{\frac{\Gamma^L(\omega;\sigma)\Gamma^R(\omega;\sigma)}{\Gamma^L(\omega;\sigma) + \Gamma^R(\omega;\sigma)}\left[Im\; G^R(\omega;\sigma)\right]\right\}. \quad (13.130)$$

Here, $G^R(\omega;\sigma)$ is the Fourier transform of the retarded Green's function $G_\sigma^R(t)$ of the single-level quantum dot in the presence of coupling to the leads. $G_\sigma^R(t)$ is defined by

$$G_{d\sigma}^R(t) = -i\theta(t)\left\langle\left\{d_\sigma(t), d_\sigma^\dagger(0)\right\}\right\rangle. \quad (13.131)$$

Let us denote the retarded Green's function of the isolated quantum dot by D_σ^R. This function was calculated in Chapter 7 (see Eq. [7.11]); it is given by

$$D_\sigma^R(\omega) = \frac{1 - \langle n_{\bar{\sigma}}\rangle}{\omega - \epsilon/\hbar + i0^+} + \frac{\langle n_{\bar{\sigma}}\rangle}{\omega - (\epsilon + U)/\hbar + i0^+} \quad (13.132)$$

where $n_{\bar{\sigma}} = n_{-\sigma}$. This is the exact retarded Green's function for the isolated single-level quantum dot.

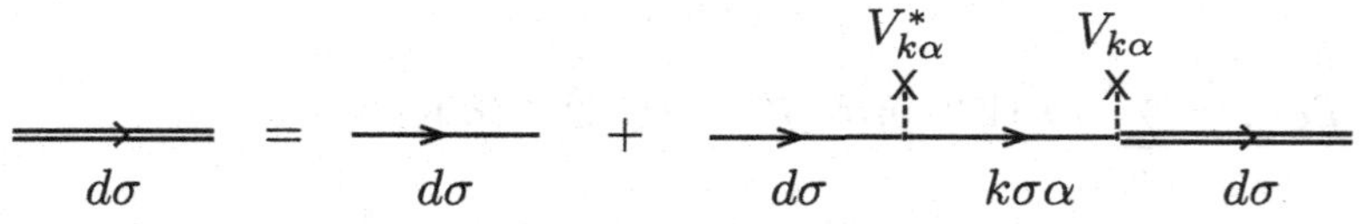

Figure 13.12 Dyson's equation for the contour Green's function of the quantum dot placed between two noninteracting metal leads.

Our next step is to calculate the retarded Green's function G_σ^R of the quantum dot in the presence of the tunneling Hamiltonian,

$$H_T = \sum_{k\sigma\alpha} \left(V_{k\alpha} c_{k\sigma\alpha}^\dagger d_\sigma + V_{k\alpha}^* d_\sigma^\dagger c_{k\sigma\alpha} \right). \tag{13.133}$$

We have assumed that the tunneling matrix elements are spin-independent; that assumption, along with the assumption of a single level in the dot, makes it superfluous to add any additional subscripts to $V_{k\alpha}$ and $V_{k\alpha}^*$.

The exact calculation of the dot's retarded Green's function is not possible, so we must resort to approximations. Our approach will be to treat the exact Green's function D_σ^R of the isolated dot as the unperturbed function $G_{d\sigma}^0(\omega)$, and then calculate a correction resulting from the inclusion of tunneling. The assumption made here is that the coupling between the dot and the leads is weak.

Dyson's equation for the contour Green's function of the dot, $G_{d\sigma}(\omega)$, is depicted graphically in Figure 13.12. We can thus write

$$G_{d\sigma}(\omega) = G_{d\sigma}^0(\omega) + G_{d\sigma}^0(\omega)\Sigma_\sigma(\omega)G_{d\sigma}(\omega) \tag{13.134}$$

where the proper self energy is given by

$$\Sigma_\sigma(\omega) = \frac{1}{\hbar^2} \sum_{k\alpha} |V_{k\alpha}|^2 \, G_{k\sigma\alpha}^0(\omega). \tag{13.135}$$

Applying Langreth's rule, Eq. (13.80), we find

$$\begin{aligned} G_{d\sigma}^R(\omega) &= G_{d\sigma}^{0R}(\omega) + G_{d\sigma}^{0R}(\omega)\Sigma_\sigma^R(\omega)G_{d\sigma}^R(\omega) \\ \Rightarrow G_{d\sigma}^R(\omega) &= \frac{1}{\left[G_{d\sigma}^{0R}(\omega)\right]^{-1} - \Sigma_\sigma^R(\omega)}. \end{aligned} \tag{13.136}$$

Replacing $G_{d\sigma}^{0R}(\omega)$ with the exact retarded Green's function of the isolated dot (see Eq. [13.132]), we obtain

$$G_{d\sigma}^R(\omega) = \frac{\hbar\omega - \epsilon - U + \langle n_{\bar\sigma}\rangle U}{(\omega - \epsilon/\hbar)(\hbar\omega - \epsilon - U) - \Sigma_\sigma^R(\omega)[\hbar\omega - \epsilon - U + \langle n_{\bar\sigma}\rangle U]}. \tag{13.137}$$

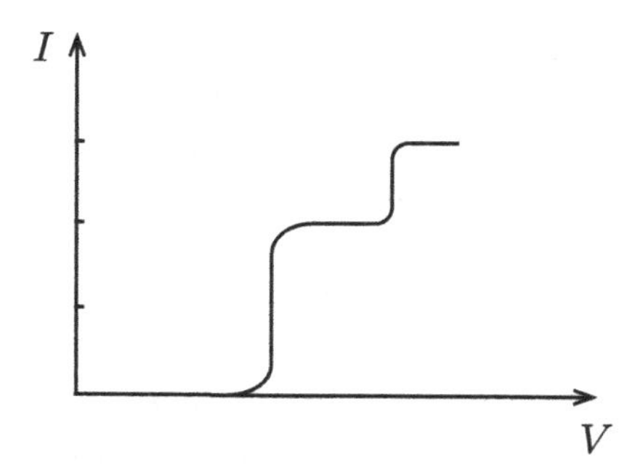

Figure 13.13 A plot of I vs V in the Coulomb blockade regime.

This is too complicated. As an approximation, we ignore the real part of Σ_σ^R, which only leads to a small shift in the energy level of the dot. If we also ignore the ω-dependence of Σ_σ^R, then, using Eqs (13.120) and (13.135), we obtain $\Sigma^R = -\Sigma^A = -i(\Gamma_L + \Gamma_R)/2\hbar$, where Γ_L and Γ_R are constants. However, the calculation of the current is still complicated because of the presence of $\langle n_{\bar{\sigma}} \rangle$ in the expression for $G_{d\sigma}^R(\omega)$; this is given by

$$\langle n_{\bar{\sigma}} \rangle = \langle d_{\bar{\sigma}}^\dagger d_{\bar{\sigma}} \rangle = -i \int \frac{d\omega}{2\pi} G_{d\bar{\sigma}}^<(\omega). \tag{13.138}$$

The lesser function is given by the Keldysh equation (13.83), which again reduces to the simple form:

$$G_{d\bar{\sigma}}^<(\omega) = G_{d\bar{\sigma}}^R(\omega)\Sigma_{\bar{\sigma}}^<(\omega)G_{d\bar{\sigma}}^A(\omega) \tag{13.139}$$

where

$$\Sigma_{\bar{\sigma}}^<(\omega) = \frac{1}{\hbar^2}\sum_{k\alpha} |V_{k\alpha}|^2\, G_{k\bar{\sigma}\alpha}^{0<}(\omega), \quad G_{k\bar{\sigma}\alpha}^{0<}(\omega) = 2\pi i f_\alpha(\omega)\delta(\omega - \epsilon_{k\alpha}/\hbar).$$

It follows that

$$\Sigma_{\bar{\sigma}}^<(\omega) = \frac{i}{\hbar}\sum_\alpha f_\alpha(\omega)\Gamma_\alpha \tag{13.140}$$

where the ω-dependence in Γ_L and Γ_R is ignored. We thus see that $\langle n_{\bar{\sigma}} \rangle$ depends on the retarded Green's function of the quantum dot, $G_{d\bar{\sigma}}^R$, which, in turn, depends on n_σ. A self-consistent solution for $G_{d\sigma}^R$ is called for, using numerical techniques. If such a calculation is carried out, using reasonable values for the various parameters, a plot of the current I versus the applied bias voltage V, as in Figure 13.13, will be obtained (Pals and Mackinnon, 1996; Swirkowicz et al., 2004; Zimbovskaya, 2008).

A qualitative understanding of how the current varies with bias voltage is gained by appealing to the so-called Coulomb blockade model. Current flow requires the addition of an electron to the dot between the metal leads. Adding an electron to the dot costs an energy of $e^2/2C$, where C is the capacitance between the dot and the leads. Since the quantum dot is of nanoscale size, C is extremely small, and $e^2/2C$ may attain a large value. There is thus an energy barrier, called a Coulomb blockade, to the flow of current through the quantum dot: no current flow takes place unless the bias voltage exceeds a threshold value. A plateau in the I–V plot results from the fact that the energy barrier to the flow of two electrons is larger than the barrier to the flow of one electron (Kastner, 1993).

The first step in the I–V plot corresponds to the addition of one electron to the dot, while the second step results from the addition of two electrons. The first step is twice as large as the second step: the first added electron may have its spin up or down, but the second added electron can be of only one spin orientation, the one opposite to that of the first added electron.

Further reading

Datta, S. (2005). *Quantum Transport: Atom to Transistor*. Cambridge: Cambridge University Press.

Di Ventra, M. (2008). *Electrical Transport in Nanoscale Systems*. Cambridge: Cambridge University Press.

Haug, H. and Jauho. A.-P. (1996). *Quantum Kinetics in Transport and Optics in Semiconductors*. Berlin: Springer.

Rammer, J. and Smith, H. (1986). *Quantum Field-Theoretical Methods in Transport Theory of Metals*. Reviews of Modern Physics **58**, 323–359.

Problems

13.1 *Evolution operator*. Derive Eq. (13.7).

13.2 *Properties of the evolution operator*. Verify Eq. (13.8).

13.3 *Ensemble average of operators*. Prove Eq. (13.11).

13.4 *Scattering matrix*. Show that $S(t, t')$ satisfies the properties given in Eq. (13.26).

13.5 *Relations among G-functions*. Derive Eq. (13.47).

13.6 *G_c in terms of interaction picture operators*. Prove the validity of Eq. (13.49) for the case when t' lies further along the contour than t.

13.7 *Langreth's rule for the advanced function*. Derive Eq. (13.72).

13.8 *The Keldysh equation for the greater function*. Derive Eq. (13.84).

13.9 *Retarded and advanced functions.* Prove the validity of Eq. (13.109) which relates the retarded and advanced Green's functions.

13.10 *Conductance.* Using the Landauer formula for the current, show that the low-temperature, zero-bias conductance, defined by $G = dI/dV|_{V=0}$, is given by

$$G = \frac{e^2}{2\pi\hbar^3} \sum_n \tau_n(E_F).$$

E_F is the Fermi energy, and τ_n's are the eigenvalues of the matrix $T(E_F/\hbar)$, where T is given by Eq. (13.128).

13.11 *Green's operator.* In the independent-electron approximation, the Hamiltonian for a system of electrons is

$$\mathcal{H} = \sum_i \left[-\frac{\hbar^2}{2m}\nabla_i^2 + V(\mathbf{r}_i) \right] \equiv \sum_i H(i).$$

$V(\mathbf{r}_i)$ is the potential energy of electron i; it includes the potential produced by the nuclei, as well as the average interaction with other electrons. Let $H|\phi_\nu\rangle = \epsilon_\nu|\phi_\nu\rangle$.

(a) Show that the retarded Green's function is given by

$$G(\mathbf{r}, \mathbf{r}'; \omega) = \sum_\nu \frac{\phi_\nu(\mathbf{r})\phi_\nu^*(\mathbf{r}')}{\omega - \epsilon_\nu/\hbar + i0^+}.$$

(b) Now define the retarded Green's operator $G(\omega)$ by

$$G(\mathbf{r}, \mathbf{r}'; \omega) = \langle \mathbf{r}|G(\omega)|\mathbf{r}'\rangle.$$

Show that

$$G(\omega) = [\omega - H/\hbar + i0^+]^{-1}.$$

13.12 *LCR system.* Consider a system which consists of a semi-infinite left lead (L), a semi-infinite right lead (R), and a molecule in the central region (C). The molecule is in contact with both leads, but the leads are not in direct contact with each other. Treat this problem using the independent-electron approximation. The Hamiltonian for the system is written by using a basis set of real atomic-like orbitals centered on the atoms. There may be more than one orbital centered on any one atom. In terms of this basis, the

Hamiltonian is a matrix of the form:

$$H = \begin{bmatrix} H_{LL} & H_{LC} & 0 \\ H_{CL} & H_{CC} & H_{CR} \\ 0 & H_{RC} & H_{RR} \end{bmatrix}.$$

Assuming that there are N basis functions in the left lead ($N \to \infty$), N basis functions in the right lead, and M basis functions in the molecule (M is finite), H_{LL} and H_{RR} are each an $N \times N$ matrix, while H_{CC} is an $M \times M$ matrix. H_{LC} and H_{RC} describe coupling between the molecule and the leads. The matrix H is real and symmetric.

The retarded Green's operator is

$$G(\omega) = \begin{bmatrix} G_{LL} & G_{LC} & G_{LR} \\ G_{CL} & G_{CC} & G_{CR} \\ G_{RL} & G_{RC} & G_{RR} \end{bmatrix}.$$

According to the previous problem, $G(\omega)$ is obtained by solving the equation

$$(\omega - H/\hbar + i0^+)G(\omega) = 1.$$

Show that

$$G_{CC}(\omega) = \left[\omega - H_{CC}/\hbar - \Sigma^{(L)}(\omega) - \Sigma^{(R)}(\omega)\right]^{-1}$$

where

$$\Sigma^{(L)}(\omega) = H_{CL}\, g_L(\omega)\, H_{LC}, \quad \Sigma^{(R)}(\omega) = H_{CR}\, g_R(\omega)\, H_{RC}.$$

g_L (g_R) is the retarded Green's operator for the isolated left (right) lead:

$$g_L(\omega) = (\omega - H_{LL} + i0^+)^{-1}, \quad g_R(\omega) = (\omega - H_{RR} + i0^+)^{-1}.$$

Appendix A
Second quantized form of operators

We present a detailed derivation of the second quantized form of one-body and two-body operators. We consider a system of N identical fermions, and follow by the case of a system of N identical bosons.

A.1 Fermions

A.1.1 One-body operators

Consider the one-body operator $H_0 = \sum_{i=1}^{N} h(i)$. Let $|\phi_1\rangle, |\phi_2\rangle, \ldots$ be a complete set of orthonormal single-particle states. H_0 acts upon the vector space $\mathrm{V}^{(N)} = \mathrm{V}_1 \otimes \mathrm{V}_2 \otimes \cdots \otimes \mathrm{V}_N$, the direct product space of the spaces of the N particles. The vector space V_i, the Hilbert space of particle i upon which $h(i)$ acts, is spanned by the basis set $|\phi_1\rangle_i, |\phi_2\rangle_i, \ldots$. For any $|\phi_\nu\rangle_i$ in this basis set, $\langle \mathbf{r}|\phi_\nu\rangle_i = \phi_\nu(i)$; for example, $\langle \mathbf{r}|\mathbf{k}\sigma\rangle_i = \frac{1}{\sqrt{V}} e^{i\mathbf{k}.\mathbf{r}_i}|\sigma\rangle_i$, where $\mathbf{r}_i$ is the position vector of particle i and $|\sigma\rangle_i$ is its spin state. The orthonormality of the basis states means that ${}_i\langle\phi_\nu|\phi_{\nu'}\rangle_i = \delta_{\nu\nu'}$, and completeness means that $\sum_\nu |\phi_\nu\rangle_i\ {}_i\langle\phi_\nu| = 1$. For the case $|\phi_\nu\rangle = |\mathbf{k}\sigma\rangle$, these relations mean that ${}_i\langle\mathbf{k}\sigma|\mathbf{k}'\sigma'\rangle_i = \delta_{\mathbf{k}\mathbf{k}'}\delta_{\sigma\sigma'}$ and $\sum_{\mathbf{k}\sigma} |\mathbf{k}\sigma\rangle_i\ {}_i\langle\mathbf{k}\sigma| = 1$. Note that an expression such as ${}_i\langle\phi_\nu|\phi_{\nu'}\rangle_j$, for $i \neq j$, is not an inner product because $|\phi_\nu\rangle_i$ and $|\phi_{\nu'}\rangle_j$ belong to different vector spaces; it is, in fact, the operator $|\phi_{\nu'}\rangle_j\ {}_i\langle\phi_\nu|$. Inserting the completeness relation into H_0, we obtain

$$H_0 = h(1) + \cdots + h(N) = \sum_{\nu\nu'} |\phi_{\nu'}\rangle_1\ {}_1\langle\phi_{\nu'}|h(1)|\phi_\nu\rangle_1\ {}_1\langle\phi_\nu| + \cdots$$

$$+ \sum_{\nu\nu'} |\phi_{\nu'}\rangle_N\ {}_N\langle\phi_{\nu'}|h(N)|\phi_\nu\rangle_N\ {}_N\langle\phi_\nu| = \sum_{\nu\nu'}\sum_i |\phi_{\nu'}\rangle_i\ {}_i\langle\phi_{\nu'}|h(i)|\phi_\nu\rangle_i\ {}_i\langle\phi_\nu|.$$

The matrix element ${}_i\langle\phi_{\nu'}|h(i)|\phi_\nu\rangle_i$ is independent of i, since the coordinates of particle i are integrated over, and it is written as $\langle\phi_{\nu'}|h|\phi_\nu\rangle$. Thus,

$$H_0 = \sum_{\nu\nu'}\langle\phi_{\nu'}|h|\phi_\nu\rangle \sum_{i=1}^{N} |\phi_{\nu'}\rangle_i\, {}_i\langle\phi_\nu| = \sum_{\nu\nu'}\langle\phi_{\nu'}|h|\phi_\nu\rangle R_{\nu'\nu}. \tag{A.1}$$

We have introduced the operator $R_{\nu'\nu}$,

$$R_{\nu'\nu} = \sum_{i=1}^{N} |\phi_{\nu'}\rangle_i\, {}_i\langle\phi_\nu|. \tag{A.2}$$

The Slater determinants form a properly symmetrized basis for the expansion of the N-fermion wave function. We consider how $R_{\nu'\nu}$ acts on an arbitrary Slater determinant $|\Psi\rangle = |\phi_{\nu_1}\ldots\phi_{\nu_N}\rangle$. If $\nu \notin \{\nu_1,\ldots,\nu_N\}$ then $\nu \notin \{P(\nu_1),\ldots,P(\nu_N)\}$, since the sets $\{\nu_1,\ldots,\nu_N\}$ and $\{P(\nu_1),\ldots,P(\nu_N)\}$ are identical (the elements of the second set are merely a permutation of the elements of the first set). Since $\nu \notin \{P(\nu_1),\ldots,P(\nu_N)\}$,

$$\begin{aligned}
{}_i\langle\phi_\nu|\Psi\rangle &= \frac{1}{\sqrt{N!}}\, {}_i\langle\phi_\nu| \sum_P (-1)^P |\phi_{P(\nu_1)}\rangle_1 \cdots |\phi_{P(\nu_i)}\rangle_i \cdots |\phi_{P(\nu_N)}\rangle_N \\
&= \frac{1}{\sqrt{N!}} \sum_P (-1)^P |\phi_{P(\nu_1)}\rangle_1 \cdots {}_i\langle\phi_\nu|\phi_{P(\nu_i)}\rangle_i \cdots |\phi_{P(\nu_N)}\rangle_N = 0.
\end{aligned}$$

The last equality follows since $\nu \neq P(\nu_i)$. Therefore, unless $\nu \in \{\nu_1,\ldots,\nu_N\}$ the action of $R_{\nu'\nu}$ on $|\Psi\rangle$ yields zero. So let us assume that $\nu = \nu_j$. Then

$$R_{\nu'\nu}|\Psi\rangle = \frac{1}{\sqrt{N!}} \sum_{i=1}^{N} |\phi_{\nu'}\rangle_i\, {}_i\langle\phi_\nu| \sum_P (-1)^P |\phi_{\nu_1}\rangle_{P(1)} \cdots |\phi_\nu\rangle_{P(j)} \cdots |\phi_{\nu_N}\rangle_{P(N)}.$$

The sum is now over the permutations of coordinates. Recall that the Slater determinant has two equivalent forms: the sum in one form is over the permutations of coordinates, while in the other form, it is over the permutations of indices. In the summation over i from 1 to N, each i belongs to the set $\{P(1),\ldots,P(N)\}$, its elements being a permutation of $1,\ldots,N$. Since the single-particle states are orthonormal, only when $i = P(j)$ will $R_{\nu'\nu}|\Psi\rangle$ be nonzero. When $i = P(j)$, $|\phi_{\nu'}\rangle_i\, {}_i\langle\phi_\nu|\phi_\nu\rangle_{P(j)} = |\phi_{\nu'}\rangle_i = |\phi_{\nu'}\rangle_{P(j)}$. Hence, the result of the action of $R_{\nu'\nu}$ on $|\Psi\rangle$ is simply to replace $|\phi_\nu\rangle$ in $|\Psi\rangle$ by $|\phi_{\nu'}\rangle$,

$$\begin{aligned}
R_{\nu'\nu}|\Psi\rangle &= \frac{1}{\sqrt{N}} \sum_P (-1)^P |\phi_{\nu_1}\rangle_{P(1)} \cdots |\phi_{\nu'}\rangle_{P(j)} \cdots |\phi_{\nu_N}\rangle_{P(N)} \\
&= |\phi_{\nu_1}\cdots\phi_{\nu'}\cdots\phi_{\nu_N}\rangle = c^\dagger_{\nu'}c_\nu|\phi_{\nu_1}\cdots\phi_\nu\cdots\phi_{\nu_N}\rangle = c^\dagger_{\nu'}c_\nu|\Psi\rangle.
\end{aligned}$$

No minus sign is needed: ϕ_ν is moved to the leftmost position, replaced by $\phi_{\nu'}$, which is then moved back to the original position of ϕ_ν. If the first movement produced a minus sign, so would the second. Since $|\Psi\rangle$ is an arbitrary Slater determinant, we conclude that

$$R_{\nu'\nu} = c^\dagger_{\nu'} c_\nu. \tag{A.3}$$

The second quantized form of $H_0 = \sum_{i=1}^N h(i)$ is therefore

$$H_0 = \sum_{\nu\nu'} \langle \phi_{\nu'} | h | \phi_\nu \rangle c^\dagger_{\nu'} c_\nu. \tag{A.4}$$

A.1.2 Two-body operators

Consider the two-body operator $H' = \frac{1}{2}\sum_{i\neq j} v(i,j)$. Given a complete set $|\phi_1\rangle, |\phi_2\rangle, \ldots$ of orthonormal single-particle states, we may write

$$\begin{aligned}
H' &= \frac{1}{2}\left[v(1,2) + v(2,1) + v(1,3) + \cdots + v(N, N-1)\right] \\
&= \frac{1}{2}\left[\sum_{klmn} |\phi_k\phi_l\rangle_{1,2}\,{}_{1,2}\langle\phi_k\phi_l|v(1,2)|\phi_m\phi_n\rangle_{1,2}\,{}_{1,2}\langle\phi_m\phi_n| + \cdots \right. \\
&\quad \left. + \sum_{klmn} |\phi_k\phi_l\rangle_{N,N-1}\,{}_{N,N-1}\langle\phi_k\phi_l|v(N,N-1)|\phi_m\phi_n\rangle_{N,N-1}\,{}_{N,N-1}\langle\phi_m\phi_n|\right].
\end{aligned}$$

In the above equation,

$$|\phi_k\phi_l\rangle_{i,j} = |\phi_k\rangle_i \otimes |\phi_l\rangle_j = |\phi_k\rangle_i\,|\phi_l\rangle_j, \quad {}_{i,j}\langle\phi_k\phi_l| = {}_i\langle\phi_k|\,{}_j\langle\phi_l|.$$

The matrix element ${}_{i,j}\langle\phi_k\phi_l|v(i,j)|\phi_m\phi_n\rangle_{i,j}$ is independent of i and j, since the coordinates of i and j are integrated over, and we write it simply as $\langle\phi_k\phi_l|v|\phi_m\phi_n\rangle$. Hence,

$$H' = \frac{1}{2}\sum_{klmn} \langle\phi_{kl}|v|\phi_{mn}\rangle \sum_{i\neq j} |\phi_k\rangle_i|\phi_l\rangle_j\,{}_i\langle\phi_m|\,{}_j\langle\phi_n| = \frac{1}{2}\sum_{klmn}\langle\phi_{kl}|v|\phi_{mn}\rangle A_{klmn}$$

$$A_{klmn} = \sum_{i\neq j} |\phi_k\rangle_i|\phi_l\rangle_j\,{}_i\langle\phi_m|\,{}_j\langle\phi_n| = \sum_i |\phi_k\rangle_i\,{}_i\langle\phi_m| \sum_{j,j\neq i} |\phi_l\rangle_j\,{}_j\langle\phi_n|.$$

A_{klmn} is a product of two operators. We rewrite it as follows,

$$A_{klmn} = \sum_i |\phi_k\rangle_i\ {}_i\langle\phi_m| \left[\sum_j |\phi_l\rangle_j\ {}_j\langle\phi_n| - |\phi_l\rangle_i\ {}_i\langle\phi_n|\right]$$
$$= \sum_i |\phi_k\rangle_i\ {}_i\langle\phi_m| \sum_j |\phi_l\rangle_j\ {}_j\langle\phi_n| - \sum_i |\phi_k\rangle_i\ {}_i\langle\phi_m|\phi_l\rangle_i\ {}_i\langle\phi_n|$$
$$= R_{km}R_{ln} - \delta_{ml}R_{kn} = c_k^\dagger c_m c_l^\dagger c_n - \delta_{ml}c_k^\dagger c_n.$$

We have used Eqs (A.2) and (A.3). The anticommutator $\{c_m, c_l^\dagger\} = \delta_{ml}$; it follows that $c_m c_l^\dagger = \delta_{ml} - c_l^\dagger c_m$. Hence,

$$A_{klmn} = c_k^\dagger(\delta_{ml} - c_l^\dagger c_m)c_n - \delta_{ml}c_k^\dagger c_n = -c_k^\dagger c_l^\dagger c_m c_n = c_k^\dagger c_l^\dagger c_n c_m.$$

We thus arrive at the second quantized form of the two-body operator,

$$H' = \frac{1}{2}\sum_{klmn} \langle\phi_k\phi_l|v|\phi_m\phi_n\rangle c_k^\dagger c_l^\dagger c_n c_m. \tag{A.5}$$

A.2 Bosons

A.2.1 One-body operators

We assume that we have a complete set $|\phi_1\rangle, |\phi_2\rangle, \ldots$ of orthonormal single-particle states. For a system of N identical bosons, the basis states are

$$|\Phi^B\rangle = \frac{1}{\prod_\mu \sqrt{n_\mu!}}\frac{1}{\sqrt{N!}}\sum_P |\phi_{\nu_1}\rangle_{P(1)}\cdots|\phi_{\nu_N}\rangle_{P(N)}. \tag{A.6}$$

n_μ is the number of times the state $|\phi_\mu\rangle$ appears in the product, i.e., n_μ is the number of particles that occupy the single-particle state $|\phi_\mu\rangle$. The one-body operator $H_0 = \sum_i h(i)$ is given by

$$H_0 = \sum_{\nu\nu'} \langle\phi_{\nu'}|h|\phi_\nu\rangle R_{\nu'\nu}, \qquad R_{\nu'\nu} = \sum_{i=1}^{N} |\phi_{\nu'}\rangle_i\ {}_i\langle\phi_\nu|. \tag{A.7}$$

It is clear that if $\nu \notin \{\nu_1, \ldots, \nu_N\}$ then $R_{\nu'\nu}|\Phi^B\rangle = 0$. Let us assume that $\nu \in \{\nu_1, \ldots, \nu_N\}$. First we consider the case $\nu' \neq \nu$. Suppose that n_ν particles $P(i_1), P(i_2), \ldots, P(i_{n_\nu})$ occupy the single-particle state $|\phi_\nu\rangle$, and $n'_{\nu'}$ particles $P(j_1), P(j_2), \ldots, P(j_{n'_{\nu'}})$ occupy the single-particle state $|\phi_{\nu'}\rangle$. In the number representation,

$$|\Phi^B\rangle = |\ldots n_\nu \cdots n'_{\nu'} \cdots\rangle. \tag{A.8}$$

Applying $R_{\nu'\nu}$ to $|\Phi^B\rangle$, we obtain

$$R_{\nu'\nu}|\Phi^B\rangle = \frac{1}{\sqrt{n_\nu! n'_{\nu'}!}} \frac{1}{\prod_{\mu\neq\nu,\nu'} \sqrt{n_\mu!}} \frac{1}{\sqrt{N!}} \sum_P \sum_{i=1}^{N} |\phi_{\nu'}\rangle_i \, {}_i\langle\phi_\nu|$$

$$\left[\cdots|\phi_\nu\rangle_{P(i_1)}\cdots|\phi_\nu\rangle_{P(i_{n_\nu})}\cdots|\phi_{\nu'}\rangle_{P(j_1)}\cdots|\phi_{\nu'}\rangle_{P(j_{n'_{\nu'}})}\cdots\right]. \tag{A.9}$$

Whenever $i \in \{P(i_1), \ldots, P(i_{n_\nu})\}$, the result of the action of $|\phi_{\nu'}\rangle_i \, {}_i\langle\phi_\nu|$ is to remove one particle from state $|\phi_\nu\rangle$ and add a particle into state $|\phi_{\nu'}\rangle$, i.e., it produces a state with $n_\nu - 1$ particles in state $|\phi_\nu\rangle$ and $n'_{\nu'} + 1$ particles in state $|\phi_{\nu'}\rangle$. For $i \notin \{P(i_1), \ldots, P(i_{n_\nu})\}$, the action of $|\phi_{\nu'}\rangle_i \, {}_i\langle\phi_\nu|$ on $|\Phi^B\rangle$ yields zero. We thus find

$$R_{\nu'\nu}|\Phi^B\rangle = \frac{n_\nu}{\sqrt{n_\nu! n'_{\nu'}!}} \frac{1}{\prod_{\mu\neq\nu,\nu'} \sqrt{n_\mu!}} \frac{1}{\sqrt{N!}}$$

$$\times \sum_P \left[\cdots|\phi_\nu\rangle_{P(k_1)}\cdots|\phi_\nu\rangle_{P(k_{n_\nu-1})}\cdots|\phi_{\nu'}\rangle_{P(l_1)}\cdots|\phi_{\nu'}\rangle_{P(l_{n'_{\nu'}+1})}\cdots\right].$$

Noting that

$$\frac{n_\nu}{\sqrt{n_\nu! n'_{\nu'}!}} = \frac{\sqrt{n_\nu}\sqrt{n_{\nu'}+1}}{\sqrt{(n_\nu - 1)!(n'_{\nu'}+1)!}},$$

we find, for $\nu \neq \nu'$,

$$\begin{aligned} R_{\nu'\nu}|\cdots n_\nu \ldots n'_{\nu'}\cdots\rangle &= \sqrt{n_\nu}\sqrt{n'_{\nu'}+1}\,|\cdots n_\nu - 1\cdots n'_{\nu'}+1\cdots\rangle \\ &= a^\dagger_{\nu'} a_\nu |\cdots n_\nu \cdots n'_{\nu'}\cdots\rangle. \end{aligned}$$

Now consider the case $\nu' = \nu$. In Eq. (A.9), whenever $i \in \{P(i_1), \ldots, P(i_{n_\nu})\}$, the action of $|\phi_\nu\rangle_i \, {}_i\langle\phi_\nu|$ on $|\Phi^B\rangle$ leaves $|\Phi^B\rangle$ unchanged, since it simply removes a particle from the single-particle state $|\phi_\nu\rangle$ and adds a particle into the same state. For $i \notin \{P(i_1), \ldots, P(i_{n_\nu})\}$, the action of $|\phi_\nu\rangle_i \, {}_i\langle\phi_\nu|$ on $|\Phi^B\rangle$ yields zero. Therefore,

$$R_{\nu\nu}|\Phi^B\rangle = R_{\nu\nu}|\cdots n_\nu\cdots\rangle = \sum_{i=1}^{N} |\phi_\nu\rangle_i \, {}_i\langle\phi_\nu|\cdots n_\nu\cdots\rangle$$

$$= \sum_{i\in\{P(i_1),\ldots,P(i_{n_\nu})\}} |\phi_\nu\rangle_i \, {}_i\langle\phi_\nu|\cdots n_\nu\cdots\rangle = n_\nu|\cdots n_\nu\cdots\rangle = a^\dagger_\nu a_\nu|\cdots n_\nu\cdots\rangle.$$

We conclude that $R_{\nu'\nu} = a^\dagger_{\nu'} a_\nu$. The one-body operator is thus given by

$$H_0 = \sum_{\nu\nu'} \langle\phi_{\nu'}|h|\phi_\nu\rangle a^\dagger_{\nu'} a_\nu. \tag{A.10}$$

A.2.2 Two-body operators

Following the same steps as in the case of fermions, we obtain the following expression for the two-body operator

$$H' = \frac{1}{2} \sum_{klmn} \langle kl|v|mn\rangle A_{klmn}$$

$$A_{klmn} = R_{km} R_{ln} - \delta_{ml} R_{kn} = a_k^\dagger a_m a_l^\dagger a_n - \delta_{ml} a_k^\dagger a_n.$$

The commutation relation $[a_m, a_l^\dagger] = \delta_{ml}$, for bosonic operators, implies that $a_m a_l^\dagger = \delta_{ml} + a_l^\dagger a_m$; hence

$$A_{klmn} = a_k^\dagger(\delta_{ml} + a_l^\dagger a_m) a_n - \delta_{ml} a_k^\dagger a_n = a_k^\dagger a_l^\dagger a_m a_n = a_k^\dagger a_l^\dagger a_n a_m.$$

The second quantized form of the two-body operator is therefore

$$H' = \frac{1}{2} \sum_{klmn} \langle kl|v|mn\rangle a_k^\dagger a_l^\dagger a_n a_m. \tag{A.11}$$

This is the same expression as in the case of fermions.

Appendix B

Completing the proof of Dzyaloshinski's rules

A Feynman diagram of order N is a sum of $N!$ time-ordered diagrams corresponding to the $N!$ permutations of $\tau_1, \tau_2, \ldots, \tau_N$. Consider the diagram $\Gamma_{N,1}$ obtained from some permutation for which, say, $\tau_{P_N} > \tau_{P_{N-1}} > \cdots > \tau_{P_1}$. The contribution of this diagram is

$$\delta g^{\Gamma_{N,1}} \propto \int_0^{\beta\hbar} d\tau_{P_N} e^{\epsilon_{P_N}\tau_{P_N}} \int_0^{\tau_{P_N}} d\tau_{P_{N-1}} e^{\epsilon_{P_{N-1}}\tau_{P_{N-1}}} \ldots \int_0^{\tau_{P_2}} d\tau_{P_1} e^{\epsilon_{P_1}\tau_{P_1}}.$$

We asserted in Section 9.8 that the sole surviving term is obtained by keeping only the contribution at the upper limit of each integral for all integrations over $\tau_{P_1}, \tau_{P_2}, \ldots, \tau_{P_{N-1}}$. To prove this assertion, consider the term obtained by keeping only the lower limits on the first d_1 integrals over $\tau_{P_1}, \tau_{P_2}, \ldots, \tau_{P_{d_1}}$, then only the upper limits on the next u_1 integrals over $\tau_{P_{d_1+1}}, \ldots \tau_{P_{d_1+u_1}}$, then only the lower limits on the next d_2 integrals, then only the upper limits on the next u_2 integrals, and so on. Only the upper limits are kept on the last u_k integrals. Let S_1 be this sequence of lower and upper limits, $S_1: u_k d_k \ldots u_2 d_2 u_1 d_1$. The diagram $\Gamma_{N,1}$, without external lines, is represented in Figure B.1, where, for simplicity, times are arranged in decreasing order from left to right (instead of top to bottom). In the figure, we lump the u_k vertices $\tau_{P_N}, \ldots, \tau_{P_{N-u_k+1}}$ together into L_1, and the vertices $\tau_{P_{N-u_k}}, \ldots, \tau_{P_1}$ into L_2. We show fermion and boson lines that connect the two lumps L_1 and L_2. The term obtained by sequence S_1 is denoted by $\delta g_{S_1}^{\Gamma_{N,1}}$. We will show that this term is cancelled out by another term that arises from a sequence in a different time-ordered diagram.

Consider the time-ordered diagram $\Gamma_{N,2}$, obtained from a second permutation which orders the times as follows: $\tau_{P_{N-u_k}} > \cdots > \tau_{P_1} > \tau_{P_N} > \cdots > \tau_{P_{N-u_k+1}}$. This diagram is the same as that in Figure B.1 except that the lumps L_1 and L_2 are interchanged. As a result of this interchange, the lines connecting the lumps have their directions reversed. In evaluating $\delta g^{\Gamma_{N,2}}$, consider the sequence $S_2: d_k u_{k-1} d_{k-1} \ldots u_2 d_2 u_1 (d_1 + 1)(u_k - 1)$. Note that, in S_2, the lower limit from

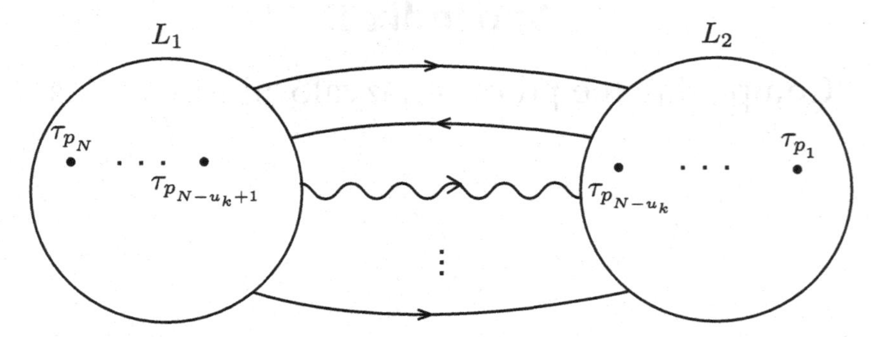

Figure B.1 A time-ordered diagram, with time decreasing from left to right, is divided into two lumps that are connected by fermion (solid) and boson (wavy) lines. The external lines are not shown.

the integration over τ_{P_N} is the one that is kept. We will show that the term $\delta g_{S_2}^{\Gamma_{N,2}}$ exactly cancels out the term $\delta g_{S_1}^{\Gamma_{N,1}}$. It is easy to check that, in both $\delta g_{S_1}^{\Gamma_{N,1}}$ and $\delta g_{S_2}^{\Gamma_{N,2}}$, time integrations produce the same denominators. We also note that $\delta g_{S_2}^{\Gamma_{N,2}}$ differs from $\delta g_{S_1}^{\Gamma_{N,1}}$ by the following:

1. A factor of -1, which results from keeping the lower limit rather than the upper limit in the integral over τ_{P_N}.
2. A factor of $\exp[-\beta\hbar(\epsilon_{P_N} + \cdots + \epsilon_{P_{N-u_k+1}})]$; this results from the fact that, in $\delta g_{S_1}^{\Gamma_{N,1}}$, we kept only the upper limits in the last u_k integrals, which produces the factor $\exp[\beta\hbar(\epsilon_{P_N} + \cdots + \epsilon_{P_{N-u_k+1}})$. In $\delta g_{S_2}^{\Gamma_{N,2}}$ we kept only the lower limit in the integration over τ_{P_N}.
3. A factor of $(-1)^J \exp[-\beta\hbar \sum_{i=N-u_k+1}^{N} \sum_{j=1}^{N-u_k} (\epsilon_{P_i P_j} - \epsilon_{P_j P_i})]$, where J is the number of internal fermion lines connecting L_1 to L_2. The reason for this factor is the following. As a result of interchanging L_1 and L_2, all lines connecting L_1 and L_2 are reversed. For each vertex τ_{P_i} in Figure B.1, $i = N, N-1, N-u_k+1$, the lines connecting τ_{P_i} to all $\tau_{P_j} \in L_2$ are reversed. Upon reversal of lines, for each fermion line directed from $\tau_{P_i} \in L_1$ to $\tau_{P_j} \in L_2$, the factor $-(1 - f_{\epsilon_{P_i P_j}})$ must be replaced by $f_{\epsilon_{P_i P_j}} = -\exp(-\beta\hbar\epsilon_{P_i P_j})[-(1 - f_{\epsilon_{P_i P_j}})]$, whereas for each fermion line directed from $\tau_{P_j} \in L_2$ to $\tau_{P_i} \in L_1$, the factor $f_{\epsilon_{P_j P_i}}$ must be replaced by $-(1 - f_{\epsilon_{P_j P_i}}) = -\exp(\beta\hbar\epsilon_{P_j P_i}) f_{\epsilon_{P_j P_i}}$. Also note that upon reversal of lines, for each boson line directed from $\tau_{P_i} \in L_1$ to $\tau_{P_j} \in L_2$, the factor $(1 + n_{\epsilon_{P_i P_j}})$ must be replaced by $n_{\epsilon_{P_i P_j}} = \exp(-\beta\hbar\epsilon_{P_i P_j})(1 + n_{\epsilon_{P_i P_j}})$, whereas for each boson line directed from $\tau_{P_j} \in L_2$ to $\tau_{P_i} \in L_1$, the factor $n_{\epsilon_{P_j P_i}}$ must be replaced by $(1 + n_{\epsilon_{P_j P_i}}) = \exp(\beta\hbar\epsilon_{P_j P_i}) n_{\epsilon_{P_j P_i}}$. Hence, the reversal of the lines produces the factor given above.

Summarizing,

$$\delta_{S_2}^{\Gamma_{N,2}} = -(-1)^J e^{-\beta\hbar\left(\epsilon_{P_N}+\cdots+\epsilon_{P_{N-u_k+1}}\right)} \exp\left[-\beta\hbar \sum_{i=N-u_k+1}^{N} \sum_{j=1}^{N-u_k} (\epsilon_{P_i P_j} - \epsilon_{P_j P_i})\right] \delta_{S_1}^{\Gamma_{N,1}}.$$

Using the definition of ϵ_{P_i},

$$\epsilon_{P_N} + \cdots + \epsilon_{P_{N-u_k+1}} = \sum_{i=N-u_k+1}^{N} \epsilon_{P_i}$$

$$= \sum_{i=N-u_k+1}^{N} \left(-\sum_{j=1}^{N} \epsilon_{P_i P_j} + \sum_{j=1}^{N} \epsilon_{P_j P_i} + i\omega_{n_1}\delta_{P_i,N} - i\omega_{n_2}\delta_{P_i,1} \right) = \sum_{i=N-u_k+1}^{N}$$

$$\left(-\sum_{j=N-u_k+1}^{N} \epsilon_{P_i P_j} - \sum_{j=1}^{N-u_k} \epsilon_{P_i P_j} + \sum_{j=N-u_k+1}^{N} \epsilon_{P_j P_i} + \sum_{j=1}^{N-u_k} \epsilon_{P_j P_i} + i\omega_{n_1}\delta_{P_i,N} - i\omega_{n_2}\delta_{P_i,1} \right)$$

$$= \sum_{i=N-u_k+1}^{N} \sum_{j=1}^{N-u_k} (-\epsilon_{P_i P_j} + \epsilon_{P_j P_i}) + i\omega_{n_1} \sum_{i=N-u_k+1}^{N} \delta_{P_i,N} - i\omega_{n_2} \sum_{i=N-u_k+1}^{N} \delta_{P_i,1}.$$

Therefore,

$$\delta g_{S_2}^{\Gamma_{N,2}} = -(-1)^J \exp\left[i\beta\hbar\omega_{n_1} \sum_{i=N-u_k+1}^{N} \delta_{P_i,N} - i\beta\hbar\omega_{n_2} \sum_{i=N-u_k+1}^{N} \delta_{P_i,1} \right] \delta g_{S_1}^{\Gamma_{N,1}}.$$

If we are interested in the self energy of a boson, then J is even and $\omega_{n_1} = 2\pi n_1/\beta\hbar$, $\omega_{n_2} = 2\pi n_2/\beta\hbar$; this gives $\delta g_{S_2}^{\Gamma_{N,2}} = -\delta_{S_1}^{\Gamma_{N,1}}$. On the other hand, if we are interested in the self energy of a fermion, then:

(a) If both external lines enter and leave L_1, then J is even, and

$$\delta g_{S_2}^{\Gamma_{N,2}} = -e^{i\beta\hbar(\omega_{n_1}-\omega_{n_2})} \delta g_{S_1}^{\Gamma_{N,1}} = -\delta_{S_1}^{\Gamma_{N,1}}.$$

(b) If both external lines enter and leave L_2, then J is even, and $\delta g_{S_2}^{\Gamma_{N,2}} = -\delta_{S_1}^{\Gamma_{N,1}}$.

(c) If one external line enters L_1 and the other line leaves L_2, then J is odd, and

$$\delta g_{S_2}^{\Gamma_{N,2}} = -(-1)^{odd} e^{i\omega_{n_1}\beta\hbar} \delta g_{S_1}^{\Gamma_{N,1}} = -\delta g_{S_1}^{\Gamma_{N,1}}.$$

(d) If one external line enters L_2 and the other line leaves L_1, then J is odd and

$$\delta g_{S_2}^{\Gamma_{N,2}} = -(-1)^{odd} e^{-i\omega_{n_2}\beta\hbar} \delta g_{S_1}^{\Gamma_{N,1}} = -\delta g_{S_1}^{\Gamma_{N,1}}.$$

In all cases, $\delta g_{S_2}^{\Gamma_{N,2}} = -\delta g_{S_1}^{\Gamma_{N,1}}$.

Appendix C

Lattice vibrations in three dimensions

C.1 Harmonic approximation

Consider a crystal having N unit cells with a basis of r atoms in each unit cell. Choosing the center of one unit cell to be the origin, the instantaneous position of atom l in unit cell n is

$$\mathbf{x}_{nl} = \mathbf{R}_{nl} + \mathbf{u}_{nl}. \tag{C.1}$$

$\mathbf{R}_{nl} = \mathbf{R}_n + \mathbf{d}_l$ is the equilibrium position of the atom, $\mathbf{u}_{nl}$ is its displacement from equilibrium, $\mathbf{R}_n$ is the lattice vector from the origin to the center of unit cell n, and $\mathbf{d}_l$ is the equilibrium position vector of atom l, measured from the center of the unit cell to which the atom belongs. The various terms are shown in Figure C.1.

Denoting the mass of atom l as M_l, the kinetic energy of the atoms is

$$T = \sum_{nl\alpha} P_{nl\alpha}^2/2M_l, \quad \alpha = x, y, z. \tag{C.2}$$

The potential energy V is assumed to be a sum of pairwise interactions between the atoms. A Taylor expansion of V about equilibrium gives

$$V = V_0 + \sum_{nl\alpha} \left.\frac{\partial V}{\partial u_{nl\alpha}}\right|_0 u_{nl\alpha} + \frac{1}{2}\sum_{nl\alpha}\sum_{n'l'\alpha'} \left.\frac{\partial^2 V}{\partial u_{nl\alpha}\partial u_{n'l'\alpha'}}\right|_0 u_{nl\alpha}u_{n'l'\alpha'} + \cdots. \tag{C.3}$$

V_0 is the potential energy of the crystal when the atoms sit at their equilibrium positions. Being merely a constant, V_0 will be ignored (we can always set it equal to zero by measuring energies relative to it). Since V is a minimum at equilibrium, the second term in Eq. (C.3) vanishes. The harmonic approximation, which we adopt here, consists in keeping only the third term in Eq. (C.3), ignoring higher order terms. The justification for this is that the displacement from equilibrium is very small compared to the lattice spacing (this is generally true for temperatures

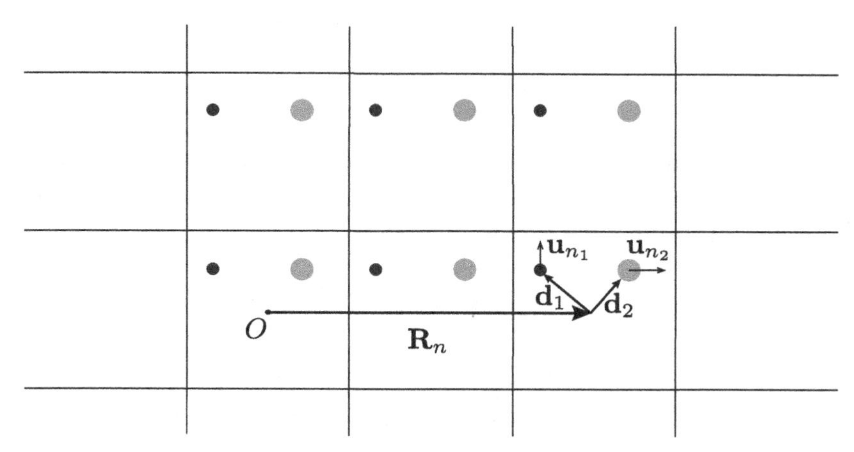

Figure C.1 A two-dimensional crystal with two atoms per unit cell. The center O of one unit cell is chosen as the origin of coordinates. $\mathbf{R}_n$ is the vector from O to the center of unit cell n. $\mathbf{d}_1$ and $\mathbf{d}_2$ are the equilibrium position vectors of the two atoms relative to the center of the unit cell to which they belong. $\mathbf{u}_{n1}$ and $\mathbf{u}_{n2}$ are the displacements from equilibrium of the two atoms that belong to unit cell n.

far below the melting point). Thus,

$$V = \frac{1}{2} \sum_{nl\alpha} \sum_{n'l'\alpha'} \phi(nl\alpha, n'l'\alpha') u_{nl\alpha} u_{n'l'\alpha'}. \tag{C.4}$$

The force constants $\phi(nl\alpha, n'l'\alpha')$ are given by

$$\phi(nl\alpha, n'l'\alpha') = \left. \frac{\partial^2 V}{\partial u_{nl\alpha} \partial u_{n'l'\alpha'}} \right|_0 . \tag{C.5}$$

C.2 Classical theory of lattice vibrations

The atoms' equations of motion are given by Newton's second law

$$M_l \ddot{u}_{nl\alpha} = -\partial V / \partial u_{nl\alpha} = -\sum_{n'l'\alpha'} \phi(nl\alpha, n'l'\alpha') u_{n'l'\alpha'}. \tag{C.6}$$

This is a set of $3Nr$ coupled, linear, differential equations. The general solution is written as a linear combination of the normal modes. This is analogous to the situation in quantum mechanics where the general solution of the time-dependent Schrödinger equation is written as a linear combination of the eigenfunctions of the Hamiltonian. Here, normal modes play the role of the eigenfunctions in quantum mechanics. We search for the normal modes of the system; their number is $3Nr$, which equals the number of degrees of freedom. In a normal mode all atoms vibrate with the same wave vector and frequency; we thus consider the following trial solution

$$\mathbf{u}_{nl} = \frac{A}{\sqrt{M_l}} \boldsymbol{\epsilon}^{(l)}(\mathbf{q}) e^{i(\mathbf{q}.\mathbf{R}_n - \omega_{\mathbf{q}} t)}. \tag{C.7}$$

The allowed values of $\mathbf{q}$ are determined by the periodic boundary conditions. $\boldsymbol{\epsilon}^{(l)}(\mathbf{q})$ is a vector to be determined, A is a constant, and the factor $1/\sqrt{M}$ is inserted for later convenience. Putting the above expression into Eq. (C.6), we obtain

$$\omega_{\mathbf{q}}^2 \epsilon_\alpha^{(l)}(\mathbf{q}) = \sum_{n'l'\alpha'} \frac{1}{\sqrt{M_l M_{l'}}} \phi(nl\alpha, n'l'\alpha') \epsilon_{\alpha'}^{(l')}(\mathbf{q}) e^{i\mathbf{q}.(\mathbf{R}_{n'} - \mathbf{R}_n)}. \tag{C.8}$$

This is a set of $3r$ algebraic equations ($l = 1, \ldots, r$; $\alpha = x, y, z$). The translational symmetry of the lattice implies that the force constants depend on $\mathbf{R}_{n'} - \mathbf{R}_n$ and not on $\mathbf{R}_{n'}$ and $\mathbf{R}_n$ separately (to convince yourself of this, think about how potential energy changes if two atoms, one in unit cell n and another in unit cell n', are displaced from equilibrium, while all other atoms in the crystal remain fixed). We define $D_{\mathbf{q}}(l\alpha, l'\alpha')$ by

$$D_{\mathbf{q}}(l\alpha, l'\alpha') = (M_l M_{l'})^{-1/2} \sum_{n'} \phi(nl\alpha, n'l'\alpha') e^{i\mathbf{q}.(\mathbf{R}_{n'} - \mathbf{R}_n)}. \tag{C.9}$$

Since the summand depends on $\mathbf{R}_{n'} - \mathbf{R}_n$ and not on $\mathbf{R}_n$ and $\mathbf{R}_{n'}$ separately, we may set $\mathbf{R}_n = 0$. Relabeling n' as n, we obtain

$$D_{\mathbf{q}}(l\alpha, l'\alpha') = (M_l M_{l'})^{-1/2} \sum_{n} \phi(0l\alpha, nl'\alpha') e^{i\mathbf{q}.\mathbf{R}_n}. \tag{C.10}$$

Since the force constants are real, it follows that

$$D_{\mathbf{q}}^*(l\alpha, l'\alpha') = D_{-\mathbf{q}}(l\alpha, l'\alpha'). \tag{C.11}$$

Using Eq. (C.9), we rewrite Eq. (C.8) as an eigenvalue equation,

$$\sum_{l'\alpha'} D_{\mathbf{q}}(l\alpha, l'\alpha') \epsilon_{\alpha'}^{(l')}(\mathbf{q}) = \omega_{\mathbf{q}}^2 \epsilon_\alpha^{(l)}(\mathbf{q}). \tag{C.12}$$

To make this notion more manifest, we define the $3r \times 3r$ dynamical matrix

$$D(\mathbf{q}) = \begin{bmatrix} D_{\mathbf{q}}(1,1) & D_{\mathbf{q}}(1,2) & \ldots & D_{\mathbf{q}}(1,r) \\ D_{\mathbf{q}}(2,1) & D_{\mathbf{q}}(2,2) & \ldots & D_{\mathbf{q}}(2,r) \\ \vdots & \vdots & & \vdots \\ D_{\mathbf{q}}(r,1) & D_{\mathbf{q}}(r,2) & \ldots & D_{\mathbf{q}}(r,r) \end{bmatrix}. \tag{C.13}$$

Here, $D_{\mathbf{q}}(l, l')$ is a 3×3 matrix whose $\alpha\alpha'$ entry is $D_{\mathbf{q},\alpha\alpha'}(l, l') = D_{\mathbf{q}}(l\alpha, l'\alpha')$. We also define the $3r$-column polarization vector

$$\epsilon = (\epsilon_x^{(1)} \;\; \epsilon_y^{(1)} \;\; \epsilon_z^{(1)} \ldots \epsilon_x^{(r)} \;\; \epsilon_y^{(r)} \;\; \epsilon_z^{(r)})^{\mathrm{T}} = (\boldsymbol{\epsilon}^{(1)} \;\; \boldsymbol{\epsilon}^{(2)} \ldots \boldsymbol{\epsilon}^{(r)})^{\mathrm{T}}. \tag{C.14}$$

The superscript T means "transpose." Equation (C.12) now becomes

$$D(\mathbf{q})\epsilon(\mathbf{q}) = \omega_{\mathbf{q}}^2\epsilon(\mathbf{q}). \tag{C.15}$$

We note that the dynamical matrix is hermitian, since

$$\begin{aligned} D_{\mathbf{q}}^*(l\alpha, l'\alpha') &= (M_l M_{l'})^{-1/2} \sum_{n'} \phi(nl\alpha, n'l'\alpha') e^{-i\mathbf{q}.(\mathbf{R}_{n'}-\mathbf{R}_n)} \\ &= (M_l M_{l'})^{-1/2} \sum_{n'} \phi(n'l'\alpha', nl\alpha) e^{i\mathbf{q}.(\mathbf{R}_n-\mathbf{R}_{n'})} \\ &= D_{\mathbf{q}}(l'\alpha', l\alpha). \end{aligned} \tag{C.16}$$

The first equality results from the definition of $D_{\mathbf{q}}(l\alpha, l'\alpha')$ in Eq. (C.9) and the fact that the force constants are real. The second equality is valid because $\phi(nl\alpha, n'l'\alpha') = \phi(n'l'\alpha', nl\alpha)$. The hermiticity of $D(\mathbf{q})$ implies that:

- The eigenvalues $\omega_{\mathbf{q}\lambda}^2$, $\lambda = 1, \ldots, 3r$, are real. Furthermore, $\omega_{\mathbf{q}\lambda}$ is real; if it were not, the displacement $\mathbf{u}_{nl}$ would be a monotonically increasing or decreasing function of time, rather than an oscillatory one.
- The $3r$ eigenvectors $\epsilon_\lambda(\mathbf{q})$, $\lambda = 1, \ldots, 3r$ can be chosen to form a complete orthonormal set; any $3r$-column vector can be expanded in terms of them.

We also make the following remarks:

(1) The definition of $D_{\mathbf{q}}(l\alpha, l'\alpha')$, as given in Eq. (C.9), along with the equality $e^{i\mathbf{G}.\mathbf{R}_n} = 1$, implies that $D_{\mathbf{q}}(l\alpha, l'\alpha') = D_{\mathbf{q}+\mathbf{G}}(l\alpha, l'\alpha')$ for any reciprocal lattice vector $\mathbf{G}$. The dynamical matrix thus satisfies the property $D(\mathbf{q}) = D(\mathbf{q}+\mathbf{G})$; consequently,

$$\omega_{\mathbf{q}\lambda} = \omega_{\mathbf{q}+\mathbf{G}\lambda}, \quad \epsilon_\lambda(\mathbf{q}) = \epsilon_\lambda(\mathbf{q}+\mathbf{G}). \tag{C.17}$$

In other words, the mode $(\mathbf{q}+\mathbf{G}\lambda)$ is identical to the mode $(\mathbf{q}\lambda)$. It is therefore sufficient to restrict the values of $\mathbf{q}$ to the first Brillouin zone: $\mathbf{q} \in$ FBZ; this exhausts all possible normal modes. Since there are N $\mathbf{q}$-points in the FBZ, there are a total of $3Nr$ eigenvalues $\omega_{\mathbf{q}\lambda}$ and $3Nr$ corresponding eigenvectors $\epsilon_\lambda(\mathbf{q})$, $\mathbf{q} \in$ FBZ, $\lambda = 1, \ldots, 3r$.

(2) The dynamical matrix satisfies the relation $D^*(\mathbf{q}) = D(-\mathbf{q})$, a consequence of Eq. (C.11). Since $\omega_{\mathbf{q}\lambda}$ is real, it follows that

$$\omega_{\mathbf{q}\lambda} = \omega_{-\mathbf{q}\lambda}, \quad \epsilon_\lambda^*(\mathbf{q}) = \epsilon_\lambda(-\mathbf{q}). \tag{C.18}$$

(3) The orthonormality of the eigenvectors (polarization vectors) means that

$$\epsilon_\lambda^\dagger(\mathbf{q})\epsilon_{\lambda'}(\mathbf{q}) = \delta_{\lambda\lambda'} \Rightarrow \sum_{l\alpha} \epsilon_{\lambda,\alpha}^{*(l)}(\mathbf{q})\epsilon_{\lambda',\alpha}^{(l)}(\mathbf{q}) = \delta_{\lambda\lambda'}. \tag{C.19}$$

(4) If we plot $\omega_{\mathbf{q}\lambda}$ vs $\mathbf{q}$ along a certain direction in the FBZ, we obtain a curve for each value of λ. Therefore, along any given direction in the FBZ, there are $3r$ curves, or branches. Of these, three are acoustic branches, while the remaining $3r - 3$ branches are optical branches. For the three acoustic branches, $\omega \to 0$ as $\mathbf{q} \to \mathbf{0}$.

The general solution of the equation of motion, Eq. (C.6), is a linear combination of the $3Nr$ normal modes,

$$\mathbf{u}_{nl} = \frac{1}{\sqrt{NM_l}} \sum_{\mathbf{q}\lambda} Q_{\mathbf{q}\lambda}(t) \boldsymbol{\epsilon}_\lambda^{(l)}(\mathbf{q}) e^{i\mathbf{q}.\mathbf{R}_n} \tag{C.20}$$

where the factor $exp(-i\omega_{\mathbf{q}\lambda}t)$ is absorbed into the expansion coefficients $Q_{\mathbf{q}\lambda}(t)$, and the factor $1/\sqrt{N}$ is inserted for later convenience. The coefficients $Q_{\mathbf{q}\lambda}(t)$ are called normal coordinates. Using the equality

$$\sum_n e^{i(\mathbf{q}-\mathbf{q}').\mathbf{R}_n} = N\delta_{\mathbf{q}\mathbf{q}'},$$

and Eq. (C.19), it is not difficult to show that Eq. (C.20) gives

$$Q_{\mathbf{q}\lambda}(t) = \frac{1}{\sqrt{N}} \sum_{nl\alpha} \sqrt{M_l}\, u_{nl\alpha}(t) \epsilon_{\lambda,\alpha}^{*(l)}(\mathbf{q}) e^{-i\mathbf{q}.\mathbf{R}_n}. \tag{C.21}$$

Since $u_{nl\alpha}$ is real and $\epsilon_{\lambda,\alpha}^{(l)*}(-\mathbf{q}) = \epsilon_{\lambda,\alpha}^{(l)}(\mathbf{q})$, it follows that

$$Q_{\mathbf{q}\lambda}^*(t) = Q_{-\mathbf{q}\lambda}(t). \tag{C.22}$$

C.3 Vibrational energy

The total energy of the atoms can be expressed in terms of the normal coordinates. The kinetic energy is given by

$$T = \frac{1}{2}\sum_{nl\alpha} M_l \dot{u}_{nl\alpha}^2 = \frac{1}{2N}\sum_{nl\alpha}\sum_{\mathbf{q}\lambda}\sum_{\mathbf{q}'\lambda'} \dot{Q}_{\mathbf{q}\lambda}\dot{Q}_{\mathbf{q}'\lambda'}\epsilon_{\lambda,\alpha}^{(l)}(\mathbf{q})\epsilon_{\lambda',\alpha}^{(l)}(\mathbf{q}')e^{i(\mathbf{q}+\mathbf{q}').\mathbf{R}_n}.$$

Carrying out the summation over n: $\sum_n e^{i(\mathbf{q}+\mathbf{q}').\mathbf{R}_n} = N\delta_{\mathbf{q}',-\mathbf{q}}$, then over l and α: $\sum_{l\alpha} \epsilon_{\lambda,\alpha}^{(l)}(\mathbf{q})\epsilon_{\lambda',\alpha}^{(l)}(-\mathbf{q}) = \delta_{\lambda\lambda'}$, the expression for T reduces to

$$T = \frac{1}{2}\sum_{\mathbf{q}\lambda} \dot{Q}_{\mathbf{q}\lambda}\dot{Q}_{-\mathbf{q}\lambda}. \tag{C.23}$$

The potential energy is given by

$$V = \frac{1}{2}\sum_{nl\alpha}\sum_{n'l'\alpha'}\phi(nl\alpha, n'l'\alpha')u_{nl\alpha}u_{n'l'\alpha'} = \frac{1}{2N}\sum_{nl\alpha}\sum_{n'l'\alpha'}\sum_{\mathbf{q}\lambda}\sum_{\mathbf{q}'\lambda'}(M_l M_{l'})^{-1/2}$$
$$\times\, \phi(nl\alpha, n'l'\alpha')\, Q_{\mathbf{q}\lambda}\, Q_{\mathbf{q}'\lambda'}\, \epsilon^{(l)}_{\lambda,\alpha}(\mathbf{q})\epsilon^{(l')}_{\lambda',\alpha'}(\mathbf{q}')e^{i\mathbf{q}.\mathbf{R}_n}\, e^{i\mathbf{q}'.\mathbf{R}_{n'}}.$$

Writing $e^{i\mathbf{q}.\mathbf{R}_n}e^{i\mathbf{q}'.\mathbf{R}_{n'}} = e^{i(\mathbf{q}+\mathbf{q}').\mathbf{R}_n}e^{i\mathbf{q}'.(\mathbf{R}_{n'}-\mathbf{R}_n)}$, using Eq. (C.8), summing over n, and using the orthonormality of polarization vectors, we find that

$$V = \frac{1}{2}\sum_{\mathbf{q}\lambda}\omega^2_{\mathbf{q}\lambda}Q_{\mathbf{q}\lambda}Q_{-\mathbf{q}\lambda}. \tag{C.24}$$

The Lagrangian L is equal to $T - V$. The canonical momentum conjugate to $Q_{\mathbf{q}\lambda}$ is

$$P_{\mathbf{q}\lambda} = \partial L/\partial \dot{Q}_{\mathbf{q}\lambda} = \dot{Q}_{-\mathbf{q}\lambda}. \tag{C.25}$$

In terms of the dynamical variables $Q_{\mathbf{q}\lambda}$ and $P_{\mathbf{q}\lambda}$, the Hamiltonian is

$$H = \left(\sum_{\mathbf{q}\lambda}P_{\mathbf{q}\lambda}\dot{Q}_{\mathbf{q}\lambda} - L\right)\Bigg|_{\dot{Q}_{\mathbf{q}\lambda}=P_{-\mathbf{q}\lambda}}$$
$$= \frac{1}{2}\sum_{\mathbf{q}\lambda}\left(P_{\mathbf{q}\lambda}P_{-\mathbf{q}\lambda} + \omega^2_{\mathbf{q}\lambda}Q_{\mathbf{q}\lambda}Q_{-\mathbf{q}\lambda}\right). \tag{C.26}$$

C.4 Quantum theory of lattice vibrations

The quantum theory of lattice vibrations is obtained by treating $Q_{\mathbf{q}\lambda}$ and $P_{\mathbf{q}\lambda}$ as operators that satisfy the following equal-time commutation relations:

$$\left[Q_{\mathbf{q}\lambda}, Q_{\mathbf{q}'\lambda'}\right] = \left[P_{\mathbf{q}\lambda}, P_{\mathbf{q}'\lambda'}\right] = 0, \quad \left[Q_{\mathbf{q}\lambda}, P_{\mathbf{q}'\lambda'}\right] = i\hbar\delta_{\mathbf{q}\mathbf{q}'}\delta_{\lambda\lambda'}. \tag{C.27}$$

Analogous to the case of the harmonic oscillator (see Section 1.2), we define two new operators:

$$a_{\mathbf{q}\lambda} = \sqrt{\frac{1}{2\hbar\omega_{\mathbf{q}\lambda}}}\left(\omega_{\mathbf{q}\lambda}Q_{\mathbf{q}\lambda} + iP_{-\mathbf{q}\lambda}\right), \quad a^\dagger_{\mathbf{q}\lambda} = \sqrt{\frac{1}{2\hbar\omega_{\mathbf{q}\lambda}}}\left(\omega_{\mathbf{q}\lambda}Q_{-\mathbf{q}\lambda} - iP_{\mathbf{q}\lambda}\right). \tag{C.28}$$

These operators satisfy the commutation relations

$$[a_{\mathbf{q}\lambda}, a_{\mathbf{q}'\lambda'}] = [a^\dagger_{\mathbf{q}\lambda}, a^\dagger_{\mathbf{q}'\lambda'}] = 0, \quad [a_{\mathbf{q}\lambda}, a^\dagger_{\mathbf{q}'\lambda'}] = \delta_{\mathbf{q}\mathbf{q}'}\delta_{\lambda\lambda'}. \tag{C.29}$$

We can use Eq. (C.28) to express the normal coordinates and momentum operators in terms of the new operators:

$$Q_{\mathbf{q}\lambda} = \sqrt{\frac{\hbar}{2\omega_{\mathbf{q}\lambda}}}(a_{\mathbf{q}\lambda} + a^{\dagger}_{-\mathbf{q}\lambda}), \quad P_{\mathbf{q}\lambda} = i\sqrt{\frac{\hbar\omega_{\mathbf{q}\lambda}}{2}}(a^{\dagger}_{\mathbf{q}\lambda} - a_{-\mathbf{q}\lambda}). \tag{C.30}$$

Using the above expressions, it is straightforward to show that

$$H = \sum_{\mathbf{q}\lambda} \hbar\omega_{\mathbf{q}\lambda}(a^{\dagger}_{\mathbf{q}\lambda}a_{\mathbf{q}\lambda} + 1/2). \tag{C.31}$$

The values of $\mathbf{q}$ are restricted to the first Brillouin zone: $\mathbf{q} \in$ FBZ. The Hamiltonian is thus a collection of $3Nr$ harmonic oscillators. The operator $a^{\dagger}_{\mathbf{q}\lambda}$ ($a_{\mathbf{q}\lambda}$) is interpreted as a creation (annihilation) operator for a particle-like entity, called a phonon, of wave vector $\mathbf{q}$, branch index λ, and energy $\hbar\omega_{\mathbf{q}\lambda}$.

Appendix D

Electron–phonon interaction in polar crystals

D.1 Polarization

Polar crystals are generally semiconductors or insulators that, at low temperatures, have fully occupied valence bands and empty conduction bands. It is possible, however, to introduce electrons into the conduction bands. For example, absorption of photons of appropriate energy leads to the promotion of electrons from the occupied valence bands to the empty conduction bands. Raising the temperature produces a similar effect. In semiconductors, doping introduces free electrons into the lowest conduction band (or free holes into the top valence band). The electron–phonon interaction in these systems is not adequately described by the rigid-ion approximation. In an optical mode, the ions in the unit cell move relative to each other, resulting in an oscillating dipole moment which, in turn, gives rise to an electric field that acts on the electrons. The electron–LO phonon interaction in polar crystals is mainly the result of this coupling of electrons to the induced electric field.

We consider the case of a cubic crystal with two atoms per unit cell. The ionic charges are $\pm e^*$. The volume of the crystal is V, and the number of unit cells is N. In the long wavelength limit ($\mathbf{q} \to \mathbf{0}$), the two ions in the unit cell vibrate out of phase, while the displacements in one cell are almost identical to those in a neighboring cell. We denote by $\mathbf{u}_+$ ($\mathbf{u}_-$) the displacement of the positive (negative) ion within a unit cell (see Figure D.1).

Due to ionic displacements, a dipole moment $\mathbf{p} = e^*\mathbf{u}$, where $\mathbf{u} = \mathbf{u}_+ - \mathbf{u}_-$, is induced in the unit cell. Since the ions are not rigid, the resulting electric field further polarizes the ions; the polarization is thus

$$\mathbf{P} = n(e^*\mathbf{u} + \alpha\mathbf{E}_{local}) \tag{D.1}$$

where $n = N/V$, $\alpha = \alpha_+ + \alpha_-$ is the sum of the polarizabilities of the ions, and $\mathbf{E}_{local}$ is the electric field at the site of the dipole. The crystal is viewed as a

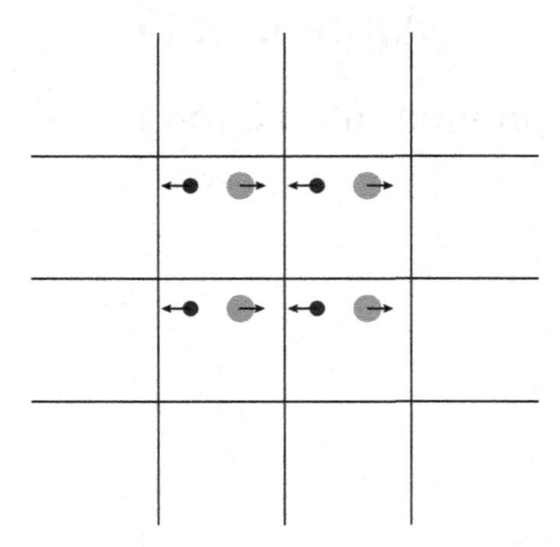

Figure D.1 In a long wavelength optical mode, the motion in neighboring cells of a lattice is essentially the same. Within a unit cell, the two ions vibrate out of phase. The dipole moment in a unit cell is $\mathbf{p} = e^*\mathbf{u}_+ - e^*\mathbf{u}_-$.

collection of dipoles, each of which occupies one unit cell. The local field acting on a dipole differs from the average macroscopic field $\mathbf{E}$ in the crystal; they are related by the Lorentz relation:

$$\mathbf{E}_{local} = \mathbf{E} + (4\pi/3)\mathbf{P} \quad (cgs), \qquad \mathbf{E}_{local} = \mathbf{E} + (1/3\epsilon_0)\mathbf{P} \quad (SI). \tag{D.2}$$

This relation is derived in standard electricity and magnetism textbooks (see, for example, [Griffiths, 1999]). In what follows we adopt the cgs system of units.

In a long-wavelength optical mode, all positive ions in adjacent cells have almost the same displacement $\mathbf{u}_+$, while all negative ions have almost the same displacement $\mathbf{u}_-$, whose direction is opposite to that of $\mathbf{u}_+$. The short-range restoring force acting on any one ion is proportional to $\mathbf{u}_+ - \mathbf{u}_-$. The equations of motion of the ions are thus given by

$$M_+\ddot{\mathbf{u}}_+ = -\gamma(\mathbf{u}_+ - \mathbf{u}_-) + e^*\mathbf{E}_{local} \tag{D.3a}$$

$$M_-\ddot{\mathbf{u}}_- = -\gamma(\mathbf{u}_- - \mathbf{u}_+) - e^*\mathbf{E}_{local}. \tag{D.3b}$$

γ is a constant related to the strength of the short-range restoring force between the ions. Multiplying the second equation by M_+/M_- and subtracting the result from the first equation, we obtain

$$M\ddot{\mathbf{u}} = -\gamma\mathbf{u} + e^*\mathbf{E}_{local} \tag{D.4}$$

where $M = M_+M_-/(M_+ + M_-)$ is the reduced mass of the two ions in a unit cell. Defining $\mathbf{w} = \sqrt{NM/V}\mathbf{u}$, a bit of algebra shows that Eqs (D.1), (D.2), and (D.4) give the following:

$$\ddot{\mathbf{w}} = b_{11}\mathbf{w} + b_{12}\mathbf{E} \tag{D.5a}$$

$$\mathbf{P} = b_{12}\mathbf{w} + b_{22}\mathbf{E}. \tag{D.5b}$$

The coefficients b_{11}, b_{12}, and b_{22} are constants. We can express these coefficients in terms of experimentally measurable quantities. In the presence of an externally applied static electric field, the ions are displaced from their equilibrium positions to new equilibrium positions, and a polarization develops even in the absence of ionic vibrations. In the static case, $\ddot{\mathbf{w}} = 0$, $\mathbf{w} = -(b_{12}/b_{11})\mathbf{E}$, and $\mathbf{P} = (-b_{12}^2/b_{11} + b_{22})\mathbf{E}$. Since, in this case, $\varepsilon(\omega = 0)\mathbf{E} = \mathbf{D} = \mathbf{E} + 4\pi\mathbf{P}$, it follows that

$$b_{22} - b_{12}^2/b_{11} = \frac{1}{4\pi}\left[\varepsilon(0) - 1\right], \tag{D.6}$$

where $\varepsilon(0) = \varepsilon(\mathbf{q} \to 0, \omega = 0)$ is the static dielectric constant.

If an external electric field with a frequency much higher than the vibrational frequencies is applied, the ions will not be able to follow the fast variation of the field; hence $\mathbf{w} = 0$ and $\mathbf{P} = b_{22}\mathbf{E}$. Since, in this case, $\varepsilon(\infty)\mathbf{E} = \mathbf{D} = \mathbf{E} + 4\pi\mathbf{P}$, we obtain

$$b_{22} = \frac{1}{4\pi}\left[\varepsilon(\infty) - 1\right], \tag{D.7}$$

where $\varepsilon(\infty) = \varepsilon(\mathbf{q} \to 0, \omega \gg \omega_{\text{phonon}})$ is the high-frequency dielectric constant. We write the solution to Eq. (D.5) in the form

$$(\mathbf{w}, \mathbf{E}, \mathbf{P}) = (\mathbf{w}_0, \mathbf{E}_0, \mathbf{P}_0)e^{i(\mathbf{q}.\mathbf{r} - \omega t)}. \tag{D.8}$$

Since $\nabla.\mathbf{D} = 0$ (there are no excess free charges in the crystal; it is neutral), it follows that

$$\mathbf{q}.(\mathbf{E} + 4\pi\mathbf{P}) = 0. \tag{D.9}$$

For a transverse optical mode, $\mathbf{q}.\mathbf{P} = 0 \Rightarrow \mathbf{q}.\mathbf{E} = 0$. Furthermore, in the electrostatic approximation, $\nabla \times \mathbf{E} = 0$ implies that $\mathbf{q} \times \mathbf{E} = 0$; hence, for a transverse optical mode, $\mathbf{E} = 0$, and Eq. (D.5a) becomes $\ddot{\mathbf{w}} = b_{11}\mathbf{w}$. We therefore conclude that

$$b_{11} = -\omega_{TO}^2, \tag{D.10}$$

where ω_{TO} is the frequency of the transverse optical phonon.

One may object to the conclusion that the electric field vanishes for the transverse optical modes, based on the electrostatic approximation, since the correct Maxwell's equation is $\nabla \times \mathbf{E} = (-1/c)\partial\mathbf{B}/\partial t$. This should be solved along with $\nabla.\mathbf{E} = 0$ (which is correct for the transverse modes) and $\nabla \times \mathbf{B} = (1/c)\partial\mathbf{D}/\partial t =$

$(\varepsilon/c)\partial\mathbf{E}/\partial t$, where ε is the dielectric constant. Writing $\nabla\times\nabla\times\mathbf{E} = -\nabla^2\mathbf{E} + \nabla(\nabla.\mathbf{E}) = -\nabla^2\mathbf{E}$, we obtain $\nabla^2\mathbf{E} = (\varepsilon/c^2)\partial^2\mathbf{E}/\partial t^2$. This gives $\varepsilon = c^2q^2/\omega^2$. Since $\mathbf{D} = \varepsilon\mathbf{E} = \mathbf{E} + 4\pi\mathbf{P}$, we see that $\mathbf{E}\approx 0$ if $\varepsilon \gg 1$. Since the optical mode frequency $\omega \approx 10^{13}s^{-1}$, we find that $\varepsilon \gg 1$ for $q \gtrsim 10^5\ m^{-1}$. Since the width of the Brillouin zone is $\approx 10^{10}\ m^{-1}$, we conclude that, except for values of $\mathbf{q}$ in an extremely tiny volume surrounding the Brillouin zone center, the electrostatic approximation is indeed satisfied.

For the longitudinal optical mode (where $\mathbf{w}$, $\mathbf{E}$, and $\mathbf{P}$ are parallel to $\mathbf{q}$), Eq. (D.9) implies that $\mathbf{E} = -4\pi\mathbf{P}$. Writing for this case $\ddot{\mathbf{w}} = -\omega_{\mathrm{LO}}^2\mathbf{w}$, it is straightforward to show, from Eqs (D.5), (D.6), (D.7), and (D.10), that

$$\omega_{\mathrm{LO}}^2 = \frac{\epsilon(0)}{\epsilon(\infty)}\omega_{TO}^2 \tag{D.11}$$

where ω_{LO} is the frequency of the longitudinal optical phonon. The above relation is known as the Lyddane–Sachs–Teller (LST) relation.

When the relation $\mathbf{E} = -4\pi\mathbf{P}$, valid for longitudinal optical vibrations, is substituted into Eq. (D.5b), we obtain

$$\mathbf{P} = \frac{b_{12}}{1+4\pi b_{22}}\mathbf{w} = \frac{b_{12}}{\varepsilon(\infty)}\mathbf{w}. \tag{D.12}$$

In the last step we used Eq. (D.7). Solving for b_{12} from Eqs (D.6), (D.7), and (D.10), and replacing $\mathbf{w}$ with $\sqrt{NM/V}\mathbf{u}$, we obtain

$$\mathbf{P} = \omega_{\mathrm{LO}}\left[\frac{NM}{4\pi V}\left(\frac{1}{\varepsilon(\infty)} - \frac{1}{\varepsilon(0)}\right)\right]^{1/2}\mathbf{u}. \tag{D.13}$$

D.2 Electron–LO phonon interaction

In order to write the electron–LO phonon interaction, we make the approximation of treating the crystal as a continuum, i.e., we express the polarization as a function of position inside the crystal. To do this, we need to write $\mathbf{u}$ as a function of position. The relative displacement $\mathbf{u}_n$ can be expanded as

$$\mathbf{u}_n = \frac{1}{\sqrt{NM}}\sum_{\mathbf{q}} Q_{\mathbf{q}}\boldsymbol{\epsilon}_L(\mathbf{q})e^{i\mathbf{q}.\mathbf{R}_n}.$$

where $\boldsymbol{\epsilon}_L(\mathbf{q})$ is the unit polarization vector of the LO phonon, and the normal coordinates $Q_{\mathbf{q}}$ satisfy the equation $\ddot{Q}_{\mathbf{q}} = -\omega_{\mathrm{LO}}^2 Q_{\mathbf{q}}$ (the longitudinal optical modes are assumed to be dispersionless: $\omega_{\mathbf{q}} = \omega_{LO}$). In the continuum description we thus write

$$\mathbf{u}(\mathbf{r}) = \frac{1}{\sqrt{NM}}\sum_{\mathbf{q}} Q_{\mathbf{q}}\boldsymbol{\epsilon}_L(\mathbf{q})e^{i\mathbf{q}.\mathbf{r}}.$$

Inserting this into Eq. (D.13) gives polarization as a function of position.

The interaction energy of an electron at position $\mathbf{r}_j$ with the polarized medium is $-e\Phi(\mathbf{r}_j)$, where $\Phi(\mathbf{r}_j)$ is the electric potential produced at $\mathbf{r}_j$ by the polarization. The polarization induces a charge density which, in turn, produces the electric potential $\Phi(\mathbf{r}_j)$. The induced charge density is

$$\rho(\mathbf{r}) = -\nabla.\mathbf{P} = -i\omega_{\rm LO}\left[\frac{1}{4\pi V}\left(\frac{1}{\varepsilon(\infty)} - \frac{1}{\varepsilon(0)}\right)\right]^{1/2} \sum_{\mathbf{q}}{}' Q_{\mathbf{q}}\mathbf{q}.\boldsymbol{\epsilon}_L(\mathbf{q})e^{i\mathbf{q}.\mathbf{r}}. \tag{D.14}$$

In the sum over $\mathbf{q}$, the $\mathbf{q} = \mathbf{0}$ term vanishes; the prime on the summation indicates that the $\mathbf{q} = \mathbf{0}$ term is excluded. The electron–LO phonon interaction is given by

$$\begin{aligned} H_{e-\rm LO} &= -e\sum_j \Phi(\mathbf{r}_j) = -e\sum_j \int \frac{\rho(\mathbf{r})d^3r}{|\mathbf{r} - \mathbf{r}_j|} \\ &= ie\omega_{\rm LO}\left[\frac{1}{4\pi V}\left(\frac{1}{\varepsilon(\infty)} - \frac{1}{\varepsilon(0)}\right)\right]^{1/2} \sum_j \sum_{\mathbf{q}}{}' Q_{\mathbf{q}}\mathbf{q}.\boldsymbol{\epsilon}_L(\mathbf{q}) \int \frac{e^{i\mathbf{q}.\mathbf{r}}d^3r}{|\mathbf{r} - \mathbf{r}_j|} \\ &= ie\omega_{\rm LO}\left[\frac{1}{4\pi V}\left(\frac{1}{\varepsilon(\infty)} - \frac{1}{\varepsilon(0)}\right)\right]^{1/2} \sum_j \sum_{\mathbf{q}}{}' Q_{\mathbf{q}}\mathbf{q}.\boldsymbol{\epsilon}_L(\mathbf{q})e^{i\mathbf{q}.\mathbf{r}_j} \int \frac{e^{i\mathbf{q}.(\mathbf{r}-\mathbf{r}_j)}d^3r}{|\mathbf{r} - \mathbf{r}_j|}. \end{aligned}$$

A change of variable $\mathbf{r} \to \mathbf{r} + \mathbf{r}_j$ shows that the above integral is simply the Fourier transform of $1/r$, which is equal to $4\pi/q^2$. We thus obtain

$$H_{e-\rm LO} = ie\omega_{\rm LO}\left[\frac{1}{4\pi V}\left(\frac{1}{\varepsilon(\infty)} - \frac{1}{\varepsilon(0)}\right)\right]^{1/2} \sum_{\mathbf{q}}{}' Q_{\mathbf{q}}\mathbf{q}.\boldsymbol{\epsilon}_L(\mathbf{q})(4\pi/q^2)\sum_j e^{i\mathbf{q}.\mathbf{r}_j}. \tag{D.15}$$

Finally, by writing $\sum_j e^{i\mathbf{q}.\mathbf{r}_j}$ in second quantized form in terms of electron creation and annihilation operators, and by expanding $Q_{\mathbf{q}}$ in terms of the phonon creation and annihilation operators, we obtain Eq. (11.46).

References

Abrikosov, A.A., Gorkov, L.P., and Dzyaloshinski, I.E. (1963). *Methods of Quantum field Theory in Statistical Physics*. New York: Dover Publications.

Alyahyaei, H.M. and Jishi, R.A. (2009). Theoretical investigation of magnetic order in RFeAsO (R = Ce, Pr). *Physical Review B* **79**, 064516.

Anderson, P.W. (1961). Localized magnetic states in metals. *Physical Review* **124**, 41–53.

Bardeen, J. (1936). Electron exchange in the theory of metals. *Physical Review* **50**, 1098.

Bardeen, J., Cooper, L.N., and Schrieffer, J.R. (1957). Theory of superconductivity. *Physical Review* **108**, 1175–1204.

Baym, G. and Sessler, A.M. (1963). Perturbation-theory rules for computing the self-energy operator in quantum statistical mechanics. *Physical Review* **131**, 2345–2349.

Bednorz, J.G. and Müller, K.A. (1986). Possible high T_C superconductivity in the Ba–La–Cu–O system. *Z. Physik. B* **64**, 189–193.

Bogoliubov, N.N. (1958). On a new method in the theory of superconductivity, *Nuovo Cimento* **7**, 794–805.

Callen, H.B. and Welton, T.R. (1951). Irreversibility and generalized noise. *Physical Review* **83**, 34–40.

Cini, M. (1980). Time-dependent approach to electron transport through junctions: general theory and simple applications. *Physical Review B* **22**, 5887–5899.

Cooper, L.N. (1956). Bound electron pairs in a degenerate Fermi gas. *Physical Review* **104**, 1189–1190.

Das, A. and Jishi, R.A. (1990). Theory of interband Auger recombination in semiconductors. *Physical Review B* **41**, 3551–3560.

De la Cruz, C., Huang, Q., Lynn, J.W. *et al.* (2008). Magnetic order close to superconductivity in the iron-based layer $LaO_{1-x}F_xFeAs$ systems. *Nature* **453**, 899–902.

Dresselhaus, M.S. and Dresselhaus, G. (1981). Intercalation compounds of graphite. *Advances in Physics* **30**, 139–326.

Duke, C.B. (1969). *Tunneling in Solids*. New York: Academic Press.

Dyson, F.J. (1949a). The radiation theories of Tomonaga, Schwinger, and Feynman. *Physical Review* **75**, 486–502.

Dyson, F.J. (1949b). The S-matrix in quantum electrodynamics. *Physical Review* **75**, 1736–1755.

Dzyaloshinski, I.E. (1962). A Diagram technique for evaluating transport coefficients in statistical physics at finite temperatures. *Soviet Physics JETP* **15**, 778–783.

Fermi, E. (1927). Un metodo statistico per la determinazione di alcune priopretà dell'atomo. *Rend. Accad. Naz. Lincei* **6**, 602–607.

Feynman, R.P. (1949a). The theory of positrons. *Physical Review* **76**, 749–759.
Feynman, R.P. (1949b). Space-time approach to quantum electrodynamics. *Physical Review* **76**, 769–789.
Frohlich, H. (1950). Theory of the superconducting ground state. I. The ground state at the absolute zero of temperature. *Physical Review* **79**, 845–856.
Ganguly, A.K. and Birman, J.L. (1967). Theory of light scattering in insulators. *Physical Review* **162**, 806–816.
Gell-Mann, M. and Low, F. (1951). Bound states in quantum field theory. *Physical Review* **84**, 350–354.
Glasser, M.L. (1981). Specific heat of electron gas of arbitrary dimensionality. *Physics Letters* **81** A, 295–296.
Glasser, M.L. and Boersma, J. (1983). Specific heat of electron gas of arbitrary dimensionality. *SIAM Journal of Applied Mathematics* **43**, 535–545.
Gorkov, L.P. (1958). On the energy spectrum of superconductors. *Soviet Physics JETP* **34**, 505–508.
Griffiths, D.J. (1999). *Introduction to Electrodynamics*. 2nd edn. San Francisco: Benjamin Cummings.
Holstein, T. and Primakoff, H. (1940). Field dependence of the intrinsic domain magnetization of a ferromagnet. *Physical Review* **58**, 1098–1113.
Horovitz, B. and Thieberger, R. (1974). Exchange integral and specific heat of the electron gas. *Physica* **71**, 99–105.
Hubbard, J. (1963). Electron correlations in narrow energy bands. *Proceedings of the Royal Society (London)* **A276**, 238–257.
Hwang, E.H. and Das Sarma, S. (2007). Dielectric function, screening, and plasmons in two-dimensional graphene. *Physical Review B* **75**, 205418.
Jackson, J.D. (1999). *Classical Electrodynamics*, 3rd edn. New York: Wiley.
Jishi, R.A. and Dresselhaus, M.S. (1992). Superconductivity in graphite intercalation compounds. *Physical Review B* **45**, 12465–12469.
Jishi, R.A., Dresselhaus, M.S., and Chaiken, A. (1991). Theory of the upper critical field in graphite intercalation compounds. *Physical Review B* **44**, 10248–10255.
Kadanoff, L.P. and Baym, G. (1962). *Quantum Statistical Mechanics*. New York: W.A. Benjamin, Inc.
Kamihara, Y., Watanabe, T., Hirano, M., and Hosono, H. (2008). Iron-based layered superconductor La[$O_{1-x}F_x$]FeAs ($x = 0.05 - 0.12$) with $T_C = 26$ K. *Journal of the American Chemical Society* **130**, 3296–3297.
Kastner, M.A. (1993). Artificial atoms. *Physics Today* **46**, 24–31.
Keldysh, L.V. (1965). Diagram technique for nonequilibrium processes. *Soviet Physics JETP* **20**, 1018–1026.
Kittel, C. (2005). *Introduction to Solid State Physics*, 8th edn. New York: Wiley.
Kubo, R. (1957). Statistical-mechanical theory of irreversible processes. I. General theory and simple applications to magnetic and conduction problems. *Journal of the Physical Society of Japan* **12**, 570–586.
Landau, L.D. (1957a). Theory of the Fermi liquid. *Soviet Physics JETP* **3**, 920–925.
Landau, L.D. (1957b). Oscillations in a Fermi liquid. *Soviet Physics JETP* **5**, 101–108.
Landau, L.D. (1959). On the theory of the Fermi liquid. *Soviet Physics JETP* **8**, 70–74.
Landauer, R. (1957). Spatial variations of currents and fields due to localized scatterers in metallic conduction. *IBM Journal of Research and Development* **1**, 223–231.
Landauer, R. (1970). Electrical resistance of disordered one-dimensional lattices. *Philosophical Magazine* **21**, 863–867.

Langreth, D.C. (1975). Linear and nonlinear response theory with applications. In *Linear and Nonlinear Electron Transport in Solids*. Vol. 17 of NATO Advanced Study Institute, Series B: Physics, ed. J.T. Devreese and V.E. Van Doren, pp. 3–32. New York: Plenum Press.

Lindhard, J. (1954). On the properties of a gas of charged particles. *Det Kongelige Danske Videnskabernes Selskab Matematisk-fysike Meddlelser* **28**, 1–57.

London, F. and London, H. (1935). The electromagnetic equations of the superconductor. *Proceedings of the Royal Society (London)* **A149**, 71–88.

Loudon, R. (1963). Theory of the first-order Raman effect in crystals. *Proceedings of the Royal Society (London)* **A275**, 218–232.

Meir, Y. and Wingreen, N.S. (1992). Landauer formula for the current through an interacting electron region. *Physical Review Letters* **68**, 2512–2515.

Meissner, W. and Ochsenfeld, R. (1933). Ein neuer effekt bei eintritt der supraleitfÃ⁄higkeit. *Naturwissenschaften* **21**, 787–788.

Migdal, A.B. (1958). Interaction between electrons and lattice vibrations in a normal metal. *Soviet Physics JETP* **34**, 996–1001.

Nambu, Y. (1960). Quasiparticles and gauge invariance in the theory of superconductivity. *Physical Review* **117**, 648–663.

Nyquist, H. (1928). Thermal agitation of electric charge in conductors. *Physical Review* **32**, 110–113.

Omar, M.A. (1993). *Elementary Solid State Physics: Principles and Applications*, revised printing. Boston: Addison-Wesley.

Onnes, H.K. (1911). Further experiments with liquid helium. D. On the change of electrical resistance of pure metals at very low temperatures, etc. V. The disappearance of the resistance of mercury. *Koninklijke Nederlandsche Akademie van Wetenschappen Proceedings* **14**, 113–115.

Pals, P. and Mackinnon, A. (1996). Coherent tunneling through two quantum dots with Coulomb interaction. *Journal of Physics: Condensed Matter* **8**, 5401–5414.

Scalapino, D.J. (1969). The electron-phonon interaction and strong-coupling superconductors. In *Superconductivity*, ed. R.D. Parks, pp. 449–560. New York: Marcel Dekker, Inc.

Shung, K.W.-K. (1986). Dielectric function and plasmon structure of stage-1 intercalated graphite. *Physical Review B* **34**, 979–993.

Stefanucci, G. and Almbladh, C.-O. (2004). Time-dependent partition-free approach in resonant tunneling systems. *Physical Review B* **69**, 195318.

Swirkowicz, R., Barnas, J., and Wilczynski, M. (2002). Electron tunneling in a double ferromagnetic junction with a magnetic dot as a spacer. *Journal of Physics: Condensed Matter* **14**, 2011–2024.

Taylor, P.L. and Heinonen, O. (2002). *A Quantum Approach to Condensed Matter Physics*. Cambridge: Cambridge University Press.

Thomas, L.H. (1927). The calculation of atomic fields. *Proceedings of the Cambridge Philosophical Society* **23**, 542–548.

Valatin, J.G. (1958). Comments on the theory of superconductivity. *Nuovo Cimento* **7**, 843–857.

Van Hove, L. (1954). Correlations in space and time and Born approximation scattering in systems of interacting particles. *Physical Review* **95**, 249–262.

Wick, G.C. (1950). The evaluation of the collision matrix. *Physical Review* **80**, 268–272.

Wilczek, F. (1982). Quantum mechanics of fractional spin particles. *Physical Review Letters* **49**, 957–959.

Yildrim, T. (2008). Origin of the 150-K anomaly in LaFeAsO: competing antiferromagnetic interactions, frustration, and a phase transition. *Physical Review Letters* **101**, 057010.
Ziman, J.M. (1960). *Electrons and Phonons*. Oxford: Oxford University Press.
Zimbovskaya, N.A. (2008). Electron transport through a quantum dot in the Coulomb blockade regime: nonequilibrium Green's function based model. *Physical Review B* **78**, 035331.

Index

Printed in the United States
By Bookmasters